COMPLETE GUIDE TO
STARGAZING

A FIREFLY BOOK

Published by Firefly Books Ltd. 2015

First printing

Publisher Cataloging-in-Publication Data (U.S.)

A CIP record for this title is available from the
Library of Congress

**Library and Archives Canada Cataloguing in
Publication**

Scagell, Robin, author
    Complete guide to stargazing / Robin Scagell.
ISBN 978-1-77085-474-1 (paperback)

    1. Stars—Observers' manuals. 2.
Astronomy—Observers' manuals. I. Title. II.
Title: Stargazing.

QB63.S33 2015   520
C2015-903765-4

Published in the United States by
Firefly Books (U.S.) Inc.
P.O. Box 1338, Ellicott Station
Buffalo, New York 14205

Published in Canada by
Firefly Books Ltd.
50 Staples Avenue, Unit 1
Richmond Hill, Ontario L4B 0A7

Printed in China

ROBIN SCAGELL is a long-serving Vice President of
Britain's Society for Popular Astronomy (www.popastro.
com). A lifelong stargazer, he has worked as an observer
and photographer, and as a journalist has edited a wide
range of popular-interest magazines. Robin is the author
of several popular astronomy books, and has contributed
to many other publications. He has been awarded the Sir
Arthur Clarke Award for Space Reporting in recognition
of his many appearances on TV and radio talking about
astronomy and space.

Published in Great Britain by Philip's,
a division of Octopus Publishing Group Ltd,
Carmelite House, 50 Victoria Embankment,
London EC4Y 0DZ
An Hachette UK Company

Title page montage: *Evening skyline* Robin Scagell/
Galaxy Picture Library; *AE Aurigae* T. A. Rector and
B. A. Wolpa/NOAO/AURA/NSF

FIREFLY

COMPLETE GUIDE TO
STARGAZING

ROBIN SCAGELL   FIREFLY BOOKS

INTRODUCTION

This guide is for anyone who wants to learn the night sky, from anywhere in the world, and to learn more about the amazing objects that are up there. You will need no prior knowledge – just add your own enthusiasm.

There's no substitute for going out under the stars and learning them for yourself. So, much of this guide is devoted to helping you to pick out which stars are visible from your location month by month, using a combination of maps of different scales and methods of plotting. With these, you can discover the appearance of the stars wherever and whenever you observe.

A special feature of the atlas section is the realistic views of the constellations that match as closely as possible what you actually see in the sky, with no labels or grid lines to clutter the page. Facing each one is a conventional map of the same area that you can use to identify the stars and constellations.

The most interesting constellations are described in detail, with illustrations that show the objects of interest in a variety of ways, from drawings that match accurately what you can see through a small telescope, to images taken through large telescopes, including the Hubble Space Telescope. Notes give you the basics of observational methods and help you to find and observe the objects, whether you have the simplest telescope or an up-to-the-minute computerized model.

Having found your way among the constellations, you will want to know more about the various objects – the Sun, Moon and planets and the much more distant stars, nebulae, star clusters and galaxies. Separate sections of the guide explain more about all the heavenly bodies you are likely to encounter, both as seen from Earth through amateur instruments and as viewed by giant telescopes and space probes. These pages will round off your stargazing experience by helping you to understand just what it is you are looking at.

A special website, www.stargazing.org.uk, accompanies this atlas to provide regularly updated links to sites giving further information and planetary positions, so you can always keep up to date. With this information, the *Firefly Complete Guide to Stargazing* can be your astronomical companion for years to come.

Robin Scagell

CONTENTS

1. THE NIGHT SKY 8

2. GETTING STARTED 20

3. EQUIPMENT FOR OBSERVING 32

4. THE MOON 44

5. THE SOLAR SYSTEM 62

6. STARS AND DEEP-SKY OBJECTS 100

7. THE SKY MONTH-BY-MONTH 122

8. STAR MAPS 196

A TO Z OF ASTRONOMY 212

APPENDIX 317

ACKNOWLEDGMENTS 318

INDEX 319

As evening falls, two astronomers are beginning their night's work. One of them, a professional astronomer, is where you would expect an astronomer to be, on a lofty mountain above much of the atmospheric murk and well away from the light-polluted cities, with a giant telescope. But the other, an amateur, is in a suburb a thousand or so miles closer to the pole, and is carrying a telescope out of a shed.

Both of them will be using the mirrors of their telescopes to catch light that has traveled across space – and both of them will this evening witness sights that no human has ever seen before. The professional is using a telescope with a mirror as big as the amateur astronomer's whole patio, in a great dome. Tonight it will be pointed at galaxies so distant that their light left them before the Solar System formed. The amateur's telescope, however, has a mirror about the size of a dinner plate and will be pointed, among other things, at the bright planet Jupiter. But the systems and methods that each of them uses have a lot in common, and while the professional will write a paper based on the night's observations that will explain just a bit more about the Universe as a whole, the amateur's observations of Jupiter will capture for posterity the appearance of that giant planet on this night.

Each telescope is a reflecting telescope – that is, it uses a mirror with a dished surface to focus the light from astronomical objects on to a detector. In both cases

this is an electronic imaging device, but the amateur will also be able to look directly into the telescope using an eyepiece – a facility denied the professional because giant telescopes are usually operated remotely from a nearby heated room. This allows the telescope itself to remain in total darkness and at the same temperature as the night air, for one of the bugbears of all astronomers is the turbulence of our own atmosphere, both at high altitude and close to the telescope. Any unnec-

▼ *An amateur astronomer* prepares to image Jupiter using a 200 mm Schmidt-Cassegrain telescope and a webcam from his backyard.

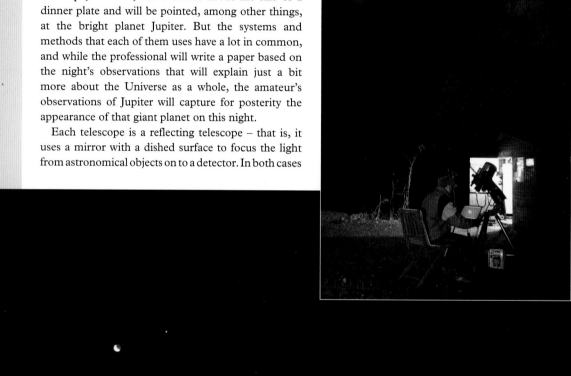

▼ *Twilight at the Teide Observatory* on Tenerife. The observatory is above the cloud layer, providing clean and usually calm air conditions that are ideal for observing.

essary source of heat within the dome will result in warm air rising through the line of sight of the telescope, causing the incoming starlight to be deflected from its true path. This is a problem for amateurs as well, who have to make sure that their telescopes are close to ambient temperature. Just putting a warm hand in front of the telescope can cause the image to shatter into a writhing mass of blobs.

So there are no warm observing conditions for most amateur observers, particularly in the winter months. Pointing the telescope through a convenient window is not an option. They must be out there in the elements if they wish to commune directly with the light that is constantly pouring down from space, mostly unwatched.

Tonight, both astronomers spend a little time just stargazing before beginning work in earnest. To the professional, the stars may be less well known than to the amateur, but at high altitude they are a dramatic sight and one well worth drinking in as the last traces of twilight drain away from the western sky. With a little imagination you can see the skies in three dimensions. The brightest object at this particular time, in the absence of the Sun and Moon, is the planet Jupiter – a brilliant, untwinkling object. It is the closest object visible – though Venus, Mars and Mercury can be closer. Astronomical distances are so great that instead of using kilometers or miles it is easiest to think of them in terms of the time light would take to travel from them.

We do the same on Earth surprisingly often – we think of a distance not in kilometers or miles but in terms of travel time. The local shop is only five minutes away on foot. Your relatives are an hour away by car. Your vacation destination is three hours by plane. In the case of astronomy we use the travel time of light, which propagates at a constant speed of nearly 300,000 km/second (186,000 miles/second).

So the light from that bright point that the planetary tables tell us is Jupiter left it some 40 minutes ago on this occasion. We are seeing it pretty much as it is at this very moment. But the light from a bright star that we also see in the same sky may have left it 40 years ago. Though it is fairly bright in the sky, it is more distant in the same ratio as the number of minutes in a year – about half a million. In reality, Jupiter is a globe of gas with no light of its own, and simply reflects the light from the Sun. But the stars are suns in their own right, and are both millions of times brighter than Jupiter and millions of times more distant.

The distance that light travels in a year is called a light year. Most of the stars we can pick out in the sky are a few tens or hundreds of light years away – just in our own neighborhood. But there are many more stars in the sky, and these merge into one another to form a band of light crossing the sky, the Milky Way.

Our professional astronomer is gazing at the Milky Way, for it is a spectacular sight. In a dark sky it dominates the view and is so clear that you wonder how you could ever have missed it before. It has subtle knots and convolutions, with some darker lanes, and here and there you can pick out brighter patches that beg to be studied in close-up. With a knowledgable eye, you can appreciate that its millions of stars comprise a great flattened disk, with us inside it. Its diameter is over a hundred thousand light years, and because we are within it the closer regions hide our view of the most distant ones, just as in a forest you see only the nearby trees and can have no idea whether you are in the middle or near one edge. Away from the plane of the Milky Way there are comparatively few stars, and the sky is much blacker.

But even the Milky Way, distant though it may be, is on our doorstep compared with the objects of tonight's professional study – distant galaxies. These are many millions of light years away, and the ones which the big

▲ *Few astronomers* *get to see the Milky Way in all its glory these days. Only those lucky enough to visit remote high-altitude observatories may witness skies completely lacking in atmospheric dust, water vapor and light pollution. This is the scene from Paranal in Chile, site of the Very Large Telescope.*

▲ **The bright stars Castor and Pollux** (left) in the constellation of Gemini appear of similar brightness to Saturn on the right. Castor, the upper of the two, is some 50 light years away, while Pollux is 34 light years away. Saturn on this occasion was 1460 million kilometers (81 light minutes) away.

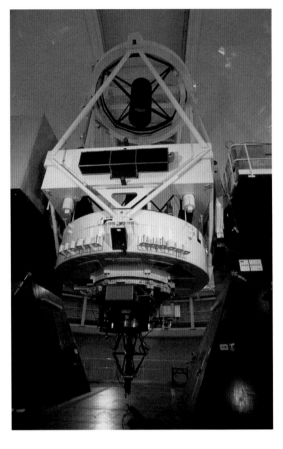

▶ *The William Herschel Telescope* on La Palma in the Canary Islands. The mirror, which is within the cream-colored circular housing below center, reflects light up to the secondary mirror within the black baffle tube at the top of the structure. This directs the light down through a hole in the mirror to the cameras and other instruments at the bottom of the telescope. The gray housing at the left is an alternative observing position.

▼ *A group of professional astronomers* using an ultraviolet spectrometer, X-Shooter, in the control room of the Very Large Telescope at Paranal, Chile. The observers control the instrument and the camera from a comfortable environment.

telescope will be studying could be thousands of millions of light years distant. Each one is a collection of hundreds or thousands of millions of stars, so they are about the same true brightness as our own Milky Way but very much more remote. This is why such a giant telescope is needed to collect the trickle of ancient light which has been traveling through space for perhaps twice the age of the Earth.

The amateur astronomer sees more or less the same regions of the sky. Jupiter still glares down, but here in the outer suburbs, even though there are trees around and fields not far away, light pollution transforms the scene. The bright stars are visible, but the Milky Way that seems so bright from the mountaintop is drowned out by the light spilling from millions of streetlights, house and office lights, car lights, floodlights on buildings, advertising signs and all the rest of the network that makes up modern life for most people.

If all those lights were suddenly to be turned off many people would probably think that the world had come to an end. The skies would again blaze with their pale light as they have done for most of the Earth's peoples until recently. Folk would wonder at the Milky Way arching across the heavens, and might even believe that the heavens have been torn apart. They would be in awe of the brilliance of Jupiter, and would gaze at the bewildering numbers of stars that have appeared, seemingly from nowhere.

Yet this is not all completely hidden, even now. The amateur astronomer has binoculars, and picks out some familiar sights. Just humble binoculars can help to bring back some stars despite the light pollution. Along the line of the Milky Way are star clusters that can be seen with just a little optical aid, even though the spectacle is very much diminished by the orange background light. The stars of constellations with familiar names, such as Cancer and Libra, that have disappeared from suburban skies, can be picked out with binoculars.

Also visible are individual stars of particular interest because their brightness varies, sometimes unpredictably. Amateur astronomers have traditionally taken these variable stars under their collective wing, because they are in a good position to keep an eye on them. With practice, it is possible to compare the brightness of the variable with that of nearby stars of known constant brightness, to monitor changes that can signal that the star is about to undergo a transformation that can be studied by professionals.

The amateur, too, can pick out galaxies, though with binoculars only the brighter and nearby ones are visible. Even so, it is fun to spot them, and to reaffirm one's acquaintance with a distant smudge of light that is a Milky Way in its own right.

The time for plain stargazing is over, and it's down to work using the two telescopes – one great, one small. But there are many similarities between them. As the Earth rotates on its axis, so the heavens appear to turn slowly above our heads. This is why the Sun, Moon, planets and stars all rise and set in the same direction, lifting above the eastern horizon and eventually setting at the western horizon. Most astronomical telescopes are arranged on mountings so as to follow this movement, which is very rapid when viewed at high magnification. In the giant dome the motors whirr softly as they drive tons of metal and glass to track the distant galaxies. The telescope must stay at the same object to within a fraction of an arc second for maybe an hour at a time. That is like pointing steadily at the pupil of the eye of someone standing five kilometers away.

In the control room of the giant telescope, the astronomer types the coordinates of the object to be photographed on to a keyboard and in the adjacent dome the monster telescope slowly moves round the sky to that point, with the slit of the dome obediently following so as to keep it in view. Shortly the target is acquired, and a view of some stars appears on a monitor. The astronomer checks these against a map of what is expected, or maybe a previously taken image, to confirm that the telescope is viewing the right part of the sky, and begins taking exposures on a state-of-the-art electronic camera cooled down to −120°C to reduce its electronic noise.

The garden astronomer's telescope, too, is driven by a small motor. But in this case the object acquisition is a bit more simple – it's just a matter of lining up the telescope on Jupiter by first using a small telescope attached to the main tube, then looking through the eyepiece to check that it is dead central, using the buttons on a handset. The amateur's camera, which then replaces the eyepiece, is a tiny affair compared with that of the professional. In fact it is a type of webcam, designed for taking live video images, and costing only about as much as a good compact camera. A view of Jupiter appears on the monitor of a laptop nearby to which it is connected.

But this is not a clear, steady image – it is constantly wriggling and writhing, as if seen at the bottom of a fast-flowing stream. This is really quite close to the truth, because the atmosphere above the telescope does indeed behave like the water in a stream as the heat of the ground seeps away into space. This is what astronomers, both amateur and professional, refer to as the "seeing." It causes star images to become bloated, and destroys planetary detail. The search for good seeing is one reason why the professional observatory is up a mountain, above the densest part of the atmosphere, and often in a more tranquil air stream than is found at ground level.

Even so, the garden telescope shows that a considerable amount of detail is visible on Jupiter – brownish belts and light blue zones, with light and dark spots within them, and a much darker spot which is the shadow on the planet of one of its large moons. The detail is there, but hard to keep track of as the atmosphere jiggles the image. But technology now comes to the rescue. At the click of a button the amateur starts to record a live video of the image on to the laptop. Later on, back indoors, software will distil a detailed image of the planet derived from the very best of the video frames.

Because Jupiter rotates in only about ten hours, its appearance changes surprisingly quickly. Within ten minutes of the first video sequence it is time to take another, as new features will have appeared at one edge, or limb, of the planet. Over a period of an hour or so, the astronomer fills the laptop's hard drive with sequence after sequence of video, each one consisting of perhaps 2000 individual color frames. The processed pictures from each sequence can then be put together in an animation, showing the speeded-up rotation of the planet. Just a few years ago, such a feat would have challenged astronomers using even large telescopes. Today, it's all in a night's work for an amateur under light-polluted skies, in a small corner of a metropolis.

Soon the resulting animation will be circulated among like-minded observers, and the views made from around the world will be compared. The images made by amateur astronomers now give round-the-clock coverage of the major planets. What an astronomer in Japan or Australia may miss, because the feature was on the other side of Jupiter at the time, an observer in Europe may detect as the planet rotates. A few hours later, observers in America will train their own instruments on the planet and will record more data. The swirling of Jupiter's clouds has never been so carefully scrutinized, and even when there are spacecraft orbiting the planet they do not give as good coverage as provided by the worldwide network of amateur astronomers.

Yet each one may pick up something that no one else has witnessed. It may be a new spot on Jupiter or Saturn, or a dust storm on Mars. You might think that there is little point in amateurs bothering these days, when spacecraft can get up close to the planets and with space telescopes able to take so much better pictures than any amateur. Yet in 2006, when a new spot appeared on Saturn, it was not initially seen by the Cassini spacecraft in orbit around the planet, but by amateurs. Cassini was on the dark side of the planet at the time, and the Hubble Space Telescope was studying much more distant objects. And it is not just planets that amateurs can keep an eye on. It could be the brightening or fading of a variable star which no professional has the funding to study, yet which is a matter of a few minutes of observation for a skilled amateur observer.

Back under the stars, eventually Jupiter gets rather too low in the sky for good results and the amateur turns to another task – photographing some deep-sky objects. These lie well beyond the Solar System. Some are within our own Milky Way, such as the nebulae, which are gas clouds from which stars are born, or are left when they die. Others are star clusters. But beyond the Milky Way lies the realm of the galaxies – and today these are prime targets, even for amateurs who are besieged by streetlights, neighbors' security lights, advertising signs, sports grounds and floodlights, each of which unthinkingly help to ruin our view of the night skies.

Imaging deep-sky objects calls for a different type of camera, this time rather more costly than the lowly webcam. Like the professional camera it is cooled to

◀ *To make an image of Jupiter* the amateur records a video sequence of 1000 frames or more, with the webcam in place of the eyepiece. Then software selects the best frames and adds them together, yielding an image which reveals a wealth of detail.

▶ *The famous "Whirlpool Galaxy,"* M51, photographed from Harrow in the London suburbs by Nick King using a 200 mm telescope. When Lord Rosse first spotted the spiral structure in this galaxy in 1845 (see page 152), he was using what was then the largest telescope in the world, in the dark skies of central Ireland. Modern technology and a telescope one-twelfth the diameter can now give an image that would have amazed Rosse.

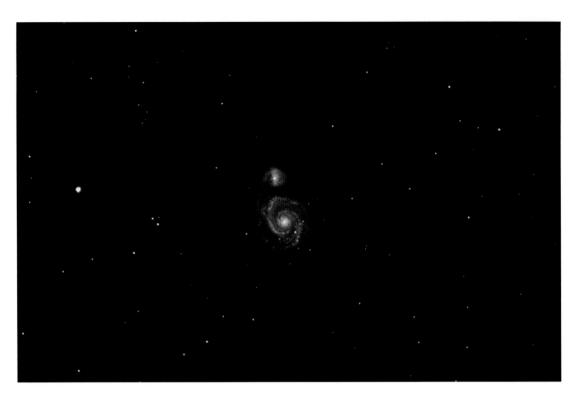

restrict the random jiggling of electrons within it, which reduce the quality of the image. But instead of using a supply of very cold liquid nitrogen, as with the professional camera, it makes use of electronic or Peltier cooling to reduce its temperature by some 30°C.

Again the procedure is to locate the target in the sky – but this time, unlike Jupiter, it is a galaxy which is completely invisible in the telescope because it is so faint and the light pollution is so bright. Probably it has not been visible from this location with such a telescope for the past few decades, since the burgeoning of city lighting. But fortunately there are stars visible in the telescope in the vicinity of the galaxy, and with some practice the observer can aim the telescope at the right spot. As before, the camera goes in place of the eyepiece and shortly an image of stars appears on the laptop.

The camera in this case is linked to the drive motors of the telescope. If the stars in the field of view start to drift across the image as a result of errors in the drive motors, the movement will be detected and a command issued to the drive motor to correct the drift. In this way, an exposure of several minutes is possible with the stars remaining in the same position. Eventually, among the stars and the general brightness from the sky, the image of a small faint galaxy will appear. Again, a fair bit of computer processing is needed, this time to remove the background of light pollution and increase the brightness of the galaxy itself.

Why would an amateur astronomer bother to photograph galaxies when there are professional photographs of these objects, some made using the Hubble Space Telescope? It is not as futile as it may seem. Around the country, people are engaged in other activities which are equally bizarre. Does the Saturday morning footballer give up because the team is nowhere near as good as those in the major leagues? Does the amateur artist lose heart because other people can do so much better? Does the angler down by the local lake pack it in because there is no hope of catching a blue marlin? In fact each is motivated to achieve a personal best, and amateur astronomers are the same.

Just seeing an elusive faint fuzzy object at the limit of visibility with a particular instrument and location is a challenge that amateurs welcome. It is sadly true that no telescope, amateur or professional, will show you the Andromeda Galaxy in as much detail and the color of the images seen in books, just by looking through it. The human eye is just not sensitive enough, and what's more will not carry on building up an image for minutes on end as a camera will. But that glimpse of the oval glow from a galaxy very like our own, and one of our nearest neighbor galaxies, is food for the imagination. Even though it is on our doorstep compared with the rest of the Universe, no individual stars are visible – they all mingle together to provide a soft glow.

But we know that there are hundreds of millions of individual stars there, many of which will be accompanied by planets. It is very likely that life will have evolved on some of those planets – just how many, no one can say with certainty – and that alien eyes are similarly peering through their own devices at our own Galaxy.

Amateur astronomy is not really about making great discoveries, though occasionally amateurs do spot things through their persistence, diligence and sheer numbers that their better equipped professional colleagues miss. As well as monitoring the planets, some amateurs, for example, will photograph asteroids or newly discovered comets and provide positions that help in determining their orbits. Others search the galaxies for supernovae –

exploding stars – allowing professionals to observe them soon after the eruption. In this way, new information is being gleaned about the distant reaches of the Universe. And one day, it may be an amateur who knows the starry skies that is the first to spot a supernova in our own Galaxy. No such supernova has been seen in our Galaxy since 1572, and when one does appear it will outshine every other star in the night sky.

But mostly what drives amateur astronomers is the realization that the night sky is a natural landscape just like our own world, a landscape to be explored and studied through telescopes. Our own Moon is a world in its own right which never fails to amaze, and although its surface has remained essentially the same for millions of years, the view is always different as a result of the changing illumination from the Sun. The planets, too, are gems when seen through telescopes, and there are glittering star clusters and faint misty nebulae and galaxies to be harvested with the eye or the camera. Occasionally we might spot a meteor or a comet. By day, the Sun itself is ever-changing.

Today we know so much about the Universe, but we also know that we have only scratched the surface. Astronomy gives us a wonderful introduction not just to seeing with our own eyes the wonders of the Universe, however dimly, but to the way we have learned about the Universe. The next few pages will give you an overview of how we know the things we do, and within the rest of this book you will get some idea of how to make the same exploration for yourself.

FINDING OUT ABOUT THE UNIVERSE

Virtually everything we know about the rest of the Universe beyond the Earth is a result of basically looking at it and contemplating what it all means. Only in a few cases within the Solar System do we have the luxury of actually going there and confirming what we believe.

Today's astronomers have many tools at their disposal in addition to telescopes on Earth and in orbit. A vast amount of information is contained within starlight. The methods they use boil down to three basic approaches:

• Making the assumption that the same laws of physics that apply on Earth apply throughout the rest of the Universe.

• Looking at the position and appearance of an object and how it changes over a period of time.

• Looking at the brightness and distribution of wavelengths of electromagnetic radiation from an object and working out what these mean.

The first assumption really can't be tested in any way other than to say that if the laws of physics do change from place to place, we don't have much hope of making sense of things. But the signs are that they do, though there may well be laws of physics that we don't yet know about. This is one great reason for studying astronomy in the first place. People will sometimes ask you, "Does it really matter that we know about the most distant galaxy? Surely we have enough problems here on Earth to worry about what's going on at the other end of the Universe!" Actually, it's by studying the extremes of time and space that we are most likely to discover the true laws that underlie the Universe, which could have a profound effect on our own lives. But this is a big topic, so for the purposes of this book we will have to accept the assumption at face value.

The second approach, that of measuring positions, is surprisingly valuable. This task started back in the 17th century, when the big challenge was to provide a means of finding a ship's position at sea. Observatories were equipped with telescopes whose job was not so much to understand the stars as to measure their positions as accurately as possible for the benefit of navigators.

This led to the discovery that stars don't remain in the same position forever. Some stars change position very slowly in the sky, which we now realize is largely a result of their closeness to our Solar System. As the Earth moves around the Sun, our viewpoint changes slightly so for the closest stars we can use the time-honored principles of parallax and trigonometry to measure their distances precisely. Parallax is what we use with stereo vision – the slight change in viewpoint between our two eyes allows us to judge everyday distances. Distance measurement is critical in astronomy, and depends on a long chain which starts right here on Earth.

Everything comes down to measuring a baseline on Earth, and working outward from there. These days our baselines are measured in laboratories and are linked to the wavelength of light, but the principle remains the same as used by early mapmakers who would pace out a known distance, measure the angles to distant objects from each end of the baseline, and use trigonometry to calculate the distances to those objects. The resulting positions provided further baselines for mapping features invisible from the original location.

Today it is possible to send radar signals between bodies in the Solar System and time the interval between transmission and their return, thus getting

▼ As viewed from Earth, the stars are practically unchanging. The constellations and patterns we see are virtually the same as those that our distant ancestors would have gazed at over their campfires. Their actual speeds through space are measured in kilometers per second, but they are so distant that very precise measurements are needed to detect a change in position of even the closest.

a very precise measurement of their distances. But finding the distances to the stars depends entirely on trigonometry again, with the best measurements coming from a satellite called Hipparcos, which measured the positions of millions of stars very precisely. Its baseline was the diameter of the Earth's orbit around the Sun.

Only the distances to stars within about a hundred light years or so are known to any real accuracy, but this still gives us a good sample of stars whose other properties we can measure. We then make the assumption that all stars with apparently similar properties are actually similar in other respects, particularly brightness.

Studying the positions of nearby stars reveals that many of them are double or even multiple. In fact, our Sun is unusual in that it does not have a stellar companion. These double stars are valuable to astronomers. If you see two stars mutually orbiting each other and you know their distance using a direct measurement of their parallax, it is a simple matter to apply Newton's law of gravity to the stars themselves. Again, this depends on measurements made here on Earth of the effect of gravity on masses, and an extrapolation out into the Universe. In this way we can discover the masses of the stars and see how their brightness changes with their mass.

The third approach to gathering information now comes into its own. In the 19th century the principles of spectroscopy were first established, laying the foundation for analyzing the actual constituents of stars just by looking at their light. The earliest experiments used prisms to split light into its component wavelengths. Newton had shown that what we see as white light is composed of all the colors of the rainbow. The different colors, we now know, are characterized by having different wavelengths, with blue being the shortest wavelengths and red the longest.

Looking at the light from the Sun, we see that the spectrum of the colors of the rainbow is crossed by dark lines, which turn out to have the same wavelengths as bright lines created when a gas is heated in the

▲ **The spectrum of the Sun** shows dark lines where certain wavelengths have been absorbed by gases in the Sun's atmosphere. The distribution of its light, with a peak in the green part of the spectrum, also reveals the Sun's temperature.

laboratory. The 19th-century astronomers realized that the presence of the same gas in the atmosphere of a star could be inferred from these dark lines in its spectrum. Looking at the spectra of stars and other objects remains at the heart of astrophysics, and shows us that the same elements and molecules exist all over the Universe. Hydrogen is by far the most common element, with helium second. All the other elements, such as those that our own bodies are made from, are actually way down the list in terms of abundance.

The importance of spectroscopy is not limited to finding out what the stars are made of. The wavelengths of light that we receive are altered by the movement toward or away from us of the object that emitted them – what is known as the Doppler shift. Think of an orange-yellow streetlight, for example, which produces light from hot sodium gas. If we were moving toward that streetlight very fast, its light instead of being yellow-orange would appear green (a shorter wavelength). If we were moving away from it very fast, it would appear red as the wavelength lengthens. Fortunately for us when driving along the road, the shift produced by everyday speeds is so small as to be unnoticeable, but in the case of stars which are moving at speeds of kilometers per second, the shift is detectable by the slight movement of the spectral lines compared with those produced on Earth. And in the case of distant galaxies, which are traveling at a significant fraction of the speed of light, a strong color change actually does take place.

Doppler shifts are incredibly useful. By measuring the position of the spectral lines, it is possible to detect even slight movements of an object that is so distant all we can see is a spot of light. This allows such things as the motions of stars and galaxies to be measured, and in turn, by applying Newton's law again, their masses.

The overall distribution of light in an object also reveals its temperature. We know from personal experience that as objects get hotter their color gets whiter and eventually bluer – think how yellow a low-wattage incandescent bulb is compared with a brighter one, for example. Exactly the same applies to stars, with the cooler and usually dimmer ones being orange and the brighter and hotter ones being noticeably blue.

The distribution of light of different colors within an object reveals not just its temperature but the process that created the light in the first place. By such means, astronomers can measure an enormous range of properties of astronomical objects, working from the

▼ Though there will be a few foreground and background stars mixed in, most of the stars in the Hyades star cluster will be at essentially the same distance so their brightness differences are real. This allows astronomers to compare the spectra of stars of different brightnesses even though the actual distance to the cluster cannot be measured directly. On the left is a photo of the cluster, while on the right is a view taken with a prism in front of the lens, which is a method of getting simple spectra of a whole region of stars.

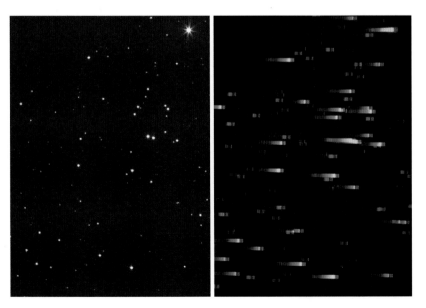

known to the unknown. It is very much like viewing a coastline at night. We see various lights around us, and even if we have never seen the view by day we use our knowledge of what sort of objects can produce light to give us an idea of the landscape. There are house lights, car headlights, streetlights and perhaps the flash of a lighthouse on land. Each has its own characteristics and even though the distant lighthouse may appear brighter than much closer house lights we know what it is from its appearance and do not assume that it is nearby because it is bright.

Similarly, when we look out to sea we might see a set of red and green lights bobbing about, so we instantly know that this is a small vessel, compared with another set of lights moving slowly but steadily which we interpret as a larger vessel. This is an example of how we use one property – the motion of the lights – to infer not just the size of the vessel but the probable true brightness of the lights. If we know how bright the lights are in reality we can judge how far away the vessels are. It helps if there is a quayside in sight where we know that all the vessels are together so we can compare the lights on a small fishing vessel with those on a larger ferry.

Astronomers use the same trick all the time. We know that stars have a wide range of true brightnesses. In general, the more distant they are, the fainter they appear, but we recognize from their characteristics that some are real searchlights which we can see over vast distances, while others are guttering candles which are hard to make out even when they are close.

Many stars occur in vast clusters, and in these cases we can be sure that the brighter stars in the cluster are truly bright compared with their fainter neighbors. This makes it possible to compare the properties of stars that are in effect at a standard distance from us. There are particular stars known as Cepheids, which are recognizable because their brightness varies in a distinctive way (see page 109). The greater the overall brightness of a Cepheid, the longer its cycle of variability. From studies of Cepheids in clusters, astronomers can say with some certainty that when they spot one, and can measure its period of variations, they know pretty well how bright it really is and from the brightness that it appears, they can calculate its distance.

ABOUT STARS

In principle a star is a fairly simple body – it is basically a globe of hydrogen gas which shines not because it is burning, like a gas flame at home, but because of nuclear reactions at its heart. These consume hydrogen, converting it to helium and releasing energy in the process. Very high pressures and temperatures are needed to start these reactions, which is why they don't occur in hydrogen gas under normal conditions on Earth. If they did, we would have a very good energy source, but currently the only way we can use this form of power is in a hydrogen bomb.

The main deciding factor that controls the life of a star is its mass. Low-mass stars – with a tenth or so of the mass of the Sun – are very dim and red and last

for billions of years. Those of high mass – which could be up to about a hundred times that of the Sun – shine brilliantly with a strong bluish light, but last only a few million years. In between is a wide range of stars with intermediate properties, and there are also stars at the beginnings or ends of their lives which have different characteristics from the general population. The high-mass stars have rather featureless spectra, with just a few lines, mostly of hydrogen, present. Those of lower mass have more complicated spectra, with increasing numbers of lines indicating the more complex atoms and molecules that can exist in their cooler atmospheres.

At one time, astronomers categorized stars by the strength of the hydrogen lines, and put them in a sequence which started with type A and ran through the letters of the alphabet. Today, this sequence has been reorganized with the same letters but based on the star's temperature. It now starts with type O stars which are the most massive, hottest and bluest, through type B, type A, then F, G, K and M. Within these classifications there are also giants and dwarfs, but these do not actually refer to the masses of the individual stars but solely their diameter compared with others of their type,

▼ **Two views of other galaxies** *that show what our own Milky Way Galaxy would probably look like if viewed from outside. Important features are the spiral arms highlighted by gas clouds and bright bluish stars, the nucleus consisting of older and redder stars, and the dust lanes that are particularly obvious when looking side-on. The Sun's location is in a spiral arm about two-thirds of the way from the center, which is at a distance of 26,500 light years. A faint halo surrounds the Galaxy, considerably increasing its diameter.*

which depends on their stage of evolution. So there are so-called dwarf stars which are much more massive than the Sun, and stars can go from being dwarfs to giants as they evolve.

There are plenty of examples of each type of star in the sky, and their colors, while not as obvious as those of, say, traffic lights, are readily visible. The colors of stars add to the spectacle of viewing clusters in particular, and these various types of star are often referred to in observing guides. Very often, astronomers measure only the color of a star, and not its entire spectrum. Although this does not give a very accurate idea of the properties of an individual star, because there will be variations from star to star, if enough of them are measured in a cluster on average the variations will even out and the true brightnesses of the stars can be estimated, and hence the distance to the cluster can be worked out. In this way a fairly accurate picture of much of the Galaxy has been established.

As in human society, the brightest and most impressive stars are usually the ones that catch the eye, but there are stars in our own neighborhood which are bright in our sky just because they are close to us. Most of the bright stars in the sky are actually among our closest neighbors, but there are also many more close stars that are invisible to the naked eye because they are so dim. The nearest star of all to the Solar System is one of these – a red dwarf which is part of the Alpha Centauri system, known as Proxima Centauri. It is about a hundred times fainter than you can see with the naked eye, and requires a reasonable telescope to show it at all. Even then, it is a matter of picking it out from many other faint stars in that direction, most of which are actually much more distant.

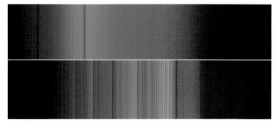

▲ *Spectra of two different types of star.* *The upper one is a B-type star, whose hot atmosphere shows mostly the lines due to hydrogen. The lower is a type M star, with a much cooler atmosphere that allows more complex molecules to exist and hence has many fine lines. Not all elements or molecules produce lines of equal strength, however.*

Red dwarf stars of type M like Proxima are by far the most common in the Galaxy: about 70% are of this type. In comparison, only about 6% of the stars in the sky are as bright as the Sun or brighter, yet these are the ones that we see and consider typical. However, more curiously still, studies of our own and other galaxies show that even if all the bright and dim stars are allowed for, plus all the dust and gas and so on that we can detect, there must still be about ten times more matter in the Galaxy. If this matter were not present, the Milky Way would have flung itself apart by now. What's more, all the other galaxies show the same state of affairs, as do clusters of galaxies. The nature of this unseen matter is one of the great mysteries of astronomy.

DISTANCES BEYOND THE GALAXY

For many years it has been known that virtually all galaxies, except the closest, have light which is red-shifted – implying that they are moving away from us. In the 1920s Edwin Hubble found that the more distant a galaxy appears – because of its size and brightness – the greater its redshift. Astronomers now believe that the Universe was created in what amounted to a "big bang," and this redshift is the main evidence for this. Alternative explanations have often been suggested, but none fit the facts as well.

Winding back the expansion gives us an age for the start of the Universe, but the whole thing needs to be calibrated so that we know just how far a particular galaxy has moved at a given speed. During the second half of the 20th century arguments raged over the distance scale of the Universe, with different methods giving wildly differing values.

If a galaxy is close enough that Cepheids can be seen within it, their periods of variability and apparent brightness make it possible to work out their actual distance. Until quite recently it was virtually impossible to pick out individual stars in any but the nearest galaxies. When it became apparent that a space telescope that would have sharper vision as a result of being above the atmosphere might be feasible, the Hubble Space Telescope was born. One of its prime missions

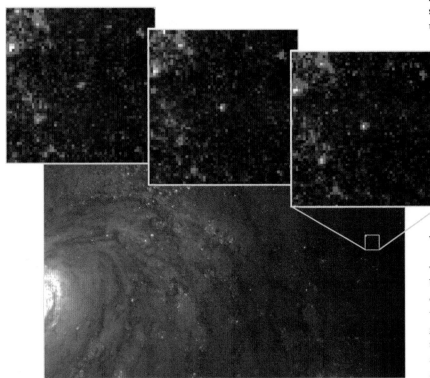

▼ *The Hubble Space Telescope* *is powerful enough to detect Cepheid variable stars in the galaxy M100 in the Virgo Cluster. Here, one of these stars can be seen varying over a period of time, yielding a true brightness for the star and hence an accurate distance to the galaxy – assuming that Cepheids in M100 behave the same way as those in our own Galaxy.*

was to detect individual Cepheids in the Virgo Cluster of galaxies, about 50 million light years away.

This it did, and in doing so gave their distances which were a vital stepping stone in the sequence of known distances. There are several methods of finding distances, the Cepheid method being one of the most accurate. But a further standard star is needed for more remote galaxies, and that has been provided by the Type Ia supernovae. These are double stars in which material leaking from one star overloads the other, causing that star to explode when its core reaches a certain critical size. Such an explosion is extremely brilliant but also probably of a uniform brightness from star to star, which makes it an ideal candidate for distance measurement.

Some astronomers, including some amateurs, devote their lives to surveying distant galaxies in the hope of finding a Type Ia supernova in them. Such supernovae are about 1000 times brighter than Cepheids, so they can be seen as far away as about 10 billion light years. When one flares up, its spectrum and the way in which its brightness varies over a matter of days and weeks is measured so as to confirm which type of supernova it is. If it is confirmed as a Type Ia, its distance can be worked out in the same way as for Cepheid stars. It is now possible to calibrate the distance scale fairly accurately, and the consensus among astronomers is that the Universe is 13.8 billion years old.

All this – the distances and masses of nearby stars, the size of the Milky Way, the distances to nearby and then more distant galaxies – has been worked out by moving from the known to the unknown. Always there are assumptions made, and sometimes these assumptions turn out to be wrong. But astronomers are fairly confident that they are on the right track.

Where does the amateur stargazer fit into all this? Barring a very unlikely chance, there is nothing that the casual observer with simple equipment can do to push forward the frontiers of knowledge. But much of skygazing is about imagination and understanding. It really helps to know where in the grand scheme of things a planetary nebula, or a Cepheid variable, or a double star fits. As you gaze, you can start to imagine the sky in three dimensions, with the planets close by, the nebulae and star clusters more remote but still within our general neighborhood, and the distant galaxies much more distant still.

These days, it is not just the light from astronomical bodies that astronomers study, but all the rest of the spectrum as well. Most of these observations are beyond the scope of amateurs, but they are crucial to our understanding of the cosmos, and are described on the next few pages.

OBSERVING AT OTHER WAVELENGTHS

Modern astronomy depends on our ability to observe celestial objects at a wide range of wavelengths. Light is just a small part of the full electromagnetic spectrum, which ranges from radio waves at one end to gamma

rays at the other. We see only a tiny part of the whole range, the colors between red and violet.

Why don't humans see all these different wavelengths? One reason is that most of them do not penetrate Earth's atmosphere – fortunately in the case of gamma rays and X-rays, which are very energetic and would destroy most life on Earth if they did get through. The low-energy radio, microwave and infrared parts of the spectrum do penetrate to varying extents, but by and large the information they carry is rather limited.

The property known as wavelength is all-important. Light has a wavelength of less than a millionth of a meter, and a system to observe such wavelengths need be only a few millimeters across – the eye being an excellent example. Radio waves, by comparison, have wavelengths over a million times longer, which is why radio and TV antennas are unwieldy arrays on rooftops. Furthermore, the short-wavelength rays of X-rays and gamma rays are so energetic that they are troublesome to focus, whereas a bit of glass does the job for light.

The human eye has evolved to make the best use of the light that does penetrate our atmosphere, seeing the full range as white. The Sun, incidentally, has to be white

▲ **The bright star** is a Type Ia supernova that appeared in 1994 in the galaxy NGC 4526, a member of the Virgo Cluster, and therefore at a similar distance to the Cepheid variable in M100. Its greater brightness is obvious, making such objects suitable for determining the distances to much more remote galaxies.

▼ **Visible light,** shown by the rainbow at the center, is just one small part of the full spectrum, which is characterized by the wavelength of the different forms of radiation. Virtually the whole range of wavelengths is used by astronomers.

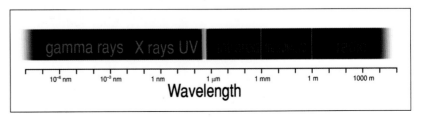

gamma rays X rays UV

10^{-6} nm 10^{-3} nm   1 nm   1 μm   1 mm   1 m   1000 m

Wavelength

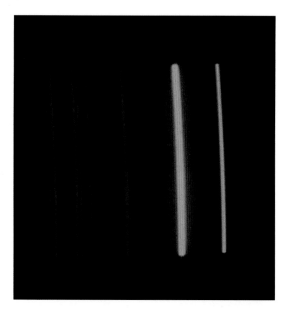

▶ *The spectrum of a sodium streetlight is typical of a hot gas – it consists of a number of narrow individual wavelengths rather than a broad band of all colors. Gases in space also have their own characteristic spectra.*

having a wide distribution of wavelengths in its light, a neon tube shines only with pink light. The orange-yellow sodium streetlights also shine with virtually a single color. Both of these are examples of the light from hot gas, each gas having its own characteristic colors. Such individual colors are not confined to optical wavelengths, but can occur in different parts of the spectrum. Cold hydrogen gas, for example, gives off radio waves with a wavelength of 21 cm, just as hot sodium gas gives off yellow light.

Radio waves are not interrupted by the atmosphere, just in the same way that a human, with long strides, can march across a plowed field that would be tough going for an ant. They will also pass through the dust and gas in the Galaxy quite readily, so the 21 cm radio emission can be detected from gas over a wide area of the Galaxy whose light just cannot penetrate the dust clouds. This has taught us a great deal about the structure of our own Galaxy that would be impenetrable at optical wavelengths.

The Doppler effect applies to radio waves and the other wavelengths just as much as to light, which is why it is possible to pick out different arms of the Galaxy from their radio output. The Galaxy is not a solid body rotating like a wheel – every star and gas cloud in it orbits the center just as the planets orbit the Sun. The different gas clouds have their own particular motions and therefore Doppler shifts, which depend on their distance from the center of the Galaxy. This has made it possible to draw a plan of the Galaxy's gas clouds, revealing a general spiral structure just as we see in many other galaxies.

rather than the yellow that it is normally depicted in books, for this very reason. Normally you cannot look directly at it because of its brilliance, but when you see sunlight reflected on water it does indeed appear white. The only reason we think of the Sun as yellow is because on those occasions when we do glimpse it as a disk, it is either very low in the sky or seen through cloud, and in each case the light is yellowed by absorption.

The smooth general distribution of light from a star is typical of an incandescent body, like a light bulb. However, there are other means of producing radiation, a neon tube being a very good example. Instead of

▶ *Radio telescopes such as the 76 m (250 ft) dish at Jodrell Bank in Cheshire, UK, can see not only through gas clouds in space, but also through terrestrial clouds. They need not be located on remote mountaintops, though locations free from electrical interference are a great advantage. However, because radio wavelengths are so long and their energy so low, a giant dish is needed.*

Gas clouds also emit radiation specific to molecules other than hydrogen, including many organic molecules – which means not that they are life-bearing, but simply that they contain carbon. These giant molecular clouds map out the spiral arms of the Galaxy.

Radio telescopes can detect radio emission from many distant galaxies as well as our own. Some galaxies stand out particularly strongly, and have either strong radio emission from their nuclei or from plumes of gas extending far outside the visible galaxy, often perpendicular to the long axis of the galaxy. Such activity is linked to the presence of supermassive black holes at the centers of these galaxies.

Black holes are much misunderstood creatures. People imagine that they are dangerous beasts, lurking round every galactic corner, ready to suck in unsuspecting stars. But in fact they are no more dangerous than the stars that they once were. "Ordinary" black holes were once just very massive stars that collapsed in on themselves at the end of their short energy-producing lifetimes. They may have masses many times that of the Sun, but nothing out of the ordinary for massive stars. When one becomes a black hole it shrinks in diameter compared with its former self, so that it becomes very dense, but the effect of its gravity on its surroundings remains unchanged. Stars and planets nearby are unaffected by its altered state. However, any gas that happens to fall toward it may now be accelerated much more before it finally enters the point of no return, so becomes very hot indeed. Paradoxically, the black hole, which has such a strong gravitational pull that neither matter nor light cannot escape from it, is betrayed by the X-rays given off by the material plunging into it. This may be detected by radio telescopes on Earth, or by X-ray satellites above our atmosphere.

There are other processes going on in the Universe which produce tell-tale signatures. Electrons whirling in magnetic fields have their own type of output, easily identified by radio astronomers. Very violent events, such as the collision of massive stars, can produce extremely energetic radiation such as gamma rays, which are detectable over vast distances. Bursts of gamma rays are fairly frequent occurrences, but they all occur so far away that they took place when the Universe was much younger. By such means astronomers can piece together the history of the Universe from the here-and-now back to a time very close to the origin of everything in the big bang.

▲ **NASA's Chandra X-ray Observatory** was launched in 1999 to study the X-rays from celestial bodies, and complements the Hubble Space Telescope and the Spitzer Space Telescope, which observe in the visible and the infrared respectively.

▶ **The WMAP satellite** (see A–Z page 315) produced this map of slight variations in the temperature of the Universe's microwave background radiation, dating from shortly after the big bang. These fluctuations are believed to have given rise to the structures in the Universe we see today.

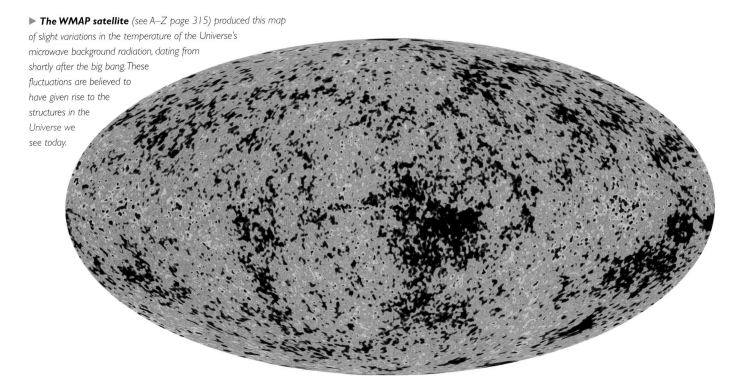

2

Learning the sky is a challenge, but the good news is that it is actually quite easy once you get started. What's more, it is something that you can do at your own pace, when the opportunity presents itself, and all you need is your eyes and a clear night sky. With a little effort you can recognize the major constellations or star patterns, and from there you can, if you want, move to the fainter ones. Soon, the sky will be as familiar to you as your own neighborhood.

Even if you live in a town you can make a good start. These days almost everyone is affected by light pollution – the glare from streetlights and other artificial forms of lighting. In towns and cities this almost drowns out the stars, but even in the largest city you can still see some stars and pick out the main constellations under the right conditions. People often say that when they see a really dark, clear sky, there are so many stars that they can't find their way around. So there is something to be said for starting with a simpler sky.

Probably the biggest hurdle that people encounter when they start to learn the sky is picking out their first constellations from the maps. One major difference between learning the sky and learning your way around a town is that the streets stay put, but the stars are always on the move. You may have been shown some well-known pattern in the past, but where is it now? Quite possibly it is not in the sky right now, or so low down that it is out of sight. Different stars are visible depending on the time of night and time of year, or where you happen to be on the Earth's surface. Those stars you learned a couple of months ago one early evening at home will be quite different from those you see after a late-night party on your foreign vacation. So the first thing to do is to find out what stars are visible at your time and place, then get your bearings so that you can locate them in the sky.

Then comes the next problem – the maps are small, but the sky is huge. You have to get the hang of the scale of the maps compared with the sky, and interpret the brightnesses of the stars as they are represented by the map symbols. There may be two or three planets in the sky which are not marked on the maps because they move around the sky at varying speeds. After a while you can pick out the difference between a planet and a star, and even work out which planet is which just by its appearance to the naked eye.

QUICK START GUIDE

This chapter explains step by step how to follow the movements of objects in the sky, and how to know which map to choose for your particular time, date and place. But if you want to plunge in at the deep end, go to Chapter 7 and choose the section for the appropriate month. Then look at either the upper or lower pair of small maps, depending on where you live. These are like tourist maps, from which you can choose a larger-scale star atlas map from Chapter 8 by referring to the page numbers around the edge. Each atlas map is shown in two ways – in a photo-realistic version that closely resembles the real sky, and, on the opposite page, in a more conventional version in which the stars and constellations are labeled. Finally, for a more detailed look at each constellation, return to your chosen month where you will find maps with targets suitable for binoculars and telescopes.

We have deliberately not shown every one of the 88 constellations in detail. This is because many of them contain little of interest, and few people spend much time looking at them, just as conventional travel guides don't bother listing every building in an area but concentrate on the more interesting ones. You may be surprised that some constellations everyone has heard of, such as Libra, don't get a special mention. But the well-known names of the zodiac are simply those that are along the path of the Sun and planets through the sky. In ancient times, people believed that the locations of the planets in the sky could affect their lives here on Earth. That has turned out not to be the case, but some of the rather ordinary suburbs of the sky have gained unwarranted fame as a result.

Telescopes and binoculars

You will probably want to find some of the sights of the sky using binoculars or a telescope. Some are easy to find, while others require more effort. People are sometimes disappointed that even an expensive telescope

▶ **The apparent path of the Sun** through the sky is called the ecliptic. The zodiacal constellations are those that lie along this path.

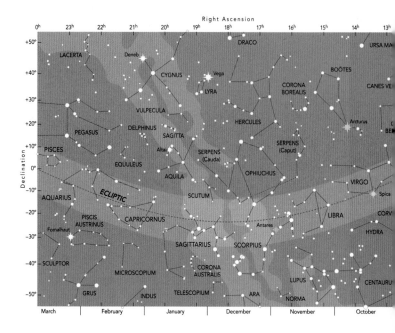

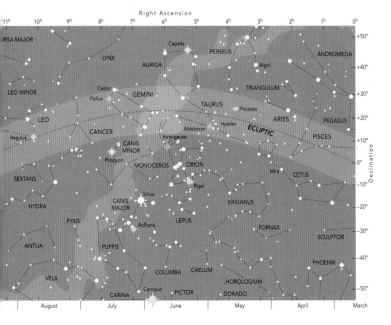

◀▲ *Two views of the Lagoon Nebula, M8, in Sagittarius. At left is an amateur photograph by Michael Stecker in California; above is a drawing by Darren Bushnall of Hartlepool, UK, using a 300 mm reflector. This telescope, though fairly large by amateur standards, shows much less detail than the photograph, and no color is visible.*

doesn't show anything resembling the dramatic color photos that you see in books. The sad fact is that the human eye, while incredibly acute and sensitive, is not good at picking out color in faint objects, while modern imaging techniques can transform a hazy blur into a vivid spectacle.

In our constellation pages we have shown objects in a variety of ways, from drawings and photographs made using amateur equipment to images made using giant telescopes or the Hubble Space Telescope. People often ask, "Are those colors real?" The answer is usually, "Yes, but they are not the colors you will see with your eyes." Though you may not be able to

see the Crab Nebula, say, in the same way that a giant telescope shows it, you can still get a sense of achievement from seeing that faint blur with your own eyes. The interest comes from tracking down an object rather than from the dramatic view.

Having said that, the sky does contain some more subtle gems. You may find from your suburban lawn a star cluster that looks like a faint sprinkling of stardust on the sky, or a pretty double star. The planets are visible even from light-polluted locations, while a sight of the Moon at first quarter through even a small telescope can be stunning, exceeding the ability of any printed page to reproduce its appearance.

These days, many telescopes have what is called a Go To facility – that is, they are equipped with motors and a self-contained computer that will, if it is properly set up and everything works as it should, find hundreds of objects for you automatically. People often believe that this is the answer to the beginner's problem of finding their way around the sky. But many experienced stargazers would argue that it is better to learn to find objects for yourself, which will give you a much better knowledge of the sky than relying on technology. And knowing in advance what you are looking for is important, even with a well set-up Go To instrument. Are you looking for a tiny, faint fuzzy blob or a scattering of stars across the sky? The catalog number doesn't tell you. But our listings of objects tell you what to expect, and what sort of

▶ *The Earth's elliptical orbit* (blue) compared with a true circle (red). The difference appears slight, but the Earth's distance from the Sun varies by 5 million km.

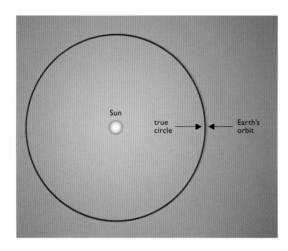

magnification to use when looking. And if you are relying on your own skill rather than electronics, we give many tips on how to find the objects from scratch.

For many people, visual observing is not enough. Some objects that are invisible to the naked eye can easily be imaged with either a digital camera or specialized CCD camera. Even in the city, some amazing results are possible using image processing that can be achieved on any modern computer. Details of methods used are on page 38, and you can find the best targets for your camera in Chapter 7.

Whether you observe with the naked eye from the city, or with a computerized telescope from a dark-sky site, you will find plenty to keep you busy in the sky. We hope that this book will be your companion for many years.

Getting your bearings

The first thing any stargazer needs to do is to establish which way is north or south. Most people have a good idea of their orientation from their home site, but if you are at a new observing site you may have problems. One way is of course to look at a map, while another is to notice the direction of the Sun at true midday. At this time it is on your *meridian* – your north–south line. This means that it is due south if you are in the northern hemisphere, or due north if you are in the southern. In the tropics it will be more or less overhead, which is not so useful. Wherever you are on Earth, the Sun and indeed the rest of the sky moves slowly as the Earth turns, rising in the eastern side of the sky and setting in the western side. Only rarely does the Sun actually rise due east, and set due west.

In practice, other factors affect the actual position of the Sun at midday. The most obvious is the presence or absence of Summer or Daylight Saving Time. When this is in force midday usually occurs an hour later, around 1 pm. Another factor is an unevenness in the Sun's apparent motion over the year, which results in it being sometimes late and sometimes early to reach the meridian. This is caused by the Earth's orbit around the Sun not being circular but being an ellipse – a slightly squashed circle, with the Sun slightly displaced from the center. Virtually all orbits, whether of planets or comets around the Sun or one star in a double star around the other, are elliptical. When the Earth is closest to the Sun, it moves faster in its orbit than when it is most distant, resulting in the varying speed of the Sun through the sky. These daily variations

▶ *A time-exposure image* of star trails taken at 44°N using a fisheye lens looking east. The Pole Star at top left remains almost stationary, while stars in the east rise at an angle to the horizon. Due south, at far right, the stars set only a short time after they have risen.

are known as the Equation of Time, and you can find them listed on the website www.stargazing.org.uk. The same factors affect the accuracy of sundials.

A final correction in the Sun's position at midday arises from your position within your time zone on Earth. We all know that the time in different countries east or west of our own is different, but for convenience we all stick to the same time within our own time zone. If you happen to be near the edge of your zone, however, you could find that time as measured by the Sun is different from that on your watch by an additional half an hour or more. These factors also affect the local times of the appearance of the sky as shown on the star maps, so if the maps always seem to be slightly in error, this could be the reason.

THE CELESTIAL SPHERE – HOW THE SKY MOVES

The changes in the Sun's position, and the overall movement of the sky, are most easily understood by thinking of the sky in terms of a huge sphere surrounding the Earth. We know that in reality the Earth goes round the Sun, and that it is the rotation of the Earth that causes the Sun, Moon and stars to rise and set. But the concept of the celestial sphere is very useful, and is the standard way of looking at such things.

A simple celestial sphere is shown at top right. It is drawn so that the Sun's apparent path through the sky – called the *ecliptic* – is horizontal and the Earth's axis is tilted to it. Imagine yourself on the Earth in the middle; you would feel that the Earth is stationary and the giant sphere surrounding the Earth appears to turn once a day along the tilted axis that sticks through the Earth, from north to south pole.

Just like the Earth, the celestial sphere has a north and south pole, and an equator. These are the extensions into the sky of the same locations on the Earth. The sky also has two hemispheres, northern and southern.

You can see that if you were at the North Pole, the sky and stars would rotate parallel to the horizon, with none rising or setting. At the South Pole they would do the same thing, but in the opposite direction. Stand on the equator and everything appears to rise and set vertically from the horizon, with objects toward the north and south performing smaller and smaller semi-circles as they get closer to each celestial pole.

Most of us, however, live at some intermediate point on the Earth's surface. We see one of the celestial poles part of the way up the sky, with all the stars rotating around it. In the Earth's northern hemisphere you can see the stars around the north celestial pole all the time, and you can also see some of the stars in the sky's southern hemisphere, below the celestial equator. But there are some stars, around the south celestial pole, that you will never be able to see because the Earth is always in the way. The opposite applies at locations in Earth's southern hemisphere. People living at the equator can see the whole sky at one time or another, while the nearer to one of the poles you live, the less of the sky's opposite hemisphere is visible to you.

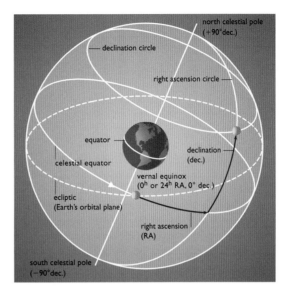

◄ **Right ascension (RA)** is measured eastward around the celestial equator from the vernal equinox (or First Point of Aries). Declination (dec.) is given by the angle north or south of the celestial equator.

Wherever you live, objects in the sky will always appear to rise along the eastern horizon and set along the western horizon. In the northern hemisphere they reach their highest point above the horizon toward the south, which is where you will find the Sun at true midday. In the southern hemisphere objects rise from the eastern horizon at the reverse angle, and reach their highest point due north.

Star trail photographs are among the easiest of astronomical photographs to take, and simply involve keeping the camera's shutter open for a period of time so that the stars trail across the chip. Photos taken in this way show the stars' movements very clearly.

The Sun's movements

The Earth's axis is inclined at an angle – 23½° – to the ecliptic (which is in actual fact the plane of the Earth's orbit around the Sun). This is what gives rise to the seasons, and from the stargazer's point of view it means that the ecliptic is inclined to the celestial equator. The Sun moves from west to east along the ecliptic during the course of the year while the Earth turns daily beneath the stars. As we measure time by the Sun, it is the average interval between successive middays that we divide into 24 hours. But because the Sun has moved slightly along the ecliptic, there is a small slippage between the time as told by the Sun, and that as told by the stars. In fact, the same stars are visible four minutes earlier each night.

After a year the Sun has moved full circle and all the four minutes have added up to get the stars back to almost exactly the same position as they were the previous year at that date. They are not in precisely the same place, which is why every four years, in what is known as a leap year, we need an extra day to keep the movements of the Sun and the stars together. Without this extra day, eventually the months and the Sun's movements would get out of step and June or December would end up occurring in the fall or spring.

When the Sun is north of the celestial equator, it is high in the sky in the northern hemisphere. It reaches its maximum altitude around 21 June, which is midsummer in the northern hemisphere. From there it moves southward until around 22 September it is exactly on the celestial equator. On this day only it rises exactly in the east, and day and night are 12 hours long all over the globe (excepting the regions around the poles where it is on the horizon). This is the *equinox,* meaning "equal nights." Three months later it is at its southernmost point, and the southern hemisphere experiences midsummer; then in March it is back again at the equator for the vernal equinox.

The ecliptic is also the approximate track of the Moon and planets, though they can move several degrees on either side of it. We are all familiar with the Sun – and therefore the ecliptic – being high in summer at midday, but look at the diagram on the previous page and you can work out that at midnight the opposite occurs, and the ecliptic is low in the sky. As a result, the Moon and planets are rather low in the sky at midsummer. Then in midwinter, the ecliptic is high in the sky at midnight and any planets that happen to be roughly opposite the Sun in the sky will appear very high up, which makes them ideally placed for observation.

Positions in the sky

The celestial sphere diagram also helps to explain the way positions in the sky are measured – the celestial equivalent of latitude and longitude. The easiest to follow is the equivalent of latitude, which is called *dec-lination.* Just as on Earth, this is measured in degrees from the equator to the pole, with the equator being 0° and the pole being 90°. Southerly declinations are usually given a minus sign. If you are at the equator, the stars directly overhead have a declination of 0°, and wherever you are on Earth, stars directly above you (in your *zenith*) have a declination equal to your latitude.

Longitude on Earth is measured in degrees west of a fixed point on the Earth's surface, which by inter-

national agreement is Greenwich Observatory near London. The equivalent fixed point in the sky is not a particular star, but the point where the ecliptic crosses the celestial equator when the Sun is on its way north-ward in March. This is known as the vernal equinox or First Point of Aries, though because of a slow move-ment of the Earth's axis, known as precession, it is no longer in Aries.

The celestial equivalent of longitude is known as *right ascension,* or RA. Though RA can be measured in degrees, it is usually measured in hours and minutes eastward from the First Point of Aries, from 0 hours to 24 hours. There is a good reason for this – it relates to the apparent rotation of the celestial sphere around the Earth. Say you start observing with the First Point of Aries on your meridian. An hour later, stars with an RA of 1 hour will be on your meridian. At that time, stars with an RA of about 7 hours and on the celestial equator will be just rising in the east. The sky is one great 24-hour clock, and the scale of RA reflects this.

Bear in mind that the separation of two stars 6 hours of RA apart depends on their declination. If you look toward the pole where all the hour lines of RA converge, two stars separated by 6 hours of RA will actually be close together in the sky. On the celestial equator, however, they will be half a sky diameter apart.

Angles in the sky

How do these measurements relate to the separation of two stars in the sky? We usually measure angles in degrees, so two objects exactly on opposite sides of the sky are 180° apart, while the distance between the horizon and the zenith is 90° and an object halfway up the sky is 45° above the horizon. Stargazers often refer to such angles, and it is very useful to have a mental idea of the appearance of angles in the sky. If you hear that Mercury, for example, will be 10° above the horizon this evening, it is worth knowing how high this is in relation to the local rooftops. Or if the Moon will be 3° from a particular star, will that be within the field of view of your binoculars, or a telescope?

The degrees of declination are the same as degrees of angular measurement. So two stars that have the same RA but are separated by 22 degrees of declina-tion, say, are a genuine 22° apart in the sky. An hour of RA is 15° only when it is on the celestial equator. Closer to the poles, an hour of RA will be a smaller angle, which is one reason why RA is usually referred to in hours and minutes rather than degrees.

It so happens that a handspan from tip of thumb to little finger, when stretched out at arm's length, is about 16° to 20°. People often use this to get an idea of scale when measuring angles in the sky. In the same way, your index finger at arm's length is about 1°. Other useful angles to remember are the separation of the two stars known as the Pointers in the Big Dipper or Plough (see page 139), which are 5.4° apart, or, in the southern hemisphere, the length of the Southern Cross, which is 6°. A typical field of view of binoculars is about 5°, and a telescope's low-power eyepiece is often about 1°.

▼ The area of the image on page 22, with the RA and dec. lines added. RA lines are marked every 30 minutes, and dec. lines every 5°. At the celestial equator, 15° of dec. is the same as 1 hour of RA. But near the pole, 15° of dec. is the same as 3 hours of RA. The map scale changes near the pole to retain the appearance of the constellations.

From time to time you may find references to positions in the sky in terms of *altitude* and *azimuth*. In this case, altitude refers to the angular distance in degrees of an object up from the horizon. In the earlier example, therefore, Mercury would have an altitude of 10°. This is not to be confused with the altitude of your observing site above sea level! Azimuth is the angle round the horizon starting from north and going through east. North is 0°, east is 90°, south is 180° and west is 270°. This applies in both the northern and the southern hemispheres. Each degree is divided into 60 arc minutes, which in turn are each divided into 60 arc seconds. Altitude and azimuth are most commonly used to refer to the positions of artificial satellites in the sky. The position where an object rises along the horizon is also often referred to by its azimuth, as this allows you to work out its direction on a terrestrial map.

GETTING TO KNOW THE SKY

Learning the sky is little different from learning your way around any new district that you happen to be in. To begin with you need to know the overall structure with a few landmarks that you can easily recognize. Then you can start to fill in the details to suit your own requirements. In some ways, learning the sky is easier because you get an instant overview anyway, as if you were seeing the district from the air.

To begin with, you need to establish which stars are visible at the time you will be observing, from your location. The monthly maps in Chapter 7 give you an overview of the sky looking either north or south, and from either the northern or the southern hemisphere. From these, you can refer to the appropriate atlas map in Chapter 8 for a more detailed view of that part of the sky.

Remember that the lower map of each pair shows the view looking toward the pole, in which case the only change from month to month is the orientation of most of the stars. These stars that never rise or set are known as *circumpolar stars*. But the upper map shows the sky looking away from the pole – south in the northern hemisphere and north in the southern hemisphere. In this direction, the stars on view change with the seasons.

The scale of these maps is a little small for picking out the constellations, so you may prefer to use the main atlas maps to which they refer instead. The key to finding your way here is to identify the brightest stars or most obvious patterns, as referred to in the captions to the monthly maps. Be aware that any bright planets in your region of sky will make a big difference to the star patterns. You can check the monthly positions of the bright planets by using the tables on the website www.stargazing.org.uk.

As you observe, the stars move slowly across the sky. After an hour or so, some stars in the west are noticeably lower, while others in the east are higher up. If you watch for several hours, a completely new sky presents itself. Alternatively, if you observe at the same time each night, the same thing happens over a period of time. In a month of observing, you will see stars that you would have waited two hours to see on the first night of observation. Should you be impatient to see stars that will be in your evening sky in three months' time, you can wait for six hours and they will be there in the early morning sky. After a year, the sky has gone full circle and you are back where you started.

An alternative to the monthly maps in Chapter 7 is to purchase a Firefly planisphere for your particular latitude. These handy star disks have a map of the sky visible from your location with a rotatable overlay that can be set to show exactly which stars will be above the horizon at any chosen date and time.

Star brightnesses

On the maps, the stars are shown by dots of different sizes according to their brightness. The true range of brightness of the stars is enormous – the brightest star in the night sky is 1000 times brighter than the faintest star easily visible with the naked eye. To make the area of its dot accurately depict its brightness compared with the faintest star on the map, it would have to be about 8 mm across, which would be impractical. This is why we have created the photo-realistic maps, in which the fainter stars are shown by less bright dots. This, and the lack of labels, makes them much more representative of the real night sky. Once you have identified the star patterns using the photo-realistic maps you can use the conventional maps to find their names.

Star brightnesses are measured on a scale unique to astronomy, which has remained in place since ancient Greek times. Put simply, the brightest stars are first magnitude while the faintest ones are sixth. Thus they are ranked like the winners in a contest, with the most prestigious having the lowest number. This is the opposite of most means of measurement, and the magnitude scale is a potential source of confusion at all levels in astronomy.

When it was required to place the ancient scale on a mathematical basis, it was decided to make five magnitudes exactly equal to a range of 100 times in brightness. To accommodate the full range of star brightnesses, the very brightest objects had to be assigned negative magnitudes. The table below gives a few representative magnitudes. Very soon you will get an idea of what each magnitude means, and how a star of, say, the second magnitude appears to the naked eye from your usual observing site, and when seen through binoculars or a telescope. This is one of the great keys to finding objects in the sky, because you

EXAMPLES OF TYPICAL MAGNITUDES			
Sun	−26.8	Typical naked-eye limit in country	5.8
Moon	−12.7	Naked-eye limit with acute vision	7
Venus at its brightest	−4.4	10 × 50 binocular limit	11
Sirius (brightest star)	−1.5	Proxima Centauri (nearest star)	11.01
Alpha Centauri (nearby star)	−0.3	114 mm telescope limit	13
Polaris (Pole Star)	2.1	200 mm telescope limit	14
Typical naked-eye limit in towns	4.5	Giant telescope limit	30+

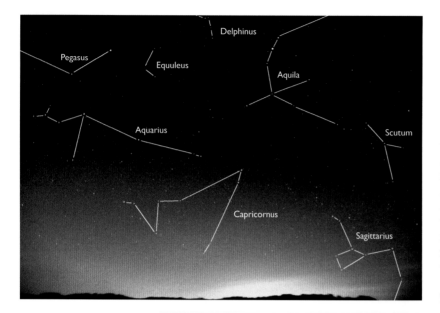

Delphinus

Pegasus

Equuleus

Aquila

Aquarius

Scutum

Capricornus

Sagittarius

▲ *The constellation*
patterns rarely look like
what they are supposed to
represent. But the use of
shapes makes it much easier
to recognize the various parts
of the sky. The lines have no
official significance, and are
shown on maps as an aid
to picking out the patterns.

need to be able to know when you have found the star
you are looking for, and by how much your optical aid
brightens its appearance.

Constellations and their names

Beginners to astronomy are often bemused or even
amused by the peculiar names of the constellations and
their meanings. What is fishy about Pisces, and what
is balanced about Libra? The star patterns themselves
rarely work on a "join the dots" basis. But the constellations are actually quite useful, even if the names have
little to do with their appearance. Say to stargazers
that a particular object is in Aquarius, and they will
immediately have an idea of which part of the sky
you are talking about, and when it is visible. But say
that it is on the celestial equator and at 22 hours RA,
and most will have to turn to a star map to see where
you mean.

THE CONSTELLATIONS							
Name	Genitive	Abbreviation	Common name	Name	Genitive	Abbreviation	Common name
Andromeda	Andromedae	And	Andromeda	Leo	Leonis	Leo	Lion
Antlia	Antliae	Ant	Air Pump	Leo Minor	Leonis Minoris	LMi	Little Lion
Apus	Apodis	Aps	Bird of Paradise	Lepus	Leporis	Lep	Hare
Aquarius	Aquarii	Aqr	Water Bearer	Libra	Librae	Lib	Scales
Aquila	Aquilae	Aql	Eagle	Lupus	Lupi	Lup	Wolf
Ara	Arae	Ara	Altar	Lynx	Lyncis	Lyn	Lynx
Aries	Arietis	Ari	Ram	Lyra	Lyrae	Lyr	Lyre
Auriga	Aurigae	Aur	Charioteer	Mensa	Mensae	Men	Table (Mountain)
Boötes	Boötis	Boo	Herdsman	Microscopium	Microscopii	Mic	Microscope
Caelum	Caeli	Cae	Chisel	Monoceros	Monocerotis	Mon	Unicorn
Camelopardalis	Camelopardalis	Cam	Giraffe	Musca	Muscae	Mus	Fly
Cancer	Cancri	Cnc	Crab	Norma	Normae	Nor	Level (square)
Canes Venatici	Canum Venaticorum	CVn	Hunting Dogs	Octans	Octantis	Oct	Octant
Canis Major	Canis Majoris	CMa	Great Dog	Ophiuchus	Ophiuchi	Oph	Serpent Bearer
Canis Minor	Canis Minoris	CMi	Little Dog	Orion	Orionis	Ori	Orion
Capricornus	Capricorni	Cap	Sea Goat	Pavo	Pavonis	Pav	Peacock
Carina	Carinae	Car	Keel (of a ship)	Pegasus	Pegasi	Peg	Pegasus (winged horse)
Cassiopeia	Cassiopeiae	Cas	Cassiopeia	Perseus	Persei	Per	Perseus
Centaurus	Centauri	Cen	Centaur	Phoenix	Phoenicis	Phe	Phoenix
Cepheus	Cephei	Cep	Cepheus	Pictor	Pictoris	Pic	Easel
Cetus	Ceti	Cet	Whale	Pisces	Piscium	Psc	Fishes
Chamaeleon	Chamaeleontis	Cha	Chameleon	Piscis Austrinus	Piscis Austrini	PsA	Southern Fish
Circinus	Circini	Cir	Compass	Puppis	Puppis	Pup	Stern (of a ship)
Columba	Columbae	Col	Dove	Pyxis	Pyxidis	Pyx	Compass
Coma Berenices	Comae Berenices	Com	Berenice's Hair	Reticulum	Reticuli	Ret	Net
Corona Australis	Coronae Australis	CrA	Southern Crown	Sagitta	Sagittae	Sge	Arrow
Corona Borealis	Coronae Borealis	CrB	Northern Crown	Sagittarius	Sagittarii	Sgr	Archer
Corvus	Corvi	Crv	Crow	Scorpius	Scorpii	Sco	Scorpion
Crater	Crateris	Crt	Cup	Sculptor	Sculptoris	Scl	Sculptor
Crux	Crucis	Cru	Southern Cross	Scutum	Scuti	Sct	Shield
Cygnus	Cygni	Cyg	Swan	Serpens	Serpentis	Ser	Serpent
Delphinus	Delphini	Del	Dolphin	Serpens Caput			Serpent's head
Dorado	Doradus	Dor	Goldfish or Swordfish	Serpens Cauda			Serpent's tail
Draco	Draconis	Dra	Dragon	Sextans	Sextantis	Sex	Sextant
Equuleus	Equulei	Equ	Foal	Taurus	Tauri	Tau	Bull
Eridanus	Eridani	Eri	River Eridanus	Telescopium	Telescopii	Tel	Telescope
Fornax	Fornacis	For	Furnace	Triangulum	Trianguli	Tri	Triangle
Gemini	Geminorum	Gem	Twins	Triangulum Australe	Trianguli Australis	TrA	Southern Triangle
Grus	Gruis	Gru	Crane	Tucana	Tucanae	Tuc	Toucan
Hercules	Herculis	Her	Hercules	Ursa Major	Ursae Majoris	UMa	Great Bear
Horologium	Horologii	Hor	Pendulum Clock	Ursa Minor	Ursae Minoris	UMi	Little Bear
Hydra	Hydrae	Hya	Water Snake	Vela	Velorum	Vel	Sails (of a ship)
Hydrus	Hydri	Hyi	Lesser Water Snake	Virgo	Virginis	Vir	Virgin
Indus	Indi	Ind	Indian	Volans	Volantis	Vol	Flying Fish
Lacerta	Lacertae	Lac	Lizard	Vulpecula	Vulpeculae	Vul	Little Fox

NAMED STARS IN GO TO CATALOGS							
Name	Designation	Name	Designation	Name	Designation	Name	Designation
Acamar	θ Eridani	Antares	α Scorpii	Hassaleh	ι Aurigae	Rasalas	μ Leonis
Achernar	α Eridani	Arcturus	α Boötis	Homam	ζ Pegasi	Rasalhague	α Ophiuchi
Acrux	α Crucis	Arkab	β Sagittarii	Izar	ε Boötis	Regulus	α Leonis
Acubens	α Cancri	Arneb	α Leporis	Kaus Australis	ε Sagittarii	Rastaban	β Draconis
Adara, Adhara	ε Canis Majoris	Ascella	ζ Sagittarii	Kaus Borealis	λ Sagittarii	Rigel	β Orionis
Adhafera	ζ Leonis	Asellus Australis	δ Cancri	Kaus Media	δ Sagittarii	Rigil Kentaurus	α Centauri
Albireo	β Cygni	Asellus Borealis	γ Cancri	Kocab, Kochab	β Ursae Minoris	Ruchbah	δ Cassiopeiae
Alcaid, Alkaid	η Ursae Majoris	Aspidiske	ι Carinae	Kornephoros	β Herculis	Rukbat	α Sagittarii
Alchiba	α Corvi	Atik	ζ Persei	Lesath	ν Scorpii	Sabik	η Ophiuchi
Alcor	80 Ursae Majoris	Atria	α Trianguli Australis	Markab	α Pegasi	Sadachbia	γ Aquarii
Alcyone	η Tauri	Avior	ε Carinae	Matar	η Pegasi	Sadalbari	μ Pegasi
Aldebaran	α Tauri	Baten Kaitos	ζ Ceti	Mebsuta	ε Geminorum	Sadalmelik	α Aquarii
Alderamin	α Cephei	Bellatrix	γ Orionis	Megrez	δ Ursae Majoris	Sadalsuud	β Aquarii
Alfirk, Alphirk	β Cephei	Betelgeuse	α Orionis	Mekbuda	ζ Geminorum	Sadr	γ Cygni
Algedi	α Capricorni	Biham	θ Pegasi	Menkalinan	β Aurigae	Saiph	κ Orionis
Algenib	γ Pegasi	Canopus	α Carinae	Menkar	α Ceti	Scheat	β Pegasi
Algieba	γ Leonis	Capella	α Aurigae	Menkent	θ Centauri	Sheliak	β Lyrae
Algol	β Persei	Caph	β Cassiopeiae	Menkib	ξ Persei	Shaula	λ Scorpii
Algorab	δ Corvi	Castor	α Geminorum	Merak	β Ursae Majoris	Shedar, Shedir	α Cassiopeiae
Alhena	γ Geminorum	Cebalrai	β Ophiuchi	Merope	23 Tauri	Sirius	α Canis Majoris
Alioth	ε Ursae Majoris	Cor Caroli	α Canum Venaticorum	Miaplacidus	β Carinae	Skat	δ Aquarii
Alkes	α Crateris	Dabih	β Capricorni	Mimosa	β Crucis	Spica	α Virginis
Almaak, Almach	γ Andromedae	Deneb	α Cygni	Mintaka	δ Orionis	Suhail	λ Velorum
Alnair	α Gruis	Deneb Algedi	δ Capricorni	Mira	ο Ceti	Sulafat, Sulaphat	γ Lyrae
Alnath, Elnath	β Tauri	Deneb Kaitos	β Ceti	Mirach	β Andromedae	Talitha	ι Ursae Majoris
Alnasl, Alnazl	γ Sagittarii	Denebola	β Leonis	Mirfak, Mirphak	α Persei	Tania Australis	μ Ursae Majoris
Alnilam	ε Orionis	Diphda	β Ceti	Mirzam, Murzim	β Canis Majoris	Tania Borealis	λ Ursae Majoris
Alnitak	ζ Orionis	Dschubba	δ Scorpii	Mizar	ζ Ursae Majoris	Tarazed	γ Aquilae
Alphard	α Hydrae	Dubhe	α Ursae Majoris	Muscida	ο Ursae Majoris	Tejat Posterior	μ Geminorum
Alphecca, Alphekka	α Coronae Borealis	Edasich	ι Draconis	Nair al Saif	ι Orionis	Thuban	α Draconis
Alpheratz	α Andromedae	Enif	ε Pegasi	Naos	ζ Puppis	Turais	ι Carinae
Alrai, Errai	γ Cephei	Errai	γ Cephei	Nihal	β Leporis	Unukalhai	α Serpentis
Alrescha	α Piscium	Etamin	γ Draconis	Nunki	σ Sagittarii	Vega	α Lyrae
Alshain	β Aquilae	Fomalhaut	α Piscis Austrini	Phad, Phecda	γ Ursae Majoris	Vindemiatrix	ε Virginis
Altair	α Aquilae	Furud, Phurud	ζ Canis Majoris	Pherkad	γ Ursae Minoris	Wazat	δ Geminorum
Altais	δ Draconis	Gacrux	γ Crucis	Polaris	α Ursae Minoris	Wezen	δ Canis Majoris
Alterf	λ Leonis	Gomeisa	β Canis Minoris	Pollux	β Geminorum	Yed Posterior	ε Ophiuchi
Aludra	η Canis Majoris	Graffias	β Scorpii	Porrima	γ Virginis	Yed Prior	δ Ophiuchi
Alula Australis	ξ Ursae Majoris	Grumium	ξ Draconis	Procyon	α Canis Minoris	Zaniah	η Virginis
Alula Borealis	ν Ursae Majoris	Hadar	β Centauri	Propus	η Geminorum	Zavijava	β Virginis
Ankaa	α Phoenicis	Hamal	α Arietis	Rasalgethi	α Herculis	Zosma	δ Leonis

Patterns are very useful to us when identifying things, and the fact is that the stars are distributed pretty much at random rather than in well-organized groups. Even if we were to give them modern names, as is sometimes suggested, they would not represent the new objects any better than they do a centaur or an eagle, for example.

The constellation names are derived from several sources. The very oldest predate history itself, and came from the Middle East. They have been added to over the years, in Greek, Roman and medieval Arabic times, and there was a flurry of additions during the 18th century. In the early 20th century their borders were regularized. All cultures had their own names, often for different patterns of the same stars, but only these classical names are in use worldwide today. In English-speaking countries it is usual to apply the classical version (such as Leo) rather than its translation (the Lion).

The sky therefore contains a motley collection of heroes, villains, creatures commonplace and fantastic, and a few items of hardware such as a lyre, a set of scales and a telescope (one of the 18th-century introductions).

Of the 88 constellations now recognized, 48 were in use in Greek and Roman times, including most of the well-known ones. Most of these refer either to animals or to mythological characters. We can picture people sitting round fires in the open, telling tales and using the star patterns to represent the characters in the stories. Today we watch television instead, but we can think of the constellation heroes as being the soap stars of the ancient past.

THE GREEK ALPHABET					
α	alpha	ι	iota	ρ	rho
β	beta	κ	kappa	σ	sigma
γ	gamma	λ	lambda	τ	tau
δ	delta	μ	mu	υ	upsilon
ε	epsilon	ν	nu	φ	phi
ζ	zeta	ξ	xi	χ	chi
η	eta	ο	omicron	ψ	psi
θ	theta	π	pi	ω	omega

These mythological characters often lent their figures to the names of the stars that comprise them. The name of the star Betelgeuse in Orion is a classic example. It is said to derive from the Arabic for "The Hand of Jauzah the Giant." Rigel is the same giant's leg or foot. Many of the names come from Arabic words, because the Arab world maintained the flame of knowledge during what are known as the Dark Ages in Europe. This is why many of them begin with "Al-," meaning "The."

In addition to the names, stars have other designations. In the 17th century the principal stars in each constellation were assigned Greek letters by German astronomer Johann Bayer. Usually the brightest star is alpha (α), the next brightest beta (β) and so on. The constellation name is turned into its Latin genitive form, as in Alpha Centauri, meaning "alpha of Centaurus," Delta Cephei ("delta of Cepheus"), and so on. So in addition to learning the constellations, the knowledgable stargazer should know the Latin genitive version of its name as well.

There are only 24 Greek letters, and in the 18th century when more stars were accurately cataloged, it became necessary to assign numbers. This was first done by the British Astronomer Royal of the day, John Flamsteed, who numbered the stars in his catalog, again constellation by constellation, beginning at the westernmost edge. These Flamsteed numbers sit alongside the names and the Bayer letters, so many stars have all three designations. The star Denebola, for example, in the tail of Leo, is also Beta Leonis and 94 Leonis. In general, names are used for the brighter stars only, though the lists of stars in Go To catalogs habitually use these rather than the Bayer letters, rather inconveniently in some cases. At least with the Bayer letter, one has an idea of where Delta Leonis might be, though not everyone may remember or realize that Zosma is the same object. Flamsteed letters apply only where there is neither a name nor a Bayer letter, and mostly cover stars of magnitudes 4 to 6.

Fainter stars still are referred to by their numbers in one of several catalogs, such as the *Henry Draper Catalog*, prefaced HD. Several commercial companies have regarded star catalogs as a source of stars which they then name after anyone who cares to pay them a fee, but these names are not recognized by any authority and there is nothing to prevent several companies selling the same star to different people. If you bought this book hoping to locate "your own" star, then the bad news is that the name is not genuinely recognized and the star may well be too faint to be shown on these maps. If the company has supplied you with a map, you may be able to locate its general area using this book, but spotting the actual star may require a fairly large telescope and some hard work linking the star's position with objects that you can find.

The planets

There are four planets that can cause confusion when learning the constellations – Venus, Mars, Jupiter and Saturn. Of these, Venus is not really a problem because it is so bright that it stands out very plainly. Jupiter is also bright, but could easily be mistaken for a bright star. Mars and Saturn are usually about the same brightness as many stars, so they can definitely create puzzlement. Mercury, the other bright planet, is only ever seen low in the twilight sky so it is usually out of the picture.

The planets always remain close to the ecliptic, and you can find their approximate positions using tables (provided on www.stargazing.org.uk). You can also pick out a planet by the fact that it rarely twinkles. This is because it has a definite disk, rather than being just a point of light as in the case of a star. It is the effect of our own unsteady atmosphere on the delicate wavefront of a star's light that causes twinkling, whereas the wavefront of light from a planet is less easily deviated.

Over a period of time, the planets move through the sky. They share in its daily movement, but have their own movements in addition. Each cycle of appearance

▲ **In March 2004,** Jupiter was in Leo (top), altering the normal appearance of the constellation (bottom).

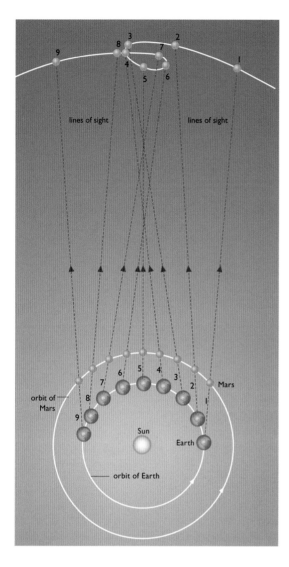

until they are eventually visible at midnight on the meridian. At this point they are opposite the Sun in the sky, and more or less at their closest to Earth and therefore at their brightest. This point is called *opposition*. From that point they move farther into the evening sky, when most people find it easiest to observe them. However, the planets are now moving away from Earth and getting smaller in the sky, so the nearer to opposition you observe them the larger they will appear in your telescope. As time goes by they start to sink into the western sky at sunset. Finally, they pass on the far side of the Sun, at what is also termed conjunction.

These movements are the same as those of the background of stars against which the planets appear, and they are the result of the Earth's annual movement around the Sun. But each planet has its own separate eastward movement along the ecliptic, so that in the case of Mars in particular, which is moving round its own orbit but more slowly than Earth, it can take a long time for Earth to catch up with it and then pull away from it; this means that Mars' apparitions are longer than those of any other planet.

Around the time of opposition, the planets farther out from the Earth stop their eastward movement and describe slow loops in the sky, known as retrograde loops. These are the result of the Earth's own faster motion around the Sun.

The Milky Way

One other feature of the night sky is shown on the maps – the Milky Way. This is a pale band of light that was commonly seen by our ancestors but which is now all but invisible to most people because of light pollution. But on those occasions when you are far from civilization and have a really clear, dark sky, it appears so bright that you wonder how you could ever miss it.

The Milky Way is actually our own Galaxy of stars seen from the inside. The term may refer to both the band of light and the Galaxy as a whole. The faint distant individual stars merge one into another as seen with the naked eye, though binoculars and telescopes help you to pick them out. Along with the stars are dust and gas, which hide the more distant stars from view. Many of the interesting objects in the sky, such as open star clusters and nebulae, lie along the plane of the Milky Way.

▲ **The Earth's faster motion** in its orbit causes the movement of the superior planets (those with orbits that lie outside that of the Earth) to appear to reverse in a retrograde loop around the date of opposition.

of a particular planet is referred to as its *apparition*. Mercury and Venus are always to be found within a certain distance of the Sun as they are closer in to the Sun than is the Earth. They start each apparition in the western twilight sky after sunset, then move away from the Sun so that they become more easily visible. After what is called their *greatest elongation* from the Sun, they move back toward it, passing between the Earth and the Sun in what is known as *conjunction* – technically, the moment when a planet shares the same longitude on the ecliptic as the Sun, though it may be above or below the Sun in the sky, and in the case of Mercury and Venus either in front of or behind the Sun. After a week or two they can be seen again in the early morning sky just before sunrise, and they then repeat their movement away from the Sun and back again.

The other planets begin each apparition in the early morning sky just before sunrise. Over the months they move westward away from it, rising earlier and earlier

▶ **Fisheye view of the Milky Way** in northern-hemisphere summer. It is shown from Cassiopeia in the north to Sagittarius in the south. The dark band along its central plane is caused by dust clouds and is known as the Great Rift or Cygnus Rift.

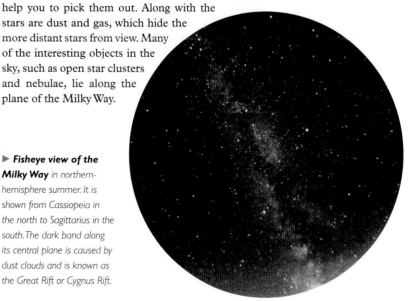

Distances in astronomy

The distances to most astronomical objects are so much greater than those in everyday use that a different style of distance measurement has been adopted. Kilometers and miles can be used for distances within the Solar System, but beyond that it is commonplace to speak in terms of *light years*. A light year is the distance that light travels in one year, and is equivalent to 9.46 million million km. The distances to stars within the Milky Way range from a few light years to thousands of light years, while the distances of other galaxies are measured in millions of light years.

THE SCALE OF THE SKY

One of the biggest hurdles encountered by beginners to skygazing is getting to grips with the difference between what you see in a book and what you can actually see in the sky. There are huge differences in scale and in brightness between the various types of object, so here are some examples of how the view changes with the scale at which you view it.

The eye often plays tricks on us – or, more accurately, the combination of the brain and the eye. Anyone who has watched the full Moon rising will appreciate this. It looks really huge, and many explanations of this have been put forward. Some books go so far as to say that the atmosphere magnifies the image of the Moon, but oddly enough it is actually slightly smaller when rising than when it is high in the sky.

There are two reasons for this. One is that when the Moon is rising it is slightly farther away from you – you are looking across one Earth-radius plus its distance, whereas when it is on the meridian and at its highest, you are closer to it than is the center of the Earth by one Earth-radius. The other reason is that far from magnifying the Moon, the effect of the atmosphere is to squash it slightly from top to bottom, the amount depending somewhat on the density of the atmosphere at the time. So the Moon is actually smaller, against all the evidence of your eye.

▼▶ **The Moon photographed over well-known London sights.** *Below, when rising over St Paul's Cathedral, it appears similar in size to the famous dome. At right, when high in the sky over Big Ben, compared with the clock face it appears tiny.*

The most favored explanation lies in the way your brain interprets distance and size. When the Moon is low down you are comparing it with foreground objects whose size you know, while when it is high up there is nothing familiar with which to compare it. The same effect applies of course to the Sun, and even to constellations, which appear larger when they are low down than when they are high up.

In fact the Moon is just about ½° across, compared with the 90° between the horizon and the overhead point, the zenith. A degree is divided into 60 minutes of arc, so the Moon is about 30 arc minutes across.

The Moon is always a good place to start your explorations because it is so easy to find in the telescope and provides a wonderful landscape of high contrast. But moving on to the planets, things become a bit more of a challenge.

The largest planet, Jupiter, can show more detail than any other planet, and you might think that a

telescope would show it with as much detail as the Moon. But instead of being 30 arc minutes in diameter, it is less than 1 arc minute across – actually only about 45 arc seconds, there being 60 arc seconds in 1 arc minute.

At low magnification in a telescope, Jupiter shows a distinct disk. If the telescope magnifies 40 times, the planet appears about the same size as the Moon does to the naked eye. But Jupiter has a gaseous atmosphere rather than a solid surface with mountains thrown into sharp relief, and the detail is much lower in contrast than that of the Moon. Many people observing it for the first time have difficulty in spotting anything other than a couple of darker belts crossing the disk.

At a magnification of 250, the planet appears about the same size as the Moon does through binoculars, still surrounded by a large area of sky. You might think that it would appear equally as crisp, but again the low contrast makes the view less dramatic. With a small telescope this view is quite dim, and even with a larger instrument your eye has to work quite hard to see real detail although the planet appears brighter.

When we turn to the stars, we can never magnify them enough to see them as anything other than points of light. Even the Hubble Space Telescope can show only a handful of stars as actual disks, and even then the detail visible is minimal. But binoculars and telescopes can reveal much fainter stars than are visible with the naked eye. A look at a fairly well-known area of the sky, the Square of Pegasus, shows what can be seen. You can see the Square on the general map on page 200, with a detailed view on page 180. It is visible worldwide on evenings in September and October.

The Square is about 20° across, which makes it easily encompassed at a glance, though you will notice that you can't see the whole of the Square in detail. Your really sharp field of view is only about 1° across – twice the diameter of the Moon – while the area we perceive well is about 40° wide.

To the naked eye, this part of the sky is not particularly starry. From a town you will be lucky to see any stars at all within the Square. Out in the country, however, you will see an increased number depending on the quality of the sky and your eyesight. A keen observer can see over 40.

However, turn a pair of binoculars on the Square even in a city sky and you will immediately see more than you can easily count. In a sky that restricts you to magnitude 4 (no stars at all within the Square) with the naked eye, you should see to at least magnitude 8, yielding around 180 stars. With a 114 mm telescope, the number should increase to many hundreds, though you can't hope to see more than a few of them at any one time. Because the binoculars can collect more light than your eye can, it can show many more stars and can also make stars that are only just visible to the naked eye appear very much brighter. The novice observer has to get used to this increase in brightness, in order to know which star is which when

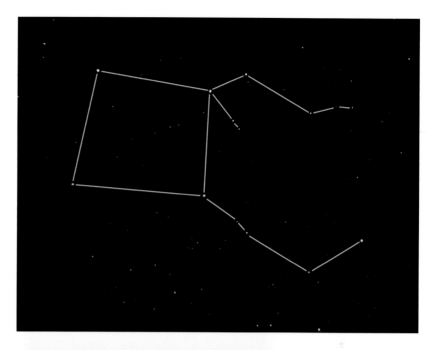

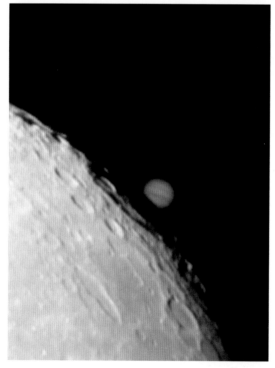

▲ *The Square of Pegasus* (outlined), showing stars down to approximately the naked-eye limit. Compare this view with the map on page 180.

◄ *Occasionally, the Moon* passes in front of Jupiter and we can compare the sizes of the two directly. Jupiter is smaller in apparent diameter than many minor craters.

comparing the naked-eye view with that through the binoculars.

The same applies when searching for deep-sky objects through binoculars or a telescope. Most of these are quite small, so often the trick is to know what you are looking for and which magnification is best for a particular object. On the monthly constellation pages of this book (see Chapter 7), we give a guide to the magnification that gives a good view of each of the objects mentioned.

So the message is, get out there and start looking. Only with practice will you really know what to expect when you look through your binoculars and telescope.

Observing with the naked eye alone goes only so far. Sooner or later, everyone wants to be able to see the heavens in greater detail, to study features on the Moon and planets and to observe deep-sky objects – the clusters and nebulae – that are barely visible, if at all, with the naked eye. This chapter is a very brief introduction to the different ways in which you can observe and record the heavens.

THE NAKED EYE

There are some characteristics of the eye that every observer should know about. The eyeball is in effect an optical instrument with an *aperture* – the hole through which light enters – a few millimeters in diameter. In bright sunlight the pupil of the eye closes down to about 1.5 mm diameter, while in the dark it opens up to between 3 mm and 9 mm diameter (the younger you are, the wider it can open). The detector that picks up the light is in effect a 120 megapixel receptor array, compared with the 12 to 18 megapixel arrays of the better digital cameras. Most of these receptors are called rods, and they are not color-sensitive. The rods simply detect light within a wavelength range of 380 nm to 760 nm (nm = nanometer, which is a billionth of a meter) – that is, violet to deep red. Color vision comes from different receptors called cones, which are concentrated toward the center of vision.

The receptors are more densely packed toward the center of vision, so we have only a small area of good vision. We are not normally aware of this, as our eye continually darts from place to place in order to fill in the details, but as you look at these words try to perceive details of your surroundings without looking directly at them. You will notice that you cannot see any great detail, and also that colors are not very obvious. Away from your center of vision the more sensitive rods are more common than cones, which means that night vision is better, being primarily provided by the rods. Your center of vision is the worst part of your eye for detecting faint objects, which is an obvious drawback for astronomers. For this reason, faint objects are best seen by using *averted vision* – that is, by using your peripheral vision without staring directly at the object.

The diameter of the pupil is not the only way, or even the primary way, in which the eye regulates its sensitivity. At low light levels the sensitivity of the rods is increased by a chemical called rhodopsin. This takes some time to have an effect, with the result that if you walk straight out of a brightly lit room into the dark you have virtually no night vision. Full sensitivity is only achieved after 30 to 45 minutes or maybe much longer. Rhodopsin breaks down in bright light, but is more strongly affected by blue rather than red light. For this reason, you should use a red light when looking at charts or making notes while observing.

Before you go out to observe, avoid using lighting with a strong blue or white content. TV screens, computer monitors, fluorescent lights and low-energy or LED lighting are bad for your night vision, but incandescent lighting has a lower blue content, particularly if you use 40-watt bulbs.

There is a blind spot in each eye, located about 17° away from your center of vision on the side away from your nose. When viewing with both eyes your brain fills this spot in, but when observing through an eyepiece you will not see stars at this point.

CHOOSING AND USING BINOCULARS

Virtually every stargazer has binoculars, even those with an array of powerful and expensive telescopes. Binoculars give a low-magnification, comparatively wide-field view of a small part of the sky. They are useful as a quick means of locating objects that are either too faint to be seen with the naked eye or are hidden by light pollution. Some of the best views of the larger star clusters and brighter nebulae are given by binoculars, and bright comets are a glorious sight. In a dark sky, the multitudes of stars in the Milky Way are breathtaking. Although binoculars are not powerful enough to show details on the planets, you can see some of the major features on the Moon. There are also many variable stars whose brightness changes can be monitored using only binoculars.

The specification of binoculars for astronomy is not as critical as might be supposed. Binoculars are described as, for example, 10 × 50 or 7 × 42, where the first figure in each case is the magnification and the second is the diameter of the main or objective lenses in millimeters. So 10 × 50 binoculars (pronounced "10 by 50") have a magnification of 10 times (10×) and objective lenses 50 mm across. A magnification of between 7 and 10 is ideal for astronomy, with 12 as an option in light-polluted skies where extra magnification helps to show fainter stars. In general, the lower the magnification the wider the field of view and the brighter the image.

There is, however, a limit to the brightness that you can use. The figure derived from dividing the magnification into the objective diameter gives the size of the circle of light emerging from the eyepieces, known as the *exit pupil*. For 10 × 50 binoculars this is 5 mm, while for 7 × 50s it is just over 7 mm. As the pupils of older people may not open wider than about 3 mm

▶ *These 10 × 50 binoculars* are a good compromise between magnification and convenience. A tripod helps to keep the image steady, though image-stabilized binoculars are available at a considerable price.

anyway, the extra light provided by 7 × 50 binoculars can be wasted. This is also the reason why it is a waste of time making, say, 3 × 50 binoculars. No one has eyes wide enough to use all the light coming from them.

The *actual* field of view – that is, the amount of sky – of specific binoculars can be between about 4° and 8°. This is not to be confused with the *apparent* field of view, which is the apparent size of the circle of light that you see when looking through and is typically between 40° and 60°. The actual field of view depends not only on the magnification (in that higher magnification generally results in a smaller field of view) but also on the nature and quality of the eyepiece in the binoculars. It is better to have a smaller, good-quality field of view than a wide one with poor definition, in which stars are not sharp. The field of view is often not specified, but may be given in the form of feet at 1000 yards. A field of view of 5° shows 261 feet of the landscape at a distance of 1000 yards.

The higher the magnification, the more difficult it is to hold the binoculars steady when viewing the sky. The magnification has no effect on the size or weight of the binoculars, which depends mostly on the size of the objective lenses. Binoculars with 50 mm lenses are heavier than 30 mm binoculars of the same type, but will show fainter stars. So the choice of binoculars depends on many factors, and it is best to test several, ideally on the night sky, before you buy. Binoculars for astronomy need not be expensive – you do not need the same level of waterproofing or robustness that a birdwatcher or mountaineer might demand – but very cheap instruments may have poor optical quality or be easily jolted out of alignment. Avoid zoom or very high-power binoculars sold cheaply, particularly through mail-order ads, even those that claim to be ideal for astronomy.

Binocular adjustments

In most binoculars there are three adjustments: the separation of the eyepieces; a center focus wheel; and a focuser on one eyepiece to allow for differences between the user's eyes. If you wear spectacles for long or short sight, you can probably remove them when viewing because the focusing range should be adequate to allow for your vision. If you use spectacles for astigmatism or extreme focusing defects, however, you may need to use binoculars with rubber eyecups that fold flat to allow you to press them right against your spectacles. Models that allow a larger eye relief – that is, the distance between the eyepiece and your eye – may be useful in this case.

It is usually easier to adjust binoculars for your eyesight by day than by night. Begin by getting the separation of the two halves exactly right for the distance between your eyes, so you are looking directly down the center of each eyepiece. Then, using your left eye only, focus on a distant object using the center focus wheel. Finally, use your right eye only and use the adjustable eyepiece (sometimes called the diopter focuser) to focus on the same object. Many binoculars

*◄ **Adjust the separation** between the eyepieces to equal the distance between your eyes.*

*◄ **Focus on a distant object** using the left eyepiece and the center wheel.*

*◄ **Adjust the focus** of the right eyepiece using the diopter focuser.*

have scales for the separation of the eyepieces and the diopter correction, so make a note of these and you can quickly adjust any binoculars to suit your eyes.

TELESCOPE CHOICE

There are so many different types of telescope available that the choice for the beginner is almost overwhelming. As well as the three main optical systems of refractor, reflector and catadioptric (explained overleaf), there are also electronically controlled mounts that claim to be able to find virtually any object in the sky after just a simple set-up procedure. Portability also affects people's choice, with many observers wanting telescopes that they can take to dark-sky sites or abroad.

The three main types of telescope are: refractors, which use lenses to focus the light; reflectors, which use mirrors; and catadioptrics, which use a combination of the two in a more compact design. The choice of mounting is as important as the optical design of the telescope, and in many cases you can get any type of telescope on any type of mounting, which adds to the range of possibilities from which to choose.

The basics of optics apply equally to all types of telescope. Each has a main mirror or lens whose job is to collect and focus the light from the object being observed. Main lenses tend to be called *objectives* or objective glasses, abbreviated OG. The aperture or diameter of the objective or mirror is its most important specification – the larger the diameter, the more light it collects. When we refer to a 150 mm or 6-inch telescope, we mean its diameter rather than its length.

The other important figure is the *focal length*. This is the distance between the lens or mirror and the focus point, and in ordinary refractors or reflectors it dictates the length of the telescope tube. It can be made short or long, and the different focal lengths have different properties. More important than the actual dimensions of the telescope is the comparison of the focal length with the diameter of the lens or mirror. Known as the *focal ratio* or *f-number*, it is the focal length divided by the diameter. So, in simple terms, a tube five times as long as it is wide has a focal ratio of 5, written f/5.

While a low f-number provides a compact instrument, there are penalties. One is that it is hard to achieve the same optical performance with short focal ratios compared with longer ones, because the optics require steeper curvature. Another is that it is harder to achieve high magnifications, because much of the magnification comes from the actual focal length of the main lens or mirror. There is a practical limit of about f/4 for reflectors and f/5 for refractors. In general terms, telescopes with small f-numbers are better suited to deep-sky observing in dark skies, while those with large f-numbers are more suited to planetary observing or observing in towns.

Eyepieces and magnification

Every telescope needs an eyepiece to make the image observable and to provide magnification. The eyepiece itself has a focal length, and the overall magnification of a telescope is found by dividing the focal length of the telescope by that of the eyepiece. So a telescope of 500 mm focal length, used with an eyepiece of 10 mm focal length, will give a magnification (or power, as it is often called) of 50. To achieve a magnification of 100, you need an eyepiece of 5 mm focal length. The same eyepieces used with a telescope of 1000 mm focal length would give magnifications of 100 and 200. In general, eyepieces are available in the range 4 mm to 40 mm.

Barlow lenses are devices that sit between your telescope and the eyepiece; they multiply the effective focal length of the telescope, usually by a factor of 2. If you are building a set of eyepieces, try to get a range that does not include factors of 2 in focal length, so that a 2× Barlow will increase the number of steps in your range. For example, if you already have a 26 mm eyepiece, a 2× Barlow will give you a 13 mm equivalent. There is little point in buying a 12 mm, as this is too close, so instead buy a 15 mm to give you a range of four widely separated focal lengths with just three purchases.

Increased magnification has two drawbacks: the field of view becomes smaller, and the image of an extended object becomes dimmer, as the same amount of light is spread over a larger area. The maximum magnification that you can use is limited by two factors – the steadiness of the atmosphere and the properties of light itself. Larger telescopes are often limited by the first of these factors to powers of about 200. Small instruments are limited by the wave nature of light, which means that only so much detail is visible through a particular aperture. A rule of thumb is that the maximum usable magnification is about twice the aperture in millimeters. A 60 mm refractor is therefore limited to a magnification of about 120.

At one time there were several different types of eyepiece available, each with its own advantages and drawbacks. Today, the Plössl type is ubiquitous as a basic eyepiece, except on the cheapest refractors, which use the simpler Huygenian eyepieces. There are also specialist eyepieces available, notably those that provide very wide fields of view. These are often very bulky and expensive, and can easily weigh and cost more than a starter telescope.

Eyepieces generally have a barrel diameter of 31.7 mm, more usually described as 1¼ inches. Some small refractors may use 24.5 mm barrels. Extra-wide-field, long-focal-length eyepieces may require a barrel of 50 mm (2 inches) in order to give the full field of view, but these can only be used on telescopes with the right size focusing mount.

It is a basic fact of optics that virtually all astronomical telescopes give an inverted view – that is, upside down. In the past, it was always said that astronomers would rather have an upside-down image than put extra bits of glass in the light path in order to bring the image the right way up. These days, however, many people spend large sums of money on eyepieces with extra glass elements that give a wide field of view. But having a non-inverted or erect image is not deemed a priority, other than with smaller telescopes that are also intended for daytime use.

Refractors

These are what most people think of as a telescope, with the main lens at the top of the tube and the eyepiece at the bottom. The smallest telescopes are all refractors, and they are available with apertures up to about 150 mm at reasonable cost. At one time, refractors were typically about f/12 or longer, but today it is not unusual to find f/5 instruments of good quality.

▶ **A range of Plössl eyepieces** in 1¼-inch fittings, from 4 mm (bottom left) to 40 mm (top right).

Refractors have always suffered from the defect known as *chromatic aberration* or false color, which means that not all colors of light are focused at the same point. The effect when viewing a planet is of a bluish halo surrounding the planet's disk, though otherwise the image in a good refractor has high contrast and brightness. Filters can reduce the false color, but only very expensive refractors using what are called *apochromatic* lenses are free from it. It is only these apochromatic refractors that are really suitable for photography, and they also have particularly wide fields of view of good definition.

A refractor will not deteriorate significantly over time, and any dust that accumulates on the lens can be easily removed with care.

When observing an object high in the sky, the viewing angle can be difficult. For this reason, refractors are usually supplied with star diagonals, which turn the image through an angle, usually 90°, and provide a much more comfortable viewing angle. However, these diagonals give a mirror image of the object, and this must be taken into account when comparing observations with those from other instruments.

Reflectors

The standard reflector type is the Newtonian, in which the eyepiece is located at the top of the tube and at right angles to it. Though this means observing at right angles to the direction of the object, the result is often a fairly comfortable viewing angle without the need for star diagonals. With long-focal-length instruments, however, the eyepiece can be inconveniently high from the ground.

The mirrors are coated on the front surface, rather than on the back as with household mirrors, so the surface coating is delicate and prone to deterioration. If the tube is capped when not in use, the surfaces should last for years. Reflecting systems are more sensitive to correct alignment than refractors, so adjustments are

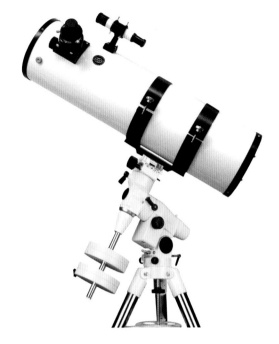

◀ *A 200 mm f/4.5 Newtonian reflector* on a German-type equatorial mount. The tube can be rotated within the cradle to bring the eyepiece to a convenient viewing angle. It has a 6 × 30 finder telescope.

provided, and the telescope may need realignment or *collimation* from time to time. However, mirrors give perfectly color-free images so they are equally suited to both visual and photographic observing. For details of collimation refer to www.stargazing.org.uk.

If you want the largest aperture for a set budget, a Newtonian reflector is the way to do it. In terms of actual observing versatility, a reflector offers the greatest chance of observing the largest number of objects, whether you live in a town or in the country. A moderately large aperture will still show more objects than a smaller one even in the town, but it will perform better in darker skies.

As the tube of a reflector is usually open at the top end, tube currents caused by local temperature variations can swirl round inside and reduce the steadiness of the image. Reflectors with open tubes avoid this, but they are susceptible to extraneous light, and the mirrors are prone to dewing and collecting dust. A reflector will not perform well until it has reached the same temperature as its surroundings. If you take it straight from a warm room into the cold night air, the image will be very disturbed until it has cooled down, and it may need an hour or so at outside temperature before it performs well. Refractors and catadioptrics do suffer from the same effect, but to a lesser extent.

Catadioptric telescopes

These are the most popular instruments available nowadays for serious amateur astronomy, eclipsing the more traditional refractors and reflectors. They include the types known as Schmidt-Cassegrain and Maksutov-Cassegrain, the latter often simply called a Maksutov.

◀ *A 102 mm f/4.9 refractor* on AZ3 altazimuth mount with manual slow motions. Notice the star diagonal, which gives easier viewing angles. This instrument has a "star pointer" zero-power finder rather than a finder telescope.

▲ *A basic 200 mm f/10 Schmidt-Cassegrain telescope* on a fork mount. This has the same aperture as the reflector shown on the previous page but has a much shorter tube. The telescope comes equipped with motors and Go To controller.

Catadioptric telescopes are basically short-focus reflectors with lenses at the top end to correct for the defects of the image. So they combine short tubes with good optical quality.

The fact that their image quality is not quite as good as that from a refractor or Newtonian reflector is outweighed for most people by their much greater compactness. This means that a 200 mm instrument is the norm, and 250 mm and even 300 mm instruments can be transported fairly easily, whereas a Newtonian of that size would require a large tube and heavy mounting. Smaller instruments are usable on a tabletop and can be carried with one hand, unlike their more traditional counterparts.

Schmidt-Cassegrain telescopes, often abbreviated to SCTs, are usually f/10, while Maksutovs are usually between f/13 and f/15. Maksutovs are particularly suited to planetary observing. They are more compact than SCTs, but require more careful manufacture and are therefore more expensive.

Another design, the Schmidt-Newtonian, is sometimes classed as a catadioptric, but this uses a small lens near the eyepiece rather than a large one at the top of the tube. Its performance is generally slightly poorer than the conventional Newtonian of the same aperture. For more details on telescope choice, refer to www.stargazing.org.uk.

Finders

Finding an object in the sky using even a small telescope magnifying, say, 30 times, is amazingly difficult compared with finding an object on land. All you see is blank sky, with no landscape features you might recognize. The solution is to use a finder, and even most Go To telescopes need a finder to help you set them up at the beginning of a session.

There are two basic types: small refracting telescopes mounted parallel to the main instrument but giving a low magnification and a wide field of view, and red-dot finders, which appear to project a red dot on the sky when you look through them, but with no magnification. These are now commonly found on budget telescopes as they are cheap to make and tend to be more usable than the very poor finder scopes that were previously common.

Aligning the finder perfectly with the main telescope is a vitally important first step, but one that many beginners ignore. It is best done during daylight by finding a distant object with the main telescope, then adjusting the finder so that the same object is precisely on its crosswires or dot. Without this step, finding objects at night can be frustrating at best and doomed to failure at worst.

To find any object in the sky, first look along the telescope tube to get the right part of the sky. This is often harder than it seems, and manufacturers are surprisingly lax in providing simple sighting devices on telescopes to help the beginner. With practice, you should

be able to point your telescope at a chosen object, which should then be well within the field of view of the finder. Then bring the object to the center of the finder's crosswires or dot and if it is properly aligned, the same object will be in the low-power field of view of the main telescope.

TELESCOPE MOUNTS

The mounting of a telescope is often crucial to its performance. Many cheap telescopes are let down by their flimsy mountings, and would be more usable if they were remounted. There are two basic types of mount – the altazimuth and the equatorial.

Altazimuth mounts

The name of these mounts refers to their two movements, one in altitude (up and down) and one in azimuth (side to side). At one time these mounts were restricted to only the very cheapest of telescopes, but today they are found on virtually all sizes, from the very smallest through to the largest professional telescopes, the advent of computer control overcoming their drawbacks.

They are usually of the fork or yoke type, in which the telescope pivots up and down between the arms or tines of a twin fork, which itself rotates about a central axis. This is simple and effective in engineering terms, but the drawback is that celestial objects move through the sky at an angle (as described in Chapter 2). This produces two problems: first, it means that the telescope has to be moved on both axes simultaneously; and second, the plane of the image rotates as the object moves through the sky. Though not a great problem for visual observers or even for short-exposure imaging, it rules out long-exposure photography.

Modern altazimuth mounts overcome the first problem by using a purpose-built computer handset.

▲ *A typical 60 mm refractor* on a yoke-type altazimuth mount and tripod.

▲ *A commercially made 200 mm f/6 Dobsonian.* This is the cheapest mounting for a telescope of this aperture.

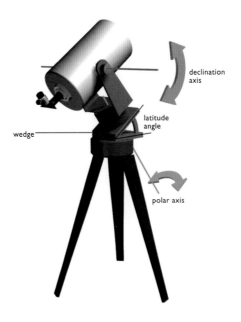

◄ *The fork mount* is basically an altazimuth mount with a wedge to incline it at the same angle as your latitude.

It contains a virtual model of the whole sky, and once it knows its orientation compared with just two fixed points in the sky it can find any other object whose position it knows. This is the Go To facility. A Go To computer, however, does not overcome the second problem – that of image rotation. It is possible to buy image derotators, which turn the camera during the exposure, but these are not commonly used.

The engineering simplicity of the altazimuth mount is exploited in the Dobsonian mount. The traditional Dobsonian is made from wood, with Teflon pads between the rotating surfaces to reduce friction. It is intended as a cheap but effective mounting and is particularly suited to the larger sizes of Newtonian telescope. Dobsonians offer the largest aperture for your money, and they are favored by deep-sky observers who are interested in using big telescopes to find faint objects. They can also be used for casual planetary observing, but are not ideal as they cannot conveniently be motorized and cannot be used for long-exposure photography.

Equatorial mounts

The drawbacks of the altazimuth mount are overcome by the equatorial mount. It again has two axes of motion, but one of them is inclined parallel to the Earth's axis. In the case of a fork mount this simply involves tilting the base at an angle equal to the latitude of the observing site using what is usually referred to as a wedge. The telescope can then be driven using one motor only to follow any astronomical object's motion through the sky. The main drawbacks of the fork mount are that it is not suited to long tubes, and that the eye-

piece becomes inaccessible when it moves between the arms of the fork.

The most common alternative to the fork mount, the German mount, overcomes both these deficiencies but has its own foibles. One is that it requires a counter-weight to the telescope, which adds to the weight of the whole assembly, and the other is that full rotation across the sky is not usually possible without some part of the assembly hitting some other part, or the whole telescope needing to be turned to the other side of the mount and rotated through 180°.

To set up an equatorial mount for ordinary observing, rather than for photography, simply set the polar axis at the correct angle for your latitude – there is usually a latitude scale marked on the mount or on the wedge. Make sure that the base is level, then point the polar axis due north in the northern hemisphere or south in the southern. This approximate alignment should be adequate for most purposes, and anything you observe should remain within the field of view for

◄ *The Meade ETX Maksutov telescopes* use altazimuth fork mounts with a computer-control option to provide a Go To facility.

▶ **An example of field rotation.** *As the constellation Taurus moves through the sky its orientation changes. The inset shows a close-up of the Pleiades star cluster at the given times. Even if you tracked the center of the cluster perfectly, a time exposure would still have star trails around the center.*

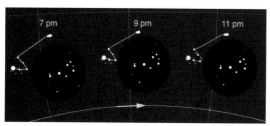

◀ **By aligning the polar axis** *of an equatorial mount with the Earth's axis, you can counteract the rotation of the Earth.*

a considerable time with only minor corrections for drift. For long-exposure photography, however, you will need much more stringent and advanced polar alignment (beyond the scope of this introductory text).

Equatorial mounts are usually provided with setting circles, which are graduated in right ascension and declination. The declination scale is fixed, but the RA scale can be rotated to suit the time of observation. The easiest way to use the scales is simply to point the telescope at a star whose position you know. If the mount is correctly polar aligned the declination scale should be correct, and you can simply turn the RA scale to read the correct position. Then you can find another object whose position you know using the circles alone. Some circles run in the opposite direction to this, in which case you need to work out the difference in RA between the known object and the one you want to observe, and turn the telescope accordingly.

Motor drives

Useful for all types of observing, motor drives are now widespread and are virtually essential for imaging. Once you have found an object it stays in the field of view. But even the most perfectly engineered and set-up mounting is likely to introduce small errors into the drive rate, so for long-exposure imaging some means of monitoring and correcting the drive rate is needed,

▶ **A lightweight Point Grey Flea 3 camera** *attached to a telescope. The unit is a fraction of the weight of even a compact camera, and gives a 648 × 488 pixel video stream in either mono or color.*

a process known as guiding. At one time this was done by eye alone, watching a star and making manual corrections to keep it on crosswires. This meant using either an off-axis guider, which directed a small part at the edge of the incoming light beam to a separate eyepiece, or a guide telescope attached to the main instrument. Today, most people use separate guide telescopes, but instead of the tedium of continually watching a star, they use autoguiders which incorporate an electronic imager to monitor errors and make corrections.

IMAGING

The word "photography" is now often replaced in the astronomical vocabulary by "imaging." The end result is still a photograph, but "imaging" tends to imply a longer process than simply pressing a shutter button. Though it is possible to take a picture through a telescope simply by holding the camera up to the eyepiece and taking a snap, the results are usually rather hit-or-miss. This is known as the *afocal method*, and it is the only method possible with compact cameras or digital cameras with non-removable lenses. With special clamps to hold the camera in place, you can get very successful photos, but there are severe limitations with this method. The camera is usually quite heavy, which may result in balance problems; operating the shutter often introduces vibration unless you can use a remote-control lead or cable release; and the image quality may be poor, particularly at the edge of the field of view.

In addition, most compact cameras are not designed for long exposures and are often restricted to an exposure time of up to about a minute. This is not a problem for the Moon and planets but is less useful for deep-sky objects.

A revolution in planetary photography occurred when around 2000 it was realized that webcams – simple video cameras that are intended for use with a computer – provide a very simple and effective camera for imaging bright objects through a telescope. They are cheap and lightweight, and produce a video stream that can be stored for later processing. One problem that has frustrated observers for centuries is the unsteadiness of the Earth's atmosphere – usually referred to as the *seeing*. This causes a planetary image to shimmer and wobble under all but the rarest of steady conditions, but the eye and brain work together to pick out the detail. A single exposure on a camera simply records one distorted image, but the beauty of a

webcam is that it can record many images per second. Software automatically picks out the best frames and adds them, creating a final image from only the best of hundreds or even thousands of frames.

This has revolutionized planetary and lunar imaging, and amateur astronomers at sea level can now take images of bright objects which far exceed what was possible even in the 20th century using large mountaintop telescopes. But commercial webcams themselves have now moved on and have become more difficult to adapt for telescope use than the early models. Instead, amateur astronomers use specialist cameras based on the same chips as those in webcams. Some units give color images directly, but the best results come from mono cameras which are used with separate red, green and blue filters to produce a color image after image processing. However, webcam-type cameras are not usually ideal for the long-exposures needed for deep-sky objects. And a computer is needed, with the attendant problems of providing a power supply during long sessions outdoors.

For long-exposure imaging through the telescope you usually need either an SLR (single-lens reflex) camera or a CCD camera. An SLR camera has a detachable lens, so the telescope can be used in place of the normal lens to provide the image. Traditionally these were film cameras, but today digital SLRs (DSLRs) are the norm. A readily available adapter is all that is needed to link telescope and camera.

As mentioned in Chapter 2, the view of most deep-sky objects through even a large amateur telescope is not a patch on a long-exposure photograph. Realizing this, many amateur astronomers now spend most of their time taking images rather than looking through the telescope. Everyday DSLR cameras can take impressive shots, far exceeding what was possible using film in the last century. They can go on building up light for minutes or hours, with the exposure time usually being limited by light pollution in all but the most remote country districts. But electronic noise or spurious speckles can often spoil the results.

For serious long-exposure photography, the cooled CCD camera reigns supreme. The cooling of the CCD chip reduces the noise level, and allows very long exposures. Image processing at the computer can remove the background light pollution, if any, and enhance faint details. As a result, it is possible to make images of faint galaxies even from city locations, while in dark skies images are possible that not long ago would have required a giant telescope.

It is not always necessary to use a telescope for long-exposure astro-imaging. Any camera that will take time exposures can be mounted piggy-back on a telescope with a driven equatorial mount, or even directly on the mount. The cradles for carrying telescopes often include a threaded stud that will carry a camera for this reason. You can take photos of constellations, or, using telephoto lenses, of the larger deep-sky objects, in this way.

TELESCOPE OBSERVING TIPS

The primary rule of telescopic observing is always to start with a low magnification. The standard eyepiece supplied with many telescopes has a focal length of about 25 mm or 26 mm, and this is the ideal eyepiece to use at the start of any observation. It gives a fairly wide field of view, so if the object you are observing is not centrally placed in the field, you should be able to spot it anyway. Furthermore, low magnifications give brighter images and are easier to focus than high powers. For the beginner, a low-power eyepiece is usually easier to look through because it has a larger exit pupil, whereas with high-power eyepieces it can be very tricky to get your eye in the right place.

Small Schmidt-Cassegrain telescopes in particular can be difficult to focus because their focusing knobs are small and require a lot of turning to make much difference. Also, unlike a Newtonian or refractor, you can't see the position of the focuser, so you have no idea which way to turn the knob. It's a good idea to find a really bright object to start with so that you can easily see which way the focusing is going. The aim is to make the blurred image as small as possible for perfect focus.

Once you have found the object you can proceed to using higher powers. Changing eyepieces takes time. If your telescope does not have a motor drive, notice how fast the object moves through the field of view and allow for this. Usually the focus position is different with the higher power, so you must refocus – which may mean that you lose the object you wanted to observe.

Higher powers mean dimmer images, smaller fields of view and quicker movement of the object through the field of view if the telescope is not driven. One advantage of higher powers, apart from the obvious one of increased magnification, is that the sky background is dimmed more than are star images, which should remain as points of light. So if you want to locate a small, faint cluster, for example, using a higher magnification should improve your chances in a light-polluted area. But a faint galaxy or planetary nebula will be dimmed equally with the background by increased magnification.

Increasing the magnification of a planetary or lunar image may show more detail but it will also increase any effects of seeing. This local turbulence can be caused by heat within the telescope or its immediate surroundings, by nearby landscape features or by higher-level atmospheric turbulence. Observing from inside a warm house through an open window is particularly bad, as the warm air pours out of the window so that you are trying to observe through a shimmering air stream. You should always take your instrument outside and let it get down to night-time temperature before beginning to observe.

People often confuse seeing with *transparency* by referring to "good seeing" when they actually mean a very clear sky. Frequently, one excludes the other. A crystal clear night is often accompanied by turbulence, whereas haze can often mean steady conditions and good seeing. Some observers find that they get the best views of the planets when they are barely visible

▲ *A single frame* from a 2-minute webcam sequence of Jupiter, made using a 200 mm reflector, shows only a few details.

▲ *Using Registax* to combine the best 900 images from 1200 frames, and with image processing, a stunning amount of detail is visible.

USING NEWTONIAN TELESCOPES

One drawback with a Newtonian reflector is that you always look at right-angles to the object you are observing. This means that to find an object you must first squint along the tube to get it roughly aligned, then look through the finder to find the actual object, but then move round to the actual eyepiece to view through the main telescope. However, the finder is not always easy to reach from the same position as the main eyepiece, and in many cases the finder is quite close to the tube so it can be hard to get your eye to it.

However, this situation also has its advantages. Most modern reflectors allow you to rotate the tube in its cradles so as to bring the eyepiece to a convenient height. You may be able to observe seated, for example. Just make sure that you don't lean on the telescope for support, or you will lose the object you are observing. The image in reflectors is always inverted, but at least it is not a mirror image as with the star diagonal that many observers use with a refractor.

Closed-tube reflectors are rarely troubled by dew, as the mirror is well down the tube. Make sure when observing, however, that you don't breathe on the front lens of the finder, and also keep your body heat away from the front of the tube, which will probably disturb the image. It is also easy to breathe on the eyepiece as you move to the finder, so take particular care not to. If the eyepiece does become misted up, resist the urge to wipe it with whatever comes to hand – see the box on eyepiece care on page 42.

Because the mirrors of a reflector will tarnish in time, make sure that you keep the dust caps on the telescope whenever it is not in use, not forgetting the cover for the focusing mount.

▲ *A Newtonian of medium size –*
here, a Russian TAL-1 – can provide a very comfortable viewing angle as the telescope can be rotated in its cradle. Following an object is just a matter of keeping one hand on the slow motion. The finder in this orientation is below the eyepiece.

through haze or cloud. The all-round skygazer chooses the best objects to observe depending on the conditions – either clear nights for deep-sky work or hazy ones for the planets.

HOW TELESCOPES PERFORM

You often need to know the limitations of your telescope. Very often these are set by the seeing conditions and the presence of light pollution, but tests do give practical values for the faintest star visible and the finest detail visible with various sizes of telescope. The *resolving power* of a telescope is a measure of the finest detail it will show, and is limited by its aperture. A small telescope will not show fine detail, no matter how much you increase the magnification – you just magnify a blur. The figure given for resolving power is the separation of the closest double star that it will show as two individual stars, but this serves as a reasonable guide to the size of detail that should be visible on planets. The table below gives figures for a range of popular telescope apertures.

RESOLVING POWERS FOR DIFFERENT APERTURES								
Aperture of telescope in mm	60	80	100	114	125	150	200	250
Faintest star visible (magnitude)	11.6	12.2	12.7	13.0	13.2	13.6	14.2	14.7
Resolving power (arc seconds)	1.9	1.4	1.2	1.0	0.9	0.8	0.6	0.5

GETTING TO KNOW YOUR TELESCOPE

There are no driving schools for telescopes, and most people have to pick everything up for themselves. On the following pages are four typical telescopes of different types, with hints and tips to get you started with each one.

Whatever the telescope, your first step must always be to align the finder with the main instrument, ideally in daylight, as described on page 36. Having done that, make sure you begin observing with your lowest-power eyepiece, which is the one with the longest focal length – usually 25 mm or 26 mm. Always observe outdoors rather than through a window, either open or closed, and don't expect much from a telescope that has just been taken from one environment to another with a very different temperature.

It also helps to have some idea of what objects are in the sky at the time of observation. The Moon is always the best object to begin your first-ever observing session, as it is easy to find and to focus on, particularly at low power, but if it is very low in the sky you will not get a very steady view. Try to get your bearings before starting, both so that you know where to find any bright planets on view, and also because if you have an equatorial mount you will need to align it with the Earth's axis.

130 mm Newtonian reflector on driven equatorial mount

These are popular beginner's instruments because they offer enough aperture to reveal detail on planets and a good range of deep-sky objects, yet are competitively priced. Anyone familiar with this instrument will also be able to use similar smaller and larger reflecting telescopes, such as a 114 mm or a 150 mm, of which there are a large number available.

Having aligned the equatorial mount (see the box on the facing page) you can now get to grips with observing. Also read the box on using a Newtonian telescope

▶ *The Sky-Watcher 130 mm is a budget Newtonian reflector which performs well and will show a wide variety of objects. The UK version has a motor drive on the RA axis, which is always worth having with even a basic equatorial mount. Details of setting up the instrument are shown in the box opposite.*

on page 40. Once you have found the object you want to observe, lock the RA and declination axes quickly while the object is still central within the field of view. Now you can track the object, either by turning the RA slow-motion cable or engaging the motor drive if fitted. The motor drive on this particular mount is engaged by flipping a metal lever on a cam which pushes the motor into contact with the large knurled drive wheel.

Newtonian telescopes usually have a rack-and-pinion focusing mount which, if it operates smoothly, can provide very easy focusing. You can feel the eyepiece move in and out as you turn the knurled focusing knob, which gives an element of feedback. Crayford-type focusers use a friction system and are usually smoother.

◄ *Once you have found* the object you want to observe, a metal lever brings the motor drive into mesh with the knurled wheel to drive the telescope. On instruments without a motor drive, you can also use this knurled wheel as a slow-motion control.

USING AN EQUATORIAL MOUNT

Equatorial mounts often cause great confusion for beginners, as they have numerous scales, clamps and bits sticking out. Take a little time to understand what everything does in daylight. In the northern hemisphere, where there is a prominent and easily found Pole Star, setting up an equatorial mount is really quite easy as long as you can find Polaris (see page 139). Southern-hemisphere observers have more of a problem, as there is no equivalent star marking the south celestial pole. The following step-by-step guide is written for the northern hemisphere, but observers south of the equator should sub-stitute south for north.

Step 1 Find your latitude to the nearest degree. Look in an atlas if you are not sure.

Step 2 Work out which is the polar axis – it is always the one which is lowest, and is attached to the tripod. Its angle to the horizontal can be adjusted, and there is a scale reading 0°–90°.

▲ *Set the polar axis scale* at the same angle as your latitude – here 50°. The threaded bolt, left, alters the tilt of the axis, while the central bolt clamps it in place. This bolt is often hard to tighten, but if it is left loose the whole telescope may overbalance.

▲ *The mount is now in alignment,* with the polar axis pointing north. The positions of the slow motions and the axis clamps are often a compromise. In this observing position, with a reflector, the RA slow motion and clamp are not easily

Step 3 Set the pointer on the scale to your latitude. The polar axis now points at the same angle to the horizon as your latitude.

Step 4 Find north (see page 22). Notice where shadows point at midday or look on a map.

Step 5 Point the polar axis in this direction. That's it. You have now aligned your equatorial mount.

But how accurate do you need to be over Step 5? The answer is, not very. As long as you are within a few degrees of the celestial pole, the equatorial mount will work perfectly well for general observing. The main

accessible though the declination slow motion and clamp are not far from the eyepiece. With the clamps loose, move the telescope within its cradles and move the counterweight along its axis until it is in balance and equally free to turn in all directions.

thing is that you should be able to observe an object by just moving one axis, either using the flexible slow-motion cable or by engaging the motor drive. Objects may drift off over a period of minutes, but unless you are observing at very high power you will not have to touch the declination slow motion very much.

Make sure that your telescope is well balanced. With the clamps on both axes loosened the telescope should not move of its own accord in any direction. To begin with lock the dec. clamp and put the dec. axis horizontal (pointing east–west). Adjust the counterbalance position so that the telescope does not swing one way or the other. Then lock the RA

clamp and make sure the telescope is well positioned in its cradle and again does not want to swing round. You should never have to tighten the axis clamps in order to prevent an out-of-balance telescope from moving.

The slow-motion cables, if fitted, can sometimes be more trouble than they are worth if they are in such a position that they obstruct the telescope's movement. If this happens you may have to remove the offending cable, or invert the telescope and move your entire observing position to the other side of the mount.

One drawback of the German-type equatorial mount is that eventually the telescope hits the tripod or central pier after it has crossed the meridian. If this happens all you can do is to rotate the mount so that the telescope is on the other side of the pier, which also means rotating the telescope through 180°. Some motorized mounts do this "renormalizing" procedure automatically when they reach a point shortly after the meridian. This can be irritating if you are in the middle of an exposure.

▲ *Oops! Slow-motion cables* and the tripod can often get in the way of the telescope. Either remove the cable or swing the telescope to the other side of the axis.

Refractor on equatorial mount

Finding an object using a refractor is a little easier than with a reflector, because you observe in the same direction as the object. The finder is usually close to your observing position. However, this assumes that you are looking straight through the telescope rather than using the star diagonal which is supplied to give you a more convenient observing position.

The alternative is to crouch or lie down, unless either you have a very tall tripod, which may be unsteady, or are observing an object low in the sky. A piece of old carpet is a useful accessory for refractor owners! Sometimes you may be lucky enough to be able to lie back on a garden chair with your eye just at the right position for observing, but this does not usually last for long as the eyepiece is always moving as you follow the object through the sky.

If you decide to use the star diagonal you must remember that your view is laterally reversed compared with the real object as well as upside down. This can easily cause orientation problems when you are star hopping or trying to identify lunar features.

Dew is a real enemy of refractor users as the objective lens is exposed to a wide area of sky, which may cause it to cool rapidly. Most refractors come equipped with

▶ **The Moon as it appears inverted** through an astronomical telescope (left) and with the addition of a star diagonal (right). Star diagonals can cause great confusion when comparing what you see with a map.

dewcaps, but these do not give complete protection. Never wipe dew from the lens as you can easily scratch the overcoating. You can buy low-power heaters for the dewcap that provide just enough warmth to reduce the chance of dewing, but many people have to rely on portable hair dryers that run from a 12-volt battery or car battery. However, do not use a household hair dryer on full power as this could cause stresses in the glass that at best will destroy the image until they cool down and at worst could permanently damage the lens.

Dobsonian telescope

The simplicity of the Dobsonian mounting means that you can very easily point the telescope at an object and begin observing with no set-up problems. For deep-sky objects, however, this requires you to know the positions of the objects quite well and to have a good finder. It is also very helpful to have a wide-field eyepiece, making it easier to find the star patterns

that you can see in the finder, but brighter and more magnified. You need experience in getting to know the difference in appearance between the two views.

Observing at higher power requires you to swap eyepieces quickly, and know the different focus positions. When you get to very high powers, the constant requirement to move the telescope can get tiresome, particularly if the movements are not silky-smooth.

Objects near the zenith in particular can get very hard to track. The trick is to arrange the base of the telescope so that you are moving the telescope in azimuth only to follow an object, with the base aligned roughly east–west rather than north–south, in which case you will be trying to turn the telescope on its turntable all the time.

The eyepiece of a Dobsonian is usually at a fixed angle, which can sometimes make it hard to reach when objects are high up.

Go To telescope

On the face of it, Go To telescopes are the answer to every beginner's dream – just let the computer do the work of finding objects. But virtually all of them require some input from the user to set them up. They need to know the date and time, and where they are, and then they need to be pointed at known stars so that they can get their bearings. The time and location data are sorted out automatically if your telescope has a GPS module in its innards, but in general this is not too hard to work out, as most people do know the time and date and where they are. The challenge comes when trying to align the telescope on known stars.

Telescopes with altazimuth mounts use a different procedure from those with equatorial mounts. Usually the same handset will work with either type, but you do need to tell it which it has to start with. Owners of equatorial mounts need to align them accurately on the pole, starting with the procedures shown on page 41 but then refining the process by using the polar-align finder, which you first need to orientate correctly for the date and time. In the northern hemisphere, you need to get Polaris in a small circle. In the southern hemisphere the procedure is more tricky as you have to find a group of three quite faint stars, which may not

---

TIPS ON EYEPIECES

Eyepieces are usually the most robust parts of your observing system, but they can too readily be abused. They can easily become dewed up, in which case observers often wipe the dew off with their fingers or put them in an inside pocket to warm up. Both practices are bad for the eyepieces, which then get scratched and dusty optical surfaces.

The best way to deal with eyepiece dew, in the absence of wired anti-dew heaters, is to keep a 12-volt portable hair dryer handy, or use the car's heater blower if you are observing away from home. Sometimes there is nothing for it but to use a different eyepiece while the affected one dries out. Canned air blowers are useful for removing dust, but they usually chill the surface as well so are best used in warm conditions rather than when observing.

be visible in light-polluted areas. Having polar-aligned, the procedures are similar to those for alt-az mounts.

If the telescope knows the direction of at least one star – usually one of the handful of the brightest stars in the sky – it can find all other objects. But each manufacturer's system for finding its reference star or stars is different. Celestron have a SkyAlign system, which requires you to point the telescope in turn to three bright stars using its motors. You don't need to know the names of the stars or even whether they are stars or planets, because the internal computer is aware of what's in the sky at its time and location. Ideally you should choose three well-separated objects that are not all in a straight line. Having done this, the telescope usually tells you that it has aligned successfully, but from time to time it does fail and you have to try again. Users generally report that this method is very easy to use and aligns well.

With Sky-Watcher's SynScan system you do need to recognize the brightest stars in the sky. The alt-az versions of the mount try to make this as easy as possible by asking the direction of the brightest star first. You then need to drive the telescope to point at that star, then choose another bright star whose name you know. With equatorial mounts, the star is chosen for you and the mount moves to the object which you then need to center in the telescope or finder. One-star alignment using the finder is often quite adequate for general observing.

Meade require you to first level the telescope and point it north, after which it has a stab at finding the first reference star. This step is automated in many models, so the telescope swings around and points up and down by itself for several minutes, but they have to rely on a built-in magnetic compass and compensate for magnetic variation to find true north. Errors in this stage can result in them not pointing accurately to their first reference star, so the automatic system is not perfect. Sometimes it is quicker to do the leveling and finding north yourself! Having aligned it on the two

bright stars which it finds, it should then be able to find any other object.

With all these systems, if the view of a particular bright star is obstructed from your site, you can choose another. It still helps to know the names of the stars even in the Go To era! Users accustomed to using apps on tablets or smart phones may find the interface with Go To telescopes very primitive. However, most Go To systems can be linked to apps using third-party connectors, allowing you to choose your target from a sky map.

Using catadioptric telescopes

Schmidt-Cassegrain and Maksutov-Cassegrain telescopes allow you to observe in the same direction as the telescope is pointing, as with refractors, but again a star diagonal is commonly used to give a better viewing angle. Remember that the image is laterally inverted. If you suspect optical problems with your telescope, try observing without the diagonal in case this is at fault.

The focusing of catadioptrics is usually by means of a knob at the back of the telescope rather than a rack-and-pinion focuser. This means that the eyepiece stays still as you focus but instead a linkage moves the main mirror up and down in the tube. Very often there is play in the linkage which means that the mirror shifts slightly as you change the direction of focusing. At high power this can be very irritating as the object can move significantly in the field of view, but you just have to allow for it.

Dew is particularly troublesome with catadioptrics as the corrector plate has a large exposed area. A dewcap is an essential additional accessory for most observers, and you should follow the same instructions as for refractors on means of coping with dew.

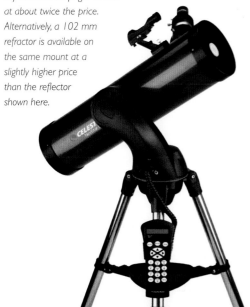

▼ *The Celestron NexStar 130 SLT* uses the popular *SkyAlign Go To system on an altazimuth mount. This 130 mm reflector has the same optical specification as the non-Go To Sky-Watcher on page 40 but at about twice the price. Alternatively, a 102 mm refractor is available on the same mount at a slightly higher price than the reflector shown here.*

USING AN OPTICAL FINDER

Check the alignment of your finder each time you observe. Finders are easily knocked when moving the instrument and you can waste a lot of time trying to find an object if the finder is slightly out. The ideal target for this is the Moon, as its brightness can make it easy to locate in the main telescope even if the finder is not accurately set, as long as you use a low power. Use a feature that you recognize – you must get it spot on. In the absence of the Moon, use the eyepiece with the widest field of view on the main telescope and aim for a bright star which is unmistakable. Aligning the finder on nearby objects is not accurate.

Most optical finder telescopes do not have illuminated crosswires so you have to rely on the sky brightness to show them. This is about the only advantage of observing in light-polluted skies.

Bear in mind that finders may have two focusing adjustments – one for the crosswires and one for the stars. Simple finders often have no focusing for crosswires, so if they are not in focus there is little you can do. The focusing for the stars is usually carried out by unscrewing a knurled ring that locks the dewcap, turning the dewcap slightly, then retightening the knurled ring. Make sure that the crosswires and the object are both in focus together, otherwise you will find that the crosswires move slightly against the object as your eye moves.

▲ *To adjust the focus* of a finder, slacken off the clamp ring, left, turn the dewcap (which carries the lens) to focus, then tighten the clamp ring to fix it. Always focus on an object at infinity.

4

To many amateur and indeed professional astronomers, the Moon is the villain of the piece. The period known as "the light of the Moon," when the Moon is high in the sky, renders observations of faint objects impossible, while "the dark of the Moon" is the joyous period when deep-sky observing can take place. But to others, the Moon itself is a fascinating world in its own right, and far from being an unchanging object is full of surprises and hitherto unknown aspects.

In fact, if you want to impress someone with the view through your telescope, you can do no better than to show them the Moon at first quarter. That

◄ *The full Moon*
photographed on two different occasions reveals changes in tilt due to libration. In the top picture Plato is well away from the northern limb while Tycho appears quite near the southern limb. In the bottom picture the situation is reversed.

familiar pale orb in the sky becomes transformed into a world in its own right, with peaks and valleys, plains and of course craters – thousands of craters, of all sizes. The Moon has been battered ever since its formation, 4.5 billion years ago. True, the Earth has been under similar attack by Solar System bodies of all sizes, but in our case most traces of the impacts have been obliterated by the action of water and wind, and also by tectonic activity and volcanism.

On the Moon there are no continental plates and erosion takes place much more slowly (over millions of years, rather than hundreds or thousands), so the features that you see now are the same as would have been seen at the time of our earliest human ancestors, a million or so years ago, or possibly even at the time of the dinosaurs, 65 million years ago. The asteroid impact that is thought to have wiped out the dinosaurs would have produced a major crater about 180 km across had it hit the Moon, but there it would be just one among many others.

Because the Moon is airless, not only do its features remain unchanged over the eons, but also the view with your telescope is virtually as good as if you were visiting it in a spaceship. Viewed with a magnification of 100, on a night when our own atmosphere is steady, you can imagine that you are 100 times closer to the Moon, and are getting the same view as the Apollo astronauts did when they were only 4000 km away from it instead of 400,000 km. The brilliance of the landscape can almost hurt your eyes, and the shadows are jet black. Although it is a wasteland, it is one that invites you to explore it, and you can do this from the comfort of your backyard.

There is something else to be said for the Moon, particularly for beginners. It provides an excellent test object on which you can hone your observing skills. Even if you can't find deep-sky objects, you can hardly miss the Moon. If stars will not come into proper focus, the Moon will help you to see where you are going wrong, because there is no doubt when you are in focus. You can practise your drawing or photography on an object that is easy to track and well lit.

MOVEMENTS OF THE MOON

As Earth's natural satellite, the Moon is always with us. Only for a few days each month, when it is more or less in line with the Sun, is it not visible at all. It orbits Earth every 27.3 days, so you might think that this is the interval between successive full Moons. But during this time the Earth has continued in its orbit round the Sun, so the Moon must move a little more in its orbit to catch up. As a result, the interval between full Moons is 29.4 days, a period known as a *lunation*.

We are so familiar with the face of the Moon in the sky that we never wonder why this is the case. If the Moon orbits us, surely we should see different sides of it? In fact, it has what is called *captured rotation*, which means that it always turns the same side toward the Earth. This is easy enough to grasp, but people often have a problem understanding that it really is rotating.

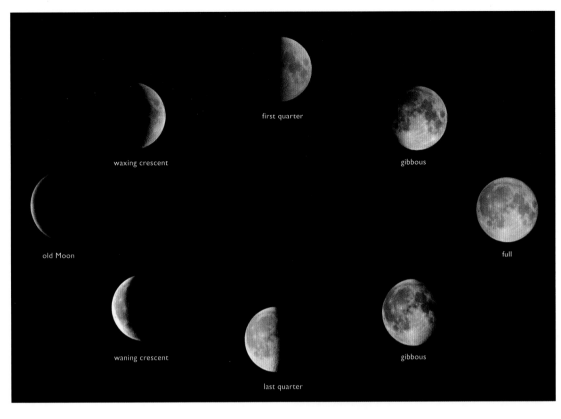

waxing crescent

first quarter

gibbous

old Moon

full

waning crescent

last quarter

gibbous

The fact is that it rotates on its axis, as seen from the Sun, every 29.4 days. Imagine standing on a fairground roundabout, facing the center all the time, just as the near side of the Moon always faces Earth. For each revolution of the roundabout you have actually rotated once yourself, so someone standing outside would have seen all sides of you, while someone at the center of the roundabout would have seen only your face.

To extend the analogy, imagine that the only illumination is a floodlight some distance away from the roundabout. As you go round, it will sometimes be behind you, sometimes to your side and sometimes shining in your eyes. The same thing happens to the Moon, and this is what causes the *phases* – the changing angle of illumination from the Sun as the Moon goes round the Earth.

The lunation cycle begins at *new Moon*, when the Moon is between us and the Sun. Only rarely does it pass directly between the two, so that we see an eclipse (see page 94) – usually, it passes above or below the line of sight. We commonly refer to new Moon, however, as the thin crescent that appears in the evening sky just after sunset. Often you can see the rest

▲ **Earthshine,** *sometimes called "the old Moon in the new Moon's arms," is caused by sunlight reflecting off the Earth on to the unilluminated part of the Moon.*

of the Moon faintly illuminated by what is known as *Earthshine*. This is caused by light reflecting off the Earth, which is nearly full as seen from the Moon when the Moon is new to us. At this point in the cycle, the Moon sets shortly after the Sun.

Each day the Moon moves eastward in its orbit and is visible later and later in the evening as a larger and larger illuminated crescent. Seven days after new Moon it is at what is called *first quarter* – because it is a quarter of the way around its orbit – and we see a half Moon. After this, its illuminated portion becomes larger and we call it *gibbous*. At this stage it is waxing, or growing in phase.

Seven days after first quarter we get *full Moon*, though for a day or two on either side of full it can look completely illuminated. At this point it is opposite the Sun in the sky, and therefore rises in the east at sunset. From then on it begins to wane, and the shadowed area starts to encroach on its landscape from the opposite edge. Each night it rises later and later until after another seven days or so it is at *last quarter*, when we again see a half Moon but with the opposite side illuminated from first quarter. Last quarter Moon is usually seen only in the morning sky, though at some times of the year it may rise well before midnight, particularly if you live at fairly high latitudes, such as in northern Europe.

After a few more days the Moon is a thin crescent again, this time rising just before the Sun in the east. It can be referred to as the *old Moon* at this stage. It then lines up with the Sun and the cycle begins again.

People often refer to "the dark side of the Moon," meaning its far side. But all sides of the Moon are illuminated at some point during its orbit. At full Moon the far side is indeed all dark, but at new Moon, when it is between the Sun and the Earth, the far side is in full sunlight and it is the near side that is dark.

In fact, because the Moon does not orbit the Earth in a circle but in an ellipse, we can see a little more than half of its surface. Sometimes we see a little more of one edge, and at other times we see more of the other edge. You can see this even with the naked eye if you are familiar with its surface. The dark area known as the Mare Crisium, near the western edge of the Moon as it is seen in the sky, is easily visible to most people. At times it is almost on the edge, while at others it is noticeably closer to the center. These variations in attitude are called *librations*, and lunar observers make use of them to see regions that are otherwise unseen.

The far side of the Moon has only been seen by Apollo astronauts and photographed by spacecraft.

OBSERVING THE MOON

Virtually any telescope will reveal details on the Moon. It does not take much magnification to show the craters and major features, which are just on the limit of visibility when the Moon is seen with the naked eye. Even binoculars show some detail, but will leave you wanting more power. The magnification limit is often set by our own atmosphere – on a night of bad seeing, a low power of 80 or so is the most you can use before the details become blurred. In these conditions, the surface appears to be constantly in motion, as if seen at the bottom of a fast-moving stream. You get an impression of what is there, but the fine details are hard to make out. On a night of average or good seeing, however, you can increase the magnification to 150, 250 or perhaps even more.

Although high magnifications can give you the impression that you are flying just over the Moon's surface, the true size of lunar features visible with an amateur instrument is still quite large. With a 300 mm telescope, for example, observing under perfect seeing conditions, you could theoretically see features about 750 m or half a mile across. Even the Pentagon in Washington, D.C., would be too small to be distinguished if it were transferred to the Moon.

Some telescopes are equipped with a Moon filter, either green or neutral color. The purpose of this filter is simply to reduce the brightness of the full Moon, which can be so bright as to leave you virtually blind when you step away from the eyepiece.

Because the Moon's phase is continually changing, the same feature seen on successive nights will appear different on each occasion. When it first appears in sunlight, on or near the shadow line or *terminator*, the feature is seen under a very low angle of illumination and as a result every slight bump or hollow in the surface casts a shadow. As the Sun rises over the feature, different details gradually come into sunlight, and when the Sun illuminates the feature from overhead all you can see are differences in brightness.

As the Moon takes almost a month to go through its cycle of phases, you might imagine that there will be little change in the illumination of the scene as you observe. However, variations in the lighting take place slowly but noticeably. If you decide to draw what you see – a traditional, satisfying and quite easy means of recording the view – you will find that as soon as you have finished all but the quickest sketch, changes

▼ **The Moon** at its most spectacular, just after first quarter.

have started to occur, particularly if you are observing around the time of first quarter.

Even if you are just Moongazing, one of the great delights of lunar observing is to watch the stately changes as the Sun rises over some great crater. To start with, only the highest peaks catch the sunlight. Soon, their lower flanks are visible, and before long the crater floor itself is crossed by the jagged shadows of the rim. After an hour or two, the whole crater is obvious. The next night you will find that the same crater is well away from the terminator, and is now blending into the mass of other craters.

Just as you learn the sky by finding a few signposts and building up your knowledge through the year, so you can learn the Moon bit by bit. It is unusual to be able to watch the sequence of phases night after night, either because of the weather or for social reasons, but you can return month after month on clear nights and pick up where you left off.

Names on the Moon

The lunar features were named in the early part of the telescopic era. The first observers did not realize what sort of a world they were looking at, and they imagined that the dark areas might be seas. These were given names using the common language of science at the time, Latin. The Latin name for sea is *mare*, pronounced "mah-ray," the plural being *maria*, pronounced "mah-ree-ah." So we have Mare Tranquillitatis, the Sea of Tranquility, and so on. One suspects, however, that romance took over from reason when Mare Nectaris (Sea of Nectars) and Lacus Mortis (Lake of Death) received their names. The area where the Ranger 7 spacecraft crash-landed in 1964, sending back the first close-ups of the lunar surface, was subsequently called Mare Cognitum, meaning "Sea that has become known." We now know that the lunar seas are huge basins left after the impact of giant asteroids, which accounts for the roughly circular appearance of many of them.

The craters have traditionally been named after astronomers and other worthies, mostly historical ones. Indeed, in many cases a prominent lunar feature is our main reminder of someone who has since sunk into obscurity.

Compass points on the Moon all changed in the 1960s when the first lunar missions took place. Prior to that time, the west side of the Moon was taken to mean the west as seen from Earth, so the Mare Orientale, for example, was on the Moon's eastern edge – hence its name. But to a lunar explorer it would be to the west, and we now refer to lunar directions as if we were seeing the surface like a map. For the same reason, lunar maps are now shown with north at the top, although astronomical telescopes invariably invert the view. For observers in the southern hemisphere, however, an official map of the Moon is now the same way up as they see it through a telescope.

The maps on the subsequent pages mark in red the locations where various spacecraft, both manned and unmanned, have either landed or crashed.

SOME LUNAR FEATURES AND THEIR TRANSLATIONS	
Sinus Amoris	Bay of Love
Mare Anguis	Sea of Serpents
Mare Australis	Southern Sea
Mare Crisium	Sea of Crises
Palus Epidemiarum	Marsh of Diseases
Mare Cognitum	Known Sea
Mare Fecunditatis	Sea of Fertility
Mare Frigoris	Sea of Cold
Mare Humboldtianum	Humboldt's Sea
Mare Humorum	Sea of Moisture
Mare Imbrium	Sea of Rains
Mars Insularum	Sea of Islands
Sinus Iridum	Bay of Rainbows
Mare Marginis	Border Sea
Sinus Medii	Central Bay
Lacus Mortis	Lake of Death
Palus Nebularum	Marsh of Mists
Mare Nectaris	Sea of Nectar
Mare Nubium	Sea of Clouds
Mare Orientale	Eastern Sea
Oceanus Procellarum	Ocean of Storms
Palus Putredinis	Marsh of Decay
Sinus Roris	Bay of Dews
Mare Serenitatis	Sea of Serenity
Sinus Aestuum	Sea of Seething
Mare Smythii	Smyth's Sea
Palus Somni	Marsh of Sleep
Mare Spumans	Foaming Sea
Mare Tranquillitatis	Sea of Tranquility
Mare Undarum	Sea of Waves
Mare Vaporum	Sea of Vapors

▼ **The craters Aristarchus** (left) and Herodotus as seen from the Apollo 15 spacecraft in low lunar orbit.

NORTHEAST QUADRANT

Though it is small, everyone's eyes are attracted to the Mare Crisium when the Moon is an evening crescent. On the floor of the Mare Crisium are two craters, Picard and Peirce, which look small at first glance. However, they are respectively 34 km and 19 km in diameter, about the diameter of major cities on Earth. To the west of the Mare Crisium is a bright ray crater, Proclus, about the same size as Picard but much more prominent. Ray craters are surrounded by bright streaks of material thrown out when they formed as a result of an impact, and the craters themselves are the brightest features on the Moon. Proclus is unusual in that the rays are asymmetric, leaving the gray plain of Palus Somni untouched. The impacting body must have come in at an oblique angle, throwing debris ahead of its track. Although oblique impacts still result in circular craters, the splash marks that we see as rays betray their direction.

The famous Mare Tranquillitatis, site of the first lunar landing, lies to the west of the Palus Somni. Its floor is crossed by the two obvious fault lines of the Cauchy Rupes (wall) and Rima (fissure). Similar features lie west of Mare Tranquillitatis, namely the Rima Hyginus, also known as the Hyginus Rille. A line of craters extends along this fault line, showing where massive ground collapses took place. Though the vast majority of the Moon's craters are caused by impacts, these are examples of volcanic features.

Northwest of Mare Tranquillitatis lies Mare Serenitatis, flanked by the eroded crater Posidonius. This comes into the category of walled plain, as it has been worn down by the impact of countless tiny meteoric bodies over eons of time and now consists of a circular wall with a flat center. Features known as wrinkle ridges cover the floor of Mare Serenitatis, and are particularly obvious under low illumination. Perhaps these helped to give the impression that the Moon's dark areas are seas, with oddly stationary waves.

The remainder of this quadrant is lighter in color, with many craters and walled plains which you can explore over a period of time.

▶ **Mare Crisium,** with the bright ray crater Proclus to its left (west) and the diamond-shaped Palus Somni.

▼ **The Hyginus Rille** at center with the crater Triesnecker and the Triesnecker rilles to its south.

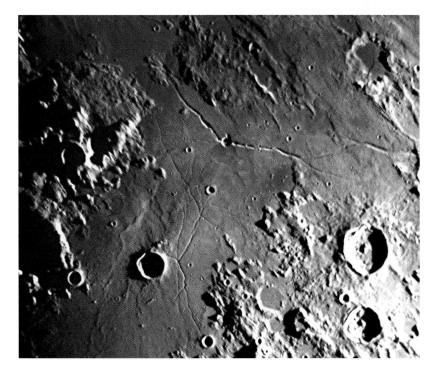

▲ **The complex interior of Posidonius** contains some fascinating details. Nearby is a wrinkle ridge on Mare Serenitatis known as Dorsa Smirnov.

NORTHWEST QUADRANT

Many lunar observers find this the most fascinating section of the Moon, as it includes some spectacular features. Much of it is covered by sea, either the vast Oceanus Procellarum or the separate and dramatic basin of the Mare Imbrium. These plains are punctuated by very striking individual craters which make the area much easier to navigate than some other parts of the Moon.

The Mare Imbrium resulted from an enormous impact comparatively late in the Moon's cratering history, though at a time when the Solar System was still young. As you gaze at it through your telescope, you can see that the lava flows from the impact have flooded earlier craters, such as Stadius at its southern edge, and the beautiful curved bay of Sinus Iridum. In fact, the impact took place some 3900 million years ago, at a time when the first primitive single-celled life forms were starting to emerge on Earth. Subsequently, a comparatively small number of impacts created such magnificent features as Copernicus and Kepler. The fact that the lava plains of Imbrium have comparatively few craters shows that the size and number of impacts must have dropped off quite sharply since the formation of Mare Imbrium itself. (Counts of craters are a standard means of assessing the age of particular landscapes on planets or their satellites.)

Copernicus is surrounded by bright rays and strings of craterlets that were formed by the debris, known as *ejecta*, that was thrown from the crater during the mas-

sive explosion that created it about 800 million years ago. It is a classic large lunar crater, with a central peak and slumped walls. Small craters, just a few kilometers across, remain as perfect bowl shapes, but the larger ones are much flatter in profile. The central peak was formed as a result of the rebound of the lunar surface, much like the central drop that you see in high-speed photos of a drop falling into milk. The lunar surface is not strong enough to support high crater walls so they slump into numerous terraces.

To the west of Copernicus lies Aristarchus, a ray crater that is the brightest feature on the Moon. Adjacent to it, issuing from a point near the older crater Herodotus, is Vallis Schröteri, a channel left after the flow of lava. This valley winds for 150 km across the surface of the Oceanus Procellarum.

Other notable features in this quadrant are the distinctive gray-floored walled plain Plato; the 130 km Vallis Alpes fault which cuts through the lunar Alps or Montes Alpes; and the lone peaks of Pico and Piton, standing 2800 m and 2300 m above the Mare Imbrium. These peaks are the first to catch the sunlight, shining like stars in the darkness.

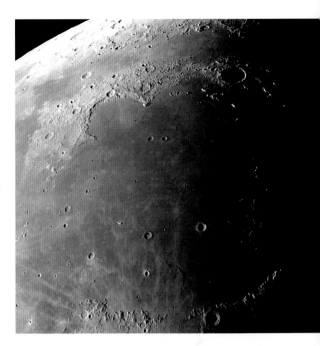

▲ *The magnificent Mare Imbrium.* Plato is at top right, with the Sinus Iridum to its left. The crater at the bottom of the terminator is Eratosthenes.

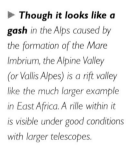

► *Though it looks like a gash* in the Alps caused by the formation of the Mare Imbrium, the Alpine Valley (or Vallis Alpes) is a rift valley like the much larger example in East Africa. A rille within it is visible under good conditions with larger telescopes.

► *The flooded walled plain Plato.* At first glance the craters on its floor are easily missed.

The pictures on these pages have north at the top, and most were taken through amateur telescopes.

▲ *Copernicus is surrounded* by the debris that was thrown out when it was formed. To its east are the ghostly remains of a flooded crater called Stadius.

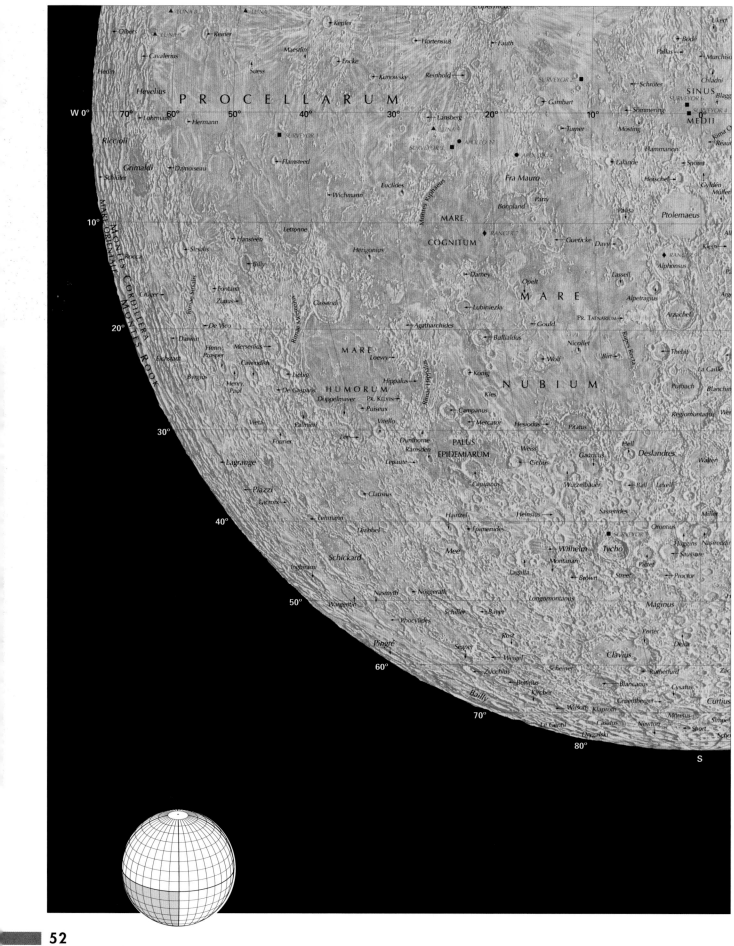

SOUTHWEST QUADRANT

This quadrant is full of interesting landmarks. The first notable feature to be seen each lunation is the north–south line of craters Ptolemaeus, Alphonsus and Arzachel, sometimes referred to as the Lunar Snowman from its sequence of decreasing size. Ptolemaeus has a dark, flat floor with a prominent crater, Ammonius, just 9 km across. Just to the west of Arzachel in the Mare Nubium is one of the most appealing features on the Moon, the Rupes Recta or Straight Wall. This is a scarp fault some 110 km long, and it appears as a dark line around first quarter or as a bright line around last quarter. Despite its name it is not particularly steep, with a slope of around 40°.

South of the Mare Nubium lies a large area of lunar highland. Here can be found one of the most curious sights on the Moon, the crater Tycho. With a diameter of 85 km, it is just a few kilometers smaller than Copernicus and as a recent crater it has many similarities (though it is in the lighter highlands rather than in a sea). When it is close to the terminator it is noticeably crisper than most of the surrounding craters, and it has the same imposing ramparts and central peak as Copernicus. But while most craters merge into the background as the Sun climbs higher over them, Tycho gets ever brighter. By full Moon it is the most

eye-catching feature on the whole Moon, as it is a brilliant ray crater.

Tycho's rays cross a large area of the Moon, though there is a darker area devoid of rays immediately surrounding the actual crater. Tycho looks just like a globe with lines of longitude radiating away from the pole, and people sometimes believe that it must be the south pole – though of course real poles do not have convenient grid lines on them!

Closer to the actual south pole than Tycho lies one of the Moon's largest craters, Clavius, with a diameter of 225 km. Only Bailly, near the limb to the west, is larger. Such a crater is so large that if you were standing in the center, you would not see the outer walls – thanks in part to the Moon's smaller diameter than the Earth, and therefore closer horizons. Clavius contains a distinctive arc of craters of decreasing size.

West of the Mare Nubium lies a roughly circular sea called Mare Humorum. At the sea's north side lies a fascinating crater, Gassendi, which has a floor covered with rilles or fault lines.

At the far western edge of the Moon is a dark oval walled plain, Grimaldi, which looks like a miniature version of Mare Crisium on the opposite limb. Grimaldi is the darkest feature on the Moon.

◀ *Clavius* (lower center) and Tycho (top left) are two of the best-known craters of the highlands area. Tycho is said to be a mere 50 million years old.

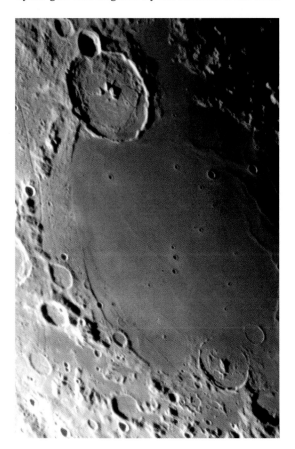

▲ *The Mare Humorum* with its great shoreline crater Gassendi. On its opposite shore is the fragmented wall of Doppelmayer. Picture taken by the Very Large Telescope.

▲ *Ptolemaeus* (top), Alphonsus and Arzachel are an easily recognized group of craters close to the center of the Moon's disk. Each crater has its own individual characteristics.

◀ **Janssen occupies the center** *of this picture, but it lies under the more recent craters Fabricius and Metius to the north and Lockyer to the south.*

▼ **The crater Petavius,** *below, with its dictinctive fault line, measures 177 km and features unusually wide walls of up to 3.4 km in height.*

SOUTHEAST QUADRANT

The first glimpse of this quadrant comes when the Moon is a thin crescent and two great craters, Langrenus and Petavius, are on the terminator. Langrenus, about 130 km across, is on the shore of the Mare Fecunditatis, which contains the unusual pair of ray craters Messier and Messier A (see page 57). Petavius, larger still at 177 km, is notable for a distinctive fault line crossing it from its central peak area to the rim, which makes it one of the most recognizable objects in this

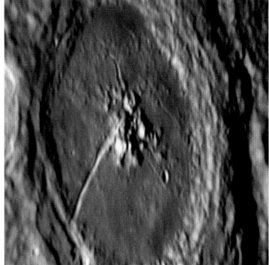

region of the Moon. Several other minor faults or rilles are also present on the floor of Petavius, but they are much narrower and more difficult to see. The walls of this massive crater are unusually wide for its size, being double in places, and rise some 3.4 km from the crater floor, while the central peaks are about 1.7 km high.

The adjacent sea, Mare Nectaris, is notable for several large craters on its shores. Fracastorius has been breached by the sea itself, while the trio of Theophilus, Cyrillus and Catharina line the northwestern shore. An arc-shaped escarpment known as the Altai Scarp or Rupes Altai surrounds the sea on the eastern side and is very conspicuous as a bright line before first quarter or by its shadows before last quarter. Its southern end is marked by the crater Piccolomini.

Much of the quadrant is occupied by light-colored highland, with numerous overlapping craters. Among the most prominent craters are the ancient ruined feature Janssen, which is one of the largest features on the Moon; Stöfler, with its floor darker than the surroundings; and Maurolycus.

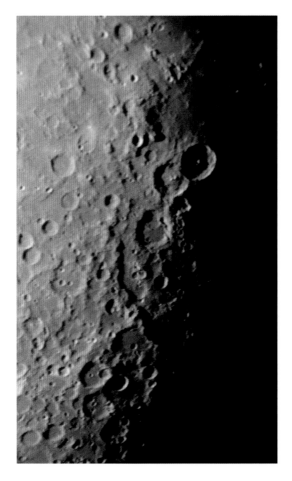

▶ **The crater trio Theophilus** *(top), Cyrillus (center) and Catharina (bottom) lie close to the Altai Scarp, which under waning illumination, as seen here, appears dark.*

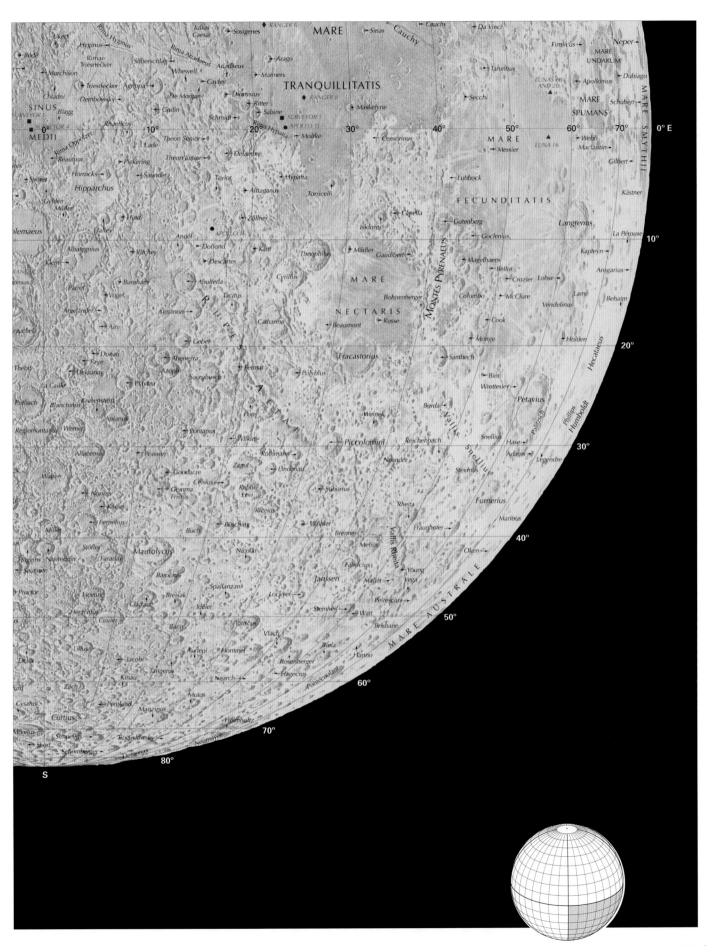

LUNAR LANDSCAPES

The Moon is the only body in the Solar System other than the Earth on which humans have stepped, a situation which is likely to remain for some time to come. Though there are hazy plans to return to the Moon, the 50th anniversary of the first Apollo landing in 1969 will be celebrated on Earth only, with no human presence on the Moon. At the time, it was widely thought that the Moon would be a prime tourist destination by now.

Instead, our lunar exploration will have to be through the telescope. It is always intriguing to compare our views through the telescope with those enjoyed by the Apollo astronauts, and the landscapes presented here compare the Earth-based views with those seen by the Apollo astronauts.

The first lunar landing was a rather brief affair, with less than 22 hours spent on the surface of the Moon and very little in the way of actual exploration. But by the time of the last lunar landing, by Apollo 17 in 1972, scientific and geological exploration was the key goal. However, the lunar landings were really politically inspired, and followed directly on from the extreme shock experienced by Americans in 1957 when they realized that the Soviet Union had orbited the first artificial satellite, and then in 1961, the first man, Yuri Gagarin. Gagarin's flight stung US President Kennedy to ask his advisers how they could upstage the Russians, and so was hatched the plan to send men to the Moon, and return them to Earth, by the end of the 1960s. It seems extraordinary, in view of the slow progress in manned spaceflight since then, that the lunar landing took place less than 12 years after the first tiny satellite had been put into orbit.

▲ *This **Apollo 8** view of part of the lunar far side shows that it is very similar in appearance to the lighter highland areas of the near side, though with very few seas.*

The triumph of the first Apollo mission showed American superiority in space, but once the point had been made the American people lost interest in the exploits of the lunar astronauts. Grand plans to reach Mars by the end of the millennium were shelved, and Arthur C. Clarke's vision of space hotels, as depicted in *2001: A Space Odyssey*, came to nothing.

We have had to rely on the efforts of the Apollo astronauts for much of our knowledge of the Moon. Their missions were carried out at a time when much of the technology we are familiar with today was in its infancy. The Apollo astronauts would have been amazed by a digital watch, let alone a smartphone or a laptop computer. But fortunately there is a rich legacy of photographs, beginning with the lunar orbits of Apollo 8 and Apollo 10.

Apollo 8

The names of the Apollo 8 astronauts, Frank Borman, James Lovell and William Anders, are not as well known as those of the Apollo 11 mission, but they were the first to voyage to the Moon and to see some sights that had never before been witnessed. Their ten orbits of the Moon allowed them to see the far side of the Moon, previously only photographed by orbiting spacecraft, and to see the Earth as just a small blue globe in space, rising over the lunar surface. The mission took place over Christmas 1968.

Apollo 10

May 1969 saw the dress rehearsal for the lunar landing. The Apollo 10 mission went through every stage of the landing mission except for the vital landing itself. Two astronauts descended in their Lunar Module to

▼ *Almost every newspaper in the world printed this photograph, taken as the Earth rose over the far side of the Moon. For an hour or so on each orbit the astronauts were completely out of touch with Earth – the first time this had happened to any humans.*

▶ The crater Moltke near
the Apollo 11 landing site,
taken from Apollo 10. The
valley in the background,
Rima Hypatia, was dubbed
"US Highway No. 1" by the
Apollo crew as it guided
them to their landing site.

▼ **Another such rima**
photographed from Apollo 10
shows that these features are
actually what are on Earth
called rifts or graben, where
two faults have allowed the
land between them to drop.

Apollo 11

The millions of people watching live on television can hardly have realized how close a shave the Apollo 11 landing was. Pilot Neil Armstrong had to overshoot the chosen site because of rough terrain, and in the end there was only 45 seconds of fuel remaining. A camera mounted externally showed some surface dust being blown away by the landing rockets.

The astronauts remained on the lunar surface for less than a day – just long enough to take a few lunar samples, set up science experiments, take photos and try to grab some sleep – though they had to try to curl up in their spacesuits with no hammocks, and neither astronaut really got any sleep. The first-ever Moonwalk lasted 2½ hours and they collected 20 kg (44 lb) of samples.

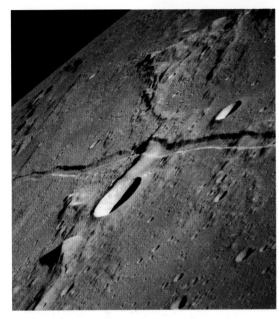

within 15 km of the lunar surface, but then ascended again and rejoined the Command Module.

One of their main tasks was to take detailed photographs of the planned landing site for Apollo 11. By this time, several robot craft known as Surveyors had already landed on the Moon, so its basic properties were known. Prior to this there had been speculation that the lunar dust, accumulated over eons through meteorite impacts, would gather in treacherous drifts. Arthur C. Clarke had written a classic science fiction novel, *A Fall of Moondust*, which centered on a lunar sea that consisted of dust many meters deep, over which future tourists could take cruises in a specially designed vessel.

The Surveyor craft had shown that, at least at the areas where they landed, the lunar surface had something of the consistency of wet sand and posed no threat. But there were major worries that the surface might be so strewn with boulders that it would be difficult to find a flat surface on which to touch down. Accordingly, the proposed landing area in the Mare Tranquillitatis was studied closely.

▲▶ **During their orbits** the
Apollo 11 astronauts saw this
view of the craters Messier
and Messier A – well known
to amateur observers because
of their curious comet-shaped
double-ray system. Many lunar
craters show rays, which are
material ejected from the impact
that created them. As in the case
of Proclus (see page 48), this
is evidence for a very oblique

impact. The craters themselves
are, however, usually virtually
circular because of the way
they were formed. Unlike the
depression caused by throwing
a stone into sand, say, impact
craters result from a much more
violent event. The incoming body
is probably traveling at several
kilometers a second. As it hits,
much of its material is vaporized
on impact, the result being an

explosion taking place some
distance below the surface,
as far as it has managed to
penetrate. The resulting crater
can be around 20 times larger
than the original body.

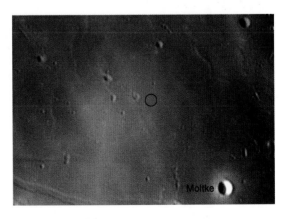

◀ **This amateur
photograph** shows the
Apollo 11 landing site.
The crater Moltke is 6.5 km
in diameter, which gives an
idea of scale. Some of the
smallest craters shown are
a kilometer or so in diameter
– still huge compared with
the size of the base of the
Lunar Module which remains
in place.

Apollo 12

Even as early as Apollo 12, the dream started to go wrong. Shortly after work began on the lunar surface, the TV camera was ruined when it was accidentally pointed at the Sun. The worldwide audience lost interest, although the mission was otherwise a success. With pinpoint accuracy, the astronauts landed within 160 m of the unmanned Surveyor 3 probe, which had itself landed in April 1967. They spent over seven hours in Moonwalks, collecting samples and performing experiments.

Apollo 13

What was becoming a mundane space mission turned into near-disaster when an oxygen tank exploded on the outward leg of the journey, resulting in ingenious repairs and a scrubbed lunar landing before the crew returned safely. Very few lunar photos were taken on this mission.

Apollo 14

The landing site originally planned for Apollo 13 was the uplands surrounding Fra Mauro crater, less than 200 km from the Apollo 12 site. This was intermediate between the comparatively smooth mare floors of Apollos 11 and 12 and the rugged highlands. It became the target of Apollo 14, and even though pilot Ed Mitchell put the Lunar Module down on a flat area, it turned out to have an uncomfortable 8° slope. As a result neither astronaut slept well during their 33-hour stay on the Moon, constantly waking from half-sleep with the feeling that the LM was tipping over.

▲ **Samples retrieved from Surveyor 3** yielded data on the effects of exposure to the lunar environment. In particular, Streptococcus bacteria were found in foam inside its camera, though some researchers believe that this may be the result of later contamination.

▶ **The consistency of the lunar soil,** known as regolith, is clear from this view. The regolith is a mixture of rock fragments churned up by meteorite impacts over billions of years.

The astronauts were helped in their sample collection by a small handcart, as they trekked about 650 meters to the rim of a nearby crater. Schoolbooks have taught us that as lunar gravity is only one-sixth that on Earth, athletic feats should be much easier on the Moon, but that does not allow for the heavy and restrictive suits. The astronauts found that an uphill walk of a few hundred meters took longer and cost them more energy than expected. However, for many the highlight of the mission came when Alan Shepard took a swing at a golf ball. Though he could only send it a few yards because of the constraints of his suit, it was a memorable event.

Apollo 15

Having had three landings in comparatively safe areas, NASA decided that it was now time to take on a more challenging site, the region of the Hadley Rille in the lunar Apennines. This is a sinuous rille, resulting from

a lava flow some 3.5 billion years ago. It and the nearby Mount Hadley provided the most spectacular landscape of any of the Apollo landings. To help in its exploration, Apollo 15 carried a battery-powered Lunar Roving Vehicle, which greatly extended the range over which the astronauts could explore. Their three-day mission involved 28 km of travel on the LRV, up to 5 km away from the Lunar Module.

▲ **Hadley Rille was formed** *by a flow of low-viscosity lava. In some places it was probably a roofed lava tube, but the roof has long since collapsed. It is some 100 km long, about 1.5 km wide and around 400 m deep.*

▲ **The crater Tsiolkovsky,** *photographed from orbit by Apollo 15, is a prominent feature of the lunar far side, and is one of the few seas on that side of the Moon. It has a particularly dark floor. The two sides of the Moon have very different characteristics, presumably because the Moon's captured rotation has led to an asymmetric interior.*

▶ **Hadley Rille** *(center) is a challenging feature for small telescopes, and requires good seeing and the right illumination for the best view.*

Apollo 16

If the landscape at Hadley Rille was dramatic, that of the highlands around Descartes where Apollo 16 landed was more of a gently undulating plain. Again, the Lunar Roving Vehicle was used for drives of several kilometers, and much useful geological work was carried out, with nearly 100 kg (220 lb) of samples being returned to Earth.

By this time, however, the American public was getting blasé about the Apollo landings, and as the initial goal of beating the Soviets had been achieved, the increasingly scientific nature of the missions was seen as not justifying the expense.

▼ **Astronaut John Young** *jumps while saluting. His heavy suit doubled his weight, but in the low lunar gravity he weighed only 30 kg (66 lb). He reached a maximum height of 42 cm and was off the ground for 1.45 seconds.*

◄▲ **The 110 km crater Gassendi** *is a favorite target for amateur observers, with its intricate network of rilles (above). The Apollo 16 photograph reveals these in exquisite detail.*

Apollo 17

The original Apollo plan called for nine lunar landings. However, by 1972 the appetite of the American public for these adventures was poor, and the remaining two were canceled. Apollo 17 was the last lunar landing, though few could have foreseen at the time that it would be over half a century before humans were next destined to venture out of Earth orbit.

Apollo 17 astronaut Harrison "Jack" Schmitt was a qualified geologist, and this mission yielded a vast amount of scientific and geological data about the Moon. The destination was near Littrow crater in the Taurus mountains, and it promised a great deal of interesting geology. The LRV was used again for a total drive distance of over 30 km, and by now the missions were becoming real explorations of the lunar surface. We can only speculate on where we would be now had the will to continue been there.

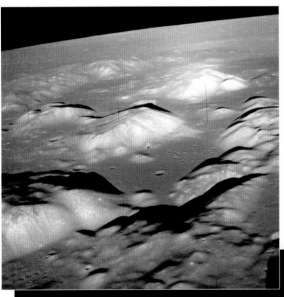

◀ **The descending Lunar Module** is in the distance near the center of this view from the Command Module. Its landing site was in the cluster of small craters just to its right, each of which is about 300 m across. The astronauts explored much of the nearby valley.

▲ **The orange soil** which Jack Schmitt discovered in one location made a dramatic change from the predominantly gray tones of the lunar landscape. At first it was thought to be evidence of recent volcanic activity, but it turned out to be caused by orange glass formed in volcanic action some 3.5 billion years ago.

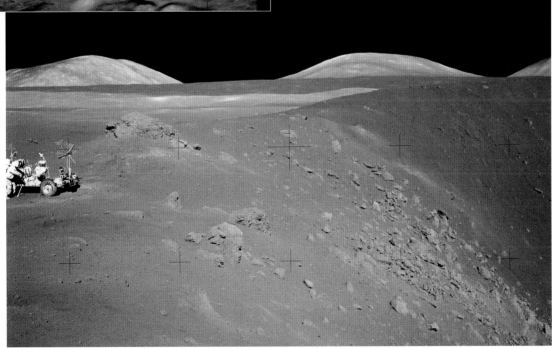

◀ **Jack Schmitt at the LRV** and the steep-sided Shorty crater, within which several patches of orange soil are to be seen.

5

THE SUN

The Sun is one of the easiest astronomical objects to observe, but it can also be the most difficult. It is easy because it is available at social hours, is unaffected by light pollution, and requires only a small telescope for good observations. It is difficult because it is blindingly bright, and there is an ever-present risk that you may unwittingly be blinded permanently.

People sometimes say, "But I often have to look directly at the Sun when I'm driving at sunset, so why is it so dangerous?" There are several reasons. One is that when you glimpse the Sun in the sky, you take only a brief glance before your reflexes make you look away, or your view dances about so that no one part of your retina receives the heat of the Sun for very long. Another is that when the Sun is low in the sky its intensity is considerably reduced. A third is that when viewed with the naked eye, the Sun's image falls on only a small part of the retina. When seen through a telescope, however, its heat is spread over a wide area and the blood vessels cannot conduct the heat away quickly enough to avoid permanent damage.

There is a tale that early observers, such as Galileo, ruined their eyesight by staring at the Sun through a telescope. But Galileo had more sense, and he was the first to realize that the simplest way to view the Sun is to project its image through the telescope on to a piece of paper or card held behind the eyepiece. This is exactly what we do today, though there are also alternative high-tech methods.

Recall those experiments with a magnifying glass focusing the Sun's image to set fire to paper and you will appreciate that there is a risk to your telescope as well as to your eyesight. Telescopes with plastic tubes or eyepiece barrels are totally unsuited to solar observation, and should not be used at all. Some instruction manuals for more solidly built instruments also advise against projecting the Sun's image to avoid damage to the optics. The danger is to the eyepiece, which could become overheated and either crack or suffer damage to the resin that cements the glass elements to each other. You can reduce the risk of damage by not leaving the telescope pointing at the Sun for long periods and by maybe using basic eyepiece types, such as Huygenian eyepieces, for solar projection. Before you project the Sun's image, make sure that the cap is on the finder telescope so that no one tries to look through it, and also to avoid damage to its eyepiece.

Unlike most other astronomical objects, where light is at a premium, in the case of the Sun the problem is that there is too much of it. This is why many telescopes have a tube cover with a small aperture that has its own separate cap. These are intended for use when observing the Sun, to reduce the amount of light reaching the eyepiece. In the case of reflectors the cap is off-center, so that the aperture is unobstructed by the secondary mirror and the spider assembly that holds it in place.

Point the telescope at the Sun by using its shadow as a guide. Use a low-power eyepiece and hold a piece of white card about 30 cm behind the eyepiece, in place of your eye. You should see a bright circle of light on the card, probably out of focus. Refocus until the disk of light is sharp – this means focusing until the image size is at its smallest. This should be the disk of the Sun.

Though there may be genuine sunspots, you may also see the out-of-focus image of any dust on the eyepiece. You can confirm which is which by rotating the eyepiece in the focusing mount. Any dust and eyepiece defects will rotate, but the Sun's image will not.

To get a larger though dimmer image, move the card farther away from the eyepiece and refocus. If the Sun is dimmed by haze or you have a very small telescope, you may need to move the card closer to the eyepiece to get a brighter, smaller view. You can experiment with higher-power eyepieces, which will probably give a view of only part of the Sun. A home-made card sunshield on the telescope or the screen will help to keep direct sunlight off the projected image. You can get large, dramatic views by projecting the image into a darkened room, but the image will move quickly and the setup is difficult to maintain for any length of time.

▲ **The Sun as seen by SOHO** is a dramatic sight compared with the bland disk seen in white light. Here, images from two separate cameras have been combined to show the Sun's surface as seen in the near ultraviolet, and the corona extending several solar radii from the surface.

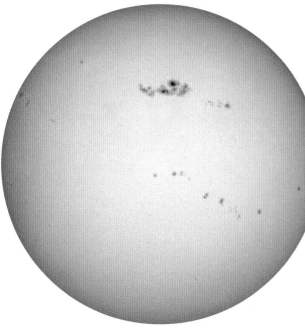

▲ **This projected image** of the Sun was photographed with a digital camera and shows it in its natural color, white. It was taken in 2001 near solar maximum when there were large numbers of sunspots visible.

Solar filters

These days, dense full-aperture solar filters are readily available which allow you to observe the Sun directly through the telescope. Do not be tempted to use other dense materials as filters because they may not have the required infrared absorption (see page 95). Solar filters must fit snugly over the aperture of the telescope – that is, over the top of the tube – where they are subject to normal sunlight rather than the focused beam. The inexpensive filters are made of thin film on which is deposited a metallized coating, and are perfectly safe as long as you follow elementary safety rules. There must be no chink of unexposed aperture, nor damage to the filter (such as by creasing, though some wrinkling is normal), and you must make sure that there is no danger of them falling off or perhaps being removed as a prank.

At one time, all small telescopes were supplied with Sun filters that screw into an eyepiece. These are universally outlawed by the astronomical community because they are known to heat up and crack, with devastating consequences. You may still come across them on telescopes that were made in the past, but never use them.

More advanced filters are available that transmit only a narrow set of wavelengths, usually the light from hydrogen atoms (hydrogen alpha). The advantage of these filters is that they will reveal the prominences on the Sun that are normally only visible during a solar eclipse, and they also show considerable detail on the Sun's surface that is not visible in white light. However, the price tag for the filter alone is about the same as that for a medium-sized telescope. Do not confuse these very narrowband hydrogen-alpha solar filters with much cheaper broadband deep-red filters designed for astrophotography.

What you can see

The Sun is a gaseous body, and what we see as the surface is actually a layer known as the *photosphere* where the maximum light is emitted. Above it, but normally too faint to be seen, is a layer known as the chromosphere. People sometimes think that haze around the Sun is the *chromosphere*, but this is actually haze caused by our own atmosphere.

Some features of the Sun's disk are a result of the photosphere being a gaseous layer rather than a solid surface. The higher regions of the photosphere are cooler, and so shine less brightly than the deeper regions. At the center of the disk we see directly into the lowest layers, but nearer the edge, or limb, we look through a greater thickness of gas and see the higher, dimmer layers. This means that the edge of the Sun appears slightly darker, an effect known as *limb darkening*. Hot clouds of gas that hang higher in the photosphere are known as *faculae*, and they are usually visible near the limb against the limb darkening.

Everyone has heard of solar flares, but these are usually wrongly identified with the prominences that are

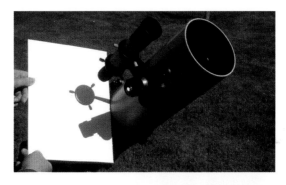

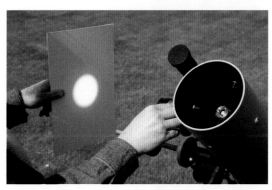

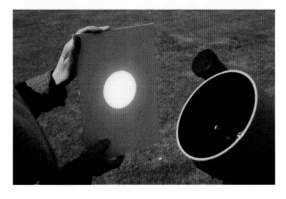

◀ **Three steps to solar observing, here using a 110 mm reflector:**

1 Cap the finder and point the telescope at the Sun using its shadow as a guide.

2 Hold a card about 30 cm behind the eyepiece to project the Sun's image.

3 Focus the image.

seen like red flames at the edge of the Sun during total solar eclipses (see page 94) or when using hydrogen-alpha filters. Flares are outbursts of solar energy, and on rare occasions they are visible as white spots or bars within sunspots. They last only a few minutes, and can give rise to streams of solar particles.

Sunspots

The most obvious features on the Sun are sunspots. These dark markings on the solar disk are usually present. They are areas where the Sun's light output is restricted to a certain extent by magnetic fields at that point in the photosphere. In fact sunspots do emit a considerable amount of light, but in comparison with the rest of the Sun's surface they appear dark. An individual spot has a dark center, known as the *umbra*, surrounded by a lighter region, the *penumbra*. If the seeing is good, you will notice that the whole solar surface has a granular structure, sometimes called a rice-grain structure, which consists of convection cells where the hot gas rises and cooler gas falls. Sunspots also have this granular appearance, but the grains are often elongated in the direction of the center of the spot.

Spots often appear in pairs or in groups. Small ones may last for only an hour or two, while big ones – usually massive complex groups – can be larger in extent than the planet Jupiter and may last for weeks. A group of sunspots is known as an *active region* or active area.

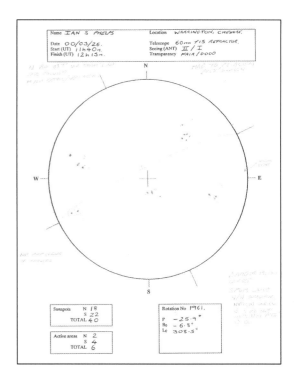

▲ *A detailed solar drawing* made using a 60 mm refractor. The Sun's equator and other particulars have been added afterward, and can be found from published tables or online, having established the east–west line from the drift of a sunspot with the telescope undriven.

The Sun rotates from east to west every 27 days or so as seen from the Earth's point of view, so spots appear to move from one side of the solar disk to the other, changing day by day. A major active region may last for more than one solar rotation. Such regions are often a source of other solar features, notably prominences and flares, and when a major active region is on the Sun, even if it has just appeared on its eastern limb, there is an increased chance of an aurora (see page 98).

The numbers of sunspots vary over a roughly 11-year cycle. At solar maximum the Sun is usually covered with spots, while at minimum there may be very few, or even none at all. Some maxima are stronger than others, and until the poor 2013–15 peak the number of sunspots at maximum has been noticeably greater than it was 100 years ago. As a new cycle begins, spots appear at high latitudes on the solar disk, though they rarely appear above a solar latitude of 35°. As the cycle continues spots are more common closer and closer to the equator, and as the cycle ends high-latitude spots from the new cycle may overlap with those at the equator from the old cycle.

The simplest way to record sunspots is to project the Sun's image on to a standard 15 cm prepared circle on your projection sheet. You can then draw in the spots and faculae (as outlines) directly, and can keep track of the movement of spots across the disk. Make a note of the date and time. At some point switch off the telescope drive if you have one, so that the image drifts across the paper. Mark the direction of drift of a sunspot

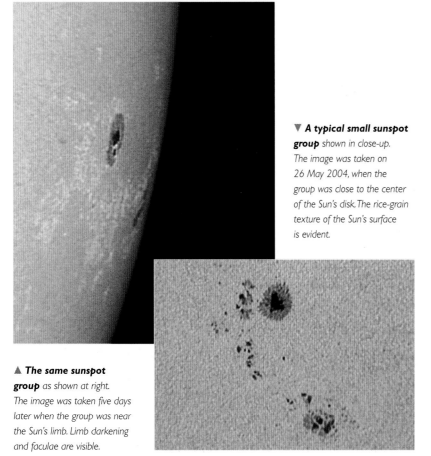

▲ *The same sunspot group* as shown at right. The image was taken five days later when the group was near the Sun's limb. Limb darkening and faculae are visible.

▼ *A typical small sunspot group* shown in close-up. The image was taken on 26 May 2004, when the group was close to the center of the Sun's disk. The rice-grain texture of the Sun's surface is evident.

so that you can get the orientation of the image the same from day to day – the drift is toward the west.

To help determine the position of the Sun's axis and for much more information on solar observing, go to the website www.petermeadows.com/.

The Sun from space

Our understanding of the Sun took a great leap forward with the launch in 1995 of SOHO, the Solar and Heliospheric Observatory. Until then, astronomers were largely dependent on photographs, taken mostly in white light, of the Sun's surface whenever it was clear at a number of dedicated solar observatories. Specialized techniques allowed the Sun to be recorded at specific wavelengths, notably hydrogen alpha. The instruments used included coronagraphs and vacuum telescopes.

A coronagraph is a telescope with a disk placed at its focus, which blocks out the bright light from the Sun so as to produce an artificial eclipse and reveal the corona, the faint outer atmosphere of the Sun. Such a device only works well at high altitudes where there is little dust or water vapor in the atmosphere to scatter light. A vacuum telescope has a tube that is evacuated of air, so as to avoid air turbulence within it as the Sun's light passes through. Vacuum telescopes on mountaintops remain the leaders in high-definition photography of the solar surface. But of course such observations

This coronal mass ejection was observed by the LASCO coronagraph aboard SOHO on 18 February 2002. It was aimed well away from Earth, but from time to time events are observed which are coming our way, and warnings are issued about possible communications disruption. The white circle at lower left marks the actual diameter of the Sun.

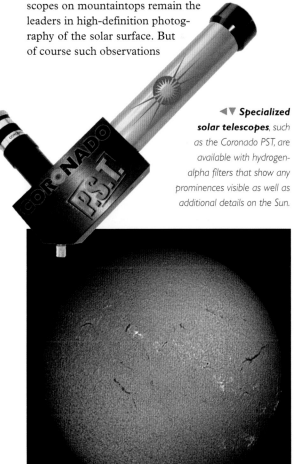

Specialized solar telescopes, such as the Coronado PST, are available with hydrogen-alpha filters that show any prominences visible as well as additional details on the Sun.

are limited by the weather and the altitude of the Sun at the observatory, and can only take place during the day.

However, the SOHO satellite has provided us with an enormous volume of almost continuous observations since early 1996. It is not in Earth orbit, but orbits the Sun at what is called the L_1 Lagrangian point. This is about 1.5 million kilometers in the direction of the Sun from the Earth, and is a location where the gravitational pulls of the Earth and Sun balance out, so allowing the spacecraft to remain in position with very little additional control. This location also has the advantage of perpetual sunlight as it is never on the night side of Earth, and is well clear of the Earth's atmosphere and radiation belts.

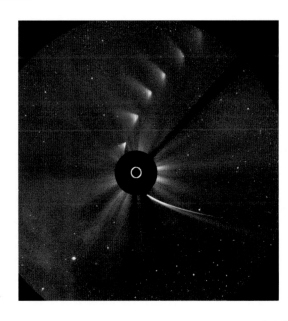

In November 2013 a large newly discovered comet, Comet ISON, plunged toward the Sun and was captured using SOHO's LASCO C3 ultraviolet camera in this time sequence, sporting a long tail just below the Sun. There were great expectations of a spectacular evening display as it re-emerged. But the comet did not survive its close approach, and the subsequent images above the Sun showed that it faded away.

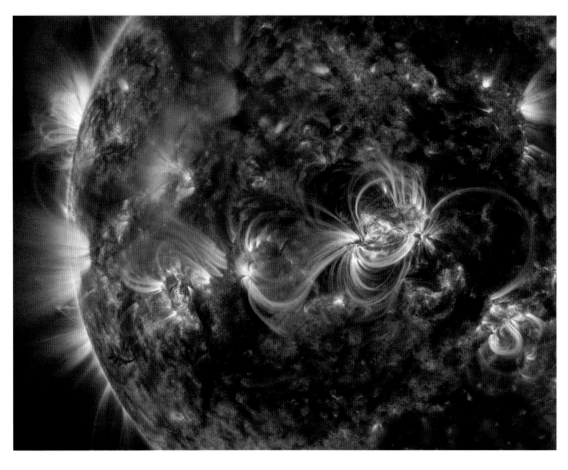

▶ **The Solar Dynamics Observatory** has been observing the Sun from a geostationary Earth orbit since 2010. Its ultraviolet camera captured this series of coronal loops across the face of the Sun, linking several active regions, in January 2015.

SOHO is equipped with several cameras which record the Sun in the ultraviolet and in visible light. Its ultraviolet cameras can show the solar disk in the light from specific gases, such as helium and iron. The views it provides are presented in colors which relate to their location within the ultraviolet spectrum – short wavelengths are blue and the longest ones are red. In general, the bluer the image, the hotter the gas that is being viewed. The images are available daily on the Internet at http://sohowww.nascom.nasa.gov/.

Another camera aboard SOHO is a coronagraph known as LASCO, with a disk obscuring the bright body of the Sun itself. It detects the outbursts of gas from the Sun which result from strong solar activity. Active areas in particular can be sources of what are called *coronal mass ejections* or CMEs. If one of these happens to be directed toward Earth, it can give rise to an aurora – the northern and southern lights – and can occasionally affect communications. Astronauts in space may also be at risk from the particles that accompany CMEs.

An unexpected spin-off from the LASCO instrument is the detection of comets, particularly small ones that are too faint to be noticed using other methods. Only when they are close to the Sun do they become bright enough to be seen – but they still remain invisible from the Earth's surface because of the Sun's glare.

Amateur enthusiasts have been at the forefront in discovering these comets by inspecting the LASCO images as soon as they are available on the Internet. Nearly 3000 such comets had been discovered by mid-2015. Many are members of what is called the Kreutz Sungrazing group of comets, which was first recognized in the 1880s when several great comets were seen near the Sun. All have similar orbits that take them within a few hundred thousand kilometers of its surface, where many of them are completely vaporized. Astronomers speculate that they are fragments of a supercomet, maybe 100 km in diameter, that broke up in the past.

▼ **Close-up of a giant sunspot** photographed on 27 April 2000 using the Swedish Solar Vacuum Telescope on La Palma in the Canary Islands. The spot's length is about three times the Earth's diameter.

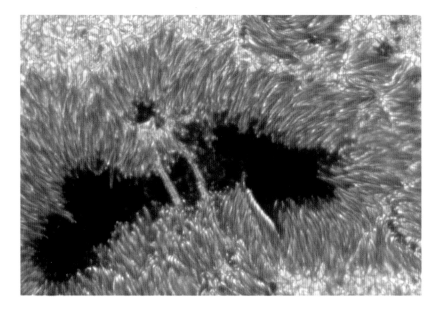

MERCURY

Mercury is not hard to find, and it is surprising how many otherwise experienced observers say that they have never seen it. However, you do need to look at just the right time and in the right place. Because the planet stays so close to the Sun and moves so quickly around it, there is a period of only about ten days at each elongation (see page 29) when it is far enough from the Sun to be seen. If the ecliptic is at a shallow angle to the horizon, it is effectively impossible to see even then.

When you do find Mercury, you may see a tiny white or pinkish disk through a telescope. The disk is small, and you will need at least a 75 mm telescope to see anything. Only elusive and vague markings are ever visible. Mercury goes through a cycle of phases similar to that of the Moon. When it is at full or crescent phase, however, it is too close to the Sun to be observed easily, and only a limited range on either side of half phase is vis-

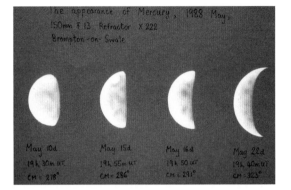

The appearance of Mercury, 1988 May,
150mm F 13 Refractor X 222
Brompton-on-Swale

May 10d.	May 15d.	May 16d	May 22d
19h 30m UT	19h 55m UT	19h 50 UT	19h 40m UT
CM = 278°	CM = 286°	CM = 291°	CM = 323°

◄ *A series of drawings of Mercury* made by David Graham over a 12-day period. They show the changes in appearance of the planet and the markings visible on the tiny disk.

ible. Occasionally Mercury passes across the face of the Sun, in what is called a *transit*. It is then visible as a tiny black disk against the Sun, smaller than many sunspots.

As a planet, Mercury is a rocky, airless world, similar in general appearance to the Moon except that it lacks the vast lava seas. It rotates on its axis exactly three times for each of its two short years, so by chance the same hemisphere is visible repeatedly from Earth, and until 1965 it was thought that its day was the same length as its year.

◄ *A photograph of Mercury* made using a webcam and a standard 200 mm Schmidt-Cassegrain telescope. Markings are evident, as is the reddish color of the planet.

▼ *Mercury is only ever visible* for a few days at a time low down in the twilight sky (here, just above the trees in the center). It is easy to miss unless the sky is particularly clear.

▶ **Mercury has several long escarpments** *such as this one, Victoria Rupes, photographed by Messenger in 2012. Shrinkage of the planet's surface has created these wrinkles, which are named after ships of discovery: Victoria was in Ferdinand Magellan's fleet. The scarp is several hundred kilometers in length.*

Visiting Mercury

It may seem odd that one of our nearest neighbors in the Solar System has been the subject of so little study, but until the arrival of the Messenger spacecraft in 2008, the only close-up pictures of Mercury were taken in 1974 by Mariner 10 – and even then it did not cover the entire planet. But Mercury is actually quite a difficult place for a spacecraft to reach. A flyby mission, such as that of Mariner 10, is comparatively easy. It is Mercury's closeness to the Sun that causes the problem. A spacecraft is drawn down the Sun's gravitational well, and slowing down so as to be able to orbit the planet requires a considerable amount of fuel. Getting there is like falling off a cliff; stopping just before you hit the bottom is the tricky bit. Spacecraft designers prefer to launch payloads rather than fuel, so Messenger used flybys of Earth, Venus and Mercury itself to adjust its orbit rather than using a lot of fuel to slow the spacecraft down once it reached Mercury. Messenger made flybys of Mercury in 2008 and 2009 before finally orbiting Mercury between 2011 and 2015.

At first glance, Mercury looks very similar to the Moon, with a jumble of craters. Like the Moon it is airless, so we see the results of impacts that took place billions of years ago. But a closer look reveals significant differences. There seems not to be the same clear-cut distinction between cratered highlands and dark seas as on the Moon.

There are maria on Mercury, but they are neither as dark nor as extensive as those on the Moon. They are also comparatively crater-free. This is probably because they formed later in Mercury's history than did the seas on the Moon.

Mercury's craters are also noticeably flatter than those of the Moon, because Mercury is a larger and denser body and has a stronger gravitational pull. So central peaks are much less likely, as these are caused by rebound of the surface following the impact that created the crater.

▶ **Images of Mercury** *taken through different color filters by the Messenger spacecraft reveal differences in surface composition. This is an oblique view of much of the planet's southern hemisphere. The main craters on the planet mostly celebrate artists, authors, poets and composers. The large basin at upper right is Rembrandt, and a smaller crater to its left, the larger of a pair, is Kipling. Mercury's craters have slumped more than those on the Moon, as a result of the higher surface gravity, and the ejecta from impact craters falls closer to the crater than on the Moon.*

VENUS

If Mercury is elusive, its companion planet, Venus, is the most highly visible. On those occasions when it is at its brightest in the evening sky, people who rarely give the skies a second glance will notice it as "the Evening Star." If this happens around Christmas time, people may ask if it is a second Star of Bethlehem; at other times of year its appearance spawns a flurry of UFO sightings. Venus is the brightest object in the sky after the Sun and Moon, and is far brighter than any other planet or star, at magnitude −4.

Like Mercury, Venus goes through a cycle of phases. It starts an evening apparition with an almost full disk when it is on the far side of the Sun. Eight months later, when it is at its greatest eastern elongation, it is at half phase. After only another two months, just before it crosses between us and the Sun, it is a thin crescent. On very rare occasions Venus transits across the disk of the Sun, but normally it passes above or below because its orbit is inclined at a different angle to that of the Earth.

The reason why Venus is so bright is that it is completely covered in cloud, which reflects sunlight very effectively. Its dense atmosphere is composed of carbon dioxide, and the pressure at its surface is about 90 times that of our own atmosphere. The surface temperature is very high – around 480°C – and this, combined with a sulfuric-acid cloud layer, makes conditions on Venus very hostile. The few spacecraft, all Soviet, to have landed there all succumbed to the extreme conditions within an hour or two of landing, and only a few direct pictures of the surface have been taken. But radar maps have been made from orbiting spacecraft, showing numerous volcanoes and other related features.

Views of Venus

Venus is not as rewarding to observe as the Moon, Mars or Jupiter. Vague markings are visible where the clouds are slightly thinned, but they are not thin enough to allow views of the surface. Venus observers use a wide variety of color filters, probably a wider range than used for any other planet, to try to make the markings

▲ **Venus seen in close-up** by the ultraviolet camera aboard the US Pioneer spacecraft, showing the cloud patterns very clearly.

more prominent. The dark markings of Venus are typically shaped like a sideways "Y" and take about four Earth days to rotate around the planet.

As the planet is so dazzling, many people prefer to observe it during the day, or at least in twilight, as Venus is then higher in the sky and the contrast between it and the sky is not so great. Venus is easy enough to find by daylight if you know exactly where to point the telescope. If it is already polar-aligned, you can use the setting circles with which most equatorial mounts are equipped. It is best to leave the telescope set up from the previous night, or at least mark the positions of the tripod legs on the ground so that you

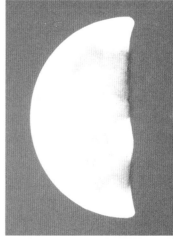

▲ **A drawing of Venus** at just over half phase, with the terminator line noticeably crooked. The drawing was made about 45 minutes before sunset.

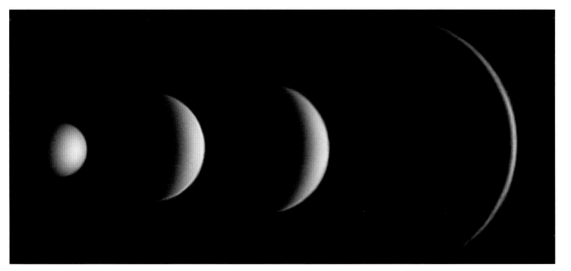

◄ **A wider range of phases** can be observed on Venus than on Mercury because it moves farther from the Sun. These photographs were taken using a 280 mm telescope and a webcam.

▶ **Venus is an unmistakable sight** *in the twilight sky, appearing before any other stars or planets are visible.*

▼ **An overall radar map** *of Venus, derived from Earth-based observations, shows that there are two major highland areas. One, named Aphrodite Terra, is close to the equator, while the other, Ishtar Terra, is near the north pole. Ishtar Terra contains a great plain, Lakshmi Planum, with one great peak, Maxwell Montes. The latter, named after the scientist James Clerk Maxwell, is one of the few features on Venus not named after a woman.*

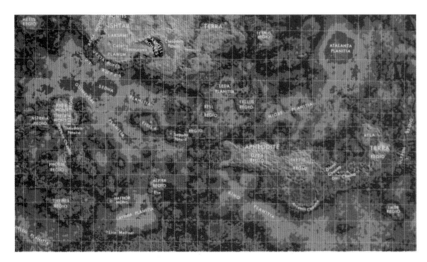

can put it back already aligned. You can then use the declination scale straight away, but will have to point the (capped) telescope at the Sun to set the RA circle. A computer mapping program will give you the correct coordinates for Venus and the Sun.

Users of Go To telescopes have a slight problem as there are no stars available for aligning the telescope. However, if you take great care over setting up the telescope's "home position" – such as pointing it north and level to begin with – you should be able to get an approximate alignment which will allow you to see Venus in the finder. Simply press "Enter" when the instrument points to each alignment star, even though you cannot see the star to center it.

As well as the markings, there are some other features that Venus observers like to look for. The terminator line – the edge of the sunlit part – is usually smooth, but on occasions it may have kinks. There may also be bright areas, known as cusp caps, around the cusps or tips of the crescent. One peculiarity is that when Venus should be at exactly half phase or *dichotomy*, with the terminator as a perfect straight line, it usually shows a lesser phase, as a slight crescent. Another interesting phenomenon is what is called the *Ashen Light*, in which the unilluminated side of Venus glows faintly when the planet is a thin crescent.

Venus unveiled

If Mercury is difficult to reach, Venus is just plain inhospitable. In size it is Earth's sister planet, but it has also been called Earth's ugly sister. Conditions on its surface are so harsh that interest in Venus as a destination for spacecraft is tiny compared with that for Mars. We can see ourselves walking on Mars, but not on Venus.

Venus is hot and oppressive. Even if untold mineral wealth were to be discovered on its surface, we would probably prefer to seek lesser deposits in the asteroid belt than attempt to recover them from Venus. Robots would have to do all the work unless new materials could be developed to prevent human explorers from being crushed and vaporized by the deep and hot atmosphere. Only the surface gravity is almost the same as that on Earth.

Venus itself has a completely different day–night regime from Earth. It rotates very slowly on its axis, once every 243 days relative to the stars, but it does so in the reverse direction compared with the Earth, Sun and most of the other planets. Coupled with its year of 225 days, this gives rise to a solar day (the interval between successive noons) of 116 Earth days. This is rather academic, as the Sun is not visible from the surface of Venus anyway. But why the cloud features rotate in just four days is a subject for speculation.

Although it is just possible for amateurs to photograph the surface of Venus, using webcams and infrared filters, most of our knowledge of the surface comes from the Magellan spacecraft which orbited the planet between 1990 and 1994.

Its onboard radar revealed a landscape of volcanoes. Planetary geologists are always particularly interested in impact craters, as the rate of bombardment over the eons throughout the Solar System is fairly well known and the numbers of impact craters reveal the age of a surface. Though Venus lacks small impact craters, because the smaller bodies would burn up in the atmosphere, there are still far fewer large impact craters than would be expected, anywhere on the planet. It appears that the entire surface of Venus is no older than about 800 million years. A planet-wide upheaval took place in the comparatively recent past, pointing to yet another major difference between Venus and Earth. Earth has plate tectonics, with deep gaps between the geological plates which happen to be water-filled, whereas the surface of Venus is one giant plate. Heat loss from the center of the Earth takes place continuously in particular areas, notably the mid-oceanic ridges and the volcanoes that surround the Pacific. But Venus seems to save its heat up for a long period and then completely resurface the planet in an orgy of volcanism. Not an environment conducive to the development of life.

But even in the absence of water, Venus has a different interior structure from that of the Earth. Planetary geologists are curious to know why, but the answer will probably be a long time in coming as few missions are being planned to Venus.

▼ **All the spacecraft** that have landed on Venus have been Soviet, the last being in 1982. These landscape views are from Venera 13 and are typical of the surface views received so far – flat volcanic rocks. The color of the top view shows roughly the color you would see if you could stand on Venus, because the clouds filter out all the blue light, giving everything a yellow cast. The brightness level is said to be "that of an overcast winter day in Moscow."

The bottom half of the panorama is reproduced with the colors that you would see if the lighting were the same color as on Earth, so we can judge the true color of the rocks. They are dark gray, like basalt on Earth.

The sawtooth at the bottom is part of the spacecraft, which continued to send data for a record 127 minutes. Because Venera landed on a tilt, the horizon appears to slope.

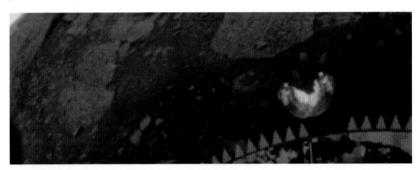

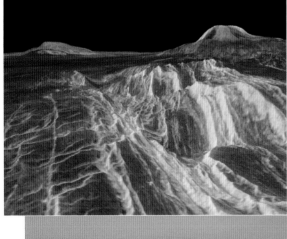

◄▼ **The data from the Magellan orbiter** has been used to construct realistic-looking three-dimensional views of the surface of Venus, as shown left. Here we see the shield volcanoes Sif Mons to the left and Gula Mons to the right. Sif is 1200 m high and 120 km across at the base, while Gula is 3000 m high. However, the vertical scale of the Magellan images has been exaggerated by up to 20 times.

Reducing this to more realistic proportions gives the view below, to which some atmospheric haze has been added – though as the true distance to Gula is over 700 km, it is questionable whether the atmosphere of Venus would be transparent enough to allow such a view in reality.

The brightness of the terrain indicates its radar reflectivity rather than its true appearance. Over the fairly brief period that Magellan recorded its data, no clear changes were noticed in the landforms. The question of whether Venus's volcanoes are currently active remains open.

MARS

Other planets may be bigger and more spectacular, but everyone wants to see Mars through a telescope. Truth to tell, it can be rather a disappointment, but a glimpse of those mysterious dark markings, once thought to be vegetation, and the glistening polar caps, so reminiscent of Earth's own, is enough to fire the imagination. No matter that all you can see is a tiny disk jumping about in the air currents over your telescope, this is Mars, possible abode of life.

Mars is often called the Red Planet, but in reality the planet is more of a pale orange color, both to the naked eye and through a telescope. Its color is most obvious at opposition, when it can become even brighter than Jupiter, and at all times it is noticeably reddish in color. Though traditionally the planet is associated with the color of blood, which has led to its links with war, you should not expect to see a strongly colored object. Binoculars show the color well, but they are not powerful enough to show anything of the disk.

Mars is a rather small planet, and it is often a long way from Earth. Because Earth and Mars are engaged in a continual race around the Sun, we only see it at all well every two years when there is a close approach and it is at opposition (see page 29). Only at these times can you hope to see much detail on the planet. At a close opposition, Mars can be as large as 25 arc seconds across, compared with 47 arc seconds for a typical opposition of Jupiter. For much of the time when it is on the far side of its orbit from Earth, however, it can be less than 4 arc seconds across. In such circumstances very little can be seen, particularly with small telescopes. So you have to choose your time carefully and be patient if you want to see much detail on Mars.

For months after its first appearance in the early morning sky Mars remains a morning object. Then for a month or two on either side of opposition it is a worthwhile target to observe. Thereafter it starts to dwindle in size, but an odd thing happens: it seems reluctant to leave the evening sky, and remains visible for many months, though tantalizingly small. In actual fact Mars lingers roughly as long in the evening sky as it does in the morning, but it is just more obvious to

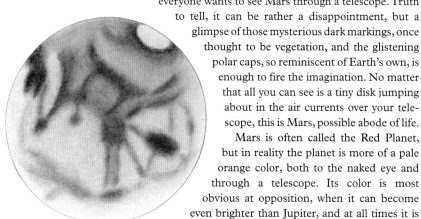

▲ **A drawing of Mars** made in September 1988, when it had a diameter of 23 arc seconds. The side of Mars on view is very similar to that of 22 August 2003, as seen below. The Solis Lacus is on the right, while Dawes' Forked Bay is on the left-hand edge. South is at the top.

▼ **The changes in size,** distance, phase and visible detail of Mars over five months from 2003 to 2004. It remained visible in the evening sky for a further eight months, eventually being half the size shown in January, before becoming lost in the twilight.

▲ **The reddish color of Mars** is obvious when compared with a blue-white star, such as Spica (the lower object).

most people. As the planet moves farther from Earth its phase changes, and at times only about 85% of the globe is illuminated.

Mars through a telescope

Photos of Mars often have their contrast enhanced to bring out the detail, and this tends to raise people's expectations of what they will see. The contrast of the features is quite low, and their outlines ill-defined, so often all you can see on Mars, even when it is close, is a slightly darker tint on part of the disk. You may even wonder if your eyes are simply playing tricks, but if you can manage a magnification of at least 70 or so, you should be able to distinguish some features in good seeing. Mars, being small, is particularly sensitive to bad seeing, and sometimes you may even see several overlapping images. If you allow your telescope and the

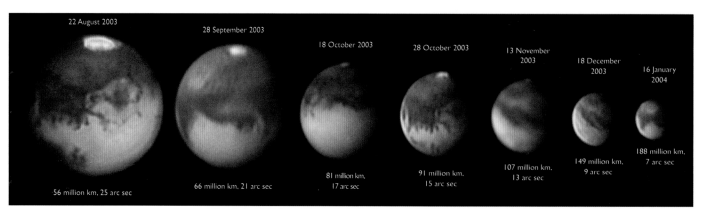

22 August 2003

28 September 2003

18 October 2003

28 October 2003

13 November 2003

18 December 2003

16 January 2004

188 million km, 7 arc sec

149 million km, 9 arc sec

107 million km, 13 arc sec

91 million km, 15 arc sec

81 million km, 17 arc sec

66 million km, 21 arc sec

56 million km, 25 arc sec

conditions to settle down, you may start to see more detail. Higher powers – the highest you can manage – will help if your telescope and the seeing are good enough, but Mars is a challenge for any small telescope.

The markings on Mars are more or less permanent, so you can compare what you see through the eyepiece with a map in order to establish which part of the planet you are viewing. But one confusing factor is the varying tilt of Mars' axis as seen from Earth. Like Earth, the axis of Mars is tilted to the ecliptic, and by about the same amount, 25°. This means that the same feature can appear at a different position on the planet's globe from apparition to apparition, depending on whether we are seeing predominantly the northern or the southern hemisphere. This makes a surprising difference to the recognizability of features. A good way round this is to use a computer-generated map of the globe, either using planetarium-type software (see www.stargazing.co.uk) or online. Such a map will show you the current aspect of the globe, though there is a risk that once you know what you should be seeing, you will let that image influence your observation.

As a day on Mars is just 37 minutes longer than our own, changes in the appearance of the planet due to its rotation are visible within a fairly short space of time. What's more, if you observe at the same time on successive nights you will see virtually the same side of the planet on each occasion. Only by observing at a considerably different time, or by observing a week or two later, will you see a different side of the planet.

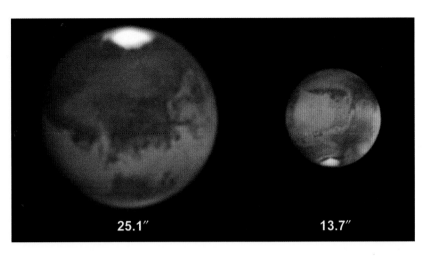

25.1″ *13.7″*

◀ *Mars, photographed at the very close opposition of 2003 (left) with a 250 mm telescope, and the distant opposition of 2012 through a 355 mm telescope, with diameters shown in arc seconds. The changed inclination of the axis makes a big difference to the features on view. South is at the top. The planet was lower in the sky in 2003 than in 2012, and was photographed with a smaller telescope, hence the difference in image quality.*

To see a globe showing the features visible on the surface of Mars at any chosen time, go to www.calsky.com, and click on Planets: Mars: Apparent View/Data.

Although the major features of Mars remain from year to year, changes in the fine details are taking place all the time as a result of dust storms on the planet. Some features have changed considerably over the years, so there is no certainty about the exact appearance. Mars has a thin carbon dioxide atmosphere and dust storms are a common feature. They usually cover just a small part of the surface, so a particular feature may appear less distinct than it should, or may even be completely absent. Dust storms tend to be more common around the Martian perihelion, which is when it is closest to the Sun. This may not coincide with Mars' closest approach to Earth, however.

The orbit of Mars is noticeably eccentric – that is to say, the ellipse of its orbit is more flattened than that of the Earth's orbit – so its distance from the Sun varies quite considerably over its year. As a result the solar heating of the surface also varies, and is at its maximum at perihelion, which occurs when it is midsummer in Mars' southern hemisphere. Occasionally, dust storms may even develop to cover the entire planet with a yellowish haze. In 1971, no surface detail was visible on the planet for several weeks.

The polar caps, too, change with the seasons. They are easily visible with even a small telescope when the planet is close, though when the seeing is bad they can appear as little more than a brightening at one edge of the disk. The northern cap is less changeable than the southern one, and the southern one can almost disappear in its summer. As the caps shrink they develop ragged edges, which appear through the telescope as

▲ **The western edge** *of the 22 km (14-mile) Endeavour crater on Mars, photographed by the Opportunity rover in August 2011. This long-lived rover landed on the planet in 2004, and has sent back thousands of panoramas and close-ups of the surface.*

▼ **A map of Mars** *from images by Damian Peach using a 355 mm telescope. This is known as an albedo map because it shows brightness variations only, unlike those made from spacecraft images which often show surface features. South is at the top.*

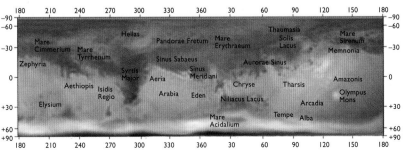

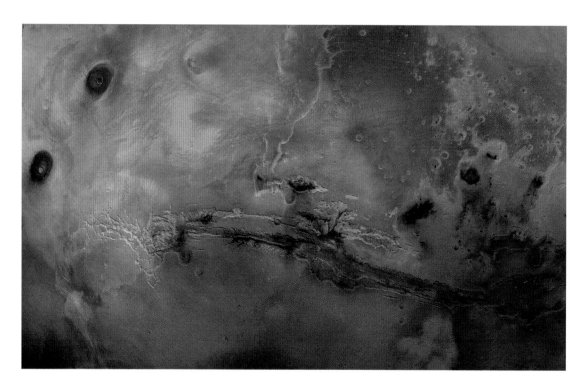

▶ *The Tharsis Ridge volcanoes* are prominent to the left of this mosaic of images made from pictures taken by the Viking orbiter spacecraft of the late 1970s. From top to bottom they are Ascraeus Mons, Pavonis Mons and Arsia Mons. The bluish haze in this area is probably clouds of water ice. To the right of Arsia Mons is the complex region called Labyrinthus Noctis, where fault lines have created depressions. This leads into the Valles Marineris, the Mariner Valleys, as there is not just one valley but several. The main gash, to the lower left, is called Coprates Chasma, which then divides into Capri Chasma and Eos Chasma.

▼ *A topographic map of Mars* using data from the MOLA radar altimeter aboard Mars Global Surveyor. High ground is coded white and low ground blue, giving Mars a somewhat misleading appearance of having snow on the peaks and water filling the low areas. The Tharsis Ridge shows up clearly to the left of the map, while the vast impact basins of Argyre and Hellas also stand out. North is at the top, unlike on the observer's map shown on page 73.

tongues or even islands of brightness. Usually only one cap is visible at a time, as a result of the planet's tilt.

Though debates rage over just how much water there is on Mars and where most of it lies, it definitely does exist on the planet. Clouds of water or carbon dioxide ice can form. They are distinguishable from dust storms by their white or bluish color and can occasionally be seen at the eyepiece.

In addition to the named features on Mars, there are one or two popular names that are occasionally used but which do not appear on maps based on spacecraft data. Dawes' Bay or the Forked Bay is one, being the distinctive double dark area of Meridiani Sinus. This is the zero longitude point on Mars' grid system, hence its official name. Another is Solis Lacus, also known as the "Eye of Mars," which in some years looks like an eye with eyebrow when seen in a telescope that inverts the view. The famous Valles Marineris canyon marks the lower lid of the eye, though this is not visible through a telescope. The view of Mars that we see at the eyepiece is essentially an *albedo* view – that is, the view is of brightness variations only and not of the actual

relief. (The term *albedo* refers to the reflectivity of a surface.) So the craters and other features are not properly visible. However, a feature once named Nix Olympica, the Olympian Snows, is now known to be a huge volcano, and has been renamed Olympus Mons or Mount Olympus. You may find both terms used.

In the late 19th and early 20th century, many people believed that they could see numerous straight markings on Mars, which were termed canals. These were mostly wishful thinking, and resulted from a tendency for the eye to join unrelated blobs. There are some linear features, but they have nothing to do with intelligent life, as was once believed.

As a planet, Mars is now dry and dusty but there are many signs that water once flowed in abundance. The wet periods were probably brief, but there is a chance that primitive life may have gained a toehold at some stage and remains there, probably below the surface.

Exploring Mars

No planet beyond the Earth has been as extensively studied as Mars. Actually we have more detailed photographs of large areas of Mars than we do of our own Moon. A succession of spacecraft since 1964 have scrutinized the planet at a wide range of wavelengths, and even though some have failed to send back data for one reason or other, our knowledge of the surface of Mars is unprecedented.

Mars has virtually the same land surface area as the Earth. Although it is only about half of Earth's diameter it has no oceans, which occupy about two-thirds of the Earth's surface. All this leaves Martian geologists with a huge terrain to study and to understand, including volcanoes, impact craters, dried-up rivers and lake beds, polar caps – and maybe even a frozen sea.

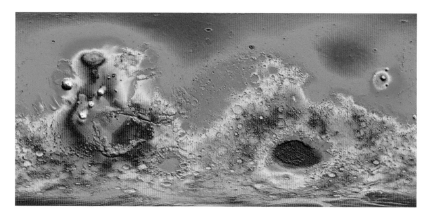

The largest volcano in the Solar System, Olympus Mons, is on Mars. It is not alone – three other major volcanoes lie in what is known as the Tharsis Ridge, and all dwarf any similar shield volcanoes on Earth, such as those of Hawaii (even allowing for its extent beneath the Pacific).

The great chasm known as Valles Marineris (see A–Z page 268, Mariner Valley) gapes for 4500 km across one entire hemisphere, roughly along the planet's equator. At first sight it looks like a giant river system, as it begins in the Tharsis Ridge and flows out into a wide basin. But it was probably created as a rift valley when the Tharsis Ridge formed, causing a great bulge in the surface. So why does it show signs of water erosion? Experts believe that although there is no liquid water on Mars today, the subsoil contains considerable quantities of water in the form of permafrost. The creation of the Tharsis Ridge volcanoes, maybe 3.5 billion years ago, could have released vast quantities of water, though some believe that carbon dioxide could flow as a liquid under the right circumstances.

▶ **The north polar cap** of Mars remains throughout the year, with a residual cap of water ice and a more temporary cap of carbon dioxide ice during the winter. The spiral structure has been put down to the tilt of Mars' axis causing unequal heating on opposite sides of any cracks that form. Ice vaporizes on one side as it heats, but is then redeposited on the other side of the valley. The south polar cap has a similar structure.

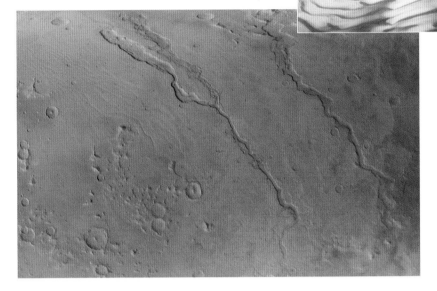

▼ **An oblique view of Olympus Mons** on Mars with late-afternoon clouds, based on images taken by Mars Global Surveyor. The base of the volcano is 540 km across, and the peak is 27 km above its base. If you could stand on the summit, the view would consist solely of the slopes of the volcano, as they would stretch beyond the horizon.

The reason why Olympus Mons was able to grow to such a height may be put down to the lack of tectonic plates on Mars. The Hawaiian Islands on Earth, for example, are constantly moving over a hotspot in the crust, so several separate shield volcanoes have built up rather than one giant as in the case of Mars. Mauna Loa is a mere 9.1 km above the sea floor.

▲ **Three dry valleys on Mars,** photographed by Mars Global Surveyor. Each valley is about 1 km deep and they range in width from 8 km to 40 km. Such features are evidence for flowing water on Mars in the past.

Despite failures, many spacecraft have sent back excellent information about Mars, though to read some newspaper articles you might think that the planet is jinxed or even populated by mysterious aliens that attack our spacecraft. After early flyby missions, Mariner 9 was the first to orbit the planet, back in 1971, when it discovered the valleys named after it and provided views of much of Mars' basic landforms. But the real flood of information came with two Viking craft in 1976, each consisting of an orbiter and a lander.

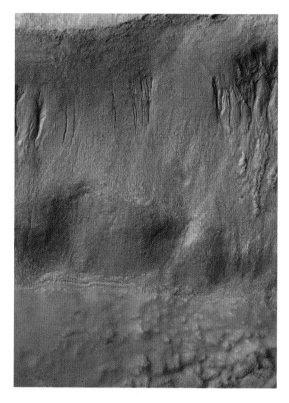

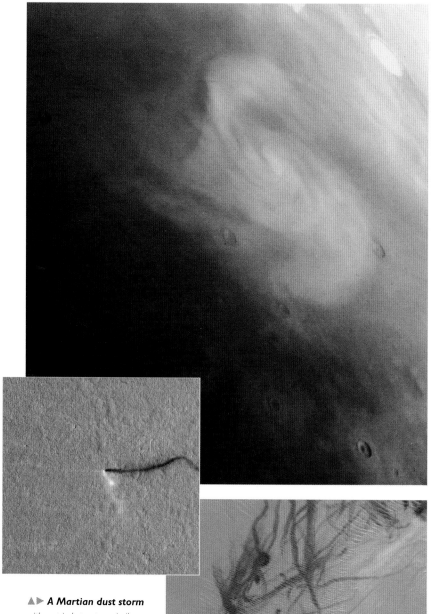

▲▶ *A Martian dust storm* with a spiral structure similar to that of hurricanes on Earth. Though the atmosphere is thin, it can pick up large quantities of the fine dust. Small dust-devils regularly cross the surface, as seen here from orbit, leaving characteristic dark tracks where they have swept the surface clear of dust (right).

▲ *Gullies are a common sight* on the slopes of craters. Here, a line of gullies starts at a particular level strongly reminiscent of a spring line on Earth, where subsurface water seeps out of a stratum. Such images suggest to some that liquid water may still flow on Mars occasionally.

There is no doubt that there is water on Mars. It has been detected in the atmosphere, in the polar caps and below the surface. The only doubts are whether it still flows occasionally. There is a great deal of evidence for erosion, particularly on the edges of cliffs or crater walls. In theory, water cannot flow on Mars today – it would either freeze or vaporize instantly, depending on the circumstances. The exact nature of the erosion is a matter for active debate.

Following Viking came a gap of 20 years before the next truly successful missions, Mars Global Surveyor and Pathfinder in 1997. Since then, there have been many successful missions, such as the orbiting Mars Odyssey, Mars Express, Mars Reconnaissance Orbiter, Mars Orbiter Mission and MAVEN, the lander Phoenix and the rovers Spirit, Opportunity and Curiosity. While the rovers have explored only a tiny fraction of the Martian surface, they have drilled, probed and analyzed the surface and sent back tantalizing clues to a time when Mars was very different, such as the presence of sedimentary rocks.

Over the next few years more Mars orbiters and landers are planned, though the date of the long-awaited robotic sample return mission remains always a few years in the future.

One question always asked by the press is, "Why so much interest in Mars?" when in fact they know the answer anyway. Mars retains a hold on everyone's imagination because of all the planets it is the only one where we could really ever feel even slightly at home. With its pale skies, wisps of cloud, occasional frosts, dunes and a hint of water, Mars is a world that we can feel is more than just a volcanic wasteland. It will never be a home from home, at least within the foreseeable future. If our planet gets crowded, it would be easier to live under the ocean, or in the wild and inhospitable regions of Earth, than on Mars.

▼ **The Mars rover Opportunity** *landed on Mars in January 2004 along with Spirit. This close-up of a rock shows the circular abrasion mark made by its drill to remove the surface layer which will have been affected by weathering, to reveal the true color of the rock itself. Both*

rovers were equipped with tools to analyze rock composition. The spherules on this site, nicknamed "blueberries" but perhaps looking more similar to rabbit droppings, are regarded as strong evidence that the area where Opportunity landed, within the outflow region of Sinus Meridiani, was once

water-covered. They appear to be hematite, an iron-bearing mineral that is associated on Earth with formation in a wet environment.

▼ **The European orbiter Mars Express** *photographed this 35 km crater near the planet's north pole filled with water ice, providing graphic evidence of the existence of water on Mars. At temperatures of around −125°C, there is little chance of the ice becoming liquid. Although surface temperatures on Mars can reach above freezing point where the Sun is overhead and Mars is at a favorable position in its orbit, the average even at the equator is about −50°C.*

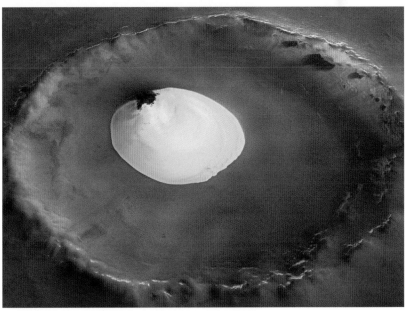

▲ **Opportunity landed close** *to a crater, nicknamed Endurance, which showed outcrops of bedrock on its flanks. Geologists were delighted by this good fortune, as the outcrops provided a ready-made cross-section through Martian rocks which would*

otherwise have required considerable excavation, for which Opportunity was not equipped. The outcrops revealed significant stratification, regarded as evidence for deposition in water. The site is thought to have once been covered with a thin layer of salty water.

◀ **The tracks of the rover Curiosity** *plow through a meter-high dune on Mars while investigating Gale crater near Mars' equator. At this time, in February 2014, it had been on the planet for 538 Mars days. The rover carries an X-ray spectrometer to analyze rock and soil samples that it finds on its journey. The color balance of this view has been shifted to show what the landscape would look like if it were on Earth, rather than in the more reddened light of Mars.*

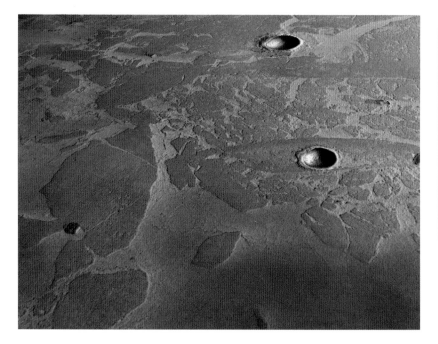

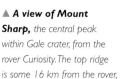

▲ *A view of Mount Sharp,* the central peak within Gale crater, from the rover Curiosity. The top ridge is some 16 km from the rover, while the pointed peak in the middle of the picture is about 100 m high. The layered rocks are characteristic of much of the Martian surface.

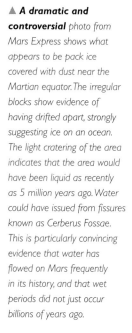

▲ *A dramatic and controversial* photo from Mars Express shows what appears to be pack ice covered with dust near the Martian equator. The irregular blocks show evidence of having drifted apart, strongly suggesting ice on an ocean. The light cratering of the area indicates that the area would have been liquid as recently as 5 million years ago. Water could have issued from fissures known as Cerberus Fossae. This is particularly convincing evidence that water has flowed on Mars frequently in its history, and that wet periods did not just occur billions of years ago.

But it invites exploration, and still holds a secret which we are eager to learn. Did life once begin there? Does it still remain there, maybe deep in some fissures? We know that life on Earth is capable of withstanding great extremes, so maybe life on Mars is clinging on.

We may not discover any life on Mars at all. In 50 years, another expedition may return from Mars to say that, yet again, they have probed deeply but in vain. Will we go on searching, or will more important matters keep us on Earth, with the realization that life is even more unique than we had thought and that we must protect our own planet first? Or will we find some strange microbes that will present us with a fascinating example of a variant on life that we had not considered, opening the door to even more curious life forms elsewhere in the Solar System? That is the excitement of Mars.

▲ *The view from the top.* Spirit saw this scene from the summit of "Husband Hill" during its exploration of the center of a crater named Gusev, 170 km in diameter. The hills in the distance are not the crater walls, though these were sighted by Spirit. Gusev was chosen because a channel system flows into it, and at some stage in the past it could have been water-filled.

▶ *A Viking Lander 2* view of Utopia Planitia, showing frost. Although the lower atmosphere does not have a significant content of water vapor, scientists believe that the water arrived on dust particles. The frost lasted for about 100 days once it had settled, even though the layer was thought to be only a few hundredths of a millimeter thick.

JUPITER

Jupiter is a favorite target for amateur astronomers. Although it is much farther away than the inner planets, Mercury, Venus and Mars, it is so large that it always presents a sizable disk when viewed through a telescope. In fact, you may be able to see its disk using binoculars, though you are unlikely to be able to see any detail.

Once you know where Jupiter is in the sky, you can keep track of its movement very easily. It takes almost 12 years to orbit the Sun once, so its movement through the sky month by month is small compared with that of the inner planets. So next year it will be in the adjacent constellation to where you see it this year, and it will be visible at roughly the same time of year. Like the other planets Jupiter marches from west to east over the months, though the Earth's daily rotation makes it move from east to west as its part of the sky as a whole rises and sets. Jupiter can also go through retrograde loops (see page 29), though it remains in the same part of the sky while doing so.

Views of Jupiter

To the naked eye, Jupiter is white and brighter than any star. Venus also meets this description, but is even brighter. If it is high in the sky late at night, it can only be Jupiter.

A glance through binoculars immediately confirms that it is Jupiter, because it is accompanied by its four brightest moons, which are easily spotted even using binocular magnifications. Not all of the moons are always visible but only very rarely can none be seen at all using binoculars.

With a telescope and a magnification of only about 20 or more, detail starts to appear. Because Jupiter is a gas planet, there is no solid surface, and all we see is the top of its opaque atmosphere. However, there are general features that remain constant over the years. To start with you should see what appear to be two darkish stripes across the planet, parallel to the equator. You will also notice that the planet is slightly flattened from a circle. This is a sign of its rapid rotation – this giant planet spins in just under ten hours.

The dark stripes are referred to as belts, and the most prominent are usually those on either side of the equator, but there are other minor belts as well. Sometimes the South Equatorial Belt fades, and Jupiter appears to have only one equatorial belt. As you gaze at the planet you should be able to make out more detail. The pictures of Jupiter in books often have their contrast and color enhanced to make the belts appear more distinct, so the planet may be a little disappointing the first time you see it. Novice viewers may not even see the belts at all, but as with all astronomical observing you have to train yourself to study the object carefully if you are to get the most out of it.

Belts and zones

To see the belts and zones in more detail try increasing the magnification, though not so much that the planet becomes dim. Even a good 60 mm telescope will start

▲▶ **Changes in the appearance of Jupiter** are obvious in this pair taken 15 years apart. Above is the Hubble Space Telescope's first color image of the planet, taken in May 1991, and to the right is an amateur image taken with a 355 mm telescope from Barbados in 2006. In 1991 the planet was in upheaval, with the Equatorial Zone appearing dark and the South Equatorial Belt having revived in color after a fade. By 2006 the colors were more normal, but the white spot near the Great Red Spot in the Hubble image had merged with others and had taken on a similar color to the GRS itself.

to show you a bit more detail as you do so, and with a magnification of about 45 the planet will appear as large as the Moon does in the sky to the naked eye. With a 75 mm telescope more details become visible, and you should be able to see that the belts are not just plain stripes but have irregularities in them. A telescope of 100 mm or more starts to show spots, both light and dark. A magnification of 90 will give you a good general view of the planet, but higher powers are helpful if you want to see the fine detail.

South Polar Regions (SPR)

South South Temperate Zone (SSTZ)
South South Temperate Belt (SSTB)
South Temperate Zone (STZ)
South Temperate Belt (STB)
South Tropical Zone (S Trop Z)

South Equatorial Belt (SEB)

Equatorial Zone (EZ)

North Equatorial Belt (NEB)
North Tropical Zone (N Trop Z)
North Temperate Belt (NTB)
North Temperate Zone (NTZ)
North North Temperate Belt (NNTB)
North North Temperate Zone (NNTZ)

North Polar Regions (NPR)

▲ *The nomenclature of Jupiter's belts and zones.*
The photograph has south at the top, as seen in a telescope, and includes the Great Red Spot.

The lighter areas are known as zones, and they and the belts have a nomenclature of their own, as shown above. There are also long-lived features that have their own names, yet which don't appear on general diagrams. Jupiter's atmosphere is one great raging storm, but there are individual disturbances which have remained for many years. The most famous storm is the Great Red Spot in the South Tropical Zone, a reddish lozenge that has been there since at least the 1870s and possibly back to the 17th century. Its color and size vary over the years – in the 19th century it was about a third larger than it is at present, and was a darker color. At the beginning of the 21st century it is hardly any redder than the belts. It sits in a lighter area known as the Red Spot Hollow. Other smaller spots may persist for several years at a time and receive their own designations.

Drawing the planet

The planet's rapid rotation becomes very obvious when you observe it for more than a few minutes at a time, particularly if you are trying to make a drawing. In a short time, the features that were in the center of the disk start to move to one side, and new features appear at the opposite edge. If you are trying to draw the planet, begin with the leading edge, containing the

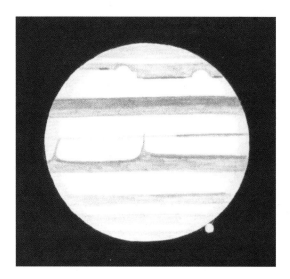

▶ *A drawing of Jupiter*
made using a 157 mm refractor. Ganymede has just emerged from behind the disk.

features that will disappear first. Observers learn to sketch the planet in outline quickly rather than laboring over subtle shadings. It is best to start with an oval outline or blank, which these days is most easily prepared by scanning an existing drawing into the computer and deleting the details before printing it out.

Sometimes you may see what looks like a particularly dark, circular spot on the planet. This is almost certainly the shadow of one of the major satellites, which will be nearby, though if it is in front of Jupiter's disk rather than to one side you may not be able to pick it out.

Jupiter in close-up

As we get deeper into the Solar System, planets and moons change. Gone are the rocky lumps, and in come the low-density but huge planets, and icy satellites. Jupiter marks the transition point. As the largest planet, it has an enormous influence throughout the Solar System, from sweeping up asteroids to killing comets. It has over 60 moons, including the largest in the Solar System and at least one which may arguably be an abode of life. And yet there are still mysteries about the planet, and no doubt surprises in store for us.

It is often said that Jupiter is a halfway house between a planet and a star. In terms of composition it is little different from a star – mostly hydrogen and helium. But the internal temperature of Jupiter is a mere 17,000°K, a far cry from the Sun's 15 million degrees K (K refers to Kelvin, the scale of temperature which starts at absolute zero, −273°C). This is nothing like enough to initiate the nuclear reactions that drive a star. If Jupiter had been about 15 times more massive, it would have started some nuclear reactions and would have been regarded as a brown dwarf star, though still not a star proper.

Jupiter has by far the most interesting surface detail of any of the giant planets – though surface is no more the right term to use for a gas planet than it is for a cloud. What we see in the case of Jupiter, as with a cloud, is a layer whose optical characteristics start to reflect light rather than transmit it. The bands and zones of Jupiter have some similarities to the weather systems here on Earth. The Sun heats the equators of both, giving rise to convection which triggers off a series of cells of rising and sinking gas. In the case of Jupiter, the rapid rotation of the planet also plays a part, smearing out the cells into bands that extend round the planet.

Time-lapse movies made over a period of days reveal very clearly that the alternating light zones and dark belts rotate in opposite directions compared with the average. In addition, the polar regions of Jupiter rotate about five minutes more slowly than do the equatorial regions. In between the belts and zones are numerous short-lived eddies, which have all the characteristics of eddies in a fast-flowing stream. So Jupiter presents an ever-changing spectacle, and one which we still don't really understand, even with the help of close-up photographs from space probes.

What causes the colors on Jupiter? There is a strong link between the overall color of a belt or zone and the height of its clouds. The white zones are the highest clouds, while the brown areas are lower. Blue regions are the lowest of all. Red features, however, are very high, so the Great Red Spot must be floating high in the atmosphere. Exactly what causes the red and brown colors is not certain. Sulfur is a strong candidate, as it is undoubtedly present and is noted for its yellows and reds. As for the Great Red Spot, phosphine or phosphorus hydride is often cited, though why this gas should be concentrated in this way is not known.

There was great excitement in 1994 when a small comet named Shoemaker–Levy 9 plunged into Jupiter. It had been a normal minor comet in orbit around the Sun when it strayed too close to Jupiter and became trapped in orbit around the planet. Comets are fragile chunks of ice, and it was torn into about 20 fragments which plunged into Jupiter in July 1994. There was great speculation as to what would happen. Nothing of this sort had ever been witnessed before, and it presented a laboratory experiment on a grand scale.

Few people really expected the dramatic spots that suddenly appeared on the planet within minutes of the comet strikes, visible with even quite small amateur telescopes. The churning up of Jupiter's clouds created dark brown spots, surrounded by circular haloes, quite reminiscent of the appearance of a lunar crater. The spots faded within a few weeks, but it was an awesome event and a reminder of the risks from cosmic collisions.

Jupiter's moons

The four large moons of Jupiter were responsible for a major change in our understanding of the Universe. They were first seen by Galileo in 1610, and his announcement immediately caused a stir. He could clearly see the moons moving from one side to the other of an object which had an obvious globe. Until then, the doctrine that the Earth was the center of the Universe held sway. But seeing another body as

the center of attraction shook this view to the core, and it spelled the end of the old notions.

Today, these four Galilean moons as they are called are still changing our ideas. When it was first learned that the closest to Jupiter, Io, is a world tortured by volcanic activity, some scientists were amazed. It is about the size of our own Moon, and by rights should be just as dead a world, having long ago cooled down from its birth. But its very closeness to giant Jupiter means that its interior is subject to great stresses as the opposite sides of it are tugged by different amounts. As a result it is heated to such an extent that it is subject to constant volcanic eruptions, with fountains of gas regularly shooting hundreds of kilometers above its surface (see Galileo Galilei and Io in the A–Z section).

▲ *A close-up of Jupiter's northern hemisphere as seen by the Cassini spacecraft in 2000. This gives a good impression of the way adjacent belts and zones rotate in opposite directions, drawing the features out into festoons and whirls.*

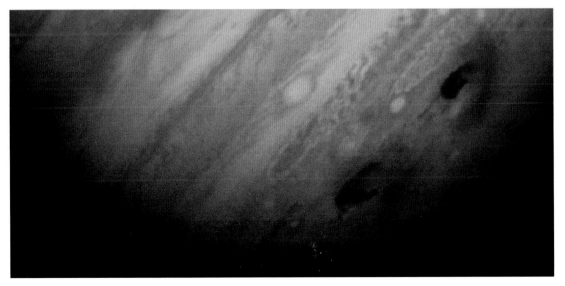

◄ *The scars left by the impact of Comet Shoemaker–Levy 9, as seen on 21 July 1994 by the Hubble Space Telescope. The spot at upper right was mostly caused by fragment G, which was probably about 2 km across and which struck Jupiter with an energy equivalent to a 6,000,000 megaton bomb.*

Next out is Europa. Though this lacks the internal heat of Io, it too provided a surprise. It has a rocky core and an icy exterior. But close-ups of its surface reveal signs of ice floes, as if they have been floating on a once-liquid sea. Far from being a solid iceball, the implication is that Europa might still have a liquid sea below an icy crust. But there is no certainty of this, nor of how thick is the ice crust. It may be a few hundred meters or a few hundred kilometers. There are clear signs of what is termed *cryovolcanism* – features which might elsewhere be due to hot lava, but which in fact are the result of similar processes involving ice instead of lava.

This is our first sight of this phenomenon in our outward trip in the Solar System, but beyond Jupiter cryovolcanism becomes an increasingly common means of shaping a landscape – far more likely than the hot volcanoes with which we are familiar. It is an indication that when we go to places with unfamiliar conditions, we may come across all sorts of features that defy immediate description and which indeed may have the appearance of being artificial. The straight features of Europa are a good example – they can look almost man-made.

As always where liquid water is suspected, there is the chance that life may be present. Europa, like Mars, is to be subjected to a detailed study in the coming years. But no one is anticipating anything more advanced than single-celled organisms. We have to remember that even on Earth, where there was liquid water, warm temperatures and abundant sunlight, although single-celled organisms seem to be present from the earliest times, they were the only life form for billions of years. We believe that conditions elsewhere in the Solar System, such as on Europa and Mars, are much more hostile.

However, we must always remember that our view of what is or isn't possible depends very much on a sample of one – our familiar Earth. As with cryovolcanism, which was unknown to us until we saw it on ice

▲ **The Galilean satellites of Jupiter,** to scale with the Great Red Spot, here seen foreshortened at the limb of the planet. From top to bottom they are Io, Europa, Ganymede and Callisto.

▶ **Parts of Europa's surface** are strongly reminiscent of ice floes which have cracked apart and drifted around on a semiliquid ocean. The largest "floes" are about 13 km across.

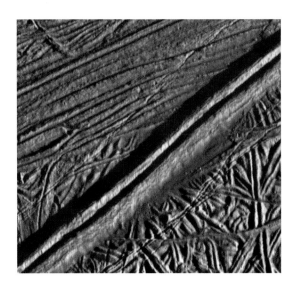

▲ **One of the curious ice ridges on Europa,** the result of cryovolcanism. Something has caused an upwelling which has created a fracture which cuts across | previously formed ridges. Different minerals have gushed up from below, creating a variety of dark stripes which criss-cross the surface of the moon.

worlds such as Europa, it is possible that molecules could find a way of organizing themselves without light and warmth. Perhaps conditions deep in Europa's suspected oceans are really ideal, but we just don't realize it.

Next comes Ganymede, the largest moon in the Solar System, and even larger than the planet Mercury. It lacks the immediate appeal of Io or Europa, but it has

its fair share of mysteries. Cryovolcanism is rife here, too, with the result that some parts of its surface seem to have been the subject of a giant rake, giving it the appearance of a Japanese Zen garden.

Callisto, the most distant from Jupiter of the four Galilean moons, is renowned for being the most heavily cratered object in the Solar System yet lacking any major relief. Despite bearing the scars of a gigantic impact, Valhalla, it has only low ridges, indicating an icy surface without any load-bearing capability. It has been pounded by impacts ever since its formation 4.6 billion years ago, with no signs of any other type of activity.

Our knowledge of Jupiter and its satellites has come from the flybys of the Voyager 1 and 2 craft in the 1970s, but particularly from the Galileo spacecraft which orbited Jupiter between 1995 and 2003. It also sent a probe into Jupiter, which relayed information on the temperature and pressure and other details as it plunged into the clouds. Contact was lost when it overheated at a depth of some 150 km.

The Galileo probe improved our understanding of Jupiter. The darker belts on Jupiter are believed to be lower in altitude than the bright zones, and there are probably three layers of clouds in all. The lowest consists of water ice, then above that is a layer of ammonia and hydrogen sulfide. The uppermost clouds are ammonia ice. But a lot remains to be discovered about the chemistry of Jupiter's clouds.

▼ **Callisto is covered by a dark layer,** which may be debris from micrometeorites. The craters show signs of erosion over possibily billions of years, but there is little hint of the cryovolcanism that is so obvious on Europa and Ganymede.

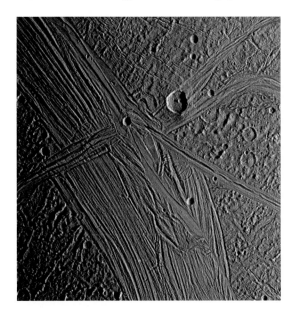

▲ **Ganymede has two separate types** of landscape: battered and jumbled dark regions, as at lower left, and brighter grooved terrain, which here crosses diagonally, with an | abrupt transition between the two. Unlike on Europa, however, the grooves have destroyed what was previously there, rather than cutting features up and separating them.

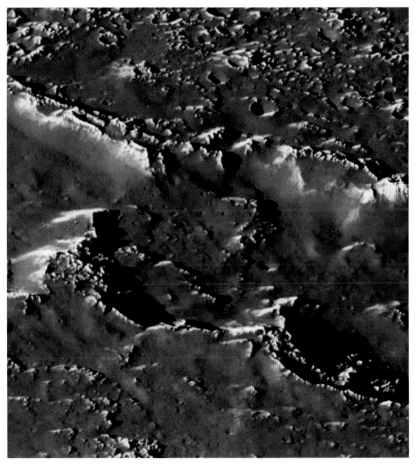

SATURN

Saturn is an amazing sight through a telescope. If we had never seen a planet with a ring round it, we would not believe such a thing was possible. Many small telescopes are sold on the basis that they will show the rings of Saturn, but in fact all but the worst telescope should be able to accomplish this feat. A magnification of 20 is all you need to glimpse the rings, and even binoculars will reveal that the planet is not a simple disk but appears elongated.

Being almost twice as far from the Sun as Jupiter, Saturn takes nearly 30 years to orbit the Sun, so it moves only slowly from constellation to constellation and remains visible at roughly the same time of year for years at a time. Its plod around the heavens gives rise to the word *saturnine*, and in ancient times it was the slowest known planet.

Views of Saturn

To the naked eye it appears as a bright, noticeably yellowish star. There are a few stars brighter than Saturn, but being a planet it rarely twinkles, so it is easy to distinguish. Though its disk is about half the size of that of Jupiter, the rings bring it up to just about the same diameter as Jupiter so even a small telescope will show it well. But if it were not for the rings, Saturn would be a great disappointment. Its disk is bland compared with Jupiter's, with just a few belts and zones and comparatively few other features. Saturn appears the same at first glance whenever you see it, but there is enough going on to interest its devotees.

One thing that clearly changes is the inclination of the planet to the Earth. While Jupiter is virtually always seen edge-on, because its polar axis is inclined at only 3° to the vertical of the ecliptic, Saturn's axis is inclined at nearly 27°. This means that as it progresses round the Sun we see it inclined at different angles. The cycle from edge-on to fully inclined either north or south takes about 7½ years, then it returns to edge-on and starts to incline at the opposite angle and show us its other pole. The rings are the most obvious sign of this change, and each year they appear noticeably different.

Of Saturn's many moons, six or seven are easily visible in amateur telescopes. They orbit in the plane of the rings, so usually they are spread out somewhat, rather than being in a line like Jupiter's Galilean moons. Only the brightest, Titan, is visible in the smallest telescopes, and you may even glimpse it in binoculars.

Observing Saturn

Many people are happy just to gaze at Saturn, for it is a spectacular sight at all times. But if you want to take your interest a step further, you could make a drawing. This is not as easy as it sounds, because whereas in the case of the other planets a simple blank disk is adequate, in the case of Saturn the changing aspect of the rings is a challenge. Observing sections of societies such as the British Astronomical Association (BAA) make blanks available to their members, and they can be obtained online for home printing for any chosen date from the Association of Lunar and Planetary Observers at http://www.alpo-astronomy.org/saturn/satfrms.html.

▲ **Saturn is the bright object above center**, here seen within the non-zodiacal constellation of Cetus. The bright star below it nearer the horizon is Fomalhaut, in Piscis Austrinus.

▶ **Even when viewed close-up,** as from the Cassini spacecraft in 2004, Saturn's disk does not usually show much detail.

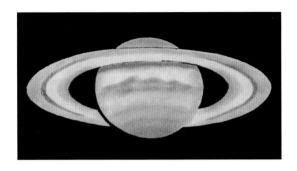

▲ *An amateur drawing* of Saturn made in August 1991 by Matthew Boulton with a 157 mm refractor and magnifications of 152 and 213. It shows a considerable amount of detail in the North Equatorial Belt. South is at the top.

Although the angle of the rings varies only slightly throughout the year, you will see a considerable difference in their appearance during the apparition as a result of the changing illumination. Around the time of opposition there are no shadows and the globe is at full phase, but on either side of opposition you may see the shadow of the globe on the far side of the rings. It is also possible to see the shadow of the near side of the rings on the globe.

The rings are actually composed of millions of tiny particles of ice and rock, all orbiting in a very thin plane, no more than a kilometer thick. As a result, when the rings are seen edge-on, they virtually disappear.

Although the globe itself is bland compared with that of Jupiter, it is by no means unchanging. There are variations in the color and intensity of the various belts and zones, and from time to time white spots appear, particularly every 30 years or so when the planet's north pole is tilted toward the Sun.

In recent years, amateur astronomers have been able to observe Saturn using CCDs and webcams. Image processing can enhance features of low contrast, and images taken from backyards with moderate aperture telescopes can reveal features that might previously have been at the borderline of visibility.

Saturn up close

Saturn, like Jupiter, is a globe largely made of hydrogen gas, with heavier gases that constitute the more complex atmosphere that gives rise to the bands around the globe. It spins on its axis in only 10.2 hours, and with light gases such a rapid spin means that Saturn's globe is very flattened along its equator. This is particularly evident if you hold a picture of Saturn on its side with the ring plane vertical.

One surprising fact is that the density of Saturn is only 70% that of water on Earth. So if such a thing were possible, Saturn would actually float in an ocean of water under Earth conditions. But in fact Saturn is over ten times the diameter of the Earth, so one just has to imagine this happening.

A joint American–European spacecraft named Cassini reached Saturn in 2004 to begin a long-term mission, taking thousands of photographs of the planets, its rings and moons. In January 2005, the Huygens probe detached from Cassini and descended through the murky atmosphere of Saturn's largest moon, Titan. After a descent of over two hours taking measurements of the atmosphere, it landed on the surface and sent back an image of a flat icy plain.

One curiosity revealed by Cassini is that it discovered Saturn's northern hemisphere to be strongly blue in color at the time of its arrival. Photographs taken by Voyager and the Hubble Space Telescope, when the

▲ *Changes in the tilt* of Saturn's rings, as recorded over 15 years. North is at the top.

◄ *An infrared false-color view of Titan,* whose diameter of 5150 km makes it larger than the Moon or the planet Mercury. In visible light, Titan is surrounded by orange smog, and only in infrared do details of its surface appear. In this 2014 Cassini view, the Sun glints off Kraken Mare, a polar sea of methane. Like Earth, Titan has a largely nitrogen atmosphere, but at a much lower temperature of only −178°C. Gases such as ethane and methane produce organic smog molecules.

▶ *The surface of Titan,* photographed by the Huygens lander on 14 January 2005. The camera was only 40 cm above the surface, so the apparent boulders in mid distance are probably ice chunks only about 15 cm across. The horizon is about 80 m distant. The color in this view is inferred from data taken through filters by photometers on the spacecraft.

▲ *A rare storm on Saturn* occurred in 2011, resulting in this stream of swirls that broke up the normally fairly bland disk. This image was taken with a 355 mm amateur telescope from Earth. Some Saturnian storms have been spotted by amateurs before they were seen by Cassini, which spends much time imaging the satellites. This photograph was taken just a few days before opposition, when the rings appear much brighter than usual because the icy particles reflect sunlight directly back in the same way as reflective road signs.

planet's equator was more nearly face-on to the Sun, did not show this. In 2004 the planet's southern hemisphere was tilted toward the Sun and from Earth that is all we could see. Cassini, however, could view the planet from different angles and was able to view the northern hemisphere, which was tilted away from the Sun. It was also crossed by the shadows of the rings and was therefore presumably considerably colder than the sunny southern hemisphere. The blue color is interpreted as being the result of a clear, cloud-free atmosphere, as on Earth and indeed on Jupiter, where blue colors indicate a lack of clouds so we are viewing deeper into the planet's atmosphere.

The color was also noticeable on amateur CCD images, which by this time had improved enormously over those made in previous years, though presumably the same colors would have been visible when Saturn had the same tilt 30 years previously. Amateur observers will be able to follow these color changes even after Cassini eventually stops working.

Cassini has also been studying the rings of Saturn in great detail, though even now it cannot resolve individual ring particles. Even the largest individual ice boulders that comprise the rings are probably no more than a few meters across, and most are much smaller. Some experts believe that the rings may be the result of the breakup of satellites that strayed too close to Saturn to be able to retain their integrity – in other words, they were pulled apart by the planet's gravitational pull. However, others argue that the rings may be the leftovers of the formation of the planet, though they would need to be replenished from time to time. While the other giant planets also have rings, they are much darker and are not generally visible from Earth. Those of Saturn are much brighter, so their surfaces seem to be fresher for some reason.

Saturn's changing tilt and the varying angles of illumination are presumably the cause of the mysterious dark ring spokes which are seen from time to time. They were first brought to prominence by the Voyager spacecraft in the 1980s, though they had been seen fleetingly by amateur observers for many years previously. In theory, spokes should not be possible in rings that consist of millions of individual particles – any structures would be smeared out within minutes by the different

▶ *The enigmatic spokes* seen in Saturn's bright ring B by Voyager 2 in 1981. Ring A is the darker, outer main ring, while the C ring is much darker still and is closer in to the planet. The only other bright ring, F, is very narrow and lies just outside ring A.

orbital speeds of the particles around the planets. This is why the early amateur observations were dismissed as optical illusions, but their existence is now thought to be linked in some way to Saturn's magnetic field.

Several of Saturn's more than 60 satellites orbit very close to or within the ring system. The gravitational pulls of the satellites have an effect on the rings, with the larger ones creating gaps in the ring system such as the Cassini Division between the main outer rings A and B, while small ones known as shepherd moons create a braided appearance of the thin outer F ring. Both the Cassini Division and the spacecraft were named after Italian-born French astronomer G. D. Cassini (1625–1712), who was the first to note the division. It is visible in telescopes larger than about 75 mm aperture.

Cassini (the spacecraft) has provided fascinating close-ups of the smaller satellites of Saturn, such as Phoebe and Dione, and in particular Iapetus. When Cassini (the astronomer) first spotted Iapetus, he noticed that he could find it only when it was on one side of Saturn and not the other. He correctly suggested that this was because one hemisphere was darker than the other, but there is a further implication – it must have captured or synchronous rotation. This means that, like our own Moon, it always keeps the same face toward its parent planet. Many moons do this, and it results from inequalities in the moon's interior. If Iapetus did not have captured rotation, there would be no link between its visibility and its position in its orbit around Saturn.

As expected, photos from Voyager and Cassini show us that Iapetus does indeed have startlingly different hemispheres. The close-up photos make it look very like a plum pudding with icing on the top, and even a central ridge where it has been turned out of a mold. But in fact the lighter color is more likely to be the true icy color of Iapetus, and it is the dark side which appears to have collected some sort of material – this is on the moon's leading edge in its orbit. The nature of this remains a mystery, along with the reason for the strange equatorial ridge.

▲ **Saturn's northern hemisphere** *was relatively cloud-free in 2004 in this Cassini image, and appeared blue for the same reason that Earth's skies appear blue – blue light is strongly scattered by a clear atmosphere. Saturn's rings are at the bottom of the picture and appear dark because Cassini at this time was viewing the non-sunlit side of them. Their shadows are projected on to the northern hemisphere.*

◄ **The leading face of Iapetus,** *showing the reddish-brown coloration of the material coating it. There is another Cassini view of the strange moon in the A–Z section on page 256.*

▲ **Hyperion,** *one of Saturn's smaller satellites, has a unique appearance, like a sponge. Dark material seems to have filled impact craters,* *which have then eroded particularly deeply at their centers. As with Iapetus, the source of the dark material is unknown.*

URANUS AND NEPTUNE

These two outer planets are rarely observed though they are both comparatively easy to locate. Uranus is sometimes just visible to the naked eye as a very faint star, but it was overlooked completely until 1781, when an amateur astronomer named William Herschel using a home-made telescope noticed that it has a disk rather than being starlike. It was the first planet ever to be discovered, rather than being known since ancient times.

Neptune was located after it was found that Uranus was deviating from the orbit that had been calculated, and it became obvious that a more distant planet was pulling on it. The planet, Neptune, was eventually tracked down in 1846. At about eighth magnitude, it is well within the visibility of binoculars and small telescopes, but it appears starlike except in larger telescopes.

Both planets are similar in composition, and while they are often referred to as gas giants, along the lines of Jupiter and Saturn, they have denser interiors, with a large proportion of water rather than hydrogen gas.

Viewing Uranus and Neptune

The best way to locate these planets is either to consult a computer sky mapping program in order to print out a map to use at the telescope, or to use a Go To telescope. Either way, you are looking for a starlike object that appears greener or bluer than the other objects. You will not see a disk at low magnifications. If you have a star map that shows stars fainter than the planet's brightness, you may be able to establish which one it is

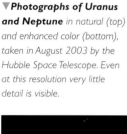

▼ **Photographs of Uranus and Neptune** in natural (top) and enhanced color (bottom), taken in August 2003 by the Hubble Space Telescope. Even at this resolution very little detail is visible.

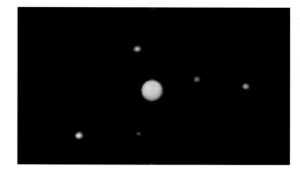

▲ **An amateur CCD photo of Uranus** and its brighter satellites, taken with a 355 mm SCT. Though the planet's disk is visible with much smaller instruments, the satellites are 14th magnitude or fainter and can only be seen in large telescopes. The planet's bluish disk is noticeable.

▲ **Neptune and its largest satellite, Triton,** photographed with a CCD on two separate nights. The planet (center) has moved between the two superimposed exposures, and Triton has also moved in its orbit around the planet. The planet's image is overexposed in order to show Triton, which is magnitude 13.5.

because it will be in addition to those shown on the map. Both planets move more slowly from night to night than the bright planets, but after only one night there is still a noticeable change compared with the background stars. Currently both planets are well away from the Milky Way, and will remain so for many years, so there are comparatively few stars in the field of view.

In binoculars Uranus is easy to spot, and you can pick it out straight away by its distinctly bluish or greenish color. No star ever appears green – it is a physical impossibility, despite occasional claims to the contrary. You will need a moderate aperture (about 150 mm or larger is adequate) with a power of 50 or more and a night of fairly good seeing to view the disk of Uranus well. It is only 3 or 4 arc seconds across – less than a tenth that of Jupiter – and is much dimmer than that of any of the bright planets. But when you do spot this tiny blue disk you are reminded of the Hubble Space Telescope photos that show virtually the same thing, but larger. There is almost no detail to be seen on Uranus, even using the HST, so your own glimpse can be quite rewarding.

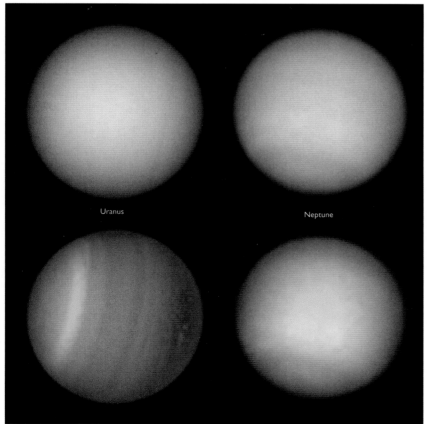

Uranus

Neptune

Neptune is also well within the reach of binoculars, but being fainter requires more of a search and less certainty that you have found it. Even with a Go To telescope you may not know which of several objects in the low-power field of view is Neptune, so again you may need to compare your view with a detailed map to be sure. You will need a telescope with a power of about 100 to show its bluish disk, of about 2 arc seconds in diameter, which like that of Uranus does not show any detail with amateur telescopes.

Probably every observer makes the effort to look at these planets very occasionally, then forgets about them for another few years. They are left in peace way out in the depths of the Solar System, largely unwatched by human eyes. By comparison, Saturn is probably being viewed by people around the clock. So spare a thought for these lonely planets and give them a friendly look from time to time. You may be surprised by how much you can see and by the strong color of their disks, which is caused by methane in their atmospheres.

As we near the edge of the observable Solar System our knowledge of the planets gets progressively less. The only spacecraft to venture that far has been Voyager 2 in the late 1980s, and the close-up photographs which it took of Uranus and Neptune and their moons will remain our best views of these lonely worlds for many years to come. The Hubble Space Telescope and its successors can provide glimpses of any changes in the atmospheres of the outer planets, but any closer studies of their moons will be a long time in coming.

Studies in detail

The close-up views of Uranus and its rings reveal straight away one extraordinary feature of the planet – its axial tilt of 98°. While the other planets have their

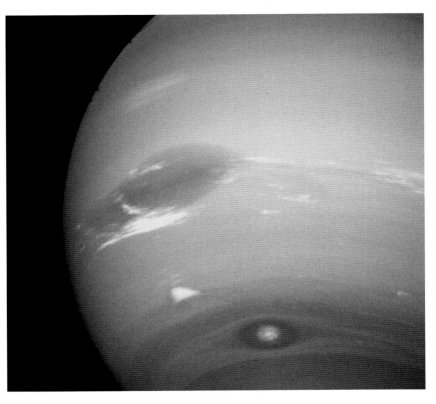

poles more or less at right angles to the plane of the Solar System, that of Uranus is tipped over almost at 90° to this, so in effect it spins on its side. With an orbital period of nearly 84 years, this means that it spends much of its time with one pole or the other pointed directly at the Sun. When Voyager flew by the planet in 1986 its south pole was pointed almost directly at the Sun and the planet was notably barren of detail. With the changing orientation to the Sun, becoming edge-on in 2007, an increased amount of banding has been observed by the Hubble Space Telescope and ground-based instruments.

Like all the giant planets, Uranus has a ring system in line with the planet's equator, though it is not visible in amateur telescopes. The orientation of the rings therefore also changes with the position of Uranus in its orbit.

The reason for the crazy tilt of Uranus, along with its retinue of moons and rings, is not known, though it hints at some extreme event, probably a collision, at a very early stage in its formation. The whole Solar System is believed to have formed from what is termed the Solar Nebula, a rotating disk of material from which eventually condensed the Sun and the planets. However, this was not a disciplined affair, and the differing tilts of the axes of several planets still bear witness to dramatic events.

The satellites of Uranus are predominantly icy. It has five major moons, though none are particularly large. We tend to think of moons as being substantial bodies, but apart from what we might call the Big Seven – our own Moon, the four Galilean moons, Saturn's Titan and Neptune's Triton – moons are either around

▲ **Voyager 2's view of Neptune,** *showing the Great Dark Spot with swirls of white methane cirrus and the triangular "Scooter" below it. Near the bottom is the Small Dark Spot. The GDS is smaller than Jupiter's Great Red Spot, but even so is as long as the Earth's diameter. It may be that these spots on Neptune are a darker color because they are deeper than the surrounding cloud layers.*

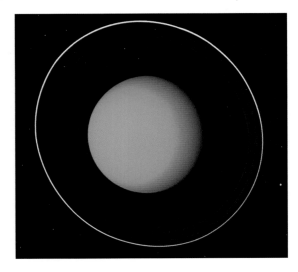

▲ **When Voyager 2 took this photo** *of Uranus in 1986 the planet was pole-on to the Sun and virtually no banding was evident. Many of its small inner satellites, plus several stars, are also shown. Compare this with the Hubble image taken in 2003 on page 88, which shows more banding as the planet's equator is more edge-on to the Sun.*

▶ **Neptune's rings** were seen by Voyager to be thin, with three strong clumps or arcs in the main ring which have been named Liberté, Egalité and Fraternité. As in the case of Saturn, there are satellites which orbit within the ring systems which result in the rings being controlled in their shape. Rings are only found within a certain radius of a planet, the Roche limit, within which the different gravitational pull of the planet on opposite sides of a large moon would cause it to break up, or to not form in the first place.

▼ **Ice volcanism** has helped to shape the surface of Triton. This region is known as the "cantaloupe terrain" from its resemblance to that type of melon and shows the ridges that presumably result from upwelling of slush along cracks or faults in the icy surface.

1000 or so kilometers in diameter, or are tiny fragments of the order of dozens of kilometers in diameter or smaller. There is no point in quoting a figure for the number of satellites of the outer planets – more are being discovered all the time, and we have to turn to the Internet for the latest figure. See www.stargazing.org.uk for links to the latest data.

As Voyager traveled out of the Solar System it reached Neptune in 1989 and found a more dynamic world than Uranus. Furthermore, it discovered that even out here, the Sun has a considerable influence on both Neptune and its largest moon, Triton. The banding on the planet is noticeable, and is presumably driven by convection as in the case of Jupiter. Neptune was found to have a Great Dark Spot (GDS), thought to be similar in some ways to Jupiter's Great Red Spot, along with smaller white spots such as the "Scooter," so-called because it raced round the planet more quickly than the others. Wispy white clouds were a feature of the planet, being reminiscent of cirrus on Earth, but in Neptune's case consisting of methane ice cystals rather than water.

Unlike Jupiter's spot, however, the GDS was moving in latitude toward the equator. Five years later, when the Hubble Space Telescope was first brought to bear on the planet, there was no sign of it and it has not been seen since though other large lighter spots have been detected.

On Triton, cryovolcanism appears yet again. It has the now-familiar linear ridges covering an icy surface, together with its own set of features not seen on the moons closer to the Sun. Amazingly, it has eruptions of dark material that form plumes several kilometers high. But what could drive such fountains on an icy world so far from the Sun? Furthermore, Triton's surface has very few impact craters, suggesting a surface that is comparatively recent and is undergoing regular activit

One suggestion is that even the slight warming from the Sun so far out in the Solar System is enough to melt nitrogen ice some way below the surface, creating geysers. But until another spacecraft visits Neptune, which will be many years in the future, this is still just guesswork.

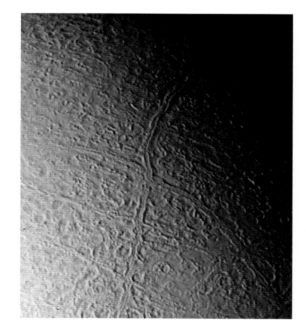

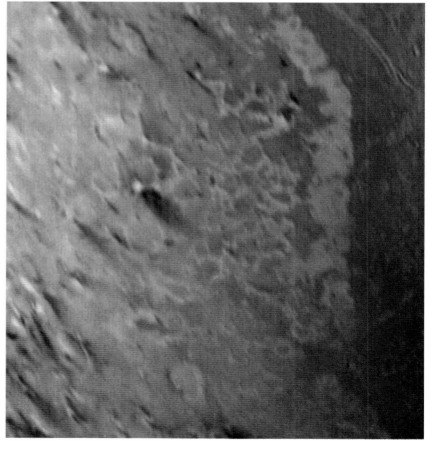

▶ **Dark streaks on Triton** probably result from plumes of material that erupt to a height of several kilometers above the surface and are then carried 100 km or more by a wind in the very thin nitrogen atmosphere. The pink color of Triton is believed to result from the action of sunlight on methane ice. These geysers have been seen in eruption even though the surface of Triton is only −235°C – the coldest in the Solar System.

PLUTO

For years, Pluto was regarded as the ninth planet, but these days more objects of similar or even larger size (though not brighter) have been discovered at the fringes of the Solar System. In 2006 it was reclassified as a dwarf planet, along with the largest asteroid, Ceres. But for many people, Pluto is still a planet and they want to find it if their telescope is large enough.

As it is 14th magnitude, you need at least a 250 mm telescope just to glimpse Pluto as a starlike object at the limit of the telescope's range. Unlike the major planets, it is far too small to show a disk – so at a glance it is not obvious which of many starlike objects of similar brightness is Pluto. This is made more difficult by Pluto's current position against the background of the southern Milky Way, where it will remain for many years. A computer-plotted star map going down to at least 16th magnitude is the best way of finding it, as even an accurately aligned Go To telescope will not show which of several faint stars in the area is Pluto. With a map, you may be able to spot which one is the inter-loper but even then you will need a second night's observation to be sure.

When the New Horizons spacecraft was launched in 2006, its target, Pluto, was still regarded as a major planet. But when it passed Pluto in July 2015, it could photograph what was now only a dwarf planet and its satellites. There are no other missions planned to visit the outer Solar System, though it might be possible for New Horizons itself to visit smaller objects if any can be found close to its trajectory beyond Pluto.

Beyond Pluto

The discovery in 2005 of a body similar in size to Pluto, but in a more distant orbit, now named Eris, prompted the realization that there may be many

▲▶ *Pluto, its major moon Charon and two additional satellites* announced in 2006, as photographed by the Hubble Space Telescope. At the distance of Pluto very little detail is discernible on the dwarf planet even in the sharpest images. The inset shows Pluto photographed by the New Horizons spacecraft. It was taken shortly before its closest approach on 14 July 2015, from a distance of less than 1 million km.

more objects at the fringes of the Solar System, some possibly even larger. These are generally known as Trans Neptunian Objects (TNOs), and their number increases all the time. Dwarf planets named Makemake and Haumea seem to be only slightly smaller than Pluto. At such a distance, even the largest telescopes currently available show nothing more than a dot of light whose properties have to be judged by making assumptions about the nature of the object.

There are suggestions that there may be yet more genuine planets, some maybe as large as Neptune. As more TNOs are located and their orbits become more precisely known, any perturbations in their orbits may reveal the presence of such objects, though this would upset current thinking about the formation of the Solar System.

The region beyond Neptune is certainly home to many thousands of smaller icy bodies and indeed to millions of comets, some of which are occasionally diverted toward the inner Solar System where they become visible as they near the Sun.

▲ *Pluto photographed using a CCD* with a 300 mm telescope. Images for two nights have been superimposed, so the dwarf planet's image has moved slightly.

ASTEROIDS

There are literally tens of thousands of asteroids – rocky bodies that orbit mostly between Mars and Jupiter. After the Sun formed some 5 billion years ago, it is believed to have been surrounded by a disk of accumulated material with the densest bodies closer in and the lighter ones farther out. The dense bodies gathered to form the inner rocky planets, while farther out the lighter material formed the giant planets and their icy satellites. Beyond Mars there was not enough rocky material to form a single planet, and what was left over remains as the asteroid belt.

The brightest asteroids are easily visible with binoculars and small telescopes, and the fainter you can observe the more you can see. To find them, you can use the same general methods as for Pluto, either using a computer-plotted star map or a Go To telescope. As asteroids are comparatively close, you need wait only an hour or two before the asteroid reveals itself by its motion, even if it is not obvious from its position on the map.

COMETS

Bright comets are few and far between. Comet Hale–Bopp in 1997 was the best known of recent years. Virtually everyone was able to see it, even from city locations, whether or not they were familiar with astronomy. But most comets are much less spectacular. Every year a dozen or so new ones are found, but only rarely are they visible without optical aid and even more rarely do they sport a noticeable tail. Most of the comets never reach the attention of the public, and details are only to be found on astronomical websites rather than in the national press.

Comets are actually fairly small icy bodies only a few kilometers in diameter. They only become easily visible if their orbits bring them into the inner Solar System, and though there are some comets that have orbits similar to those of asteroids, most have very eccentric elliptical orbits and spend most of their time in the depths of the outer Solar System where they originate. The time a comet takes to orbit the Sun is known as its *period*, and this ranges from a few years to many thousands of years. Only for a few weeks at a time are they close enough to the Sun to become observable, when the increased solar heating results in their ice turning to gas. At this point the cometary body or nucleus becomes surrounded by a halo known as the *coma*, with tails of gas or dust or both stretching away from the nucleus.

The nucleus itself is too small to be observed directly. The starlike point often seen within the coma is really the brightest part of the gas emission, and is known as the *pseudo-nucleus*.

Almost all known comets are in orbit round the Sun, but those with short periods of a few years are rarely bright enough to be noteworthy. Only the unpredictable comets that are unknown until they suddenly arrive in the Sun's vicinity are likely to become spectacular, the only exception being Halley's Comet, which has a period of about 76 years and is not due back until the middle of the 21st century.

▼ *Comet Hale–Bopp* in July 1996, when it was a binocular object. It shows the typical appearance of a comet – a hazy blob, with no tail.

▲ *By March 1997,* Comet Hale–Bopp was a spectacular sight in the night sky and was visible even from city centers. The white tail is dust and the blue tail is gas.

▶ *A collection of asteroids* that have been photographed in close-up by a variety of spacecraft, together with dwarf planet Ceres, photographed in 2015. Asteroids are not all the same: there are several types, categorized by their reflectance. That of Mathilde, for example, is only about 4%, compared with 42% for Vesta.

1 Ceres

4 Vesta

21 Lutetia

253 Mathilde

243 Ida

433 Eros

951 Gaspra

· 2867 Steins

◄ *The nuclei of six comets*
that have been photographed
in close-up by spacecraft,
at a scale about 30 times
that on the facing page. In the
background is the 15 km long
nucleus of Halley's Comet,
then counterclockwise from the
bottom are Comet Hartley 2
(2.2 km), Comet Churyumov–
Gerasimenko (4 km), Comet
Borrelly (8 km), Comet Tempel 1
(7 km) and Comet Wild 2
(5 km). The sizes of the nuclei
are shown in their correct
proportions. Jets of gas can
be seen erupting from their
surfaces except for Comet
Tempel 1, though a faint coma
was seen around the nucleus
from a greater distance.

When a bright comet appears it can best be observed with ordinary binoculars. But the more run-of-the-mill faint comets require large binoculars or telescopes. Astronomy magazines and websites give either maps of their path, predictions of their position from day to day, or just the details of the orbit (known as *orbital elements*) which you can insert into computer sky mapping software to produce your own predictions. A set of predictions is known as an *ephemeris*. For more details see the website http://www.minorplanet center.org/iau/Ephemerides/Comets/.

There are six orbital elements that precisely define the orbit of a comet or indeed any other body, such as an asteroid. The time and date of perihelion (the closest passage to the Sun) are known as T, which defines the position of the body in time. The size and shape of the orbit are given by q, the semimajor axis, and e, the eccentricity of the orbit. Finally, the body's direction in space is given by three parameters called the longitude of perihelion, ω, the longitude of the ascending node, Ω, and the orbit's inclination to the ecliptic, i. In addition, the magnitude of the body may be given by two figures, H and G. These figures are given for all comets by the International Astronomical Union's Central Bureau for Astronomical Telegrams, which is the world clearing-house for discoveries.

Studies of asteroids and comets

While the nature of the outer planets is scientifically interesting, asteroids and comets are of much more immediate interest to mankind, if only because their orbits can occasionally cross that of the Earth with the potential for a collision – so the more we know about them, the better.

The first mission to a comet was the European spacecraft Giotto, which photographed the nucleus of Halley's Comet in close-up in 1986. Since then there have been NASA missions to Comet Borrelly in 2001, Comet Tempel 1 in 2004, Comet Wild 2 in 2005 and Comet Hartley 2 in 2010. In addition, the European Rosetta mission arrived in orbit around Comet Churyumov–Gerasimenko in 2014, and released a probe named Philae to the surface of the comet.

Each nucleus has its own characteristics. All are of roughly similar dimensions and have pitted surfaces, though part of Tempel 1 is remarkably smooth. It was expected that comet nuclei would be icy, yet they seem to have comparatively warm surfaces. Coupled with the finding that they are among the blackest objects in the

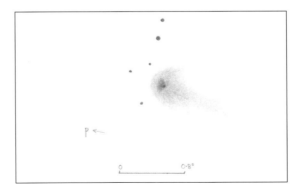

▲ *An amateur drawing of Comet NEAT* (C/2001 Q4).
It appeared in 2004 at third magnitude in the twilight evening sky.
Drawings are often able to show a wider brightness range than
photographs, and here the starlike pseudo-nucleus is visible within
the coma.

▲ **Face to face with Eros.**
This view from the descending NEAR Shoemaker spacecraft is only about 33 m across – about the same size as a tennis court. The largest boulder is some 7.4 m across.

zone between Mars and Jupiter. Eros, however, is a near-Earth asteroid, which means that it can come within 195 million km of the Sun. Some NEAs are also Earth-crossers, which means that their orbits cross that of the Earth, but Eros is not one of these.

When NEAR Shoemaker reached Eros it began a series of orbits around the asteroid, sending back photos of the body in great detail. Eventually it was directed to a touchdown on Eros, though pictures then ceased as it was not designed for a soft landing. Another spacecraft that has made a touchdown on an asteroid is the Japanese-built Hayabusa, which returned some 1500 grains of material from a small asteroid named Itokawa. The dust turned out to be similar to meteorites.

ECLIPSES

Eclipses of the Sun and Moon are usually well covered in the press these days as they can be seen without specialist equipment. But as always there are misconceptions about what will happen, and in the case of solar eclipses the information is sometimes downright misleading.

It is a great coincidence that the Sun and Moon happen to be about the same size in the sky, around 30 arc minutes. Solar eclipses occur when the Moon goes directly across the face of the Sun, while eclipses of the Moon take place when the Moon passes through Earth's shadow.

As the Moon goes round the Earth once a month, you might expect there to be an eclipse of some sort every couple of weeks, but of course this is not the case. The reason is that the tracks of the Sun and Moon through the sky (or more correctly, the orbits of the Earth around the Sun, and the Moon around the Earth) are tilted with respect to each other. So only when the Sun and Moon happen to be close to the place where the two tracks cross can there be an eclipse. The tracks cross every 173 days, or just under six months, so for a week or two on either side of the date when the Sun and Moon are close to their crossing points there is a chance of an eclipse of either the Sun or Moon. Whether the event will be visible from your location is another matter, as you must be within the Moon's shadow on the Earth (about 3500 km across) to witness any sort of a solar eclipse. To see a lunar eclipse, however, it simply needs to be night-time at your location. So from any location, lunar eclipses are far more common than solar ones.

A moment's thought will tell you that the Moon must be new for there to be a solar eclipse, and full for there to be a lunar one (see page 45). During a lunar eclipse the Sun and Moon are exactly opposite each other in the sky.

Solar System, with a reflectivity similar to that of coal, this implies that they have a thick dust layer and may indeed be more dust than ice. Samples of dust captured close to Comet Wild 2 by the Stardust mission also surprised scientists by showing signs of having a hot rather than a cold origin.

Several asteroids have been visited by spacecraft during flyby missions and two spacecraft have even orbited asteroids and touched down on them. Galileo passed the asteroids Gaspra in 1991 and Ida in 1993 while on its way to Jupiter. It discovered that Ida has a small companion asteroid, now named Dactyl. An increasing number of asteroids have been found to have companions as a result of variations in their brightness which can only be explained by assuming that there are two bodies orbiting each other.

NEAR Shoemaker was a dedicated asteroid mission which passed asteroid Mathilde in 1997 on its way to Eros, which it reached in 2000. Mathilde is a main-belt asteroid, which spends all its time in the main asteroid

▶ **The total eclipse of 11 August 1999,** *as seen from Cornwall, UK. Prominences surround the Sun. The corona can be seen to a considerable distance from the edge of the Sun.*

Observing solar eclipses

Solar eclipses are seen as total by observers exactly within the shadow of the Moon, but partial by those near the edges. During the partial phases, the Sun appears to have a bite out of it, as the silhouette of the

Moon begins to cover the brilliant solar disk. At this stage, you can only observe the Sun by using the same methods as at any other time – by projection or by using the correct dense solar filters, either over the aperture of the telescope or with the naked eye. At one time, people tended to use any material that reduced the visible brightness of the Sun, such as smoked glass, but we now know that such materials may allow infrared light to be transmitted. This can damage the eye just as readily as the white component, but as the Sun looks visually dim there is no reflex action to make you look away. Color filters come into the same category, along with black plastic bags. Only use approved solar filters to look at the Sun with the naked eye.

When the Sun is totally eclipsed, however, its bright surface is hidden by the Moon and it is safe to view the spectacle, but only for the duration of the total stage itself.

Observing lunar eclipses

Fortunately there is no direct danger associated with observing lunar eclipses. During the event, the Earth's shadow slowly encroaches on the Moon and then moves away again. If the Moon passes through only the outer edge of the Earth's shadow, the *penumbra*, it may just appear slightly dimmer than usual at one edge. It may partially pass through the black part of the shadow, in which case we see a *partial eclipse*, or entirely into the shadow, which gives a *total eclipse*.

During totality, the Moon often turns a dark shade of red, orange or brown. This is because sunlight is bent and absorbed by the edge of the Earth's atmosphere, with the red component being most likely to reach the Moon. The darkness or lightness of a total lunar eclipse is always an unknown quantity beforehand and adds to the interest and spectacle of the event. Through a telescope, the lunar features usually remain dimly visible.

Eclipse phenomena

A loyal band of followers trek round the globe to make sure that they are in just the right spot to see the Moon completely hide the Sun for just a few minutes. Total eclipses can last no longer than 7½ minutes but most are around three or four minutes long. The differences

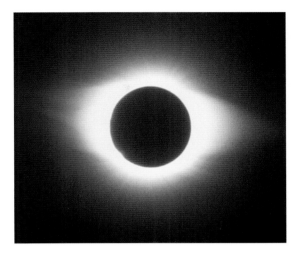

◄ The total eclipse of 24 October 1995, here seen from India, took place near solar minimum and the equatorial streamers are much more prominent than those from the Sun's polar regions.

◄ Another view of the 1999 solar eclipse, but with a long exposure to reveal the Sun's outer corona. This was close to solar maximum, and the corona is equally active in all directions. The photograph was taken from Bulgaria where skies were particularly clear.

are caused by the changing diameters of the Sun and Moon as a result of the elliptical orbits of the Earth and Moon. The longest total eclipses occur when the Moon is closest to the Earth and the Earth is close to aphelion, its farthest point from the Sun, which occurs in July.

There is comparatively little scientific value in observing an eclipse, as the views of the corona which it affords can now be made more continuously from space. But the sheer spectacle is enough to attract people from all over the world. They are drawn not just because of the

▼ Stages of the lunar eclipse of 9 November 2003. The Earth's shadow encroaches from the west as seen in the sky. At totality the Moon appears reddish, usually with a brighter segment if the Moon does not pass centrally through the Earth's shadow.

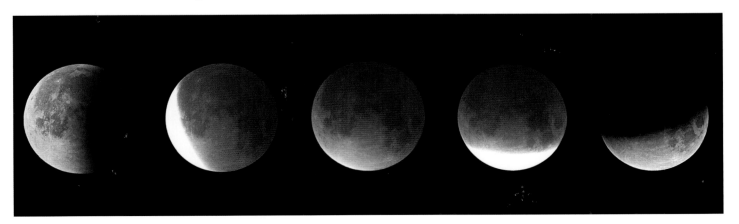

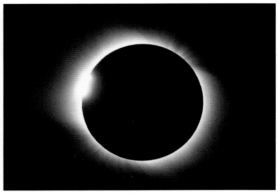

◀ *Another solar minimum eclipse,* on 29 March 2006, photographed from Turkey. Several separate exposures have been combined to show detail from the inner to the outer corona, evening out the wide range in brightness. Lunar details are even visible, being illuminated solely by light reflected from the Earth.

▲ *The "diamond ring" effect* does not always occur, but during the total solar eclipse of 1 August 2008, visible from China, it was seen at both the beginning and end of totality.

unique spectacle but because every eclipse is different. The state of the Sun's activity is a major variable factor. When the Sun is near a maximum in its 11-year cycle of activity, its corona tends to be active around the entire disk, while at minimum it is more concentrated along the line of the equator.

At the beginning and end of totality, just as people are warned to keep their eyes averted, can occur some of the most spectacular events, known as Baily's Beads and the "diamond ring" effect. Both are caused by the Sun's bright disk, or photosphere, shining through deep valleys in the Moon's limb, with the effect of the "diamond ring" being caused by a single point of light on the surrounding corona.

The landscape during a total solar eclipse can be as fascinating as the event itself. Though the sky around the Sun is fairly dark, it is by no means pitch black and only the brighter stars or planets are visible. Nearer the horizon the sky appears more reddish, as if there is a sunset all round the horizon. Close to the moment of totality, as the Sun is only a thin crescent, people occasionally see what are called shadow bands on the ground – rapidly moving light and dark bands. These have the same cause as the twinkling of the stars, that is turbulence in our lower atmosphere, but they appear only briefly when the sky is very clear.

OCCULTATIONS

From time to time, the Moon passes directly in front of a star, temporarily hiding it from view. This is known as an *occultation*, and precise timings of such events are of some use in keeping accurate track of the Moon's movement. Of course, the Moon is passing in front of stars all the time, but stars fainter than about magnitude 8 are virtually unobservable visually when so close to the bright Moon.

▶ *During an annular eclipse* the corona and prominences are not visible as the photosphere remains visible throughout, though Baily's Beads may appear. Because so much of the photosphere is visible, the same viewing precautions must be taken as for the partial stages shown here on either side. This eclipse was visible on 3 October 2005 from Spain.

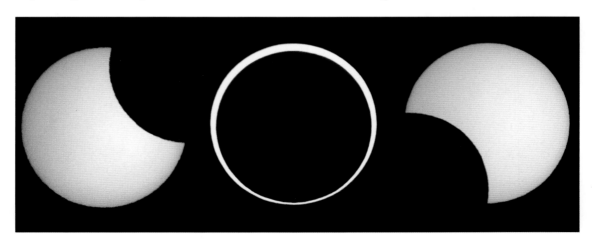

▶ **A wide-angle view** of the total eclipse of 26 February 1998 as seen from Curaçao in the Caribbean reveals the range of sky colors. Venus was visible well below the Sun, and Jupiter could be seen close to the Sun.

When a star is occulted, it disappears instantly because the Moon has no atmosphere to dim the star's light before it is occulted. This is why timings can be so precise. It is easiest to observe occultations by the dark limb of the Moon, as you can watch the star until it disappears and there is no glare to worry about. But reappearances or occultations by the bright limb are subject to greater errors because you do not know precisely where the star will reappear, and the bright limb creates glare, which may mask the instant of reappearance. Lists of observable occultations from any location worldwide are available from the International Occultation Timing Association website at www.lunar-occultations.com/iota/iotandx.htm. Occasionally, the planets may be occulted by the Moon, and occultations of stars by planets or even asteroids are also possible.

METEORS

Watching for meteors – popularly known as shooting stars – is the astronomical equivalent of fishing. You never know what will happen next, and there can be long periods when nothing happens. But there is always the chance that in the very next instant you will get the big one that makes it all worthwhile.

Meteors can occur at any time. They are tiny grains of interplanetary dust that burn up in a sudden trail of light when they collide with the Earth's atmosphere at a height of about 80 to 100 km. The bodies that cause meteors are called meteoroids, and they are generally very small, low-density objects, more like grains of instant coffee than grains of sand.

Meteors mostly come from the tails of comets, so the shooting star that you see dashing through the sky is the destruction of a tiny piece of material that probably originated billions of years ago in a distant star. It then became part of the formation of the Solar System and found itself in the body of an icy comet. Eventually it was released from the comet on a passage close to the Sun and has been orbiting for millions of years until its vaporization before your eyes.

◀ **A Geminid meteor below Orion.** Photographing meteors is a matter of leaving the shutter open and hoping that one will appear. Only the very brightest will record successfully.

The orbits of comets are strewn with the dust released over a long period of time. This dust spreads outward from the path of the comet as a result of various forces, such as the pressure from particles emitted by the Sun and the absorption and re-emission of sunlight. Eventually the dust becomes randomly spread throughout the Solar System, and it is this dust that gives rise to the *sporadic* meteors – that is, meteors that occur at random.

If the Earth passes close to the spreading cloud of dust from a comet, however, there will be a greatly increased rate of collisions and we see a *meteor shower*. In a shower, meteors appear to come from a particular point in the sky because of the effects of perspective. This point is known as the *radiant*, and the shower of meteors is named after the constellation in which it lies. One of the most famous and regular showers, for example, is the Perseids. Each year, in the second week of August, increased numbers of meteors appear to radiate away from a point in the constellation of Perseus. The shower can be seen from anywhere that Perseus is above the horizon, with the greatest numbers visible when it is high in the sky. A list of the major meteor showers is given below.

The number of meteors visible is given by the Zenithal Hourly Rate (ZHR), and this is the figure usually quoted in the media when talking about what you might see. For the Perseids, it is around 60–100 meteors an

hour. But in practice you will almost never see as many, because as its name implies the ZHR is the number that you would see if the radiant were in the zenith (directly overhead), in perfect conditions, and for a single observer looking attentively at as much of the sky as possible. This never applies to the Perseids, because the radiant is never overhead during night hours, so the numbers seen will be reduced. Skies are rarely perfect, and it is easy to miss the fainter meteors, which are often more plentiful than the bright ones. Most people would be happy to see a half or even a quarter of the stated number each hour, though a group of people observing in good conditions could see more meteors an hour than the ZHR figure would suggest.

Meteors are more common during the second half of the year than the first half, for reasons that are not fully known. Also, meteor rates increase toward dawn, because at that time the side of the Earth you are on is heading directly into the path of the meteors. Patience is often necessary if you want to see a shooting star, even on the night of a good shower.

In addition to meteors, occasionally much brighter objects dash through the sky. Anything brighter than magnitude −5 at its brightest is termed a *fireball*, and if it explodes it may be called a *bolide*. Fireballs may occur during a regular meteor shower as particularly bright ordinary meteors, but those that appear outside normal shower times, or do not come from the expected radiant, may originate within the asteroid belt rather than from a comet. Instead of being composed of rather crumbly dust grains, fireballs are likely to be chunks of rock or even metal. If they survive their descent, they are called *meteorites*. Thousands of meteorites fall to Earth daily, but only a tiny fraction of that number is ever recovered because most fall in the sea or over unpopulated territory. When a very bright fireball appears, people are asked to describe as accurately as possible its path through the sky in the hope that its possible drop point can be found and a meteorite recovered. Very often, however, people believe that a run-of-the-mill bright meteor or fireball is lower than it actually is, and associate a nearby chance noise with its impact.

AURORAE

Although aurorae – also called the Northern Lights or Southern Lights – are a feature of the polar regions, from time to time they can be seen from lower latitudes. They are visible as colored glows in the sky, often with streamers or rays, and are the result of interactions between streams of particles from the Sun and Earth's magnetic field. They are most common at times of enhanced solar activity, such as sunspot maximum (see page 64), but are not out of the question at other times. The appearance of a large sunspot may herald an aurora, but there is no certainty, and sometimes aurorae occur when there is no obvious major solar activity.

Auroral activity is most intense in a band surrounding the Earth's geomagnetic poles, which are considerably offset from the geographic poles. The geomagnetic north pole is in Canada, and the geomagnetic south

METEOR SHOWERS						
Shower	Maximum	Normal limits	Rate at maximum	Radiant RA	dec.	Remarks
Quadrantids	Jan 04	Jan 01–06	100?	15^h 28^m	+50°	Blue meteors with trains
Lyrids	Apr 22	Apr 19–25	10–15	18^h 08^m	+32°	Bright meteors
η-Aquarids	May 05	Apr 24– May 20	40	22^h 20^m	−01°	Broad maximum and multiple radiants
α-Scorpids	Apr 28 May 12	Apr 20– May 19	10	16^h 32^m 16^h 04^m	−24° −24°	Multiple radiants – long activity April to July
δ-Aquarids	July 28 Aug 06	July 15– Aug 20	20	22^h 36^m 22^h 04^m	−17° +02°	Double radiant – southern component is richer
Perseids	Aug 12	July 23– Aug 20	75	03^h 04^m	+58°	Rich shower, bright meteors with trains
Orionids	Oct 21	Oct 16–30	25	06^h 24^m	+15°	Fast meteors, many with trains; associated with Comet Halley
Taurids	Nov 03	Oct 20– Nov 30	10	03^h 44^m	+14°	Slow meteors, some fireballs
Leonids	Nov 17	Nov 15–20	20	10^h 08^m	+22°	Trains, rates declining after major storms in 1999–2002
Puppids- Velids	Dec 08 Dec 25	Late Nov– Jan	15	09^h 00^m 09^h 20^m	−48° −45°	Two of several radiants
Geminids	Dec 13	Dec 07–15	120	07^h 28^m	+32°	Many fireballs
Ursids	Dec 22	Dec 17–25	10	14^h 28^m	+78°	Increased activity in certain years

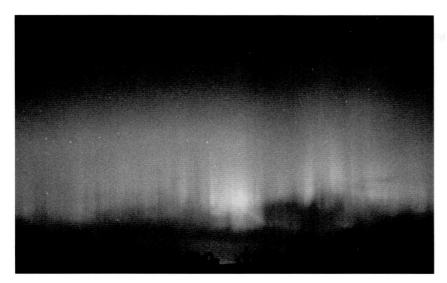

◄ An aurora seen from southern England in April 2000. It shows the characteristic curtain appearance that is more typical of those seen nearer the poles. The color changes from green to purple with increasing height.

pole is near Vostok in Antarctica. So observers in Alaska, Canada or Tasmania are more likely to see the aurora than those in Europe or South America at similar latitudes. In Europe, people living in Scotland or Norway get far more opportunities to see the aurora than those in London or Paris, while in Rome only the most dramatic events are visible.

An aurora may consist of little more than a pale glow along the poleward horizon, but in the more dramatic events there are noticeable rays or streamers. During a major display you may see the classic curtain-like appearance of these streamers, with the lower ends finishing in definite curves. Sometimes the rays can appear to come from a point high in the sky, an effect known as a *corona* (not to be confused with the solar corona visible during a total solar eclipse).

The glow of an aurora occurs at an altitude of more than 100 km in the Earth's upper atmosphere, in a region normally considered to be space. In particular, oxygen molecules glow either green or red, with the red tending to be at higher altitudes than the green. As a result, observers in midlatitudes tend to see red aurorae rather than green, and only during a major display does the green become more obvious. Other colors are caused by other atmospheric gases, such as nitrogen. The colors are visible to the naked eye when the display is bright, but faint glows are easy to confuse with light pollution. A time exposure of 30 seconds or so with a sensitive digital camera will often reveal whether an aurora is in progress, as it shows the color much more strongly. Another clue is that light pollution inevitably shines on the underside of clouds, whereas clouds show dark in silhouette against an aurora. The auroral streamers change in intensity and position over a period of seconds, so photographs are best restricted to exposure times of less than 30 seconds, if possible, to avoid smearing of the detail.

There is no good way of predicting well in advance when an aurora will take place, although aurora alerts based on the release of particles from the Sun are now sent out by email to anyone interested. There are also continuous online monitors of geomagnetic activity and these will tell you whether it is worth even looking out of the window for enhanced auroral activity. For a listing of websites worldwide that provide such information, go to www.stargazing.org.uk.

The website www.spaceweather.com issues alerts by email and provides links to solar activity sites.

NOCTILUCENT CLOUDS

At one time these beautiful high-altitude clouds were hardly ever mentioned in observing guides because they were very rare events. But they are on the increase, and are now visible from lower latitudes than ever before, though exactly why no one is sure. Noctilucent clouds appear around midsummer at latitudes higher than 50° north or south. Almost all the reports received come from the northern hemisphere because there is very little populated territory below 50°S, so here we deal only with the northern-hemisphere appearance.

At first glance, noctilucent clouds (NLC) look like cirrus clouds in the late twilight, but they are visible long after sunset and only toward the northern horizon, where the sky is still light from late May to early August. They are silvery or bluish in color, and any ordinary clouds are seen in silhouette against them. NLCs are water vapor clouds, and they may be the result of condensation on tiny particles of dust from volcanoes, meteors or aircraft vapor trails. They are about 80 to 85 km above the Earth's surface, where normal clouds do not form. Because they are rather faint, NLCs are not seen during the day.

▼ Noctilucent cloud seen from a London, UK, golf course. The star is Capella, which is usually near the horizon when noctilucent clouds are seen from the northern hemisphere.

6

A CLOSER LOOK AT STARS

Each star that we see in the sky is a sun in its own right. Our Sun is a star of average brightness, so we can picture the others as being similar self-luminous globes, each one brilliant and a great source of heat and light in its surrounding neighborhood. But not all stars are the same as the Sun. Some are hotter and brighter, while many more are cooler and fainter. Inevitably we tend to see the bright ones rather than the faint ones, and while it is often said that the Sun is an average star, this does not mean to say that it is a typical star. As many as 90% of all stars are fainter than the Sun. Furthermore, fewer than half the points of light we see in the sky are single stars: about 55% are double or multiple stars, with two or more companions. So the Sun is anything but ordinary, though it is roughly midway between the brightest and faintest, and also between the most and the least massive.

How stars and planets are born

Stars are born in very specific parts of the Milky Way – what are called *starbirth regions*. These lie along the plane of the Milky Way, among the dark clouds that are so obvious on deep wide-angle photographs, such as the Cygnus Rift (see page 29).

Only a comparatively small number are readily visible from Earth, the best known being the Orion Nebula and the Eagle Nebula. Others can be picked out better on infrared or radio surveys, which are not so strongly affected by the obscuration by dust clouds.

Starbirth regions are known as molecular clouds because they contain not just hydrogen but also numerous other molecules which involve carbon, oxygen, nitrogen and other atoms. Over 125 different molecules, some quite complex, have been identified in interstellar space. However, only about 0.1% of the molecules in a cloud are anything other than hydrogen. But there are also what are called dust grains mixed

▶ A Hubble view of part of the Trifid Nebula (see page 167). What look like horns on some space monster are two aspects of star formation. The left-hand one is a jet from a star that has formed within the dark cloud. The stalk to its right may be the remains of part of the cloud that is being evaporated by the intense light from a brilliant star nearby. The stalk points toward the star and is protected from evaporation by a knot of gas at its tip, which may contain a newborn star yet to emerge.

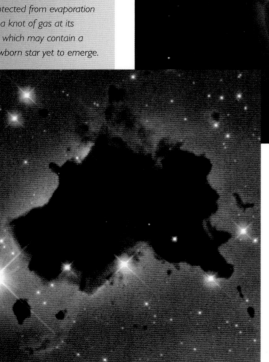

◀ A close-up of the nebula IC 2944 shows dark blobs of gas silhouetted against the bright nebula. These are known collectively as Bok globules, and the largest here have a combined mass about 15 times that of the Sun. Eventually these may collapse and form into new stars. The radiation from recently formed stars is causing the smaller clouds in the area to evaporate, so they may never form stars.

in with the hydrogen – about 1% by mass. These dust grains consist of carbonates and silicates, maybe with a coating of plain old water ice. It may seem a bit academic to consider such tiny proportions of materials, but they will become very important later in the story.

Just what makes a star begin to form is uncertain. An area of increased gas density will have a slightly greater gravitational pull on the surrounding gas, which will be attracted toward it. Perhaps the explosion of a nearby supernova will trigger off a wave of star formation as it compresses the gas cloud.

Once such sites of star formation have arisen, it can take less than a million years from just a swirl of gas to the early stage of the birth of a star, which is termed a *protostar*. The gas molecules at the core of the developing star jostle each other to the extent that they begin to heat up, which means that they move faster. At this stage the heat is powered by gravitational contraction alone.

The core of the developing star heats up, and when it reaches some 10 million degrees K it is sufficiently hot that hydrogen nuclei may collide with such force that they merge to create helium nuclei of slightly lower mass than the original components. The difference in mass is released as energy in the form of gamma rays. Although this has nothing to do with the sort of combustion we are used to, the process is usually referred to as hydrogen burning. In massive stars a different reaction occurs, but the result is the same – the release of energy. In a well-used phrase, a star is born.

In real stars, however, there is another important feature to consider – their rotation. As an object contracts in size, its speed of rotation increases. The usual example is that of a skater spinning with arms outstretched, who spins faster when they pull in their arms close to their body. So the slightest swirl of movement in the initial cloud will become a significant rotational speed when the star contracts. The resulting star will therefore be spinning, as indeed is our own Sun. Newly born stars may rotate in as short a time as just a few hours. This can cause them to throw off material which in the case of a massive object may form into new stars, or simply form a disk around the star. As more material falls in toward the protostar, it may add to the mass in the disk. The rapidly spinning star may also generate a considerable magnetic field around it, which can play a part in its development.

The newly formed star is still immersed in a gas cloud, but the furnace within soon clears the gas away from its immediate vicinity. Starbirth regions contain what are called T Tauri stars (after the first one to be discovered) and the Hubble Space Telescope in particular has given us fascinating glimpses of the regions around these stars. Like very young persons, very young stars find it hard to contain themselves, and produce a copious outflow of gas. Jets of gas known as bipolar flows can emerge from the poles of emerging stars, and when these hit the surrounding gas a curious type of nebula called a Herbig–Haro object can be the result.

What is left of the disk of material surrounding the star following these outflows may now itself condense to form planets, making particular use of the heavier elements and molecules. The whole disk will be rotating in the same direction and plane as the original star, so we expect the star's planets to do likewise. This view may well prove in time to be rather simplistic. The theory of planetary formation has been based on explaining how our own Solar System was formed, but we are slowly learning that what applies to the familiar usually doesn't apply to the unfamiliar.

What are called protoplanetary disks can indeed be seen surrounding stars in starbirth regions, and since 1995 an increasing number of other stars have been observed to have planets (see the panel on page 102). So far, none of these planetary systems can be observed in sufficient detail that we can see the direction in which the star and its planets rotate. In fact only large planets can be detected at all, and many of these turn out to have most unexpected orbits compared with those of our own comfortable and neat Solar System. We expect that the heavier materials, notably the silicates which make up rock, will reside close to the star to form terrestrial planets, while

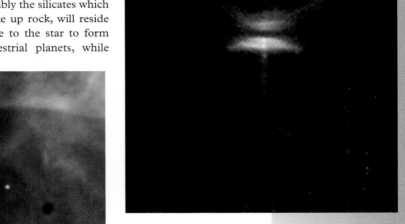

▼ *A star caught in the act of formation* by the Hubble Space Telescope. This edge-on view of Herbig–Haro object HH30 shows a dark disk of matter which hides the star itself, but which is illuminating the gas on either side of the disk. Red bipolar flows shine out on either side of the star.

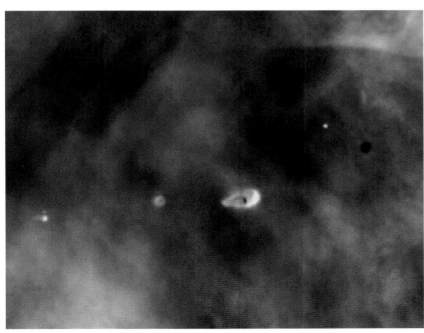

◄ *Suspected protoplanetary disks* in the Orion Nebula. A nearby bright star illuminates most of the disks, so they shine brightly, though one on the right is more distant from the star and is silhouetted against the background with a low-mass red star at its center. This disk is about 7.5 times the diameter of the Solar System.

HOW EXTRASOLAR PLANETS ARE DETECTED

The task of finding a planet around a star brings out lots of analogies – "like seeing a moth next to a searchlight" is a common one. This is a fair comparison, because even a large planet of the size of Jupiter, which is about a tenth of the Sun's diameter, only reflects the light of its star rather than being self-luminous. However, rather than trying to pick out a feeble light so close to a brilliant one, another way of detecting the moth would be to measure the brightness of the searchlight very accurately and wait for the moth to fly in front of it. In the case of a relatively large planet passing in front of a star this will make a detectable difference. This is known as a transit, like the passages of Venus or Mercury in front of the Sun.

This is exactly how the majority of extrasolar planets have been detected, notably by the Kepler spacecraft. Another technique is to look for changes in the motion of the star through the sky. As a planet orbits a star it will tug the star slightly one way and then the other. Gravity is a mutual force – even a small object has a gravitational pull. Though the Earth is pulling you toward its center, you are also pulling the Earth toward you, very slightly. When it is a planet doing the pulling, the effect on the star's motion through the sky is measurable.

The motion of a star across the sky is called its *proper motion*, and only for comparatively nearby stars is this detectable, and then only by precise measurements of its position over a long period of time. However, a star's movement toward or away from us, known as its *radial velocity*, is easier to measure from the Doppler shift of its spectral lines. A star moving in a straight line will have a constant radial velocity. But if there is a planet swinging about it in such an orbit that it tugs the star slightly toward or away from the Earth, the radial velocity of the star will change periodically.

Many extrasolar planets are now being detected, over 800 being found during 2014 alone. They come in all sizes and distances from their primary stars, including some Earth-sized planets in orbits where water could exist as a liquid. In some cases, details of their atmospheres are revealed as the planet transits its star. However, we are still a long way from seeing direct images, or even detecting signs of life.

▼ *An artist's impression* of a planet orbiting the yellow Sun-like star HD 209458, 150 light years from Earth. It was first found by the radial velocity method, but subsequently was observed in transit by the Hubble Space Telescope, which detected sodium in the planet's atmosphere. The planet is regarded as a "Hot Jupiter" – just 70% of the mass of Jupiter, but only about 6 million km from the star. Its surface temperature must be about 1100°C – quite unlike that of giant planets in the Solar System.

▼ *A cluster of brilliant, massive stars* shines at the center of nebula NGC 3603. The radiation has cleared the region of gas, but surrounding it is the nebula from which stars are still condensing. The blue gas is believed to have been ejected from the bright star nearby.

the more volatile materials will remain farther out where they are safe from the excesses of the star's early years, and will form gaseous or icy planets. Just how our theories of planet formation will fare when more and more solar systems are analyzed remains to be seen. So the heading of this section should really be "How stars and planets may be born."

Stars in middle age

Our Sun is a good, sturdy, middle-aged star. There is a great deal to be said for middle-aged stars when you depend on them so absolutely. To the best of our knowledge, it has been shining virtually constantly for billions of years. It forms a standard by which other stars are judged – we generally compare other stars with the Sun in terms of their brightness, mass and so on. But not all stars are so reliable, as astrophysicists have discovered.

By human standards, stars live for a very long time, so the starry sky that we see today is essentially the same as that seen by the very earliest human eyes a few hundred thousand years ago. Stars take millions of years to evolve. A star's future is pretty well mapped out for it as soon as it forms, because its subsequent evolution depends largely on its initial mass. The Sun is roughly midway between the largest and the smallest stars – just right for Goldilocks, as it were.

Actually, Goldilocks is associated with the Earth's position in the Solar System as well. The Earth orbits at just the right distance from the Sun that water can remain liquid on its surface. Our two nearest neighbors in space are examples of planets that lie just outside this zone – Mars is too cold, and Venus is too hot, but the temperature on Earth is just right. This region around a star is known as the habitable zone, or sometimes, whimsically, the Goldilocks zone.

It should not be a surprise that this is the case. If the Solar System did not happen to have a planet in this zone, life would presumably not have emerged on it, and there would be no one to write books about it and equally no one to read them. We need not marvel that the Earth is somehow amazingly well adapted to suit our kind of life.

The most massive stars shine brilliantly from the day they are born. A star twice the mass of the Sun shines about 14 times as brightly, while one of 15 solar masses shines a staggering 30,000 times as brightly. But these pay the price and they use up their fuel at 30,000 times the rate, which means that they have a lifetime of only about 5 million years compared with the 10 billion years of the Sun.

So while middle age for humans occurs at roughly the same sort of age for everyone, in the case of stars it depends very much on their mass. Astronomers refer to stars in the middle period of their lives as being on the *main sequence*. This means that they have settled down from their adolescent years to shine steadily, with a plentiful supply of fuel still remaining. The main sequence is a natural progression from the hot, blue stars with short lifetimes down to the faint, red stars with very long lives. Plotting all the stars in the sky on to a graph of brightness compared with temperature shows that most stars fall on this line. Such a graph is referred to as a Hertzsprung–Russell diagram (see A–Z page 253).

Stars more massive than the Sun have hotter cores, which enable a different sequence of nuclear reactions to take place. These reactions are known as the carbon-nitrogen cycle, and they have as by-products carbon, nitrogen and oxygen – three elements that are essential to life as we know it. Combinations of these three, plus hydrogen, give us the overwhelming proportion of the molecules in our bodies. Massive stars kindly produce much of the carbon dust and the organic molecules that are found in the molecular clouds permeating the spiral arms of the Galaxy. These materials eventually accumulate again to form new planetary systems, which in time may become populated with life-forms consisting of organic molecules containing the carbon. Everything you see around you, including your own body, is made from material that was forged inside stars.

Red giants

Eventually, stars will have used up most of the hydrogen available to them in their cores, and although they are still predominantly hydrogen, they have no way of using the material that lies in the cooler regions beyond the core. Hydrogen burning still continues in a shell around the core, but inside that, where no fresh hydrogen can penetrate, a crisis is brewing. Gravity starts to take over again and the core contracts. Just as in its early life, the core gets hotter again and a new cycle of reactions can start to take place, this time burning first helium, then carbon, rather than hydrogen. Different elements are produced in the process. Whereas hydrogen is the simplest element of all, consisting of just one proton and one electron, other elements combine greater numbers of protons, neutrons and electrons, and individual atoms are much heavier than, say, hydrogen or oxygen. The heaviest element that is produced by normal burning is iron.

At this point the star will have left the main sequence, and will have changed its character completely. It now has a swollen outer atmosphere but a brighter core. Its diameter increases dramatically but the temperature of its visible surface actually goes down, because the energy is spread over a much greater surface area. As always, a lower temperature means a redder color, so the star is now a red giant. Each stage of burning a different fuel results in a repeated core collapse, so the star may change its nature several times, particularly in the case of the more massive stars.

During this stage in a star's life, when it is trying to come to terms with a depleted fuel supply, it may become one of many different types of variable star. The brightness of these stars fluctuates in some way, either periodically or irregularly.

As far as stars are concerned, the terms "giant" and "dwarf" confusingly refer solely to the stage in a star's life rather than to its actual mass or diameter. A star becomes a giant when it starts to expand from its comfortable life on the main sequence. So even a star much more massive than the Sun may still be a dwarf star until it expands at the end of its life and becomes either a blue giant or a red giant. A blue giant is basically a

◀ **The Hubble Space Telescope** *has taken one of the few photographs of the actual disk of a star – in this case the red supergiant Betelgeuse, which has a disk less than 0.1 arc seconds across. This view was taken in ultraviolet light, in which the visible disk is larger than the orbit of Jupiter around the Sun. The yellow area is believed to be a bright area on the surface of the star.*

massive star that has left the main sequence and is on its way to becoming a red giant, though sometimes the term is wrongly used just to mean a rather massive blue star.

Red giant stars are huge. Our Sun, when it becomes a red giant in 5 or 6 billion years' time, will grow to as much as 100 times its present diameter. Its exact size is a matter for debate, but it is a question of whether it will engulf Mercury, or expand even farther out and surround the Earth. If it swells by 100 times it would reach between Venus and the Earth, but its actual size could be considerably less. Either way, it is academic, not only because we will not be there to see it happen, but because the Earth will have been fried long before the Sun becomes a red giant. As it does so it will increase in temperature by a factor of perhaps three. Maybe our distant descendants will flee to the outer planets, as the habitable zone shifts outward and a moon such as Europa becomes a lush watery world.

The most massive stars of all become not just red giants, but red supergiants. These too can alternate from being an extended red giant to a smaller blue supergiant, depending on what fuel they are burning. Red supergiants are truly colossal, and can swell to 500 times their original size. The only stars which show actual disks in the sky that can be seen by existing instruments are red supergiants, but even these are at the limit of detectability.

One of the most famous red supergiants is Betelgeuse in Orion. The photograph taken by the Hubble Space Telescope on page 103 shows that it does not have a sharp outline like our own Sun, but is surrounded by a huge gaseous atmosphere which just fades away at the edges. Its size depends on the wavelength of light that

▲ **Sirius** *(see also page 128) consists of a normal A-type star with a white dwarf companion. Visually the A-type star is by far the brighter of the two, but in this photo taken by the Chandra observatory in X-rays, Sirius B is the brighter of the two objects because, being much hotter than Sirius A, it emits more X-rays even though it is a faint star.*

you view it in. If it were to replace the Sun, its red light radius would reach out as far as the asteroid belt, but its ultraviolet radius would be beyond Jupiter. However, the gas at these distances is so rarefied that, overall, the star has an average density less than that of air.

One of the features of very red stars is that most of their energy is emitted in the infrared. Incandescent light bulbs have the same characteristic – they waste a lot of energy as heat, as anyone who tries to remove one that has only just gone out will agree. In the case of Betelgeuse, only 13% of its output is visible light.

Betelgeuse is also noted for its light variations. Sometimes it can rival Rigel, the brightest star in Orion, which is magnitude 0.2, while occasionally it is reported as faint as magnitude 1.5. These changes take place with an approximate period of 420 days, which puts Betelgeuse into the class of semiregular variable stars.

Death of Sun-like stars

At some point, no amount of rearrangement of its layers will keep the star going. The end is nigh. The mass of the star now makes a great difference to the eventual result.

A star with the mass of the Sun undergoes a series of explosions in its core that drive off the surrounding layers of gases, which spread out far into the space surrounding the star. While the disks of the stars themselves are far too small to be seen, the shells of gas that surround them may spread over a light year or more. Such objects are easily visible with telescopes, and it so happens that the typical nebula that results is a few arc seconds across in the sky and appears greenish or bluish in color. This is just how the newly discovered planets Uranus and Neptune looked through telescopes to 19th-century astronomers, so these ghostly shells acquired the description "planetary nebula," a name

▶ **The constellation of Orion** *is a starbirth region about 1500 light years away. Many of its stars are massive, blue objects, such as Rigel at lower right. The Orion Nebula is below the Belt of three stars, while to the upper left is the red giant star Betelgeuse. A filter has been used to spread the star images and make their colors more obvious, but the difference between Rigel and Betelgeuse is easy to spot with the naked eye.*

which has stuck though it leads to some confusion. For example, a shell surrounding a star that is on the point of turning into a planetary nebula is called a proto-planetary nebula. But this is not to be confused with the protoplanetary disk that represents a planetary system in formation, which is seen at a very early stage in a star's evolution.

There is a weird zoo of different types of planetary nebulae. The classical appearance is of a circular shell, like that of the Ring Nebula in Lyra (see page 161), but there are many other types. Many of them exhibit strange features such as what are called "FLIERS" – standing for "fast low-ionization emission regions." These are extensions on either side of the nebula that seem to be moving very rapidly, so are younger than the rest of the shell. Experts are still debating what causes these picturesque features.

At the center of the planetary nebula is the exposed core of the star, now incapable of shining as it once did because it has run out of fuel and is simply cooling down over billions more years. It is tiny in comparison with its former self, so its feeble glow is emitted over a small surface area – the opposite of a red giant – and is therefore hot but dim. This is a white dwarf.

A star therefore goes from one extreme to the other. White dwarf statistics are astounding. A typical white dwarf might be about the size of the Earth, yet it packs in all the mass of a star like the Sun. This means that its material is a million times denser than the Sun – in other words, about a ton per cubic centimeter (the size of a sugar lump).

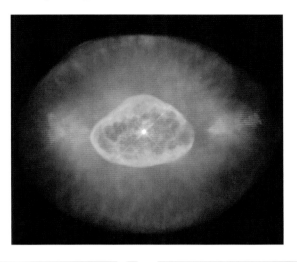

◄ *The Blinking Planetary nebula* *in Cygnus, also shown on page 176. This Hubble photo shows the enigmatic red "FLIERS" which have been shot from the central star.*

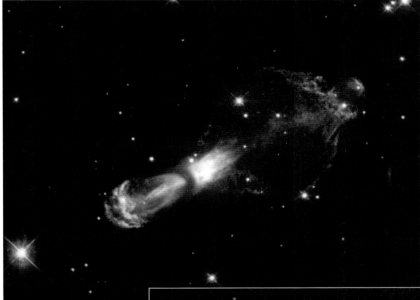

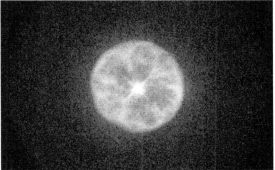

▲ *The planetary nebula IC 3568 in Camelopardalis has the appearance of a slice of lemon in this Hubble Space Telescope photo. It is a classical circular planetary nebula with* *some structure within its outline, and with a faint outer shell that was probably ejected prior to the central shell but has now faded. The nebula is circa 0.4 light years across.*

▲ *The Calabash Nebula is a protoplanetary nebula – a planetary nebula in the stages of formation. The star itself is hidden within the dark band at the center, but is ejecting a bipolar flow of gas shown here in yellow. Where it hits gas between the stars it shines blue. The evolution of this object into a fully fledged planetary nebula will take centuries to unfold.*

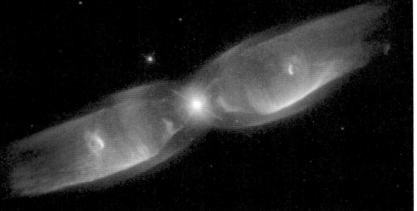

◄ *The jets from planetary nebula M2-9 strongly resemble a rocket exhaust. Its central star is a double, which may have helped to constrain the shape of the jets. The object is about 2100 light years away and the nebula measures about ten times the diameter of Pluto's orbit around the Sun.*

Massive star death

Stars much more massive than the Sun, as always, have to be different. Though they may attempt to throw off mass as red giants and planetary nebulae, if their cores end up having more than 1.4 times the mass of the Sun there is no chance of a peaceful old age as a white dwarf. The only solution is a catastrophic one and the star explodes with enormous force as a supernova. (The plural of supernova among the astronomical fraternity is supernovae, though some popular books do not believe that ordinary people can understand this, and write supernovas instead.)

The intensity of the explosion is such that the star can briefly emit as much energy as all the other stars in the galaxy put together. If this seems an extravagant claim, such a thing can be witnessed in other galaxies. Some astronomers, including a number of amateurs, patrol large numbers of galaxies night after night in the hope of catching a supernova shortly after it explodes. These searches are largely automated these days, though just occasionally an amateur astronomer who happens to be looking at a galaxy visually may notice a star that should not be there and beat everyone else to the discovery. Sometimes a supernova appears almost as bright as the rest of the galaxy.

Such an intense explosion is capable of forcing elements together to form new ones. Massive stars are already hot enough during their main-sequence lifetimes to forge some elements heavier than iron. Copper, zinc and lead are among such elements. But a supernova explosion is one of the few places in the Universe where the temperature is high enough that the heaviest elements, such as gold and silver, can be created. Massive stars are the main source of all but the commonest elements, and without a supernova explosion to spread them throughout the gas and dust

British supernova discoverer Tom Boles photographed this supernova (arrowed) in galaxy NGC 2906 in May 2005. The Type II supernova is a substantial proportion of the brightness of the whole galaxy.

from which the Sun was formed, our world would not exist. The silicon and aluminum that make up much of the Earth's crust, and the calcium in our bones, probably originated in massive stars which then went supernova.

The supernova obligingly spreads these vital materials throughout its immediate neighborhood, and, as we have seen, may trigger the formation of new stars. The expanding shells from supernovae are visible in certain parts of the sky as supernova remnants, of which one of the best known is the Veil Nebula in Cygnus (see page 113).

This type of supernova is Type II, as distinct from the Type Ia supernovae referred to on page 17. Those are also stars at the end of their lifetimes, but instead of being massive superstars they are quite ordinary stars which were initially somewhat more massive than the Sun, and therefore are white dwarfs just on the borderline of being 1.4 times the mass of the Sun. Their only crime is to be part of a double star with a less massive companion which is just going through its red giant stage.

As the red giant starts to swell, material pours on to the white dwarf and upsets its delicate balance. It then explodes as a supernova, possibly destroying itself and leaving the remains of its former companion star careening across the galaxy as it is released from its orbit. Such runaway stars are known, whose proper motions are unlike the other stars in their neighborhood.

The two types of supernova can be distinguished by their spectra. Type I supernovae lack any hydrogen in their spectrum and show just heavy metals, while Type IIs also include hydrogen. The shape of their light curves over a period of time – the way they rise to a maximum over a matter of just a few hours, then tail off over months – also help distinguish between the two.

Astronomers estimate that a galaxy such as our own should have a supernova every 40 or 50 years on average. However, the last supernova to be seen in our Galaxy was in 1604, which was witnessed by Kepler.

▼ Supernova 1987A lies at the center of this ring of gas, photographed in 2004, 17 years after the star exploded. The star itself is invisible in this view, but the pink bubble in the center is material expelled by the star. The brighter ring is gas that it threw off about 20,000 years previously and which has been caused to glow by the radiation from the supernova.

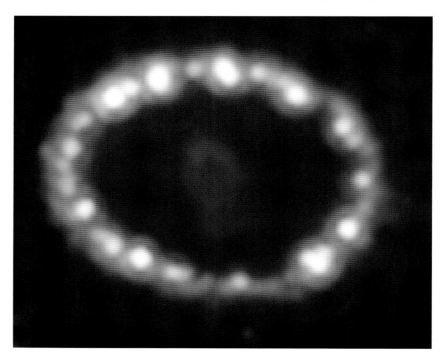

Probably there have been others but we can only see a fairly small part of the Galaxy, with much of it being hidden by dust clouds. In the Andromeda Galaxy, which is nearby and of similar size to our own, only one supernova has been observed, and that was in 1885. The only other supernova to be seen in our own Local Group of galaxies was in 1987, in the Large Magellanic Cloud. There may have been others in the Andromeda Galaxy, and indeed in the other large spiral in the Local Group, M33, but they were unobservable. The appearance of a nearby supernova will trigger frenzied activity in the world's observatories and could happen at any time.

The aftermath of a supernova

What remains of a massive star after a supernova explosion may be either a neutron star or a black hole. A neutron star is even denser than a white dwarf. Instead of containing the mass of a star within the size of a planet, it comes down to the size of a large city (or at least, a sphere with a diameter that of a city, in other words maybe 10 km to 20 km). Whereas in a normal atom there are protons, electrons and neutrons held apart by force fields, a neutron star is so compressed that these forces are overcome and the electrons and protons are forced together to form just neutrons. Curiously, the more massive the neutron star and the greater its gravitational field, the smaller its diameter.

In the same way that a star rotates faster as it shrinks from being a cloud of gas, so the neutron star speeds up yet further. It may now spin once every few seconds, or even many times a second. This seems incredible, and yet there is plenty of evidence for just this state of affairs. Radio astronomers have detected hundreds of radio sources that pulsate very regularly with periods of a second or so, with some that have shorter periods still, of the order of milliseconds. These are *pulsars*, spinning neutron stars with pairs of radio jets streaming

from their magnetic poles, which do not coincide with their rotation poles (see A–Z page 286). When these jets sweep across the line of sight of the Earth, radio astronomers detect a pulse. In this way they can measure the rate of spin of tiny objects that are perhaps thousands of light years away.

But a very big star may not even collapse to a neutron star following the supernova explosion. Its core may shrink so much that nothing in the Universe has enough energy to escape its huge gravitational field. A light beam that tries to get out will just be bent round by gravity. This is a stellar-mass black hole.

There are many myths about black holes. For one thing, they aren't black – or at least, their surroundings are not black, even though no light can get out of them. Material close to the black hole is accelerated so much by the intense gravitational field that it radiates strongly in X-rays and other short wavelengths. This is the key to identifying a stellar black hole – an intense source of X-rays that appears otherwise invisible.

Another myth is that black holes suck up everything in their neighborhood. While it is true that once you stray too close to a black hole there is no getting away from it, they have no more powerful effect on the nearby stars once they are a black hole than they did as ordinary stars. If our Sun were to turn into a black hole overnight, for instance, the Earth would continue to orbit it as if nothing had happened. The Sun's gravitational field would remain the same, but it would now be concentrated into a very tiny region of space.

▲ *A visual impression of a stellar-mass black hole,* to the right, in orbit with an aging star which is losing material. The material flows from the star on to the black hole where it heats up and emits X-rays, though the black hole itself is invisible at optical wavelengths.

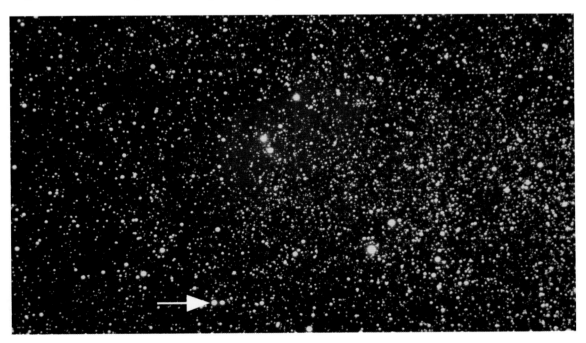

◀ *One of the best candidates* for a stellar-mass black hole is an invisible companion of the star HDE 226868, arrowed. The object, also known as Cygnus X-1, emits copious X-rays and its orbit suggests that it has a mass of between 6 and 15 solar masses. Only a black hole is thought to be able to meet these requirements.

OBSERVING STARS

Although the word "astronomer" means "star namer," amateur astronomers probably spend less time observing stars than any other type of object. Individual stars in particular are usually mere stepping stones to other, more interesting objects. There is little difference in appearance between one star and another other than its brightness. In terms of visual spectacle an ordinary star does not have the same appeal as a planet, say. However, there are some things to be learned from studying even individual stars. They can be used to test a telescope or the quality of the seeing (see page 39), for example, and the colors of stars tell you something about them.

To all intents and purposes stars are simply points of light. Though many of them are larger than the Sun, they appear as points of light because they are so far away. The distances to stars are usually given not in kilometers but light years (see page 30). Consider the Sun, our nearest star. Its light, traveling at 300,000 km/s, takes just eight minutes to reach the Earth from its distance of 150 million kilometers. If we were to travel away from the Sun at the speed of light, within a month it would appear so small that even a 250 mm telescope would no longer show it as a disk. To reach the nearest star, however, we would have to travel for over four *years* at the speed of light. This star, Proxima Centauri, is 4.23 light years away.

So if the Sun were as distant as even the nearest star it would be far too tiny for an amateur telescope to see as anything other than a point of light. The disks that you see on photographs of stars are caused by the spreading of light on the image-sensing medium used to take the picture, and have nothing to do with the actual diameters of the stars. Not even the Hubble Space Telescope has the power to show the nearest star as a disk. In fact, it can reveal the disks of only one or two of the very largest nearby stars, and these are rare exceptions.

View a star through a telescope and it does not actually appear point-like. There are two main reasons for this – the resolving power of the telescope (see page 40) and the seeing. A small telescope, which has limited resolving power, will show a star as a tiny disk, known as the *Airy disk*. The size of this disk is set by the wave properties of light. In the case of a 50 mm telescope, the Airy disk is easily visible with a magnification of 100. The image of an object that is not a point of light can be thought of as being composed of large numbers of overlapping Airy disks, one from each point on the object. The larger the telescope the smaller the Airy disk, so the image from a larger telescope is effectively

composed of finer dots. Unlike an image in a book, however, all the dots merge into each other rather than being seen separately.

The quality of the atmospheric seeing also affects the size of a star image. Unless the atmosphere is totally still, which is never the case, the effect of seeing is to enlarge the star image. This seeing disk, as it is called, is constantly on the move, and under extreme circumstances it can look like a sparkling writhing amoeba, with projections appearing and disappearing as you watch. With a telescope larger than about 150 mm aperture, the Airy disk is often smaller than the smearing effect of the seeing disk and is not easily visible. Typically, fairly good seeing produces disks about 1 arc second across. Two or 3 arc seconds is not unusual, while a quarter of an arc second is particularly good. Visual observers learn to make the most of the odd moment of better seeing.

The quality of the telescope can also affect the size and shape of a star image's disk. Triangular or other non-circular images are caused by poorly manufactured optics, or optics that are pinched in their mounting cells, causing them to be distorted. Sources of local heat, such as uneven cooling of the telescope components or even someone standing right below the line of sight, can produce distorted, moving images.

So observing a star image can tell you a great deal about the telescope and the atmospheric seeing conditions, though not very much about the star itself. The main details that you can find out about a star are its brightness and whether it has a companion star (in which case it is known as a double or binary star). Its color may reveal its star type; this is particularly the case with the brighter stars because their colors are easier to see.

Because stars have point-like images, it is possible to beat light pollution to a certain extent. Magnify a point of light and it remains a point of light of the same brightness, but the background will become darker with increased magnification because the light is spread over a larger area. So you can improve the contrast between a star and the background by using a higher power. In practice, the seeing can enlarge the star image as well, but the technique is useful when observing any stars or star clusters.

Double stars

Most double stars lose out in the popularity stakes to nebulae, clusters and galaxies. Through the telescope you just see two stars side by side, often with one much fainter than the other so that it is hard to spot, and you might be forgiven for feeling, after a while, that if

▼ **Increasingly poor atmospheric seeing** degrades a star image from a perfect Airy disk (left) to a larger and less well-defined blur. The rings seen around the Airy disk are very subtle and are known as diffraction rings.

 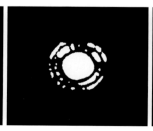   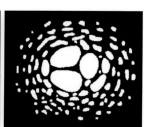

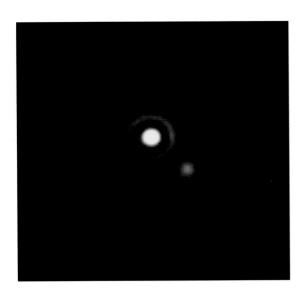

▲ *The star Izar, Epsilon Boötis,* is a double with a separation of 2.8 arc seconds. The main star is of type K and has a magnitude of 2.4. The A-type secondary star is at magnitude 4.9. Izar is a favorite among double-star observers because of its attractive contrasting colors.

you have seen one double star you have seen the lot. But there are exceptions – some doubles are famous for their contrasting colors, while others have several components, which makes them a more interesting sight. Tracking down double stars is a good occupation for a night when the seeing is good but the transparency (see page 39) is not perfect.

True double stars consist of a pair of stars that are physically close to each other – usually within a distance comparable with those of the planets round the Sun. They mutually orbit each other with periods that range from days up to thousands of years. Those stars that orbit within days are so close that they cannot be seen separately, and can only be observed using special techniques. There are also some stars that appear double, but where the two stars just happen to be in the same line of sight. These are known as *optical doubles.*

Although some amateurs make detailed measurements of double stars with a view to improving our scientific knowledge, most are content simply to view them from time to time. Observing a double star that is close to the resolving power of your telescope is a good way to confirm its quality, and it also tests your own observing ability at high magnification.

The important features of a double star to consider are the brightnesses of the two components and their separation. The easiest doubles are those where the two stars are of more or less equal brightness and have a large separation compared with the resolving power of your telescope. As the magnitude difference between the two increases, or the separation decreases, the doubles become more tricky. A difference of three or four magnitudes is enough to make a star with well-separated components hard to see except at high magnification. Separations are usually measured in arc

seconds. Another measurement that is usually given is the *position angle*. This refers to the angle in the sky of the line from the brighter to the fainter star, with north being 0° and east being 90°. The scale therefore goes counterclockwise. The brighter star is usually referred to as A, the secondary B, and subsequent components of a multiple star C and so on.

Variable stars

Some stars are variable in brightness, for one reason or another. In some cases – known as *eclipsing binaries* – the orbital plane of a double star happens to be edge-on to the Earth, so we see mutual eclipses of the stars. Algol in Perseus is the best known example of this type. More often, the variations result from a single star pulsating in brightness because it has reached an unstable stage toward the end of its life. Some of these variable stars are known as *Cepheids*, after the first star of this type to be identified, Delta Cephei. These stars vary in brightness over a matter of days or even weeks, with a characteristic slow fall and more rapid rise. They have the important property that the longer the period of variation, the brighter the star. So if a star is seen to vary in this way, its true brightness is known once its period has been measured, and hence its distance can be found from its apparent brightness.

There are many other types of variable star, some of which vary regularly, while others are unpredictable. The variations in the irregular types are often caused by pulsations in an old star, but others, known as *eruptive variables*, are subject to flickers or flares. Many eruptive variables are young stars, but flares are also a feature of some red dwarf stars. Another cause of variability is the rotation of a star with numerous spots or brighter regions.

Cataclysmic changes generally occur when double stars are so close together that material falls from one on to the surface of the other, causing an explosive outburst, just as when fat drops on to a barbecue. The most dramatic and best known of these stars are *novae*. A nova is a star that appears apparently out of nowhere, and in rare cases it may for a few weeks be one of the brightest stars in the sky. In fact a nova is a faint star whose close companion star has started to evolve into a red giant, resulting in a tremendous flare-up. Most novae are fairly faint, but occasionally one reaches naked-eye visibility.

▼ *The light variations of Delta Cephei,* which pulsates regularly with a period of 5 days 9 hours, are easily followed with the naked eye. A comparison chart is given on page 177.

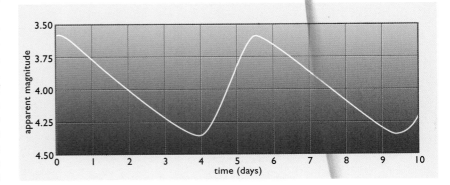

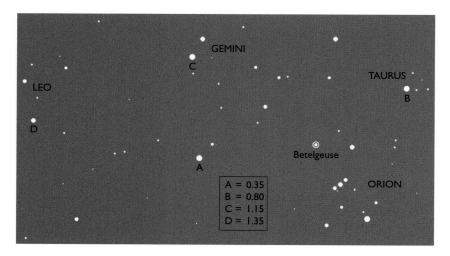

GEMINI

C

LEO

TAURUS

B

D

Betelgeuse

A

ORION

A = 0.35
B = 0.80
C = 1.15
D = 1.35

▲ *A comparison chart*
for estimating the magnitude
of the irregular variable star
Betelgeuse in Orion. Decide
whether the star is brighter or
fainter than the comparison
stars shown, and by how much.
Then use the comparison star
magnitudes to estimate the
magnitude of Betelgeuse.

Making variable-star estimates

The eye is quite good at deciding whether one star is brighter than another, and by how much. So it is possible to make reasonably accurate estimates of the brightness of a variable star by comparing it with nearby reference stars of known brightness. Ideally there would be a good range of stars of similar brightness within the same field of view as the variable star, making it easy to decide which one it is equal to. In practice, however, the comparison stars are usually more widely spaced in brightness, and may not be in the same field of view. Typically you have to estimate where in the brightness gap between the two comparisons the variable lies – say two-fifths or three-quarters of the way between them. With experience, observers can recognize a brightness difference of a fraction of a mag-

nitude, and can estimate the brightness of a star that is brighter or fainter than the comparisons.

The comparison stars used when making estimates of variables must be chosen with care, avoiding stars with strong colors or that may themselves be variable, and ensuring that the stars have accurately determined magnitudes. So it is best to make estimates using comparison charts published by organizations such as the American Association of Variable Star Observers or the British Astronomical Association's Variable Star Section. Mostly these charts are for stars that require a telescope or at least binoculars, but there are a few that can be estimated using the naked eye alone.

Star clusters

Stars are often born together and in some cases they may remain in a group for a considerable period of time. Star clusters are useful to professional astronomers because they contain stars that share a common origin and are clearly all at roughly the same distance. To amateur astronomers they are appealing because of their beauty. A field of view full of stars is a sight to behold, and the occasional presence of a strongly colored star or two adds to the effect. Many clusters are easy to spot even from light-polluted skies which blot out most diffuse nebulae and galaxies.

There are two general types of cluster – the loose or *open cluster*, and the *globular cluster*. Open clusters range from a few scattered stars to a group of several hundreds all within a fairly small region of sky. All the stars are kept in a compact group by their mutual gravitational forces, though over a period of tens of millions of years they may slowly spread apart and eventually separate. They are found within the spiral arms of our Galaxy, so they are mostly seen along the line of the Milky Way, and may contain quite young stars.

Globular clusters, however, consist of thousands or even hundreds of thousands of stars in a generally spherical body. Their distribution is also quite different – they surround the nucleus of the Galaxy but at a considerable distance from it. Most of the stars are old, and there is no trace of gas or dust within the cluster. Although many globular clusters happen to lie in the same line of sight as parts of the Milky Way, there are plenty outside the general plane of the Milky Way. Because they surround the nucleus, they are a particular feature of the evening skies of the middle of the year, when the center of the Galaxy is prominent. By contrast, there are virtually no globulars in December skies.

Some nearby open clusters are easily visible with the naked eye, the Pleiades and the Coma Berenices star cluster being good examples. Binoculars improve their appearance, but many telescopes have too small a field of view to do them justice. Binoculars bring many other clusters within range, such as M35 in Gemini or M25 in Sagittarius. Though these may be visible with the naked eye under good conditions, binoculars reveal individual stars and low powers on telescopes reveal a glittering array of stars. Yet other clusters such as the

▶ *The Pleiades cluster,*
M45, is the brightest and
best known star cluster,
and is visible to the naked
eye. It can be seen in all
conditions, and binoculars
show it particularly well.

▼ *A drawing of globular*
cluster M3 in Canes Venatici
made using a 150 mm
reflector with a magnification
of 150.

▲ *The globular cluster M92,* in the constellation of Hercules, is about 28,000 light years away. The cluster is about 100 light years across and contains several hundred thousand stars.

Wild Duck Cluster and the Jewel Box or Kappa Crucis Cluster are too small to be resolved into stars with binoculars, so telescopes with medium powers are really needed to show the cluster well.

With the naked eye or small instruments, many star clusters are indistinguishable from the other deep-sky objects, such as nebulae and galaxies, because the stars in them are too faint to be seen individually. Globular clusters in particular can also look like distant comets. In the 18th century, a French astronomer named Charles Messier listed many such objects that he encountered while searching for comets, so that he could eliminate them quickly from his searches. His listing remains in use, and many of the brighter objects are known by their Messier numbers, either as, for example, Messier 35 or just plain M35.

The brighter globular clusters are visible with binoculars (and in some cases to the naked eye) as hazy stars, but are best seen with telescopes. The brighter ones, such as Omega Centauri or M22, are resolved into stars with low powers and are a spectacular sight through a medium-sized telescope with a power of about 50 or 75. The fainter globulars require a similar or greater magnification to reveal them properly, and may be hard to pick up at all in light-polluted skies with small telescopes.

Photographs usually make the cores of globular clusters appear brilliant, with all the stars combining together. This is not the case through a telescope, however, and you can always pick out individual stars

► *The region around the Cone Nebula* in the constellation of Monoceros illustrates all three types of gaseous nebula. The pink nebulae are H II regions, while the dark intrusion known as the Cone, at bottom, is a dark nebula. The lack of stars at the bottom edge indicates that the dark nebula extends throughout this area and is hiding more distant stars. Within the cluster NGC 2264 at top are dust clouds shining as blue reflection nebulae; they are illuminated by nearby stars.

even at the center of the brightest globular clusters, though there may be a background glow of stars that are too faint to see. This is because photographs enlarge star images, whereas your eye sees them as points of light of increasing brightness. Few photographs do justice to the appearance of a great globular cluster in a good sky.

NEBULAE

The word *nebula* is Latin for "cloud." The ones referred to are not those that ruin our view of the sky, but those that appear as misty patches when seen through a telescope. The plural is *nebulae*, pronounced "neb-you-lee."

There are many small cloudy patches in the sky, and at one time all were referred to as nebulae. By the mid-20th century, however, their true physical nature became clear and now those that are actually galaxies are referred to as such. True nebulae are gas clouds of one type or another, though at the telescope it is hard to tell one type from another just by looking.

▶ *One of the brightest planetary nebulae,* NGC 3132 or the Eight-Burst Nebula. It lies in the southern constellation of Vela.

As we have seen, there are two basic types of nebula, one occurring at the beginning of a star's life, and the other at the end. Those at the beginning are known as diffuse nebulae, and they are clouds of gas and dust – the raw material from which stars form. They can be either bright (*H II regions* or *reflection nebulae*) or *dark nebulae*. Those at the end of a star's life are *planetary nebulae* or *supernova remnants*, and they are shells of gas thrown off at the end of a star's life. Some are small and still surround the dying star, while others may cover a wide area of sky and may be the remains of an exploded star.

Gas and dust nebulae

Most of the gas and dust in the Galaxy remains dark, and is only visible in silhouette against brighter objects, such as star clouds or bright nebulae. Few dark nebulae are visible with telescopes, though they become more obvious using photography, if only because certain regions have fewer stars than others because many background stars are hidden.

A star near a cloud of gas or dust may provide some illumination. The main way in which this occurs is a form of fluorescence – the gas emits light of its own when it receives a blast of ultraviolet light from a nearby hot star, often one that has recently formed from the gas cloud itself. Hydrogen glows pink, but in some cases oxygen and nitrogen may glow green. Nebulae of this sort are called H II regions – pronounced "H-two" – from the nature of the hydrogen molecule of which they mostly consist.

Strong gas flows from red giant stars distribute quantities of carbon particles – essentially soot – through the space between the stars. The carbon grains gain a coating of water ice – which is, of course, nothing more than a combination of hydrogen and oxygen. So gas, ice and soot are the major constituents of the material between the stars.

Starlight shining on this material illuminates it, and just as the light scattered by woodsmoke appears blue, so does the illuminated dust in space, though the color is not usually visible to the eye at the telescope. These nebulae are less common than H II regions, and are known as reflection nebulae.

Planetary nebulae

As already mentioned, planetary nebulae are shells that surround dying stars. These objects are much more compact than H II or reflection nebulae, and sometimes show a distinct blue-green color when seen through a telescope. Photographs taken with the Hubble Space Telescope, for example, show beautiful colored rings or tubes of gas. Usually the naked eye is not sensitive enough to see color in nebulae. However, though most planetary nebulae are fairly faint they are usually quite compact, and their surface brightness is quite high. When looking for color, it is the surface brightness rather than the overall brightness that counts, so in a medium-sized telescope, say 150 mm to 200 mm, on a good night, the color can be easily visible.

Supernova remnants

Only a few supernova remnants are observable with the eye at the telescope, the most spectacular being the Veil Nebula in Cygnus, though more can be photographed fairly easily using cameras mounted on telescopes. But there is also one virtually unique supernova remnant that can be observed using small telescopes, and that is the Crab Nebula (see page 187). This is a one-off, and there is no other object quite like it.

The light from nebulae

Many nebulae are quite faint, but fortunately there is one major difference between their light and that of other celestial bodies, such as stars. To see the difference, astronomers use the technique of *spectroscopy*. This involves splitting the light from the object into its component colors, using either a prism or a diffraction grating (a flat surface with very fine grooves). We are all familiar with CDs or reflective paper that is embossed with fine pits or lines; they split light in exactly the same way.

White light from the Sun, stars and glowing light sources such as light bulbs consists of a rainbow of all colors from deep violet (short wavelengths of light) to deep red (long wavelengths). This is known as *continuous emission*. A reddish star has more red than blue light, and a bluish star vice versa, but they both contain all the colors of the rainbow.

A nebula, however, usually emits only very specific colors of light, and this is known as *line emission*. Hydrogen, for example, emits a few individual colors or wavelengths of light. The brightest is deep red, with the next brightest being blue-green and the next being blue. These colors combine to make a hydrogen nebula appear pink. Our eyes are not very sensitive to red light, so visual observers tend to see only the blue-green and blue colors, but film in particular is more sensitive to the red, which is why it is possible to photograph some nebulae that are virtually undetectable to the eye. This is one of the few remaining advantages of using film for astrophotography.

Planetary nebulae and supernova remnants also have line emission, though often of different colors. Only reflection nebulae have continuous emission.

The importance of all this is that when we are trying to observe nebulae we are usually looking through a certain amount of light pollution. While some sources of light pollution consist of continuous emission, many have either line emission or a strong color bias. This makes it possible to manufacture filters that cut out some of the light pollution (known as light-pollution rejection or LPR filters) while allowing through the general bands of color that are emitted by nebulae. These are broad-band filters, but you can also get narrowband filters, which allow through only the light from the nebula. These are known by the type of gas that emits those colors. An O III filter, for example, allows through only the light from oxygen while blocking everything else.

Observing diffuse nebulae

Some H II regions are bright enough to be visible even in poor skies. The very brightest, such as the Orion Nebula and the Eta Carinae Nebula, are visible to the naked eye in a fairly dark sky and with binoculars even in a light-polluted sky. For the best views, however, the darker the conditions, the better.

There is a wide range of sizes and brightnesses of nebulae, so different approaches are needed depending on the object. While a bright and large nebula such as the Lagoon Nebula, M8, is easily visible in binoculars or a small telescope, other large nebulae, such as the North America Nebula or the Rosette Nebula, are large and faint. Though they may require eagle eyes and perfect skies to be seen visually, they can be photographed quite easily with time exposures using

ordinary cameras mounted piggyback on driven telescopes. Film and CCDs are more sensitive to the red light of hydrogen than is the eye, so they can pick out large nebulae from the light pollution, even from suburban skies. Most digital cameras, however, include an IR filter which reduces their red sensitivity.

The smaller and fainter nebulae require a telescope, usually at low power, and clear skies. This is an occasion where a specialized filter is worth trying on the eyepiece. Simple color filters are not generally of much use for nebulae, as their effect is too general – they cut down the light of the nebula as much as they do the sky background. But LPR filters or specialized narrowband filters such as an O III filter can help, though because they absorb light they are unsuitable for use with small telescopes.

Don't expect too much from LPR filters – they cannot magically transform a city sky into a country sky. But they can make gaseous nebulae more visible, even in what would normally be regarded as very dark skies. Rarely do you find a site where there is no background light at all. Narrowband filters will not cope with really bad light pollution, but they can have a magical effect where the object in question is on the borderline of visibility.

GALAXIES

The Universe is populated by a wide variety of galaxies. Their sheer numbers become obvious when you look at one of the Hubble Deep Field photographs, in which the Hubble Space Telescope was used to observe the same region of sky

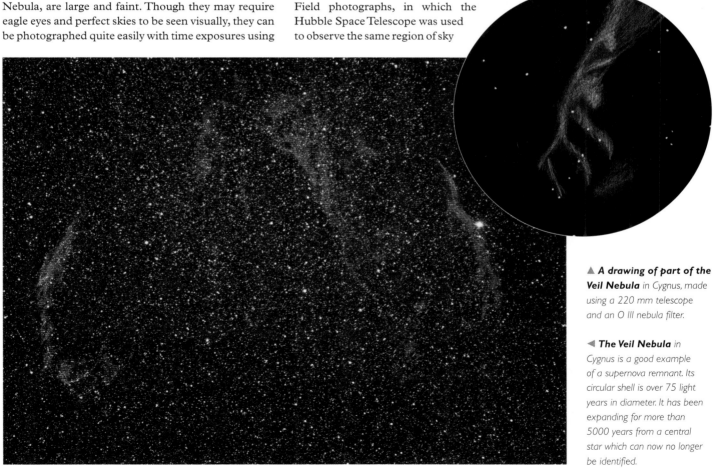

▲ **A drawing of part of the Veil Nebula** in Cygnus, made using a 220 mm telescope and an O III nebula filter.

◀ **The Veil Nebula** in Cygnus is a good example of a supernova remnant. Its circular shell is over 75 light years in diameter. It has been expanding for more than 5000 years from a central star which can now no longer be identified.

for days at a time, so as to detect the faintest objects possible. The result shows a region of sky which is virtually empty of stars, but full of galaxies at all different distances, and of a variety of types. Our imaginations are stretched beyond their limits just comprehending the numbers of galaxies, and when we consider that each of them probably has a similar number of stars to our own Galaxy, we realize that the Universe is vast, and must contain a enormous variety of possibilities. The tragedy is that as a naturally inquisitive species, which thrives on exploration, these galaxies must remain forever as just tantalizing lights in the sky.

Though it is easy enough to write a science fiction story in which the action takes place "In a

◀ *The brightest elliptical galaxy, M32,* is a companion to the Andromeda Galaxy, M31. Though only a dwarf elliptical galaxy, its closeness makes it easy to see.

galaxy far, far away…," we know that in reality it will always remain impossible to travel among the galaxies. The theory of relativity tells us that nothing can accelerate to a speed faster than light – to do so would require an infinite amount of energy. Fair enough, we say, we will invent something that warps space, as in *Star Trek*, without knowing how to do it in practice, so that we can travel at ten times the speed of light, or maybe a hundred or a million times the speed of light. That would enable us to reach the Andromeda Galaxy, our nearest large neighbor, in a mere 2½ years. But to get to the Virgo Cluster at a million times the speed of light would take 50 years, and by the time the offspring of the original voyagers returned home (assuming that they could avoid the time paradoxes that result from traveling at such a speed) their descendants would be living in a very different world. And those tiny dots on the Hubble Deep Field would take thousands of years to reach.

At this point we have to turn to parapsychology and "remote viewing" – and for a species whose members find it hard enough to locate even their closest companions within a shopping mall, let alone over galactic distances, that seems unlikely.

Though there are great clusters of galaxies containing maybe 1000 members, and superclusters with maybe 10,000 individual galaxies, galaxies are scattered throughout the Universe. Surveys that plot the positions of millions of galaxies show that they tend to be distributed in a sort of honeycomb structure, with strands or walls of galaxies or galaxy clusters, and voids between them.

Types of galaxy

Although all galaxies are great collections of stars, they fall into various categories as a result of their appearance. The main types are *elliptical*, *spiral* and *irregular*, but there are many further subdivisions such as *lenticular* and *ring* galaxies. Until fairly recently it seemed that irregular galaxies were the most common, but in recent years, as telescopic power has increased, it is now believed that faint dwarf elliptical galaxies consisting of a fairly small number of stars are the most abundant. There may also be lone stars in the space between the galaxies, but their numbers are probably not significant.

Elliptical galaxies include the largest of all galaxies, though they come in all sizes. They are, as the name implies, elliptical in shape, and generally have a very smooth gradation of brightness from the center to the edge with no internal structure visible in amateur telescopes, except sometimes an almost starlike nucleus. Some appear virtually spherical, while others are elongated. It is difficult to tell the true shape of an elliptical galaxy just by looking at it. If it appears in the telescope as a sphere, it could be a genuinely spherical galaxy, or it could be a flattened sphere (like a tangerine) seen end-on, or it could even have three axes of different lengths, such as a flattened American football or rugby ball, seen at such an angle that it appears spherical. Other elliptical galaxies, however, do appear elongated.

Elliptical galaxies contain virtually no gas or dust, and they appear yellowish or even reddish in photographs because many of their stars are old rather than being young. Blue stars always burn themselves out quickly,

▶ *A typical spiral galaxy, NGC 4414.* The central regions are tinged yellow because they contain older stars, while the spiral arms contain many massive blue stars. The spiral arms are not solid structures – stars constantly orbit through them. They are thought to be regions where matter is temporarily denser, rather like the sudden concentrations seen in dense flocks of flying birds.

but they spew gas and dust into the space surrounding them as they become supernovae. However, there comes a point where all that is left are large numbers of stars nearing the end of their lifetimes, which tend to be red. The very largest elliptical galaxies are found at the heart of large galaxy clusters, the giant galaxy M87 being a good example.

Spiral galaxies are undoubtedly the most popular among observers, as they have the classic and beautiful spiral arms. The nucleus of a spiral galaxy is virtually indistinguishable from an elliptical galaxy and has the same yellowish color. However, there is a flattened disk through its center which displays some sort of spiral structure. The disk contains gas and dust, and has bright young blue stars as well as pink H II regions when photographed in detail. Some galaxies have a nucleus that is bar-shaped rather than spherical, in which case they are known as *barred spiral galaxies*. Lenticular galaxies are elliptical, but they have a disk that appears to have no obvious spiral structure.

Although photographs of spiral galaxies may appear to show individual stars, we only ever see the general mass of stars rather than individual ones. Occasionally there may be a massive cluster of stars that looks like a single star, but in fact even the Hubble Space Telescope can only just distinguish individual stars in most galaxies. The only exception is when a star in a galaxy explodes in a supernova explosion, in which case it may briefly appear as a star within or near to the galaxy. Some amateur astronomers spend many hours searching for these events, usually using CCD cameras on

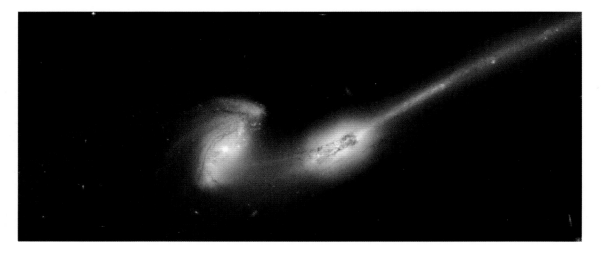

▲ *The irregular galaxy NGC 6822* is a member of the Local Group of galaxies. The numerous pink gas clouds and blue stars are evidence of the rapid rate of star formation typical of an irregular galaxy. Like other galaxies we are viewing it through a field of foreground stars in our own Galaxy, which cover the entire area. This galaxy is in Sagittarius, which is rich in star fields.

▶ *A pair of interacting galaxies* known as "The Mice" because of their long tails of stars drawn out following a close encounter. Notice how the tails are blue, as the interaction has triggered off intense star formation in the tails.

automated telescopes, as supernovae are of great interest to professional astronomers. Bear in mind that most of the stars seen in the field of view of a galaxy are much closer ordinary stars within our own Galaxy.

Irregular galaxies tend not to be as large as ellipticals or spirals, but they do contain a vast amount of gas and dust and may have giant star-formation regions. The two Magellanic Clouds which lie close to our own Galaxy are often quoted as good examples of irregular galaxies, though they do show some of the characteristics of spirals. Farther afield, however, we see much more irregular shapes.

There is a Local Group of galaxies of which our own Milky Way is a major member. The Magellanic Clouds are companions to our own Galaxy, and they are in orbit around it. Radio observations reveal a long stream of hydrogen with the Magellanic Clouds at one end, and presumably drawn off it by gravitational interactions with our own Galaxy. At some time in the future the Magellanic Clouds may merge with our own Galaxy.

Other members of the Local Group include the Andromeda Galaxy M31, the Triangulum or Pinwheel Galaxy M33, and the dwarf elliptical companions of the Andromeda Galaxy (see table in the A–Z section, page 265). Many galaxies exist in similar small groups of 30 to 50 members. Our own is on the fringes of the Virgo Cluster of galaxies, and there are other nearby groups, such as the one that contains M81 and M82 in Ursa Major (see page 139).

Gravitational interactions, also known as tidal interactions, probably play a major role in shaping galaxies. Some galaxies show much more obvious evidence than the Magellanic Stream, with long tails dragged off them. Others show what happens when one galaxy collides with another – a relatively common occurrence. The galaxy NGC 5128, also known as the radio source Centaurus A, is a good example (see page 157).

It seems that gravitational interactions and mergers are responsible for the different types of galaxies we see. When a galaxy collides with another, the gas and dust are the main casualties. Individual stars are actually small targets and there are few actual collisions, though there will be many close encounters and changes of orbit. However, the shock waves caused by the impact of gas clouds trigger off a wave of intense star formation. Perhaps the impact of one galaxy with another results in the formation of spiral structure in galaxies.

Some galaxies grow very large at the expense of the smaller ones that merge with them, and the rapid star evolution that takes place eventually uses up all the available gas. We are left with a huge mass of aging stars with no gas and dust – in other words, a giant elliptical galaxy.

Some large galaxies have nuclei which are strong sources of radio waves or visible light. These are known as *active galactic nuclei*, and seem to harbor giant black holes. Instead of having formed from a single collapsed massive star, these black holes have the mass of millions of stars. The intense radiation is caused by material spiraling around the black hole. There may be jets of energy coming from these nuclei. Sometimes these jets are visible as lobes of radio emission on either side of the galaxy, and sometimes they may be visible as bright points of light, depending on the orientation of the jet as seen from Earth. This is now thought to be the explanation of the quasars, which are brilliant starlike points seen far away in the Universe.

It seems that many large galaxies, including our own, have supermassive black holes at their centers, though in the case of our own Galaxy the object is quiescent and is simply a moderate source of radio waves compared with those at the hearts of some other galaxies. Which came first – the black hole or the galaxy? Did galaxies form around black holes, or did black holes form at the centers of large galaxies? This is the sort of question which is currently a hot topic for research among astronomers.

Dark matter and dark energy

Another question to which there is currently no good answer is that of the nature of dark matter. When astronomers analyze the motions of stars within galaxies, it becomes clear that there is far more material in them than literally meets the eye. If there were only the amounts of stars, gas, dust, black holes, dead planets, particles and so on that we know about or can surmise, galaxies would simply fly apart. Something is holding them all together, and holding the groups and clusters of galaxies together. In the absence of anything obvious, this is referred to as *dark matter* or missing matter. It is not a trivial amount, either – we can only account for about 4% of the matter in the Universe.

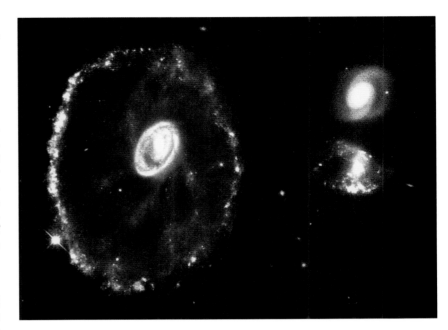

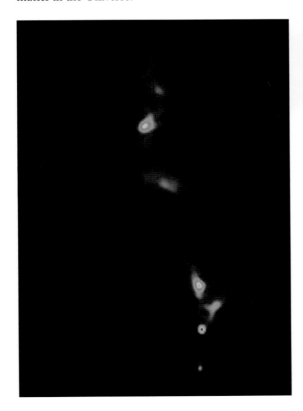

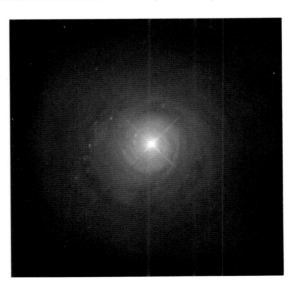

▲ *A ring galaxy,* known as the Cartwheel Galaxy, caused by the passage of another galaxy through its center. Probably one of the two galaxies at right was the culprit. It has caused a ring of star formation to spread outward through the galaxy.

◄ *The quasar 3C48,* one of the nearest, is close enough for the Hubble Space Telescope to photograph the surrounding galaxy. When first discovered, these radio sources appeared to be lone brilliant objects, as the intensity of the nucleus over-powered the light from the core.

▶ *NGC 5548* is a Seyfert galaxy, distinguished by its particularly bright nucleus. It is typical of a galaxy with an active galactic nucelus, and is believed to have a super-massive black hole at its center.

◄ *A close-up radio view of 3C48* reveals a giant jet from the quasar, which is the faint blob at the bottom end of the jet. The picture was made using the Very Long Baseline Interferometer, a worldwide network of radio telescopes that provides very detailed observations of radio sources.

In recent years, the search has widened as a result of observations of distant supernovae, which are accelerating faster than the big bang theory would suggest. Something known as *dark energy* has been put forward to account for this expansion. As matter and energy are equivalent, according to Einstein's theory of relativity, this dark energy could in itself account for a large proportion of the dark matter.

OBSERVING GALAXIES

Apart from the Andromeda Galaxy and the two Magellanic Clouds, which are visible with the naked eye, most galaxies are rather small and faint. The brighter ones are visible with binoculars, but most require a telescope and clear, dark conditions. As with all deep-sky objects, the larger the aperture and the darker your skies, the better the view you will get. But under the right conditions even a 110 mm telescope will reveal hundreds of galaxies.

Being quite small, most galaxies take some finding unless you have a Go To telescope. It helps if there are some bright stars fairly near to aid star-hopping, but in the case of the Virgo Cluster, which contains a large number of galaxies, there are few stars in the area. The galaxies might appear in the telescope, but it is hard to tell which you are looking at. The trick is to choose an easily repeatable route into the cluster, as described on page 149, and work from there.

Once you have found a galaxy, notice its shape and any stars nearby. Spiral galaxies are often not obvious even though their photographs show spectacular arms. With a small telescope, frequently all you can see is the nucleus of the spiral, which looks like an elliptical galaxy. Only in good conditions and with practice can you see the spiral arms, much fainter and well outside the nucleus. The Andromeda Galaxy, M31, is a case in point. It is visible with the naked eye by averted vision in average skies, and with binoculars even from city centers. But what you usually see is the nucleus only. In dark skies the galaxy extends way beyond this, and structure becomes visible, notably the sharp edge to the nearer spiral arm, which is caused by a line of dust clouds. Some galaxies are spindle-shaped, and edge-on spirals in particular can be almost needle-like in their appearance.

Look for subtle variations in the brightness of the galaxy, and give your eye time to get used to the field of view. Although in theory the eye cannot build up light over time like a CCD, some observers claim that they can see more the longer they keep looking at an object.

Narrowband filters suitable for observing nebulae will not be of any use with galaxies, as the light from a galaxy is essentially white light rather than line emission from a nebula. Light-pollution filters, however, can be of value, though the effectiveness of a particular brand of filter will depend on your local conditions and the particular sources of light pollution. City centers have a predominance of white light, with essentially the same color distribution as the galaxy, so using a filter will simply cut down the light from both sources. But in suburban areas with low-pressure sodium (orange) or mercury (blue-white) lighting, an LPR filter could help.

The magnitude of a galaxy quoted in catalogs is not a very good guide to its visibility, as this value gives

▲ **A drawing of the spiral galaxy NGC 253** *in Sculptor, made using a 500 mm Dobsonian reflector.*

▶ **M109 in Ursa Major** *is a good example of a barred spiral galaxy.*

*▲ **An amateur photograph** of the Andromeda Galaxy, M31, from Wales, UK. It was taken with a CCD camera through a 200 mm reflecting telescope with a CCD autoguider attached to a separate telescope. Exposures were made through red, green and blue filters with a total exposure time of 2 hours 30 minutes.*

the total light output from a considerable area. The appearance is quite different from that of a star of the same brightness, whose light is concentrated into a point. Also, face-on spirals are more difficult to observe than edge-on spirals. This is because the light from the latter is contained within a narrow needle shape rather than spread over a larger area, as in the case of a face-on spiral.

OBSERVING DEEP-SKY OBJECTS

Observing deep-sky objects, whether clusters, nebulae or galaxies, requires quite different techniques from those used for objects within the Solar System. Deep-sky objects are often quite faint and harder to find. Light pollution is frequently a major problem. While urban observers may struggle to see a comparatively large and bright object such as the Crab Nebula, those in dark areas may face similar problems trying to find much fainter objects. Just finding the object is a challenge in itself. If you can't see it, is the object too faint for your telescope, or is it too small or too large for the magnification you are using, or is it hidden in the light pollution – or are you looking in the wrong place to start with? Owners of the smaller Go To telescopes need not feel smug, because although they may have been pointed in the right direction, all the other questions still apply.

OBSERVING TIPS

All deep-sky objects by their nature are best viewed in a very dark sky with a complete absence of extraneous light and with a telescope in perfect condition. These are criteria that rarely fully apply, but you can do your best to make the most of what you have.

Make sure that your telescope is up to the job. Clean optics make a lot of difference, but it is better to put up with a little dust than to risk scratching a delicate surface. Keep covers on optics until they are needed and try to avoid dewing up. If dew starts to form on a lens or corrector plate, cover it straight away or blow warm air on it gently to dry it off. A portable hair dryer is a useful accessory if you have a 12-volt power supply handy. Dust falling on dewed-up optics is likely to stick.

Discussion of the cleaning of optics is beyond the scope of this book; suffice to say that it should only be carried out with care and when strictly necessary. Eyepieces as well as objectives and mirrors should be kept dry and dust-free.

Any dust on the optics will reduce the contrast between the object you are looking for and the sky background. Other causes of low contrast are poor baffling and reflections within the telescope tube. These are not easily altered with commercial instruments, but you can at least make sure that a refractor or SCT has a good dew cap, which should be matt black on the inside. In the case of a reflector, any worn black paint or shiny parts at the top of the tube should be attended to. But most importantly, try not to observe with extraneous light shining on the telescope or into your eyes. If possible move the telescope to a darker spot or set up a screen to block out the light – even a blanket on a clothesline or draped over a ladder will help.

Choose the correct eyepiece for the object you are trying to observe. Our observing notes in the constellation entries in Chapter 7 give suggestions for the minimum power that gives a good view, but in general for diffuse nebulae you will need a low power, for galaxies a medium power and for planetary nebulae a high power. Star clusters can be either low- or high-power objects depending on their size. The smaller star clusters and globular clusters generally require a fairly high power, though they can usually be found with a lower power. Make sure you are comfortable when observing and that the eyepiece is at a convenient position.

Wait until the objects are as high in the sky as possible, and until conditions are as clear and dark as possible. Cold air is usually clearer than moist, warm air. Observing after midnight may help as some lighting does tend to be switched off then. Wait until the sky is properly dark – that is, the Sun is at least 12 degrees or, better still, 18 degrees below the horizon. In summer at high latitudes, such as in northern Europe, it never gets fully dark. Wait until the Moon is out of the way.

Avoid looking at any bright lights or TV or computer screens for up to an hour before observing, and use only a dim red light for looking at your star map. The maps in Chapter 8 are designed to be used with such a light. Practise using averted vision (see page 32). Tapping the telescope can help to reveal a tiny faint object – the eye is good at detecting movement.

FINDING DEEP-SKY OBJECTS

If you have an ordinary telescope with a finder, get used to the inverted image given by the finder and its field of view. The 5 × 24 finders on cheap telescopes are virtually useless for finding deep-sky objects, and even the better quality 6 × 30 finders are of limited value

▶ **The galaxies M31 and M32** *as drawn using a 76 mm refractor. The larger object is M31. Compare its extent with the photograph on page 119, in which M32 is the small circular fuzzy object below the nucleus of M31. The scale is in arc minutes.*

where there is light pollution. Non-magnifying finders or sighting devices (see page 36) can be useful when you know roughly where the object you are searching for is to be found.

It is often easy to locate an object in binoculars yet hard to find the same object in the finder. This is because the finder generally inverts the view and has worse optical performance. The key to finding deep-sky objects is either to have a good finder, or to know the performance of the finder you have so that you can allow for its deficiencies. Some observers draw a circle on a clear plastic overlay for their star map that shows the field of view of their finder, which can help when trying to recognize star patterns.

The most popular deep-sky objects are usually those that are easy to find, either because they are bright or because they are conveniently located near to bright stars. Some comparatively bright objects can be overlooked simply because they are more tricky to locate. So the key to finding deep-sky objects is to refine your finding skills. Go To telescopes can help greatly, but only if they are accurately set up in the first place. But without Go To, you need to develop your star-hopping skills.

Star-hopping involves starting from a star that you can find easily and jumping from there, one field of view at a time, to the object you want to find. If your finder shows few stars, you may have to use your lowest power and move the telescope one field of view at a time from a known star to the chosen object. An alternative method, if you have an equatorial mount reasonably accurately aligned, is to locate a star with the same or similar declination as the object you want, then sweep the telescope in right ascension only to find the object. If there is a star with the same right ascension as your chosen object, you can do the same but this time sweeping only in declination. Even if you have an altazimuth mount, this procedure will work when the object is close to the meridian.

All these methods require persistence and practice, but eventually you will find that you can pick out many deep-sky objects quickly while owners of Go To telescopes are still fumbling with their key pads or changing their batteries. Nevertheless, there are times when all Dobsonian owners wish that they could have a bit of technical help in finding a particularly elusive object, so there is something to be said for both systems. Users of Dobsonians and other telescopes on altazimuth mounts must take into account that the axes of their telescopes are not usually aligned with the RA and dec. grid lines, so the orientation of the sky probably differs somewhat from the star map.

Drawing deep-sky objects

Many observers like to make a drawing of what they see at the eyepiece as a simple and quick record of the object. A clipboard with a red light attached is a useful accessory, and most observers favor a selection of pencils of different densities. Sketches are invariably made in negative form, with dark shading and black

◀ **Reflection nebula NGC 7023,** *the Iris Nebula, in Cepheus, photographed from southern France using a CCD and 155 mm refractor.*

stars, for simplicity, though it is easy enough these days to scan the drawing into a computer and invert the tones so as to get a more realistic result. It is usually necessary to emphasize the contrast of the view, because many objects are barely visible against the background. Making a drawing teaches you to look carefully at the object, and many people return to drawings they made with previous instruments, with younger eyes or in darker skies, to see how their perceptions have changed.

Star clusters are tricky to reproduce accurately, and many people settle for putting in the main stars and then stippling in the remainder, taking care to preserve the general appearance and shape of the object.

As with all astronomical observations, it is essential to make a note of the date, time, instrument, magnification and seeing conditions.

Photographing deep-sky objects

Although simple methods, using compact digital cameras and webcams, can give very good photographs of objects in the Solar System, most deep-sky objects require a different level of complexity. Time exposures, often minutes or even hours in total duration, are the norm. Basic webcams can easily pick up the brighter stars, and can give very good images of double stars, but that is about their limit without modification.

Successful deep-sky photography requires a sensitive system and long exposures, which means using a well-aligned equatorial mount with a precise guiding system. Digital cameras, particularly digital SLR cameras which allow you to use long telephoto lenses,

can be mounted piggyback on a driven telescope to obtain nice images of the larger and brighter deep-sky objects. Photographing planetary nebulae and galaxies, however, is mostly the preserve of the cooled CCD (see page 39) on a medium- to large-aperture telescope; it must be either manually guided or controlled using an autoguiding system with a CCD to correct the telescope's drive rate.

Deep-sky nomenclature

Most of the bright deep-sky objects are included in Messier's catalog of 110 objects, which have M numbers. The majority of other deep-sky objects observable with small- and medium-sized telescopes have either NGC or IC numbers. These refer to the *New General Catalogue* published by J. L. E. Dreyer in 1888, and to its *Index Catalogue* extension. Most of the Messier objects also have NGC numbers. While M numbers are assigned more or less at random throughout the sky, they do not appear south of declination −35°, which was the practical southern limit of Messier's observations. The NGC numbers are assigned in approximate order of right ascension, and include a wide range of objects, bright and faint, throughout the sky, so adjacent numbers may refer to objects widely separated in location and type.

The other major catalog is a compilation by British astronomer Sir Patrick Moore of 109 bright deep-sky objects that are easily observable with small- and medium-sized telescopes. It is known as the *Caldwell Catalogue*. The Go To lists of Meade telescopes in particular use Caldwell numbers.

7

On the following pages is a guide to the appearance of the sky for two months at a time, with more detailed information on the 50 most important constellations. The four maps for each pair of months provide a quick reference to the stars in your sky at any time. First choose the map that best matches the location you will be observing from, either northern or southern hemisphere. The exact view depends on your latitude, so notice the lines at the bottom of each map, which indicate the position of the horizon for various latitudes. The farther south you are, the more you see of the southern part of the sky and the less you see of the northern.

Individual horizons are shown for a range of latitudes in each hemisphere. However, the maps are still usable beyond those ranges, but some stars in the lower halves of the maps will be higher or lower in the sky than shown.

The maps for each hemisphere are in pairs, showing the view looking south and the view looking north on a particular date. Because of the movements of the sky, each pair of maps shows the sky at different times depending on whether you are viewing in the evening or the early morning, so choose the date and time closest to when you are observing. Bear in mind that these times do not allow for Daylight Saving Time (Summer Time), so the maps show the sky for an hour later than the times given when this is in force. So if you are observing at 10.30 pm in summer, use the map for 9.30 pm.

Each map shows the horizon along the bottom edge and the overhead point at the top. The edges of each semicircle are due east and west, which means that the semicircles join each other. Any constellation that crosses the edge will be split between the two halves of the map. Each map covers a wide area of sky, and a constellation that appears small on the map will cover quite a large area in reality.

Once you have chosen the best map for your observing date and time, note the page numbers at the edge of each chart to find out which of the main star maps in Chapter 8 will give you a larger-scale view of the same area. Remember that the planets move from month to month so they are not shown on the maps. A bright planet could change the pattern of the constellations, but you can check the positions of the planets for each month using the tables on www.stargazing.org.uk.

If you are observing from the southern hemisphere, bear in mind that your view of star maps 3 to 6, of the equatorial regions of the sky, will be with north at the bottom, so they are upside down from your point of view.

In the monthly guides, not every constellation is included because some have few objects of interest within them. In each case there is a range of objects to look for, mostly with small to medium telescopes, with illustrations for each one. Some of these illustrations were made with amateur telescopes both large and small, while others are taken through some of the largest telescopes in the world, to give you some idea what the faint misty object you are seeing looks like in depth. The descriptions are intended for observers under average modern conditions.

The descriptions of objects often refer to telescopes as being small, medium or large, and in this book the terms refer mostly to instruments used by amateurs. We have adopted a general classification in which small means telescopes from 50 to 100 mm aperture, medium means 110 to 200 mm, while large means any telescope bigger than 210 mm – usually up to about 450 mm in practical terms. The terms low, medium and high magnification are also used, but they are harder to define precisely as they depend on the instrument being used. For many catadioptric telescopes at f/15, the lowest power usually available is 75× to 100×, which for a small refractor would be medium to high power. But in general, low power means 25× to 75×, medium means 100× to 200×, and high means greater than 200×.

The tables of information about the objects described give a visual magnitude (M_V), magnification and distance for each object. Each of these figures is only a guide. The magnitude and magnification figures are intended to give some idea of what you are looking for, the magnification quoted being a suggestion for getting a good view of the object, rather than the minimum that will show it. Distances are also unreliable – different sources often give widely differing values, so don't be surprised if you see other values quoted elsewhere. Establishing a good distance scale is one of the most challenging aspects of modern astronomy.

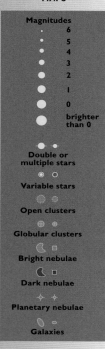

◄ **Comparing the hemisphere map** *with the sky. Chart 1 from page 123 is shown here with a photograph of the same region of sky taken from North Carolina (latitude 35°N) and shows Orion, Canis Major, Canis Minor and surrounding constellations.*

JANUARY–FEBRUARY

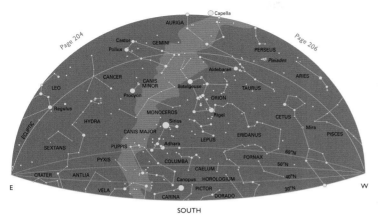

Page 204 · Page 206

Page 196 · Page 198

The northern-hemisphere sky

Orion dominates the view at this time of year. Its bright belt of three stars is a brilliant pointer to other constellations. To the upper right they point toward the bright star Aldebaran in Taurus, and farther on to the Pleiades star cluster. To the lower left they indicate Sirius, the brightest star in the sky, in Canis Major. Above Orion, roughly overhead, is the pentagon of Auriga, while to the east of this is a pair of stars known as Gemini, the Twins. Directly below these lies the star Procyon in Canis Minor.

A line from Gemini through Auriga takes you to Perseus, which lies in the Milky Way. Farther on from this is the W-shape of Cassiopeia. Setting in the west are the stars of Andromeda and, low down, the Square of Pegasus. Notice that the ecliptic is high in the sky at this time of year, so there may be bright planets in Gemini, above Orion or in Taurus.

Evening
1 January at 11.30 pm
15 January at 10.30 pm
30 January at 9.30 pm

Morning
1 October at 5.30 am
15 October at 4.30 am
30 October at 3.30 am

The southern-hemisphere sky

Although the Milky Way stretches overhead, this is its faintest part and it is not easy to see. But the brilliant stars of Orion act as a great signpost. The three belt stars point downward and to the left to the star Aldebaran in Taurus, and continuing the line, to the unmistakable star cluster of the Pleiades. Take the line from the belt in the opposite direction and you come to the brightest star in the sky, Sirius, in Canis Major, overhead. Below Orion lies the pentagon of Auriga with its brightest star Capella at its base.

At right angles to the line of the belt, down to the right, lie the two bright stars of Gemini, the Twins. Farther on, rising in the east, are the stars of Leo. The ecliptic runs from Leo, through the faint stars of Cancer, then through Gemini and below Orion to Taurus, so the patterns in this part of the sky may be distorted by any bright planets that happen to be around.

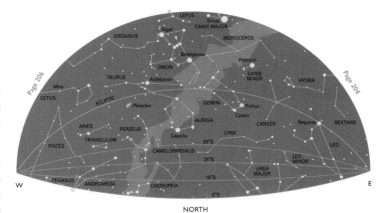

Page 206 · Page 204

Evening
1 January at 11.30 pm
15 January at 10.30 pm
30 January at 9.30 pm

Morning
1 October at 5.30 am
15 October at 4.30 am
30 October at 3.30 am

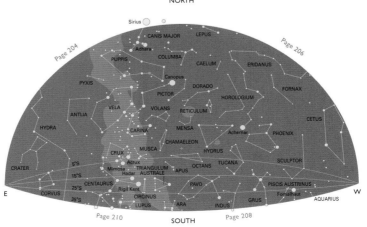

Page 204 · Page 206 · Page 210 · Page 208

ORION

There is no constellation more brilliant than Orion. When it is in the sky, it burns through any light pollution and haze, and being slap-bang on the celestial equator it is visible the world over. In the northern hemisphere its stars are symbolic of the winter sky; in the southern hemisphere they mean hot summer nights.

The three stars of Orion's belt, equally spaced in an almost straight line, have no counterpart anywhere in the sky, and provide an unmistakable signpost. The area is also the closest stellar nursery to Earth, and within its borders are good examples of stars at virtually every stage in their lives. Most of the bright stars we see in Orion share a common origin in the starbirth region, though they have subsequently spread over a wider area of space.

▼ **Orion is on the celestial equator** and is visible the world over in January and February – here, seen from Arizona over a saguaro cactus.

The figure we call Orion is a hunter, who lives in Greek legends. Before that he was a giant in the folk tales of other cultures. He faces a raging bull, Taurus, and holds up a shield, marked by the line of stars Pi¹ to Pi⁶ Orionis, made from the skin of a lion that he has slain. From his belt hangs a glowing sword which includes the Orion Nebula. At his heels are two dogs, the constellations Canis Major and Canis Minor, while below him in the sky cowers a hare, the constellation Lepus.

In a good dark sky the whole area seems to glow, and long-exposure photographs show that it is indeed a mass of faint nebulosity. To its east is a great glowing arc, known as Barnard's Loop. This is not readily visible to the eye, but is easily photographed using a red-sensitive CCD. It is probably the remnant of a supernova, a star that has exploded at the end of its life. This may seem odd if Orion is a stellar nursery, but the most massive stars, the sort that end up as supernovae, go through their life cycles in a matter of maybe 10 million years – so they are exploding while other stars in the region are still being born.

Alpha Orionis (Betelgeuse)

Although Betelgeuse (pronounced "Bet-el-jooz" rather than "Beetlejuice") is the Alpha star of Orion, it is usually noticeably fainter than Beta Orionis, Rigel. However, it is a variable red supergiant, and can vary between magnitudes 0.2 and 1.3 or maybe even fainter. This is a wider brightness range than any other bright star, and it is worth making an estimate using the chart on page 110 whenever you observe.

Betelgeuse's red color is very obvious in comparison with the other main stars in Orion, which are mostly bluish. It is considerably closer, at 500 light years compared with the estimated 1500 light years of the main Orion starbirth region, and there are suggestions that it may not be part of the Orion complex. It will someday become a brilliant supernova in its own right, but this is not likely for millions of years.

Sigma Orionis

This is a delightful multiple star. Even a small telescope will immediately show three stars, two being of sixth magnitude alongside the main third-magnitude star. If you look more closely and with a higher magnification you should see that on the opposite side of the main star is a closer ninth-magnitude companion. The main star itself is also a close double, though this is a difficult object. Other nearby stars add to the spectacle.

M42 and M43, Orion Nebula

What we see as the Orion Nebula is really only the tip of the iceberg. The whole of the Orion region is occupied by a great dark cloud of hydrogen and molecules that hides more distant regions of the Galaxy. Only the region where stars are actually forming and illuminating the cloud is visible. Radio and infrared measurements show that there are many other stars within the cloud that are hidden from view.

The Orion Nebula rivals the Eta Carinae Nebula as the brightest gaseous nebula in the sky, and it is bright enough to be visible with the naked eye as a misty star in a good sky, and with binoculars in a light-polluted area. This is a favorite target for all astronomers, as it offers not only a visual spectacle and a suitable subject for sketches, but also an unbeatable area for photography. Although photographs on film show it as red, digital cameras and the eye are not so red-sensitive and they tend to show it as whitish. Some people see the central regions as greenish, because there is also a strong green component to the light.

The region is wreathed with nebulosity, and the outer wings of the nebula contain delicate details visible even in very small telescopes. A dark central bay in the nebula is popularly called the Fish's Mouth from its shape.

At the end of the Fish's Mouth lies the multiple star Theta[1] Orionis. This consists of stars that have been born from the nebula within the past million years or so. Although you are unlikely to spot a new star popping into view, changes do take take place in this area from time to time. The four main stars of Theta are known as the Trapezium from their shape, but there are at least two other tenth-magnitude stars in the pattern visible with medium-sized telescopes using high magnification. The Trapezium is simply the brightest part of a cluster of maybe 1000 stars, many of which are visible only in infrared, which is not absorbed as much by the gas and dust in the area as is the visible light.

To the north of the Fish's Mouth lies M43, which would be a popular target for observation in its own right if it were not overshadowed by the larger and brighter M42. Farther north still, surrounding a group of stars of which 42 Orionis is the brightest, is the fainter reflection nebula NGC 1977. Go yet farther north and you find a small cluster of stars known as NGC 1981.

Horsehead Nebula and Flame Nebula

The Horsehead Nebula is undoubtedly one of the top ten illustrations in astronomical books, but as a visual object it is elusive. Those who live under pitch black skies might disagree, but for most people it is lost in the murk. This does not prevent them from looking, and

▲ **Sigma Orionis** is one of the most celebrated double stars in the sky, with a number of blue stars in the same field.

◄ **The Orion Nebula,** M42, with M43 just above the main nebula and NGC 1977 at the top.

Object	Type	M$_v$	Magnification	Distance
M42/M43	Nebula	4.0/9.0	7×	1350 light years
Horsehead	Dark nebula	–	100×	1500 light years
Flame	Nebula	–	Photographic	1350 light years
Trapezium	Multiple star	–	100×	1600 light years

THE ORION STAR NURSERY

Although we can see signs of the star-forming regions of Orion with the naked eye and binoculars, to really see what is going on requires special techniques. One method is to take pictures using CCDs which are particularly sensitive to the light of hydrogen gas, which glows strongly at a wavelength of 656.3 nm. The huge arc of Barnard's Loop stands

The regions that glow red, however, are only part of the massive molecular cloud in this area, as only the outside of the cloud glows. Some idea of the extent of the dark gas comes from looking at the numbers of stars seen across the whole region. In some areas there are fewer stars because the molecular clouds hide stars beyond. Compare, for example, the numbers

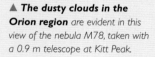
▲ The dusty clouds in the Orion region are evident in this view of the nebula M78, taken with a 0.9 m telescope at Kitt Peak.

out straight away, together with a fainter roughly circular region surrounding the star Lambda Orionis, sometimes called the Head of Orion.

These hydrogen clouds shine because they are illuminated by the massive stars of what is called the Orion OB1 association. An association of stars is a group of stars with a common origin, though they might not appear close together in a cluster. Many of the bright stars of types O and B in this area are members of the Orion OB1 association and over 50 such stars have been identified.

of stars on the left of the Lambda Orionis cloud with those on the right. The giant molecular clouds in the Orion complex stretch for maybe several hundreds of light years and have a mass of hundreds of thousands times the mass of the Sun. They extend over the whole area of Orion and into the neighboring constellation of Eridanus to its west.

Stars have been forming in this region for about 12 million years. Barnard's Loop is evidence for at least

▶ A far infrared view of the BN/KL complex itself made using the European Southern Observatory's 3.6-meter telescope. Several stars are in the process of forming within the cloud. The obscuration caused by the dust in this area is 60 magnitudes.

one supernova that has happened since then. But there is another clue to such activity in this area. Three separate stars have been found in different parts of the sky, all with unusually high velocities through space, whose paths can be traced back to a point very close to Orion's belt. These are what are known as runaway stars, thought to be the

partners of stars which exploded, thus releasing them from their orbits to career across the Galaxy.

The stars are Mu Columbae, in Columba to the south of Orion, 53 Arietis in Aries, and AE Aurigae in Auriga (see page 135). They all seem to have been ejected from this region a few million years ago. At one time it was thought that they

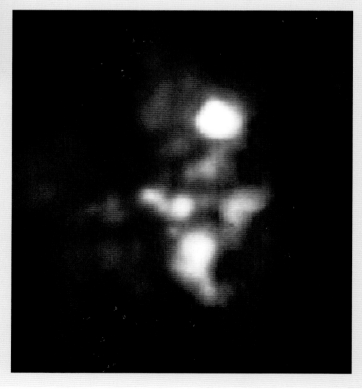

◀ *The region of the Trapezium* photographed at near-infrared wavelengths using the Very Large Telescope. This cuts through much of the nebulosity and reveals many more stars than can be seen in visible light. However, the protostars deep inside the nebula are still not visible. The rectangle marks the BN/KL complex.

▲ *A deep view of the Orion region* taken by Alex Mellinger, who combined a number of images taken on film and CCD into a mosaic showing the red hydrogen-alpha emission.

could all have been released in the same event, but recent measurements make this unlikely.

A glimpse inside the Orion Nebula itself comes from infrared imaging. Close to the Trapezium is a region known as the BN/KL complex, which was first studied by astronomers named Beckwith, Neugebauer, Kleinmann and Low in 1966 and 1967. They used infrared imaging to locate intense sources of radiation within the area, believed to be stars in the process of formation.

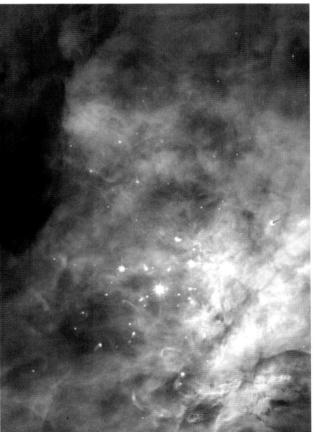

▲ *Hubble's close-up of the Trapezium area* shows many of the stars in the cluster. Interesting tails can be seen on stars that are in the process of being formed; they are caused by strong gas outflows.

▲ *In this view with north at left,* the bright star to left of center is Zeta Orionis. Below it is the Flame Nebula, and to its right is IC 434, with the Horsehead in the middle of the nebula. The star at top is Sigma Orionis.

there is great satisfaction in actually locating this famous object. It usually requires a large telescope, though it has been glimpsed in medium-sized instruments.

The Horsehead is actually a dark nebula within a strip of nebulosity called IC 434 running south of Zeta Orionis. To the east of Zeta lies a brighter part of the nebula, known as NGC 2024, which is also called the Flame Nebula from its ragged appearance in photographs. A hydrogen-beta filter is a considerable help in seeing the faint glow of IC 434 and the bay within it that we call the Horsehead. In fact, this type of filter is often referred to as the "Horsehead Filter" because this is one of the few objects for which it is actually useful. Others include the California Nebula in Perseus and the Flaming Star Nebula in Auriga.

Although these objects are a challenge to visual observers, they are fairly easy targets for astrophotographers using red-sensitive CCDs and small telescopes on a driven equatorial mount. Exposure times of ten minutes or so should show them quite easily.

CANIS MAJOR

The celestial tableau centered on Orion also includes two dogs, larger and smaller. Canis Major is the larger dog, and its brightest star, Sirius, is the brightest in the night sky. It hardly needs instructions to be able to find it – the three stars of Orion's belt point southeastward directly to it, and it is visible the world over. Canis Major contains a number of other bright stars that even without the help of Sirius would make it a major constellation. Its stars make the shape of a rather odd stick dog, which lies just to one edge of a rather faint part of the Milky Way, and is accompanied by Lepus, the Hare.

Alpha Canis Majoris (Sirius)

From our point of view, Sirius is top dog in our little corner of the Galaxy, outshining the Sun by more than 20 times. But as stars go, it is not particularly special: it is only bright because it lies so close to us – a mere 8.6 light years, just twice the distance of the nearest bright star, Alpha Centauri.

Exactly why Sirius is regarded as the Dog Star is unknown. It may have been called this long before the constellation itself got its name. To the ancient Egyptians it was a crucial star, as its appearance in the dawn sky before sunrise heralded the vital flooding of the Nile. During northern-hemisphere summer it is in the sky at the same time as the Sun, which is why hot, sultry summer days are known as Dog Days, when the heat of Sirius is supposed to be added to that of the Sun.

Sirius twinkles like no other star, merely because it is so bright. It can even give rise to UFO scares, with turbulence in our atmosphere making it appear to be sending out rays and sparks.

Accompanying Sirius in the sky is another star, a tiny white dwarf known as Sirius B. It is not a faint star, at eighth magnitude, but its closeness to Sirius makes it very hard to spot without using special techniques. This is the closest of this unusual class of stars, which have diameters comparable to those of planets but virtually as much mass as a star, making them exceedingly dense.

Open cluster M41

Just 4° south of Sirius is the bright open cluster M41, which is easily visible with binoculars and can be resolved into stars with a low magnification on a telescope. It is made all the prettier by three or four orange stars of sixth or seventh magnitude.

◀ **Sirius as viewed in close-up** with the Hubble Space Telescope. Sirius B is the tiny dot to the upper left of the main star image. The spikes are caused by diffraction of light in the optical system and by CCD artifacts.

▶ **A view of M41** taken with a 0.6 m telescope at Kitt Peak.

Object	Type	M$_v$	Magnification	Distance
Sirius	Star	−1.44	Naked eye	8.6 light years
M41	Open cluster	4.6	10×	2300 light years

LEPUS

The constellation of Lepus suffers from being so close to its illustrious neighbors as it contains some relatively bright stars in its own right, notably Arneb, Alpha Leporis, at magnitude 2.6.

Globular cluster M79

This may look like a fairly ordinary globular cluster, though considerably better than some; but it is unusual. Virtually all the other bright globular clusters are round the other side of the sky, and for a short time this is the only globular cluster in the Messier catalog in the entire visible sky. The reason is that M79 lies even farther from the center of the Galaxy than our own Sun, whereas the other globulars are all to be found surrounding the center of the Galaxy.

This may be because our Galaxy has poached it from another dwarf galaxy that it has swallowed up. The galaxy in question is known as the Canis Major dwarf galaxy, but you will search for it in vain. It was discovered in 2003 by analyzing the locations of red giant stars. At one time it may have had a billion stars

◄ **Globular cluster M79** is fairly easy to locate by taking a line through Alpha and Beta Leporis and going an equal distance beyond. It is visible in small telescopes. This image was taken using a 0.9 m telescope at Kitt Peak.

of its own, but these are now in the process of being subsumed within our own Galaxy. It seems to have contributed three other globular clusters and some open clusters to the Milky Way.

Object	Type	M_v	Magnification	Distance
M79	Globular cluster	7.7	75×	42,000 light years

CANIS MINOR AND MONOCEROS

The constellation of Canis Minor, the lesser dog, is noted for containing the star Procyon, but little else. Like the greater Dog Star, Sirius, Procyon is close, at 11.4 light years, and with our Sun, Alpha Centauri and Sirius it makes a little group within the Galaxy that we might consider our own neighborhood. Curiously, Procyon has a white dwarf companion like Sirius, though fainter and even more difficult to see.

Sirius, Procyon and Betelgeuse in Orion are sometimes called the Winter Triangle, a description that applies only in the northern hemisphere, to match the Summer Triangle of Vega, Deneb and Altair.

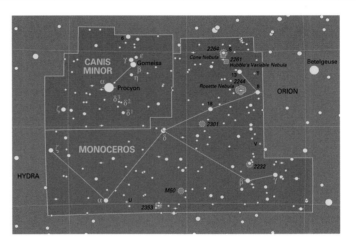

Monoceros is not a well-known constellation, and many stargazers would be hard pressed to identify its pattern of fourth-magnitude stars. This unicorn inhabits the region to the east of Orion and south of Canis Minor. Though it may be unimpressive as a constellation, it lies within the Milky Way and therefore contains a number of interesting objects. Beta Monocerotis, which with Gamma lies between Sirius and the belt stars of Orion, is a beautiful triple star with three bluish stars.

Open cluster M50

As Monoceros is poor in stars, this cluster is probably easiest to find by locating Gamma Canis Majoris, just to the east of Sirius, and then moving 7° north. The

◄ **Open cluster M50** taken with a 0.6 m telescope at Kitt Peak.

Object	Type	M_v	Magnification	Distance
M50	Open cluster	6.5	50×	3000 light years
Rosette	Nebula	–	Photographic	5500 light years
NGC 2261	Reflection nebula	10	175×	2500 light years

▲ The Rosette Nebula *is a spectacular sight in photographs and records easily on cooled CCD cameras.*

Hubble's Variable Nebula, NGC 2261

Though the photograph below was taken by the Hubble Space Telescope, the nebula is named after the man himself. Edwin Hubble, pioneer observational astronomer, took a great interest in this nebula from 1916 onward, when he noticed that its shape can vary from month to month as seen in photographs. At first sight the object looks exactly like a comet, with a bright head and a fan-shaped tail. Although it has been seen in small telescopes, under ideal conditions the nebula requires a medium-sized instrument and a fairly high magnification. As is always the case, the larger the instrument and the better the conditions the more will be visible, particularly as the nebula fades away into darkness at its northern edge.

Virtually all deep-sky objects show the same appearance even over an observer's lifetime, so this behavior is very unusual. At a distance of 2500 light years, and with a size of about 1 light year, the movements required would have to be impossibly rapid to be real. However, as it is a reflection nebula around the young star R Monocerotis, it is likely that its flame-like flickering is the result of the movement of gas clouds very close to the star, which cause shadow effects on the nebula. The star itself is hidden by the dark nebulosity in the region.

Some amateurs equipped with CCD cameras have photographed the nebula on different occasions and created intriguing animations showing its variations. This method shows up the changes very effectively, as the eye is sensitive to very slight movements.

cluster is easily visible in binoculars as a blur, and small telescopes will show a group of stars of which the brightest are eighth and ninth magnitude, with outliers that make it hard to decide on the boundaries of the cluster.

Rosette Nebula

Another of the sky's top book illustrations, the Rosette is very easy to photograph. Though its location is readily found, all that most people can see visually is a small cluster of stars known as NGC 2244, more scattered and less visually impressive than M50. But a telephoto lens and a CCD with red filter will yield a result with an exposure of minutes, though a longer exposure and good guiding are needed for really impressive results. The circular appearance of the Rosette makes it look like a supernova remnant, but it is an H II region from which the cluster of stars has been born.

▶ Hubble's Variable Nebula *reveals itself in photographs as a reflection nebula by its blue color. This photograph was made with the Hubble Space Telescope, while the drawing was made from Hertfordshire, within 40 miles of the center of London, using a 25 cm Dobsonian telescope and a magnification of 184. The object is about a light year across.*

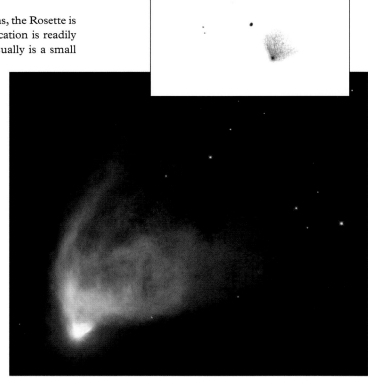

PUPPIS

This most northerly part of the old Argo Navis (see page 143) extends close enough to the celestial equator for much of its area to be visible to northern observers. The brightest parts, however, are really best seen from the southern hemisphere. Puppis lacks bright stars and a clear pattern. Its most readily recognizable section is a group of stars, which includes its brightest star, Zeta Puppis or Naos, to the southeast of Canis Major. There are, however, numerous scattered stars of third and fourth magnitude.

Open clusters M46 and M47

Follow a line from Beta Canis Majoris through Sirius and you arrive at this pair of Messier objects. The brighter of the two, M47, is visible with the naked eye under good conditions, though most people will need binoculars in average skies. The fainter M46 is also said to be a naked-eye object, but if there is light pollution around, it can be a challenge even through binoculars. The pair make an interesting contrast – both cover a similar area of sky, but M47 contains less numerous but brighter stars, while M46 is more suited to a low-power telescope view as it has numerous stars of about tenth magnitude.

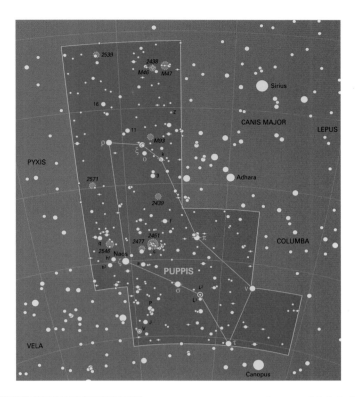

Open cluster NGC 2451

Another naked-eye cluster, NGC 2451 lies midway between Zeta and Pi Puppis, and surrounds a third-magnitude orange star. Though good in binoculars, it lacks fainter stars so the telescopic view is less impressive. It is one of the ten closest clusters to us, but a much fainter cluster known as NGC 2451B lies at twice its distance.

◄ *A Digital Sky Survey view* of open cluster NGC 2451. Though there are faint stars across the field, most of the cluster members are of magnitudes 5 or 6.

▲ *The open clusters* M46 (left) and M47 compared. The field of view is 2½°, so both will fit within the field of view of binoculars.

Object	Type	M$_v$	Magnification	Distance
M46	Open cluster	6.0	10×	5400 light years
M47	Open cluster	5.2	10×	1600 light years
NGC 2451	Open cluster	2.8	7×	850 light years

GEMINI

The two leading stars of Gemini – Castor and Pollux – are not a matched pair, but they have been regarded as twins since ancient times. Mythologically they are the sons of Leda, Queen of Sparta, but to two different fathers – a process that does not normally result in twins. The two stars do have subtly different colors, Castor being bluish and Pollux yellowish. A good way to remember which is which, once you know the sky a little, is from their capital letters: Castor is closest to Capella while Pollux is closest to Procyon.

Two loose lines of stars mark the figures of the twins themselves. Their feet dabble in the Milky Way, where there are several interesting deep-sky objects. The ecliptic also runs through Gemini, so its pattern may be altered by the presence of a planet. Although Gemini is a northern constellation it can be seen from both hemispheres – though for southern-hemisphere observers it is quite low.

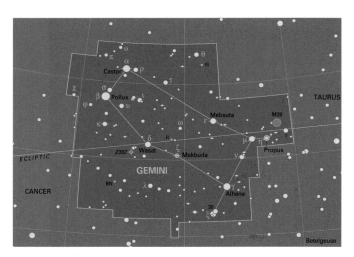

Multiple star Alpha Geminorum (Castor)

Despite being labeled Alpha, Castor is the fainter of the twins. Nevertheless, it consists of no fewer than six individual stars, of which three are visible using small telescopes. The two main components are over 4 arc seconds apart, and a 60 mm telescope should easily separate them. Look carefully and you should see a ninth-magnitude star about 60 arc seconds from the main pair. Each of these three stars is a close double star, though the components cannot be separated visually, so Castor is six stars in one.

▲ **Castor as photographed** through a 200 mm telescope is split into two stars of magnitudes 1.9 and 2.9 with a ninth-magnitude red star nearby.

▶ **Mu and Eta Geminorum** and open cluster M35, showing stars down to 11th magnitude.

Open cluster M35

The two stars Mu and Eta Geminorum are easily spotted at the western end of Gemini. They act as signposts to the cluster M35, just to their northwest. M35 is visible to the naked eye in good conditions or binoculars in poorer skies, and is one of the larger open clusters in the sky, so binoculars will easily show some individual stars. A telescope reveals masses of stars from magnitude 9 downward.

▲ **Supernova remnant IC 443** is a challenge to visual observers, but can be photographed relatively easily even with wide-field photography. This view was made using only a 76 mm telescope and a CCD camera at Kitt Peak National Observatory in Arizona.

Object	Type	M_v	Magnification	Distance
Castor	Double star	1.6	200×	52 light years
M35	Open cluster	5.3	25×	2800 light years
NGC 2264	Open cluster	3.9	25×	2800 light years
NGC 2392	Planetary nebula	9.2	100×	3000 light years
IC 443	Supernova remnant	–	10×	5000 light years

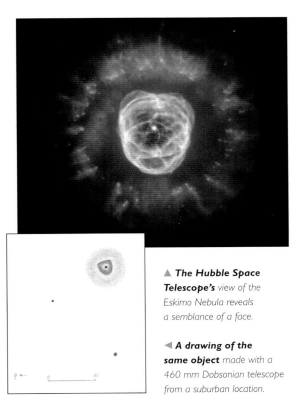

▲ **The Hubble Space Telescope's** *view of the Eskimo Nebula reveals a semblance of a face.*

◀ **A drawing of the same object** *made with a 460 mm Dobsonian telescope from a suburban location.*

NGC 2392, the Eskimo or Clown Face Nebula

One of the brighter planetary nebulae, the Eskimo is fairly easy to locate from a little triangle of stars near Delta Geminorum, Wasat. With a small telescope you can see it looking like an eighth-magnitude star, which with some magnification becomes an oval shell with maybe a blue-green tinge. With a large telescope and high magnification, some detail becomes visible, with the outer shell looking a little like a fringe surrounding a face, hence the nebula's name.

Supernova remnant IC 443

One of the few supernova remnants that can be observed with amateur instruments lies almost midway between Mu and Eta Geminorum, though it is not shown on the map. Even the brightest part is not an easy object to find visually, requiring a fairly large instrument and clear skies, together with an O III filter, but using a CCD with red filter it is quite easy to photograph. It is also known as the "Jellyfish Nebula," from its shape.

The supernova itself probably exploded some 30,000 years ago, though a wide range of ages has been suggested. Searches for the pulsar that should exist near the center of the supernova remnant were intitially in vain, until in 2000 a group of high-school students found a point source some way from the center using data from the Chandra X-ray Observatory satellite (see A–Z section, page 227). The source was accompanied by a trail, and projecting its motion back along the direction of travel yielded a timescale of 30,000 years for the movement of the object.

Some of the nebulosity closer to the star Eta Geminorum, the bright star in the photograph on page 132, may be the results of previous supernovae in the area.

Cone Nebula and cluster NGC 2264

Though they lie in Monoceros, and are shown on the map on page 129, these objects are best found by starting from the stars of Gemini rather than Monoceros.

Orion is well known as a starbirth region, and several other constellations visible at this time of year contain evidence of stellar nurseries. The Cone Nebula is a good example. The Cone is hard to observe visually, and the cluster NGC 2264 is rather unrewarding under most conditions, though it has earned the nickname of the Christmas Tree Cluster because of the triangular pattern made by the stars when viewed with south at the top. However, the area is a treat when seen in deep photographs. A conventional view is shown on page 111, but below is a photograph taken in the infrared by the Spitzer Space Telescope, which is ideally suited to observing objects such as this.

The Christmas-tree shape, which covers the whole area of this photograph, has disappeared, but instead there is a smaller central cluster of newly born stars with a shape resembling the spokes of a wheel. In keeping with the snowy feel to the area, this has been dubbed the Snowflake Cluster. The stars are about 100,000 years old and have yet to move away from the area where they were born.

▼ *An infrared view of* ***the Cone Nebula*** *and surrounding region, from the Spitzer Space Telescope. Though the cone is dark at optical wavelengths, it merges into the background in this view. The colors are entirely false, as the picture is a composite of four images taken through different filters and the color range has been shifted far into the infrared.*

AURIGA

Though not as well known to the general public as nearby constellations such as Orion or Taurus, Auriga is a favorite with astronomers because of its richness in interesting objects. It lies in the northern Milky Way so it is ideally placed for northern-hemisphere observers, but it is also accessible quite low down for most southern observers.

Auriga himself is a charioteer, though curiously he is saddled with a goat and her kids, which he carries in his left arm. The brilliant yellowish star Capella, sixth brightest in the sky, marks the goat and two of the adjacent triangle of stars mark the kids – known as the Haedi. The traditional pattern of stars of Auriga is a pentagon, but the most

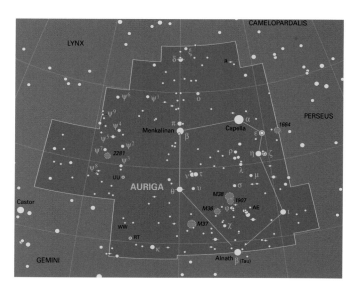

southerly of these stars is actually spoken for by neighboring Taurus, where it has a respectable existence as Beta Tauri. So although the pentagon remains, there is no Gamma Aurigae and the star in question is exactly on the border between the two constellations.

The Haedi and Epsilon Aurigae

Some books say that all three stars adjacent to Capella are the Haedi, but old constellation drawings show only two kids, which would be Eta and Zeta. It is a coincidence that two of the three stars are eclipsing binaries (see page 109) – a fairly rare type of variable star. Zeta is a common-or-garden eclipsing binary, with a period of 2.7 years. Its brightness drops from 2.7 to 3.0 for 40 days when its dimmer companion star partially hides it.

But Epsilon, the nearest of the three stars to Capella, is a different matter altogether. Its companion star takes 27 years to orbit it, and the eclipse reduces the star's magnitude from 2.9 to 3.8, which is quite noticeable, for over a year. Such a long eclipse can only be caused by an enormous object, far larger than any known star. The eclipsing body would have the same diameter as Saturn's orbit around the Sun, yet be invisible. Probably it is a cloud of dust, but there is great uncertainty. The last eclipse of Epsilon Aurigae lasted from 2009 to 2011.

Open clusters M36, M37 and M38

A line of open clusters adorns the center of the pentagon of Auriga. In one of the numerous quirks of astronomical nomenclature, they are numbered from east to west in the order: M37, M36, M38. All three are at roughly the same distance of between 4100 and 4400 light years, but there are noticeable differences between them. All are easily visible to the naked eye in good skies and with binoculars in poorer skies, though a small telescope is needed to see the individual stars.

The richest of the three is M37, and it has inspired earlier authors to wondrous prose: "A magnificent object, the whole field being strewed as it were with

▶ *A close-up of the Haedi and Epsilon Aurigae* (top). *In some listings, Eta and Zeta (the bottom pair) are named Haedus I and Haedus II.*

Object	Type	M$_v$	Magnification	Distance
M36	Open cluster	6.3	50×	4100 light years
M37	Open cluster	6.2	50×	4400 light years.
M38	Open cluster	7.4	50×	4200 light years
Flaming Star	Nebula	–	Photographic	1455 light years

sparkling gold-dust," according to the 19th-century observer Admiral Smyth. In a small- or medium-sized telescope it is certainly a beautiful sight, and it even shows through light pollution. M36 is a slightly poorer relation in terms of numbers of stars: it is smaller but with brighter and indeed younger stars. Nearby M38 is the largest of the three clusters, but less rich in bright stars. A smaller and fainter cluster, NGC 1907, lies nearby; it is just visible with small telescopes, consisting of stars mostly fainter than 11th magnitude.

AE Aurigae and the Flaming Star Nebula

Photographers of the M38 region may find that a nearby star has a red haze around it. This is the sixth-magnitude blue variable star AE Aurigae, which is illuminating a patch of hydrogen gas, known as the Flaming Star Nebula. AE Aurigae is an escapee from the region around the Orion Nebula, farther south, and is thought to have been flung away when a companion star exploded as a supernova, releasing AE Aurigae from its orbit. Visually the nebula is very difficult to see, though a hydrogen-beta filter may help. The star itself is an interloper in the region, and just happens to be passing through. In 10,000 years' time it will have moved on and the nebula will fade from view.

◀ **Cluster M37,** *photographed using a CCD through a 130 mm refractor from Corringham, Essex, UK, by Eddie Guscott with a total exposure time of 75 minutes.*

▲ **The Flaming Star Nebula,** *photographed from the same location and equipment by Eddie Guscott. This location is within 40 miles of central London.*

▶ **The central part of Auriga,** *showing the three clusters M37 (lower left), M36 (center) and M38 (upper right). The distinctive parallelogram of stars at center right is a good guide to the location of AE Aurigae.*

MARCH–APRIL

Evening

1 March at 11.30 pm
15 March at 10.30 pm
30 March at 9.30 pm

Morning

15 November at 6.30 am
1 December at 5.30 am
15 December at 4.30 am

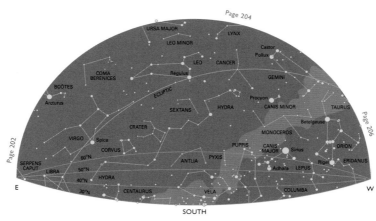

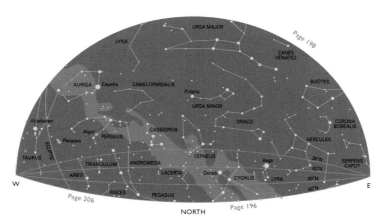

The northern-hemisphere sky

The key constellation in spring skies is the crouching lion of Leo, visible in the mid sky looking south. The backward question mark known as the Sickle, with bright star Regulus at its bottom, is a very distinctive feature because there are few bright stars nearby. Over to its west are the two bright stars of Gemini, following Orion down to the western horizon as the evening progresses. Sirius is still visible as a very bright star quite low in the southwest. Below Leo is the faint and sprawling Hydra, with its lone bright star Alphard.

The Big Dipper or Plough is virtually overhead. Follow the curve of its handle down to the east and you will find brilliant orange Arcturus in the otherwise unremarkable constellation of Boötes; carry on farther to the southeast to locate Spica, the brightest star in Virgo.

Evening

1 March at 11.30 pm
15 March at 10.30 pm
30 March at 9.30 pm

Morning

15 December at 4.30 am
30 December at 3.30 am
15 January at 2.30 am

The southern-hemisphere sky

Though the Milky Way runs high in the south, looking north the skies are much more barren. Leo stands out clearly, with its bright star Regulus at the top of a hook of stars known as the Sickle. To the northwest you can see the two bright stars of Gemini, while Orion is still easily spotted setting in the west, with brilliant Sirius and Canis Major following it down as the evening progresses.

Don't confuse Gemini with another pair of stars, Alpha and Beta Centauri, which are almost exactly opposite in the sky, in the southeast, and nearly the same separation but each about a magnitude brighter. In addition, they point upward to the Southern Cross, now getting high in the sky. Over to the southwest, high up, is the bright star Canopus, the second brightest in the night sky after Sirius.

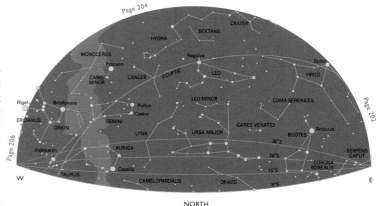

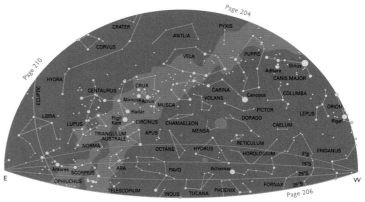

LEO, LEO MINOR AND SEXTANS

One of the most ancient constellations, Leo occupies a central place in the evening skies around March, April and May. Though just north of the celestial equator it can be viewed the world over, but only from the northern hemisphere does its actual resemblance to a crouching lion become obvious. Its most prominent feature is the reversed question mark of stars, which is widely referred to as the Sickle from its shape. This marks the lion's mane, and the rest of its body is marked by three other stars with Denebola as its tail. The word "deneb" crops up in a few star names, as it means "tail." Leo Minor, to its north, and Sextans, to its south, are comparatively recent inventions.

Being well away from the Milky Way, Leo is devoid of nebulae and clusters within our own Galaxy, but in their place we can see beyond to other galaxies. There are a few relatively bright galaxies within its boundaries. Leo is a zodiacal constellation, so the planets may be found there, and from time to time the Moon or a planet may be close to or even pass in front of Leo's brightest star, Regulus. A series of conjunctions, or close approaches, between Regulus and Jupiter in 2 BC is one theory behind the story of the Star of Bethlehem.

Double star Gamma Leonis (Algieba)

Many doubles are famed for their beauty because they have contrasting colors, or because they are widely separated, but Algieba's stars are quite close, at just over 4 arc seconds, and both are yellowish in color. Nevertheless, their brilliance and closeness makes them a very popular pair, and because they are so close slight differences in their colors are often seen. The individual stars are types K and G, so some color difference is to be expected. A fairly high magnification is needed for the best views of this double.

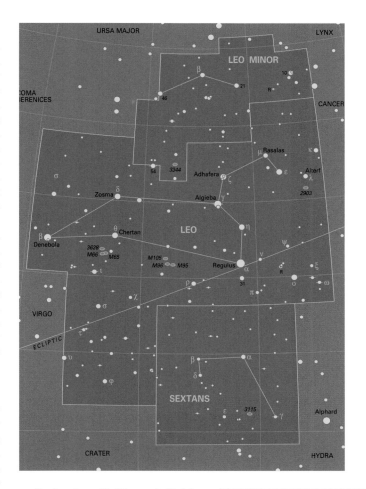

Galaxies M65 and M66

A number of galaxies are within the reach of binoculars on a good night, but they are usually small and unremarkable. It always helps to have a pair of them, because this provides confirmation that you have seen a genuine tiny misty object rather than just maybe a faint star. These two galaxies in Leo are easily found by first locating third-magnitude Theta Leonis, Chertan. About 2½° to its south is a vertical line of three fifth- and sixth-magnitude stars. M66 is at the apex of an equilateral triangle with these stars, with M65 nearby. Both are noticeably elongated spiral galaxies, and a small telescope shows them easily, with some detail visible using larger apertures. Having found these galaxies, you may spot the more elongated spiral NGC 3628 to the north.

▲ **Double star Gamma Leonis** (Algieba) presents two golden-yellow stars close together.

◄ **The trio of galaxies** M65 (lower right), M66 (lower left) and NGC 3628 (top).

Object	Type	M$_v$	Magnification	Distance
Gamma Leonis	Double star	2.0	200×	126 light years
M65	Galaxy	9.3	75×	41 million light years
M66	Galaxy	8.9	75×	32 million light years
NGC 3628	Galaxy	9.5	75×	36 million light years
M95	Galaxy	9.7	75×	33 million light years
M96	Galaxy	9.2	75×	35 million light years
M105	Galaxy	9.3	100×	34 million light years
NGC 3115	Galaxy	9.2	75×	33 million light years

Galaxies M95, M96 and M105

Having found M65 and M66, you can either move exactly 8° to the west to find M105 in the same field of view, or you can move 9° east of Regulus to find this group of galaxies. They are slightly fainter than M65 and M66, so binoculars may not show them, but they can be easily located using a low power on a small telescope. M95 is a good example of a barred spiral galaxy, though with a small telescope only the nucleus is visible and the bar requires a larger aperture. M96 is a spiral and M105 an elliptical galaxy.

Lenticular galaxy NGC 3115

The brightest example of a lenticular galaxy, this appears as a smooth spindle in a telescope, and indeed it has the nickname of the Spindle Galaxy. It is relatively easy to find by moving just over 3° eastward and very slightly northward from fifth-magnitude Gamma Sextantis. It has a fairly bright center, so it is easily observable with medium-sized telescopes. This is the only bright object of any note in Sextans.

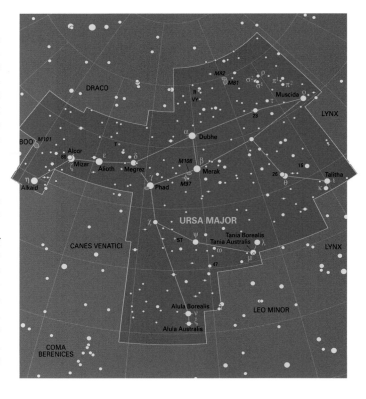

◀ **Barred spiral galaxy M95,** *photographed with a 0.9 m telescope at Kitt Peak.*

(inset) **A drawing of the lenticular galaxy NGC 3115** *made with a 222 mm reflector.*

URSA MAJOR

Call it the Big Dipper or call it the Plough, the pattern of the seven brightest stars in the constellation of Ursa Major is the best known in the northern hemisphere. These stars are by no means the brightest in the sky, but they are visible all night from North America and Europe. They are also easy to recognize, and once you have spotted the familiar grouping you can draw a line through the two right-hand stars, known as the Pointers, to point roughly to the Pole Star. They are to the northern hemisphere what the Southern Cross is to the southern hemisphere.

The seven stars are just a small part of Ursa Major. The group was known as a bear from ancient times, though like the lesser bear Ursa Minor, this one has an amazingly long tail. According to Greek legend the bear was originally a beautiful girl named Callisto, who was seduced by Zeus. For her sins she was turned into a bear and thrown into the heavens, and her tail was drawn out in the process.

To the pioneer Americans it was the handle of the dipper used to ladle water, while to the peasants of Britain centuries back, the tail of the bear resembled the handle of the wooden plough that nowadays is mostly seen on the walls of pubs of the same name. Neither appliance is in common use today, but the names live on.

Galaxies M81 and M82

This is a delightful pair of galaxies, not just because they are fairly bright but also because they have contrasting shapes that show up even using small telescopes. Users of Go To telescopes should have no trouble finding them, as they are large enough to be seen in a low-power field of view. To find them in binoculars, take a diagonal line between the stars Gamma and Alpha and you will spot them the same distance beyond Alpha, just short of a pair of stars.

M81 is a classic spiral galaxy, though with most telescopes all you see is the oval haze of the central regions. M82 by contrast is termed a peculiar galaxy. In a telescope it looks quite different from M81, as a short, stubby bar, and photos reveal signs of what looks like

▲ **The galaxy M82** *as seen with an optical telescope, left, and with the Spitzer Infrared Telescope, right. A huge infrared halo surrounds the galaxy as a result of dust being emitted by the vast number of hot stars in the starburst which this galaxy is undergoing. In the infrared image, the light from stars has been removed by subtracting the view seen through the shorter wavelength filter so as to improve the visibility of the infrared feature. The optical photo at left shows the supernova (arrowed) which appeared in January 2014 – the closest Type I supernova for over 40 years.*

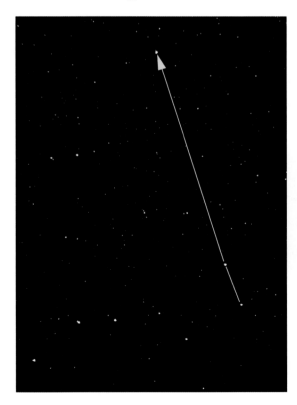

▲ **The stars of the Big Dipper** *make an ideal guide to the Pole Star, as shown here. The line misses Polaris by a few degrees.*

▲ **Though different sizes** *in a photograph, the galaxies M81 (left) and M82 are similar sizes when seen in light-polluted skies because the spiral arms of M81 are hard to see.*

Object	Type	M_v	Magnification	Distance
Mizar	Double star	2.3	50×	86 light years
M81	Galaxy	6.9	75×	12 million light years
M82	Galaxy	8.4	75×	13 million light years
M101	Galaxy	7.9	75×	22 million light years
M97	Planetary nebula	9.9	100×	2000 light years

an explosion at its center. It is what is known as a starburst galaxy, and star formation is taking place there at a great rate. One theory is that this activity was triggered by an inter-action with M81 some 100 million years ago.

Galaxy M101

Very few galaxies actually do show spiral arms when seen through a telescope. Quite often, as when view-ing M51 (see page 152), you imagine that you can see the spiral arms, but that is only because you know they are there. But M101 is one of the small band of galaxies that are indeed large and bright enough to reveal their arms in amateur telescopes.

The drawing (at left) was made using a large reflec-tor at a dark site, but telescopes of 300 mm or larger should show the spiral arms under the right conditions. In a dark sky, M101 seems difficult to miss even when viewed with binoculars, but because of its fairly low surface brightness it can be very hard to find under light-polluted conditions, even with a Go To telescope.

▲ **A drawing of galaxy M101** made using a 450 mm Dobsonian with a power of 102×.

▲ **The Big Dipper or Plough** (top). The inset (bottom) shows the telescopic view of Mizar and Alcor. The two stars of Mizar are separated by 14 arc seconds.

Double star Zeta Ursae Majoris (Mizar)

Mizar is the middle star in the handle of the Big Dipper or Plough. With the naked eye you can see a fourth-magnitude star nearby, Alcor. Through a small telescope, you can see that Mizar is a true double star in its own right.

M97, Owl Nebula

Though this is a well-known planetary nebula, many amateur astronomers have never actually seen it because it is particularly large and faint. The photo-graph (at left) shows why it got its name: two dark areas within the disk of the planetary look like the eyes of an owl. You need at least a 250 mm telescope and no haze to be able to see these with your own eyes.

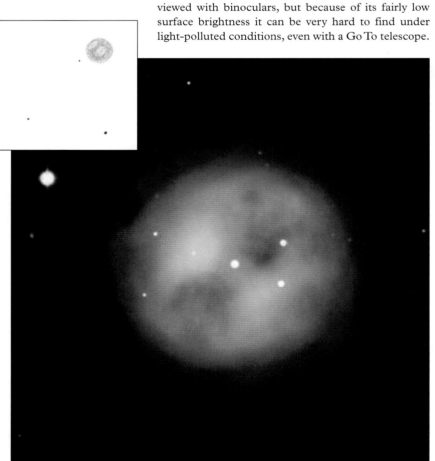

◄ **A photograph with a 0.9 m telescope** shows the green color of M97 and the two voids that resemble the eyes of an owl. Seen in a 450 mm reflector in good skies (inset), the "eyes" are visible, but they have been seen even with 250 mm telescopes.

CANCER

Everyone has heard of the constellation of Cancer as it is in the zodiac, but this is no guarantee that it has bright stars. In fact, only two of its stars are brighter than fourth magnitude. Nevertheless, Cancer is an ancient constellation containing some interesting objects, which of course may include the Moon and planets. It is a northern constellation but is easily visible from southern-hemisphere locations.

Star cluster M44, the Beehive or Praesepe

To the naked eye, M44 appears as a little misty patch just to one side of the line between the stars Gamma and Delta Cancri. It is visible in clear skies even if there is a certain amount of moonlight. Binoculars reveal it as a large cluster of stars from sixth magnitude downward, and in a telescope at the lowest power the eyepiece is full of stars. In longer-focal-length instruments, however, not all of the cluster will fit inside the field of view. The cluster is about 590 light years away, putting it somewhat more distant than the Pleiades. The word *praesepe* is Latin for "manger," referring not to the manger of Christ's birth but to the stars Gamma and Delta, known as the *aselli* or donkeys.

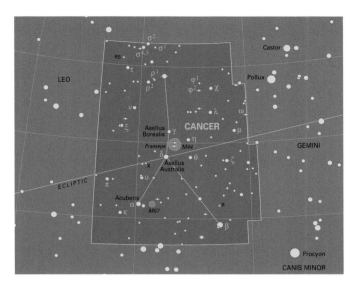

▼ **The open cluster M44,** known as the Beehive or Praesepe, flanked by the two donkeys.

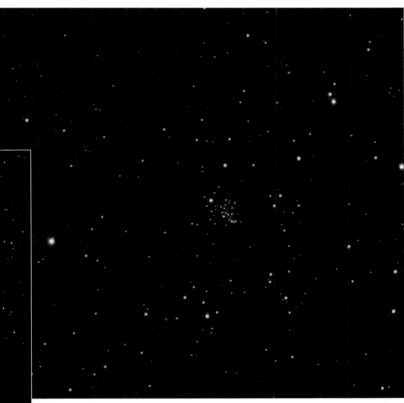

Star cluster M67

Easily found by star-hopping from Alpha Cancri, this is a much more distant and hence smaller and fainter cluster than M44. A medium-sized telescope is needed to do it justice, as most of its stars are tenth magnitude or fainter. M67 is a particularly old cluster at around 4 billion years, compared with 730 million years for M44.

▲ **In contrast to M44,** M67 is a challenge to binocular users as it is much smaller and fainter.

Object	Type	M$_v$	Magnification	Distance
M44	Open cluster	3.7	7×	590 light years
M67	Open cluster	6.1	25×	2700 light years

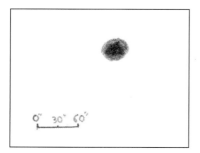

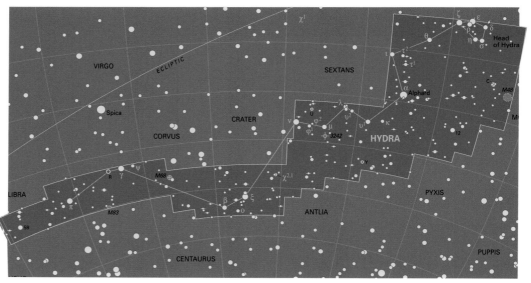

▲ **The Ghost of Jupiter,** as seen through a 75 mm refractor. This bright planetary is easily visible in small apertures.

▼ **M83 is sometimes referred to** as the "Southern Pinwheel" for its resemblance to M33. Though smaller and fainter than that galaxy, it is easy to see even with binoculars under the right conditions.

HYDRA

The largest constellation of all, this sea snake wriggles over a great swathe of the midsouthern sky, squeezing between several other constellations on its way. Its head is in the northern hemisphere, and while all of it can be seen by most northern observers, some parts of it are so low that they are effectively unobservable. Hydra covers nearly seven hours of right ascension, but for all its size it is a barren area of sky and has fewer than its fair share of deep-sky objects.

The head of Hydra is its most recognizable pattern, a group of six third- and fourth-magnitude stars lurking below Cancer. The brightest star, Alpha Hydrae, is in an isolated position to the southeast of the head, and its name, Alphard, does indeed mean "Solitary One" as there are no other nearby bright stars.

Below Spica, one of the wriggles in Hydra's body is marked by the star R Hydrae, which is often invisible as it is a Mira-type star (see page 188), varying between magnitudes 4 and 10 in about 390 days.

Galaxy M83

A classic face-on spiral galaxy, this is a very difficult object for northerly observers, but when it is high up and in dark skies it is a splendid sight, being one of the few galaxies whose spiral arms are visible with amateur telescopes. The lack of nearby bright stars makes M83 a little hard to find. It is at one apex of a large triangle with Pi and Gamma Hydrae, with a short line of fifth- and sixth-magnitude stars to its west, and another star to its east. These stars should be visible in the finder to help you locate the galaxy itself.

If the galaxy is low down and hard to find in the murk, you are likely to see only its nucleus. But the better the skies, and the higher it is, the more detail you should see, even with medium apertures. Wait until it is as high in the sky as possible for the best views.

Planetary nebula NGC 3242, the Ghost of Jupiter

As if to prove that not every object in Hydra is out of reach of northern observers, the Ghost of Jupiter planetary is easily visible even in small instruments. It gets its name from its similarity to the planet itself – in terms of size and shape but not brightness or color, hence the ghostly appearance. The object is eighth magnitude and is large enough to be found with little trouble, once you have the right part of the sky. Find Alphard and then locate third-magnitude Mu Hydrae. From there, the Ghost of Jupiter is just 2° south and a little to the west, and many observers notice a blue color. Larger telescopes reveal a fainter outer shell.

Object	Type	M_v	Magnification	Distance
M83	Galaxy	7.6	25×	15 million light years
NGC 3242	Planetary nebula	7.8	100×	2500 light years

VELA

In Greek times this area was occupied by an enormous ship, Argo Navis – the very same ship in which Jason and the Argonauts set sail. But it proved to be too cumbersome for celestial navigation, so it has now been divided up into its component parts – the Sails (Vela), the Keel (Carina) and the Stern or Poop (Puppis). This was done after the stars had been assigned their Greek letters, so each constellation now has letters missing. Vela, for example, has no Alpha or Beta. Its brightest star, Gamma Velorum, however, has at least two claims to fame. It is a wide double star visible as a pair in a telescope at low power, and the brighter component is a particularly hot type of star known as a Wolf–Rayet star. Such stars have a temperature of about 60,000°C, compared with the Sun's modest 5700°C, and Gamma Vel is the brightest example.

Vela is regarded as a southern-hemisphere constellation, though it is visible from the southern United States. Like many southern constellations, it lacks a real pattern, and some of the stars making up the polygon that traditionally marks the sail's outline are unlettered and no brighter than the stars which are not part of the outline. Vela is on the fringe of the Milky Way in an area that is rich in stars of fourth and fifth magnitude.

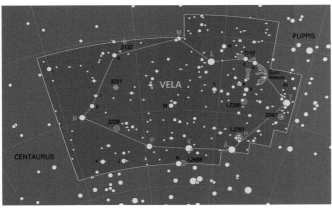

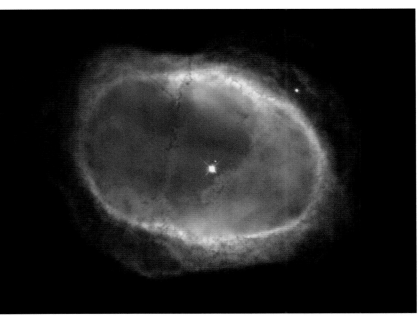

polluted areas. The name comes from the loops within it, which are visible on photographs. Visually it appears as an oval disk somewhat larger than Jupiter, with a prominent central star. This star, however, is not thought to be the one responsible for the nebula. That honor goes to another much fainter object, visible on the spectacular Hubble Space Telescope image below, very close to the apparent central star. NGC 3132 is also sometimes called "the Southern Ring Nebula" because of its similar easy visibility, though the ring is not as pronounced as with M57, the Ring Nebula in Lyra.

◀ NGC 2547 is a widely scattered star cluster best seen with binoculars.

▼ Compare this view of NGC 3132, the Eight-Burst Nebula, photographed with the Hubble Space Telescope, with the amateur photograph that is shown on page 112.

Open cluster NGC 2547

Just 2° south of Gamma Velorum lies this open cluster, which can be located with binoculars. It is quite large so requires a low power in a telescope, and consists of a wide scattering of stars mostly of eighth magnitude and fainter.

NGC 3132, the Eight-Burst Nebula

A fine planetary nebula even for small telescopes, NGC 3132 is bright enough to be visible from light-

Object	Type	M$_V$	Magnification	Distance
NGC 2547	Open cluster	4.7	25×	1960 light years
NGC 3132	Planetary nebula	9.4	100×	2000 light years
Gum Nebula	Supernova remnant	–	7×	1300 light years
NGC 2736	Supernova remnant	–	60×	815 light years

The Gum Nebula and the Vela supernova remnant

Virtually the whole western part of Vela is wreathed with faint nebulosity – the Gum Nebula. It covers about 36° of sky and therefore qualifies as the largest nebula in the sky as well as the closest supernova remnant. It was not recognized until 1952 when it was found on photographs by an Australian astronomer, Colin Gum. He cataloged several different parts of the nebula, but the whole thing was eventually named after him.

The nebulosity is probably about 1300 light years away, and is about 840 light years in diameter – a huge area. So what is causing gas spread over hundreds of light years to glow? One possibility is the hot stars Gamma Velorum and Zeta Puppis, both of which lie within it. However, they should not be powerful enough to cause such a vast nebula to shine. The supernova that gave rise to the Gum Nebula probably exploded around a million years ago – a very long time for its energy to continue to power the nebula.

Also within the Gum Nebula, however, lies the smaller Vela supernova remnant. At its center lies a pulsar which flashes every 9.3 milliseconds both at radio and optical wavelengths – one of the fastest known. The pulsar is slowing down at a constant rate as it ages, which allows astronomers to estimate its age at 11,000 years, which is therefore probably the age of the supernova remnant as well.

It is possible that the energy from the Vela pulsar is sufficient to power not only the Vela supernova remnant but also the Gum Nebula as well.

Can these nebulae be seen visually? Parts of the Gum Nebula certainly can, under the right conditions. One part was seen by John Herschel in the 19th century and is known as NGC 2736, or Herschel's Ray or the Pencil Nebula (not shown on the map). As its name implies, this is a very thin and straight nebula. The Vela supernova remnant can also be seen visually with the right conditions, though it is considerably fainter than its northern counterpart, the Veil Nebula, which is a comparatively easy object when using an O III filter. But photography is the best way to observe these objects.

▼ *Vela is a mass of nebulosity* as seen through a camera equipped with a filter that transmits the light of hydrogen alpha. The whole area is covered by the Gum Nebula, which also extends farther to the northwest. The Vela supernova remnant, referred to on the map as the Gum Nebula, comprises the fine strands of nebulosity just above center.

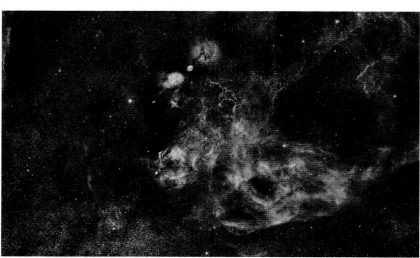

▲ *The central regions of the Eta Carinae nebula.* Compared with the picture on page 146, this shows the brightest part of the V-shape, with what Sir John Herschel called the Keyhole Nebula above the point of the V. Eta is the bright orange-tinged star just to the left of the waist of the Keyhole.

▼ *A close-up of the top part of the Keyhole Nebula* taken by the Hubble Space Telescope, in the same orientation as the picture above. This shows the roughly circular region which marks the top half of the Keyhole, together with the bright wisp of gas to its left. Eta Carinae is beyond the left-hand edge of the view. To the right of the Keyhole is a strand of gas that is being evaporated by strong ultraviolet light from another bright star out of the field of view to the lower right. The colors in this image are not the same as those of a normal photograph, as the image was constructed from separate exposures through five different filters, including one in the near infrared.

ETA CARINAE – WAITING FOR THE BANG

It is always the big stars that grab the attention, and Eta Carinae (see page 146) is a case in point. At over 100 solar masses, it is near the limit above which no star can remain together. It shines with a brightness 4 million times as bright as the Sun, though because it is hidden within a cloud of its own making we see it as only

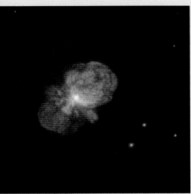

▼ **The Homunculus** surrounds the star Eta Carinae itself, seen by the Hubble Space Telescope in approximately natural color. The deep red material around it was ejected in the 19th-century eruption and is faster moving than the twin-lobed bubbles of gas. Astronomers do not really understand the mechanism that has caused the nebula to expand in this bipolar fashion, but it may be the result of the star being a binary.

magnitude 5. Fireworks are to be expected from this region at some point in the future, but whether they occur this year, this century, this millennium or even longer is just guesswork. But a star in such a fragile state is a target for close study, so that when something does happen, we have a better idea of the conditions that led up to the event.

In the 1843 eruption, Eta flared up to become one of the brightest stars in the sky and faded only slowly over a period of years. Prior to that event it had been at about magnitude 4, and is shown on several old maps at this brightness. By the beginning of the 20th century, however, it had sunk down to about eighth magnitude. Then in the 1940s it began to brighten slowly, and is currently at about magnitude 4.5.

During the 20th century visual observers noticed that the star did not appear as a point source, and by mid century it started to take on a vaguely human shape, and was referred to as the Homunculus or "little man." This appearance is easily seen in amateur telescopes: the nebula itself appears strikingly orange. It is now shown in great detail in photographs from the Hubble Space Telescope. Pictures taken only 17 months apart show a noticeable change in the size and appearance of the Homunculus, giving a good estimate of its rate of expansion, about 650 km/s or 2.4 million km/h.

Within its cocoon, Eta is probably a blue star with a binary companion which orbits it every 5.52 years, judging by the regular changes in its spectrum. It is a source of both radio waves and X-rays, and its intensity

▶ **Another part of the Eta Carinae Nebula,** this time viewed in the infrared with the Spitzer Space Telescope. Here we see the lower left-hand part of the nebula, though the image is rotated counterclockwise compared with the other views. At infrared wavelengths, the sharp left-hand edge of the bright V-shape appears dark at the top of the picture, while the dark void on the visible-light picture here appears bright. Eta Carinae is outside the field of view at the top, but it is illuminating and evaporating numerous fingers of gas in this part of the nebula.

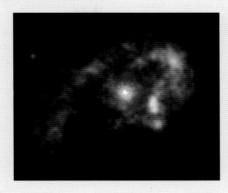

◀ **The Chandra X-ray Observatory** has provided this view of the Homunculus of Eta Carinae, showing a wider area than the Hubble shot. The star is a strong source of X-rays, and is surrounded by a horseshoe-shaped ring, two light years across, which may have been ejected from the star in an earlier eruption.

at these wavelengths changes simultaneously on this timescale. What are termed eclipses lasting three months took place in 2009 and 2014. But scientists studying the complex data from these events are not convinced that we are really seeing eclipses like those of a normal eclipsing binary star such as Algol (see page 193). It is possible that the X-rays are caused by the collision of the strong outflows of gas from the two massive stars which comprise Eta Carinae and that there is some kind of beaming effect.

As with most giant stars about to reach the end of their lifetimes, Eta is in a stellar nursery, the Eta Carinae Nebula, but which in fact covers a larger region of space. Infrared telescopes can peel away some of the nebulosity which surrounds the area, to reveal some of the young stars which have this unstable star as a troublesome neighbor.

What will happen when Eta Carinae finally goes bang? It could be

a brilliant supernova, which could shine at about magnitude −6 or even brighter, making it the brightest object in the sky apart from the Sun and Moon, and visible during the day. Some suggest that it could even become a hypernova, which is a theoretical phenomenon that has never been definitely observed. The star's core collapses to become a black hole, accompanied by a prodigious energy output, particularly in gamma rays. Such an event taking place comparatively nearby has been suggested as the reason for a mass extinction on Earth 450 million years ago. The gamma rays could have removed a considerable part of the ozone layer, leaving much life on Earth unprotected from ultraviolet radiation from the Sun. Fortunately, at a distance of about 8000 light years, Eta Carinae is too far away to have this effect, and there are now no stars of sufficient mass near enough to be a problem.

CARINA AND VOLANS

The Keel of the ship Argo occupies one of the richest areas of the southern Milky Way. Like Vela, it is crowded with stars, and Carina includes the second-brightest star in the sky, Canopus, which is at the extreme western end of the constellation. You really need to be south of the Equator to see this part of the sky in its full glory. Volans, the Flying Fish, is a comparatively recent addition to the sky.

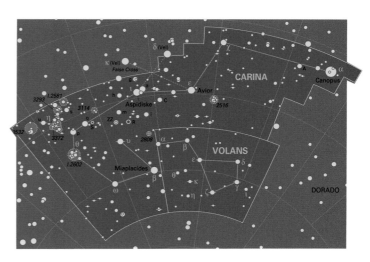

False Cross asterism

Vela and Carina have together created much confusion by forming, between them, a rival to the Southern Cross. This False Cross is actually larger and a somewhat different shape, and in particular it fails to indicate south as does the true Cross. Once you have seen the Southern Cross it is unmistakable, so it is unlikely that many have been seriously misled by these competitors – but it is worth recognizing the difference to prevent yourself being caught out. The stars in question are Kappa and Delta Velorum, and Iota and Epsilon Carina.

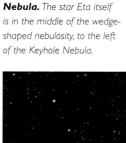

▼ **The Eta Carinae Nebula.** *The star Eta itself is in the middle of the wedge-shaped nebulosity, to the left of the Keyhole Nebula.*

Eta Carinae Nebula

The nebula that surrounds the star Eta Carinae is the largest and brightest in the sky. It lies in a star-studded region, and the whole area attracts the eye

▲ **The "False Cross"** *of stars in Vela and Carina is different in many details from the real one (see page 158).*

straight away, even in poor skies. It lies on the opposite side of the Southern Cross from Beta Centauri, at virtually the same declination, seemingly balancing the star in the sky.

Binoculars reveal a large hazy area with numerous stars, while in a telescope at low power there is a wealth of detail in the nebula, notably an L-shaped dark lane. The nebula is a starbirth region and numerous young stars and little clusters are visible surrounding the nebula.

The star Eta Carinae itself is currently fairly insignificant, at sixth magnitude. It lies within a dark zone of

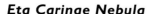

Object	Type	M$_V$	Magnification	Distance
Eta Carinae Nebula	Nebula	1.0	10×	10,000 light years
NGC 2516	Open cluster	3.8	25×	1300 light years
NGC 3114	Open cluster	4.2	10×	3000 light years
NGC 3532	Open cluster	3.0	25×	1300 light years
IC 2602	Open cluster	1.9	25×	479 light years

the nebula known from its shape as the Keyhole, though this patch has varied somewhat since it received its name in the mid 19th century. Eta is noticeably orange and is embedded in a bright nebulosity that is visible at high magnification. In 1843 Eta astonished observers by flaring up to magnitude −1, outshining even Canopus. It may do so again (see page 145). Recent observations with the Hubble Space Telescope indicate that it is actually a double, with one very massive component at 100 to 150 times the mass of the Sun, and the other at a mere 30 to 60 solar masses.

Open cluster NGC 2516
Within a binocular field of Epsilon Carinae lies this splendid open cluster, visible with the naked eye under good conditions. Its brightest star is a fifth-magnitude red giant. Binoculars show many of its brighter stars and a haze of the fainter ones that are visible with a telescope.

Open cluster NGC 3114
Sweep just to the west of the Eta Carinae Nebula with binoculars and this beautiful cluster appears. It is a naked-eye object in good skies, easily resolved into stars with binoculars or a small telescope.

▲ *Open cluster NGC 2516,* as seen on the Digital Sky Survey made using large Schmidt telescopes. The field of view is 1°.

▶ *Binoculars will give a view* of open cluster NGC 3532 similar to this photo, which has a field of 3.5° and shows stars to magnitude 10.

▲ *A wide-field view of NGC 3114,* covering 4° of sky and showing stars fainter than magnitude 14.

Open cluster NGC 3532
Every star cluster is different, and this one, celebrated as the richest open cluster, is neither the brightest nor the largest. To the naked eye it is a misty patch northeast of Eta Carinae. It lies in the same binocular field of view as Eta, and appears as a mass of stars of seventh magnitude and fainter. You will need a low power and a wide-field eyepiece on a telescope to do it justice, as it covers nearly 1°. The cluster includes several bright orange stars of type K, and there is a curious darker band across its center where there are few stars.

Open cluster IC 2602
Nicknamed "the Southern Pleiades," this cluster, like the Pleiades themselves, is a test of eyesight – there are seven or eight stars visible with the naked eye, spread over a similar area to the Pleiades. However, this cluster's stars are in general fainter. Like the Pleiades, it is best seen in binoculars because of its large size.

▼ *This view of "the Southern Pleiades"* shows a similar area to that of the original Pleiades, as seen on page 186.

MAY–JUNE

Page 202

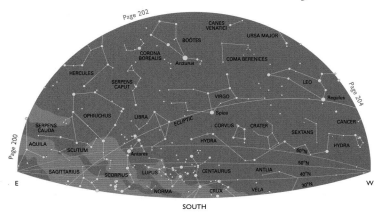

SOUTH

Page 200
Page 204

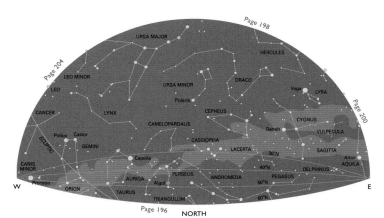

NORTH

Page 198
Page 196
Page 204
Page 200

Evening

1 May at 11.30 pm
15 May at 10.30 pm
30 May at 9.30 pm

Morning

15 January at 6.30 am
1 February at 5.30 am
14 February at 4.30 am

The northern-hemisphere sky

On May evenings the sky looking south is particularly barren of bright stars. This is because the Milky Way lies low on the horizon and in the main part of the sky we are looking at right angles to the plane of the Galaxy. There is only one really bright star, yellowish Arcturus, in the otherwise rather faint constellation of Boötes. Below this, in mid sky, is Spica, in Virgo, the rest of which extends in a straggling Y-shape toward Regulus, the brightest star in Leo, to the west. Any planets will be in mid sky, in Virgo or Leo. The ecliptic extends toward the southeast where the reddish Antares in Scorpius is rising.

Looking north, the Big Dipper or Plough is virtually overhead, while over in the northeast the bright star Vega is a taste of the stars of summer, with Deneb and Cygnus lying below it.

Evening

1 May at 11.30 pm
15 May at 10.30 pm
30 May at 9.30 pm

Morning

14 February at 4.30 am
28 February at 3.30 am
15 March at 2.30 am

The southern-hemisphere sky

At this time of the year, the northern skies are particularly devoid of bright stars other than Arcturus, in Boötes, which is a kite-shaped pattern hanging downward from it. High overhead is Spica, the brightest star in Virgo, which stretches down to the west where you will find Regulus in Leo. This also makes the line of the ecliptic, which continues east from Spica to the bright reddish star Antares, at the heart of Scorpius, the Scorpion, with the curve of its sting in the Milky Way.

If the northern skies are barren, the far southern section of the Milky Way is on full view, with large numbers of stars through Vela, Carina, Centaurus and Crux. Many of these stars are quite hard to organize into patterns, the great exception being the Southern Cross (also known as Crux) and the two adjacent stars Alpha and Beta Centauri. Below and to the left of these stars is the well-defined triangle of Triangulum Australe.

Page 202
Page 204
Page 198
Page 200

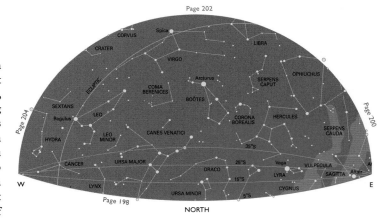

NORTH

Page 202
Page 210
Page 200
Page 208
Page 210

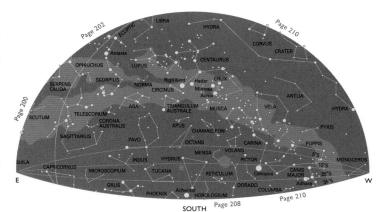

SOUTH

VIRGO

The Virgo region of sky is unimpressive to the eye, or even to binoculars. But to astronomers it is one of the most important areas of the whole heavens, for it contains the great Virgo Cluster of galaxies. But the constellation itself occupies a much wider part of the sky, and it is the second-largest constellation, after its neighbor to the south, Hydra.

The brightest star is Spica, a Latin name meaning "ear of wheat," which is what the maiden carries in her hand. The Virgin referred to is not the Christian one, but instead either the Greek goddess of justice or the corn goddess. The ecliptic runs the length of Virgo, so a bright planet may often rival Spica for attention. Virgo straddles the celestial equator, so is equally visible from either hemisphere.

This part of the sky is well away from the Milky Way, so it is comparatively barren in stars. This is fortunate, as it means that the Virgo Cluster is unobscured by the dust and gas clouds that would otherwise restrict our view. Our own group of galaxies, which also includes the Andromeda Galaxy and the Magellanic Clouds, is regarded as a minor outlier of the Virgo Cluster, the center of which lies some 55 million or so light years away. It contains around a thousand galaxies in all.

Exploring the Virgo Cluster

A star map makes the Virgo Cluster look as if one could hardly fail to stumble upon galaxies, but in fact they are not so easy to find, as the individual galaxies are quite small and faint. In a dark, clear country sky, the brighter members are just visible with binoculars if you know exactly where to look. Small telescopes will show the brighter galaxies if conditions are dark enough, as they are mostly eighth and ninth magnitude, but finding them can be a challenge. Owners of small Go To instruments may think that they will have no problems, but they may well find that unless their instrument is accurately aligned and the skies are truly dark, they will be rewarded with blank areas of sky, or uncertainty as to which object is which.

With medium and large telescopes on a good night, however, the situation changes. The galaxies are all there, and you can hunt for fainter members to your heart's content. Simply pointing the instrument in the right direction will start to bring them in. But even so, if you want to be able to identify the galaxies and not get lost, you need to learn a route.

Of several such routes, possibly the simplest is to start with the star Denebola in Leo in the finder and then move eastward to find the fifth-magnitude star 6 Comae Berenices, 6½° away. As the map shows, you

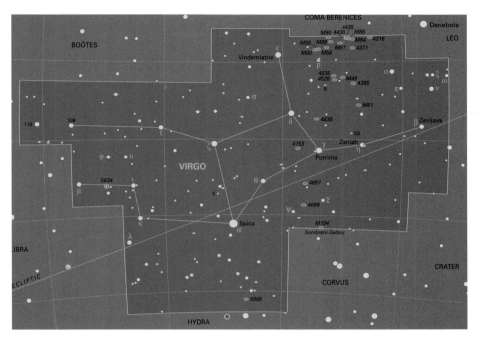

Object	Type	M_v	Magnification	Distance
M49	Galaxy	8.4	75×	53 million light years
M58	Galaxy	9.7	75×	65 million light years
M59	Galaxy	9.6	75×	51 million light years
M60	Galaxy	8.8	75×	54 million light years
M84	Galaxy	9.1	75×	56 million light years
M86	Galaxy	8.9	75×	54 million light years
M87	Galaxy	8.6	75×	54 million light years
M89	Galaxy	9.8	75×	51 million light years
M90	Galaxy	9.5	75×	40 million light years
M104	Galaxy	8.0	75×	33 million light years

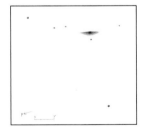

▲ *Through a 450 mm Dobsonian,* the Sombrero Hat galaxy shows a distinct straight edge on one side.

◀ *A photomosaic* of the central region of the Virgo Cluster, with M87 at center left and M86 and M84 at upper right.

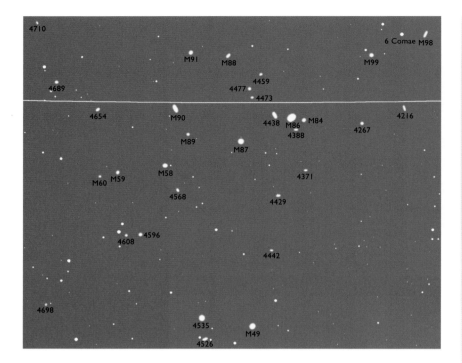

▲ **The Virgo Cluster** *mapped with stars to magnitude 9.5 and galaxies to magnitude 11, adapted from the SkyMap computer mapping program.*

▼ **The Hubble Space Telescope image of M104** *reveals the dust lane across the center of the galaxy.*

are now on the edge of the cluster (more properly known as the Virgo–Coma Cluster because it extends into Coma Berenices). Less than a degree to the southeast is M99. From here you can move 2° southeast to find two of the brightest galaxies in the cluster, at ninth magnitude, M84 and M86. They are both elliptical galaxies, and you should be able to see that M86 is larger and more elongated.

Continuing in the same direction you come to M87, a giant elliptical that is at the heart of the cluster. From there you can find M89 just over a degree to the east. Center on this elliptical, and with a low power you should find a more noticeable galaxy only 40 arc minutes to the northeast. This is M90, the first spiral galaxy along this route, and you will see that it looks more concentrated. Return to M89 and follow a route southeast to find M58, M59 and M60, a spiral and two ellipticals.

There is a second chain of galaxies farther south. To find these, return to M87 and move 4° south to the elliptical galaxy M49, then hop from there to the other galaxies in the chain, NGCs 4526, 4535 and 4365. There is one other bright galaxy in Virgo worth locating, and that is M104, the Sombrero Hat. To find it, locate Spica and then move 11° west. It looks like an oval with a strangely flat edge, resembling the wide-brimmed Mexican hat that gives it its name. Photographs show it as an almost edge-on spiral with a dark dust lane, which is what produces the straight edge. It is one of the easiest galaxies in Virgo as it is somewhat closer and probably not a true member of the cluster.

▲ **Close to the center of M87** *lies the optical jet, which is noticeably blue. This color is due to what is called synchrotron radiation, which is created when electrons are forced to spiral in a magnetic field.*

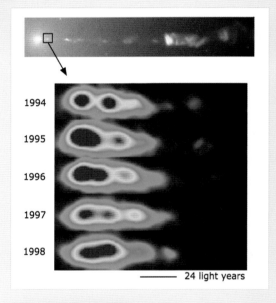

1994

1995

1996

1997

1998

———— 24 light years

▲ **Details within the M87 jet,** *here photographed by the Hubble Space Telescope, move outward along the jet year by year. This rapid movement appears to be faster than the speed of light, and is known as a superluminal velocity.*

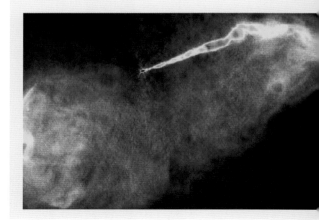

M87 AND ITS SUPERFAST JET

The giant elliptical galaxy M87 is one of the largest galaxies known, with somewhere between 800 billion and 1000 billion stars. Galaxies this big are thought to have grown at the expense of others, which have been unfortunate enough to collide with them and been stripped of all their gas and dust. A deep exposure photograph (above) shows it to be surrounded by globular clusters, some of which are giants in their own right.

Although the scale is different, you can imagine the nucleus and halo of our own Galaxy appearing like M87, with a bright more-or-less spherical core and the globulars scattered symmetrically around it. But in the case of our own and other spiral galaxies, the spiral arms spread outward from the nucleus.

However, a shorter exposure (above left) reveals a curious situation at the center of M87. A bluish jet is sticking out of the core. Although it may look insignificant, it is actually about 5000 light years in length. What could produce a jet of this sort, which must require a prodigious amount of power maintained over a long period of time? And how is it constrained to flow in such a narrow jet?

◀ A radio image shows the jet looking very similar to the optical version. But at a certain distance from the nucleus the constraint on the narrowness of the beam is lost and the electrons seem to be carried away by a separate gas flow within the galaxy.

◀ Giant elliptical galaxy M87, at the heart of the Virgo Cluster of galaxies. As with most elliptical galaxies there is a smooth increase in brightness toward the center with no hint of spiral arms or other structures. However, the fuzzy blobs surrounding it are globular clusters. Other galaxies are visible farther out.

observations. It is visible not just at optical wavelengths, but over the entire range from radio waves to X-rays. There are several lumps or knots along its length. Astronomers were amazed to discover that they could actually see changes in the positions of these knots over a matter of a year or so, at both radio and optical wavelengths, even though the source is some 55 million light years away. The movement implied that the speed of the jet was six times the speed of light!

The problem with this is that according to the special theory of relativity, accelerating a particle from a standing start to beyond the speed of light should be impossible. However, it was soon realized that because the jet is coming toward us rather than being seen side-on, there is a geometrical effect taking place.

This is not the only such jet to be observed, however. Other galaxies have jets that stream for over a million light years. And as well as the comparatively tiny bipolar jets from newborn stars, similar features are seen on some quasars – the enormously energetic nuclei of some distant galaxies. Astronomers believe that such astrophysical jets have much in common, but on vastly different scales.

The jet in M87 provides a convenient test bed for theories and

▲ A Hubble Space Telescope close-up of the M87 jet and the nucleus of the galaxy. This gives a strong impression of perspective with the jet coming toward us, but the change in

width must be real rather than due to perspective. However, the jet clearly emerges from a disk surrounded by spiral structures as gas feeds the central supermassive black hole.

The jet is actually moving at close to the speed of light. The nearest blob of gas to us has to both move toward us and then emit its light, while the light from the blob behind has already started on its journey and is only a short way behind it, so we receive the light from both of them in what seems to be an impossibly short space of time. It is a bit like someone traveling round the world by plane and airmailing letters home from each destination each day. The letters arrive daily, with the same frequency with which they were sent. But on the final day, the traveler may get back home just hours after the previous day's letter, thus spoiling the arrival rate of the information.

The fact that the jet is moving toward us also means that for similar reasons it is much brighter than a beam moving away from us. There is little sign of an opposing beam, though there probably is one.

The exact mechanism that produces the beam is poorly understood, but is undoubtedly caused by the supermassive black hole at the center of M87. The evidence for this comes from Hubble Space Telescope observations of the nucleus of the galaxy, which show a disk of matter at the very center, with the jet emanating from it. Doppler measurements of the disk show that the opposite sides of it are whirling toward or away from us at 550 km/sec, within a few light years of the center. The mass that could cause such high velocities is estimated at about 3 billion times that of the Sun. Compare that with the supermassive black hole thought to reside at the center of our own Galaxy, which is a mere 4 million solar masses.

Such a disk is known as an accretion disk, because it is fed from matter that accretes on to it from the rest of the galaxy, which in effect feeds it. This maelstrom has a strong magnetic field associated with it which beams material outward in a direction perpendicular to the disk, though the exact mechanism is still a matter for study. The poleward direction probably offers the best escape route for the material being dragged in by the intense magnetic field. Probably much more goes into the black hole itself, never to re-emerge.

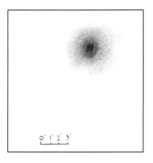

▲ **Galaxy M94** *as seen through a 76 mm refractor.*

▼ **With a 76 mm refractor,** *the two components of M51 appear as hazy blobs.*

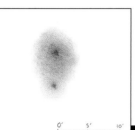

CANES VENATICI

These hunting dogs were placed in the sky comparatively recently, in 1687 by Johannes Hevelius, and they are there to help Boötes, the Herdsman, to the east. The brightest star in Canes Venatici has a curious history. Alpha is also known as Cor Caroli, meaning Charles' Heart. There are two stories about how it got this name. One says that it was named after the English King Charles II, who was restored to power on 29 May 1660, when the star was said to be particularly brilliant. But another story holds that it commemorates his predecessor, the unfortunate Charles I, who was beheaded in 1649. The star is indeed slightly variable, so who knows what may happen if it chooses to flare up again! Cor Caroli is also an easy double star.

Globular cluster M3

A bright globular cluster, M3 is one of the best in the northern sky. It is said to be the original object that Charles Messier mistook for a comet. But when he realized that he had been fooled, he set about drawing up the catalog of misty objects that now bears his name. You will need about a 100 mm telescope to turn the circular haze into actual stars.

◀ **A CCD image** *taken with a 0.9 m telescope reveals interactions between the two galaxies that comprise M51. It is just visible with binoculars.*

▲ **When photographed** *using a large telescope, M3 shows a burnt-out center. Compare this with the drawing on page 110, which was made with a 150 mm reflector.*

Galaxy M94

This is a spiral galaxy, with tightly wound arms. Even a small telescope, however, gives a hint of its spiral nature. It is easy to find even without a Go To telescope because it forms a triangle with the two brightest stars in Canes Venatici.

Galaxy M51

One of the most famous galaxies in the sky, M51 is also known as the Whirlpool Galaxy. It is the object that was first seen to be in the shape of a spiral, by Lord Rosse in 1845, though it was not known to be a galaxy at the time.

Lord Rosse used a giant 72-inch (1.8 m) telescope, but today amateurs with telescopes as small as 300 mm can repeat the observation. This is simply because today's telescopes, with mirrors made of glass and coated with aluminum, are much more reflective than the polished metal mirror used by Rosse. Even a 75 mm telescope will show the companion galaxy, NGC 5195.

M51 is most easily found by star-hopping from the end star of the handle of the Big Dipper or Plough.

Object	Type	M$_v$	Magnification	Distance
M3	Globular cluster	3.2	25×	33,900 light years
M51	Galaxy	8.4	50×	26 million light years
M94	Galaxy	8.2	75×	16 million light years

COMA BERENICES

None of the stars of Coma is much brighter than fourth magnitude, but the constellation is striking. It contains a notable star cluster, looking to the naked eye much like other clusters do in a telescope. It is best seen in binoculars, and most telescopes are of less use because the cluster is so large, at about 5° across. The group is about 250 light years away and is simply known as the Coma Star Cluster. There is another Coma Cluster, of galaxies, but these are all 11th magnitude or fainter.

The stars of the Coma Star Cluster are meant to represent the hair of the Egyptian Queen Berenice, but the name itself is not ancient and dates from the 16th century.

There are numerous galaxies within Coma, some of which at the southern end of the constellation are part of the great Virgo Cluster. Like the other Virgo galaxies, if you do not have a Go To telescope they are best located by starting at Denebola in Leo, moving east to 6 Comae, and moving from one to the other from there. Unlike many of the galaxies in the main part of the Virgo Cluster, those in Coma are mostly spirals rather than ellipticals, M85 being an exception.

M64, the Black Eye Galaxy

Coma has some galaxies that are closer than the Virgo Cluster, of which the brightest is M64. It is conveniently located near to the star 35 Comae, which you can find by star-hopping from Alpha Comae. The galaxy, which is about 18 million light years away compared with the 55 million of most of the Virgo Cluster, has a notable dark lane that supposedly lends it an eye-like appearance. However, this lane is not easily visible with small telescopes or under poor conditions, and it often looks like an oval smudge.

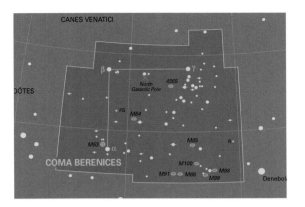

Galaxy NGC 4565

Edge-on spiral galaxies are popular with deep-sky observers because they have a greater visual appeal than the featureless ellipticals and some of the blander spirals. This is one of the best and most popular, and even a small telescope will reveal it as a spindle with a central bulge. A medium to large telescope in a good sky will show a central dust lane as well. The object is tenth magnitude but is easier to see than some of the Messier objects in the Virgo Cluster because of its more condensed nature. It is easily found from the stars of the Coma Star Cluster.

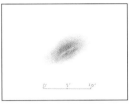
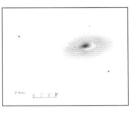

▲ **Through a 75 mm refractor** (top), M64 shows a darker spot within its body. With a larger instrument (bottom), in this case a 450 mm, the galaxy takes on a more eye-like appearance.

▶ **The Hubble Space Telescope's view of M64** reveals in startling detail the dark dust lanes. The reason for its strange appearance is uncertain. The dust lanes rotate in a different direction from the outer disk, which is mostly gas with little dust. Furthermore, the two disks are not exactly in the same plane. The obvious explanation is that two spiral galaxies are in the process of merging, but some astronomers have cast doubt on this. Even the distance to M64 is uncertain. It lies between us and the Virgo Cluster, so it is probably gravitationally attracted in that direction, which makes it difficult to use its redshift as a means of working out its distance.

◀ **The edge-on spiral galaxy NGC 4565** is a dramatic sight through a large Dobsonian.

Object	Type	M$_v$	Magnification	Distance
M64	Galaxy	8.5	75×	18 million light years
M99	Galaxy	9.9	75×	50 million light years
M100	Galaxy	9.3	75×	54 million light years
NGC 4565	Galaxy	9.6	75×	40 million light years

▶ **The spiral galaxy M100,** with two dwarf companions. This view made with a 0.9 m telescope at Kitt Peak shows that the spiral arms are quite faint, while the nucleus is fairly bright and also has a hint of spiral structure.

Galaxy M100

This spiral galaxy is one of the largest spirals in the Virgo Cluster. It lies close to 6 Comae, a popular route into the Virgo Cluster, but although it is listed as being brighter than the nearby spirals M98 and M99, its light is spread over a larger area, making it harder to see. M98 is almost edge-on, while M99 is also face-on.

However, M100's main claim to fame is that it has helped to pin down the whole distance scale of the Universe. Because it is a large and face-on spiral, it was a good candidate for finding individual stars using the Hubble Space Telescope. The picture on page 16 shows the changes in brightness of a Cepheid variable star within the galaxy that has helped to establish its distance at 54 million light years. Even so, the error in this measurement is given as 6 million light years either way, so further observations are still needed to refine the value.

DRACO

Dragons are meant to be fearsome, but the celestial version is rather faint, if large. It winds across a swathe of the northern sky around Ursa Minor, between the two bears. Its most prominent feature is its head, which is represented by four stars above Hercules, the hero who slew the dragon in mythology. Another star in Hercules makes a quite distinctive diamond shape with Draco's three brightest stars. The brightest star in Draco is actually Gamma, in the head of the dragon, and the lettering of the stars in Draco is quite chaotic, following neither their order along the dragon nor the normal bright-to-faint sequence. This is probably the celestial dragon's most fearsome aspect.

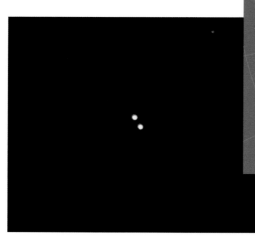

▶ **The wide double star Nu Draconis** is a pretty sight even with low-powered binoculars.

Double star Nu Draconis

The faintest star of the head of Draco, Nu Draconis is a pretty double star seen with binoculars. Two virtually identical white stars look like car headlights separated by just over a minute of arc. If you need to know what 1 arc minute looks like, this star shows you.

Object	Type	M$_V$	Magnification	Distance
Nu Dra	Double star	1.2	10×	108 light years
NGC 6543	Planetary nebula	8.1	250×	3600 light years

Planetary nebula NGC 6543

This was just another planetary nebula, though a fairly bright one, until the Hubble Space Telescope photographed it and revealed it as one of the most distinctive objects in the sky – the Cat's Eye Nebula. Through a medium-sized telescope it is a bluish disk, almost as bright as but much smaller than the Ring Nebula in Lyra. The first Hubble versions showed it as a red eye, but in a photo published in 2004 it is blue, with numerous shells surrounding the star.

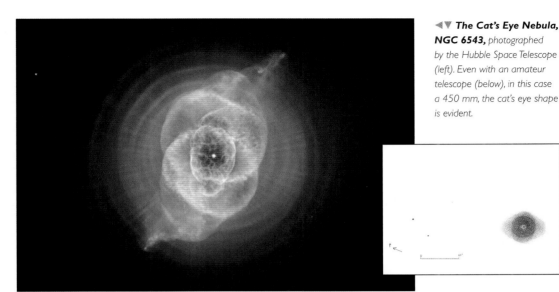

The Cat's Eye Nebula, NGC 6543, photographed by the Hubble Space Telescope (left). Even with an amateur telescope (below), in this case a 450 mm, the cat's eye shape is evident.

URSA MINOR

Ursa Minor owes its fame entirely to its brightest member – the Pole Star, Polaris. Though this is not a particularly bright star (contrary to popular myth), its location close to the sky's north pole has made it the most significant in the sky for hundreds of years. Since late Roman times it has been the Pole Star, though only since medieval times has it been particularly close. To mariners in particular, a glimpse of Ursa Minor and the Pole Star has told them the direction of north and their latitude. Even today, many telescope mounts have small telescopes built in that allow you to use Polaris to align your equatorial mount correctly. But beware of Kochab, the star at the other end of Ursa Minor, which is almost equal in brightness to Polaris. Choose it by mistake and you will be way off.

The rest of Ursa Minor is a fairly faint and undistinguished constellation that roughly resembles its more illustrious neighbor, Ursa Major, with a rectangle and a tail of stars that curves off to end in Polaris itself. In North America it is also known as the Little Dipper, though no one in Britain calls it the Little Plough. Because Ursa Minor is always at more or less the same altitude in the sky from any particular location in the northern hemisphere, many people use it as a guide to the transparency of the sky. You can look for its stars at any time of night or year and judge how much they have been affected by the absorption of the atmosphere.

Alpha Ursae Minoris (Polaris)

Polaris is probably observed more than any other star in the sky because so many people use it to align their mountings. Few pay it any attention once it has served its purpose, but it has its merits. Nearby lies a sixth-magnitude star, a member of an attractive circlet of stars sometimes known as the "Engagement Ring," which has Polaris itself as the diamond. Polaris is a Cepheid-type variable (see page 109), but its brightness changes are very slight.

Object	Type	M$_v$	Magnification	Distance
Polaris	Star	2.0	Naked eye	431 light years

◀ Polaris and the "Engagement Ring" circlet of stars, with stars shown to 12th magnitude. The field of view is 3°.

CENTAURUS

Despite the fact that it lies below declination −30° and is therefore largely unobservable from Europe, Centaurus is an ancient constellation. Over the past 2500 years, a slow movement of Earth's axis, known as *precession*, has resulted in its stars slipping southward, so Centaurus is now regarded as a southern constellation. However, almost all of it is visible from the most southerly extremities of the United States.

Centaurus is a difficult constellation to get to grips with because much of it consists of scattered second- and third-magnitude stars with no clear pattern. But its leading stars are unmistakable – a brilliant pair usually known as Alpha and Beta Centauri rather than by their names of Rigil Kent and Hadar. They are known as the Pointers because they indicate the nearby Southern Cross – not that this famous group needs much additional identification.

From Alpha and Beta a couple of straggling lines of stars spread northward and westward. Though the northern extremities of the constellation are rather barren, the southerly end of Centaurus is in a rich part of the Milky Way which contains a number of clusters, though none is particularly bright or rich.

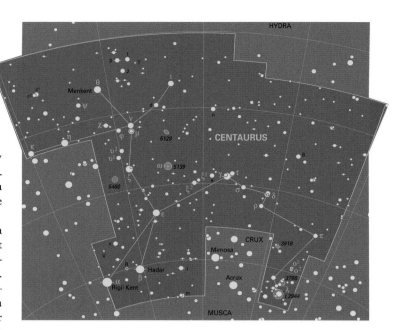

Alpha Centauri

This is the nearest bright star to the Sun, a mere 4.31 light years away, and the third brightest star in the sky. But unlike the Sun, it is a multiple star. Even a small telescope shows that it is a double, with one star somewhat brighter than the other. The brighter of the two is a type G star like the Sun, and is very similar in true brightness. Its companion, however, is a fainter type K star that orbits it every 80 years. Despite science fiction stories, no planets have been detected around the star. If there were any, the inhabitants would see a night sky very similar to ours except that there would be an extra star at first magnitude in Cassiopeia – our Sun.

There is a third member of the system, Proxima Centauri, which is slightly nearer to Earth, at 4.23 light years away, and is therefore the closest star to the Sun. However, it is a dim red dwarf and is only 11th magnitude, so it is not readily seen. Nor is it close to Alpha itself – it is 2° away, well outside the field of view of most telescopes.

▶ **This deep view of the sky around Alpha Centauri** (overexposed at top left) shows the vast difference in brightness between it and Proxima, marked by the arrow.

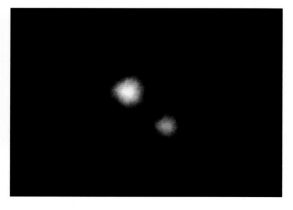

▲ **A close-up of Alpha Centauri** shows that the two stars are slightly different in brightness and color. They are separated by 14 arc seconds.

Object	Type	M$_v$	Magnification	Distance
Alpha Cen	Double star	−0.3	50×	4.31 light years
Proxima Cen	Star	11.0	–	4.23 light years
Omega Cen	Globular cluster	3.7	25×	16,000 light years
NGC 3918	Planetary nebula	8.4	100×	3000 light years
NGC 5128	Galaxy	7.0	75×	12 million light years

▲ **Omega Centauri's resemblance** *to a comet is shown in this 1986 photo of Halley's Comet, the turquoise object above center, and Omega itself, at bottom right. NGC 5128 is visible to the right of the comet.*

Globular cluster Omega Centauri

Normally, a designation like Omega Centauri would be given to a star, but in this case the fourth-magnitude star that you can see in the sky looks distinctly more fuzzy than the others. You can use the stars of the Southern Cross to point to it. Turn binoculars or a telescope on this star and it is clearly a globular cluster, though unwary observers using low powers might think that they have discovered a new tail-less comet. It is not just any old globular, however, but the brightest in the sky. It is a must-see object for anyone who lives far enough south to observe it well – which means anywhere south of the Mediterranean, or the southern states of the United States. A small telescope will resolve some stars, and the larger the aperture the more amazing the view. In a medium or large telescope under good conditions the sight is generally described using superlatives such as "awesome." The cluster contains around a million stars, and it is one of the nearest globular clusters.

NGC 3918, the Blue Planetary

Several planetary nebulae are bright enough to appear blue, but many observers comment on the blueness of this one. Find it by using Crux to indicate a little triangle of fifth- and sixth-magnitude stars. It is fairly small so a considerable magnification is needed to show a fuzzy disk with little internal detail.

Galaxy NGC 5128

If you are observing Omega Centauri, move just under 5° to the north and you will spot this peculiar galaxy, which, at seventh magnitude, is one of the brightest in the sky. It is noticeable in binoculars, often in the same field of view as Omega but very much smaller, appearing as a small fuzzy glow. A small telescope shows it as a circular patch of light, but with a medium-to-large telescope you can see a dark lane bisecting the center. One nickname for this object is "the Hamburger Galaxy."

This is one of the most thoroughly observed galaxies in the sky as far as professional astronomers are concerned. The dark lane seems to be the remains of a spiral galaxy that has collided with the main elliptical galaxy, and it is a major source of radio waves, known as Centaurus A.

◀ **It may be known as the Blue Planetary,** *but this Hubble view of NGC 3918 emphasizes different colors.*

▼ **Giant galaxy NGC 5128** *(Centaurus A), photographed by the Very Large Telescope in Chile.*

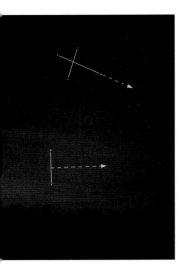

▲ **How to use the Southern Cross** and the Pointers to find the south celestial pole.

CRUX AND MUSCA

There can be few people who have not heard of the Southern Cross. The Latin version of the name is simply Crux, meaning the Cross. As we have become used to these stars on the flags of several nations, it may seem odd that the group has not always been a separate constellation. Until the 17th century, however, the stars were regarded as part of Centaurus, though they have not been visible from Europe since ancient times because of precession. It was when European navigators first saw the pattern that it became regarded as a cross, and it is now generally taken as a sign that you have arrived in the southern hemisphere, despite being visible from as far north as the southern tip of Florida in the United States.

The Southern Cross can be used as a quick means of indicating the south celestial pole. Northern observers have the luxury of a bright star, Polaris, within a degree of the north celestial pole. But there is no such bright star in the southern hemisphere, so observers must make more of a guess. The long axis of the cross points roughly north–south, but by itself this is not enough. Bisecting the line between Alpha and Beta Centauri provides another line, and the two intersect about 3° from the true pole.

Crux is the smallest constellation of all, but its location in the Milky Way provides it with several interesting objects. Alpha Crucis is a double star that can be resolved using small telescopes, the separation of the two stars being 3.9 arc seconds. Gamma is a red giant star, appearing noticeably orange compared with the other three bright stars of the cross, which are all type B.

Musca, the Fly, is one of those constellations devised in the 16th century to fill in an otherwise unclaimed area of sky. The globular cluster NGC 4833 is an easy object to find just north of Delta Muscae, though it requires a medium aperture to resolve some stars.

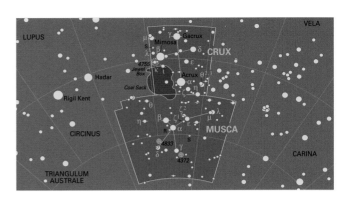

NGC 4755, the Jewel Box or Kappa Crucis Cluster

Binocular observers, having heard that this is the most beautiful cluster in the sky, may wonder if they are looking at the right spot, because all they see is a knot of stars with none of the haze that signifies a rich cluster. The object is 1° from Beta Crucis, and is also known as Kappa Crucis after the cluster's brightest star, which is of fifth magnitude. With a telescope and a moderate power, however, the object becomes transformed. The stars are mostly bluish, with the exception of an eighth-magnitude red giant near the center. The color contrast has led to the cluster's popular name.

The Coalsack Nebula

If the sky is sufficiently clear that the Milky Way is visible, it becomes obvious that to the east of the Cross is a dark region, apparently devoid of stars. This is a dark nebula of the same type as those of the Cygnus Rift (see page 29). Although there is at least one star of naked-eye visibility within its borders, the effect is nevertheless striking. It is about 600 light years away.

▼ **A photograph of the Jewel Box Cluster** with a large telescope shows the contrasting colors of the bright stars.

▲ **The Coalsack is a dark cloud** in an otherwise bright part of the Milky Way, close to the Southern Cross.

Object	Type	M_v	Magnification	Distance
NGC 4755	Open cluster	4.2	50×	7600 light years

JULY–AUGUST

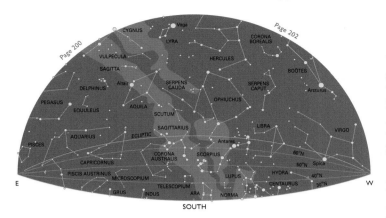

SOUTH

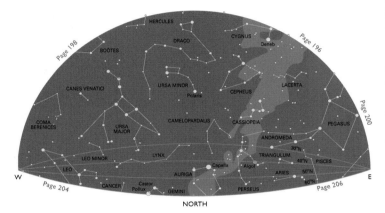

NORTH

The northern-hemisphere sky

Your summer signposts are the three stars which comprise the "Summer Triangle": brilliant Vega, almost overhead; Deneb, farther to its east; and Altair, in mid sky, which is flanked by two lesser stars. Having found Deneb, look for the "Northern Cross" of Cygnus, which lies between Vega and Altair. Also spot Arcturus, the same brightness as Vega but with a yellower color, halfway down to the western horizon. Between Vega and Arcturus lie the wedge shape of the Keystone of Hercules and the semicircle of Corona Borealis.

This is the best time of year to view the Milky Way, which stretches virtually overhead in summer, leading down toward the south where you will find the zodiacal constellations of Sagittarius and Scorpius. Look for "the Teapot" of Sagittarius and the reddish star Antares, also flanked by two others, which is the heart of the Scorpion.

Evening

1 July at 11.30 pm
15 July at 10.30 pm
30 July at 9.30 pm

Morning

1 April at 5.30 am
15 April at 4.30 am
30 April at 3.30 am

The southern-hemisphere sky

The Milky Way dominates the winter heavens, with its brightest parts virtually overhead at this time of year. Even in very poor skies you can make out its path, marked by the Southern Cross on its side in the southwest, Beta and Alpha Centauri in mid sky, then Sagittarius and Scorpius overhead. Look for the shape of Scorpius, with reddish Antares at its heart, a line of stars as its claws, then a curve of stars ending in a distinctive pattern that marks its sting.

Over to the north is a very large triangle of stars – Altair in mid sky, flanked by two lesser stars, bright Vega lower down, and Deneb scraping the horizon from many locations. Look toward the west to find Arcturus, the same brightness as Vega but more yellowish. Between the two lie Hercules and Corona Borealis.

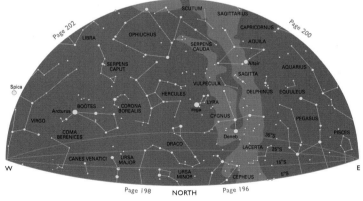

NORTH

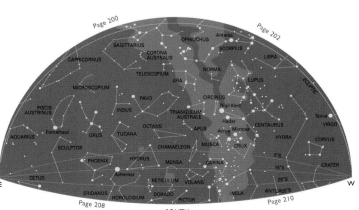

SOUTH

Evening

1 July at 11.30 pm
15 July at 10.30 pm
30 July at 9.30 pm

Morning

1 April at 5.30 am
15 April at 4.30 am
30 April at 3.30 am

159

SAGITTA AND VULPECULA

These two tiny constellations have a popularity beyond their size because of their location within crowded starfields of the Milky Way. The arrow shape of Sagitta, lying midway between Albireo in Cygnus and Altair in Aquila, is easily found when the skies are dark enough to show the Milky Way properly, though binoculars are needed in bright skies. Vulpecula is less obvious but is easily located from Albireo.

Planetary nebula M27, the Dumbbell Nebula

Even binoculars will show this bright and large planetary nebula. Though it is in Vulpecula, many observers locate it by using the stars of Sagitta – it is at a right angle to the arrow from the faint end star, Eta. The nebula itself is magnitude 7.3, though it appears fainter than a star of this magnitude because its light is spread over an area about a quarter of the size of the full Moon. Its name comes from its double-lobed appearance in a telescope.

▲ **There is no mistaking the Coathanger** – *its alignment of stars is easy to spot even though they are between magnitudes 5 and 7.*

▼ **Cluster M71 in Vulpecula** *is now generally considered to be a loose globular cluster rather than an open cluster.*

▲ **This view of M27** *was taken with a digital camera on a 127 mm refractor, with an exposure of just 60 seconds.*

Open cluster M71

Point a telescope midway between Gamma and Delta Sagittae and you will find the cluster M71. This remote object seems to be midway between an open cluster and a globular cluster; astronomical authorities differ as to how it should be classified, but in a telescope it looks decidedly like a globular cluster as it is hard to resolve into stars.

The Coathanger

This well-known and attractive asterism gets its name from its shape, though it is also known as Brocchi's Cluster. The stars that comprise the asterism are all at different distances so it is nothing more than a chance alignment. The asterism is best viewed in binoculars.

Object	Type	M$_v$	Magnification	Distance
M27	Planetary nebula	7.3	50×	1200 light years
M71	Globular cluster	8.2	100×	13,000 light years

LYRA

Vega, to the west of Deneb, is one of the brightest stars in the sky, at magnitude 0. With Deneb and Altair it forms what is known to northern-hemisphere observers as the Summer Triangle, a large asterism (star pattern) that actually persists well into the early winter and is useful for helping to locate objects.

The constellation itself is compact, mainly consisting of a triangle of stars to the east of Vega and another pair of similar brightness to the south, toward Albireo in Cygnus.

Beta Lyrae is an unusual variable star. It is an eclipsing binary (see page 109) with individual components so close together that each star is distorted into an egg shape. This is not visible in a telescope, but amateur astronomers can confirm it for themselves by monitoring the brightness changes of the star. It varies between magnitudes 3.3 and 4.4 in 12.9 days, and when its light curve is plotted, the results are much smoother than would be expected from a more normal eclipsing binary such as Algol.

Epsilon Lyrae, the Double Double

Young observers can easily see with the naked eye that the northernmost star of the triangle beside Vega is a double star. The two components are separated by 208 arc seconds, which is 3½ arc minutes. Those with less keen eyesight can see this easily with binoculars or a finder scope. But look at Epsilon through any but the smallest telescope and you should see that the two components of the star are themselves double. The periods of the two double stars are 1000 and 600 years, so even over a lifetime of observing you will not notice any great change; nevertheless, the Double Double is a showpiece of the sky and features high on the list of objects to observe at star parties.

Object	Type	M$_v$	Magnification	Distance
Epsilon Lyr	Double star	4.0	200×	160 light years
M57	Planetary nebula	8.8	100×	2300 light years

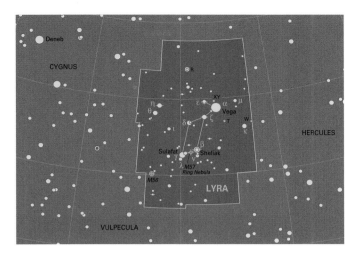

M57, the Ring Nebula

This planetary nebula is another favorite, not least because it is so easy to find, roughly midway between Beta and Gamma Lyrae. Point the telescope between these stars, possibly slightly nearer to Beta, and it is in the center of the field of view. However, you need a moderate power to see it well – around 100 is fine. It appears as a ghostly ring some 1.4 arc minutes across – about twice the size of Saturn as seen in the telescope. The central star is about 15th magnitude and you need a large amateur telescope to see it.

▶ **The best donut in the sky,** *the Ring Nebula displays strong colors in this Hubble Space Telescope photograph. But visual observers rarely report seeing color in this object.*

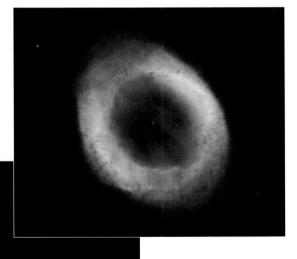

◀ **The components of each** *of the two doubles of the multiple star system Epsilon Lyrae are separated by 2.3 and 2.6 arc seconds, so they are a test for a 60 mm telescope.*

161

HERCULES

This is a large, faint and sprawling constellation in the mid northern sky. Its most recognizable feature is a pattern of four stars known as the Keystone, which lies midway between the more easily spotted semicirclet of Corona Borealis and the bright star Vega. Once you have found the Keystone, look for the extensions southward to Beta Herculis (Kornephoros) and Alpha Herculis (Rasalgethi). The Keystone is also your signpost to the two globular clusters, M13 and M92, which are the main attractions of the constellation.

Hercules is an ancient constellation, featuring in many of the Greek legends about the heavens. But the poor chap is depicted upside down in the sky, this being the only way he can rest his foot on the head of the dragon, Draco, that he slew.

Globular cluster M13

Sometimes called "the Great Hercules Cluster" or similar, M13 is the brightest globular cluster in the northern half of the sky (though southern observers have been heard to sneer at it in comparison with the great southern globulars). It is a fine sight in a medium-sized telescope, appearing as a ball of stars. Find it by looking about two-thirds of the way up the western edge of the Keystone. In small instruments and binoculars it is a hazy circular blur, but the larger your telescope, the more it turns into individual stars. Photos made with large telescopes usually make the center appear white as all the star images run together, but through even a large telescope you see only individual stars as pinpoints of light. The cluster has around a million stars in all.

Globular cluster M92

Though less rich than M13, M92 is still an enjoyable sight. It is easy to spot in binoculars in a good sky, though city observers will struggle as it may be difficult to distinguish M92 from a star, being quite a compact object. It makes a triangle with the two northernmost stars of the Keystone. It is easy to compare M92 with

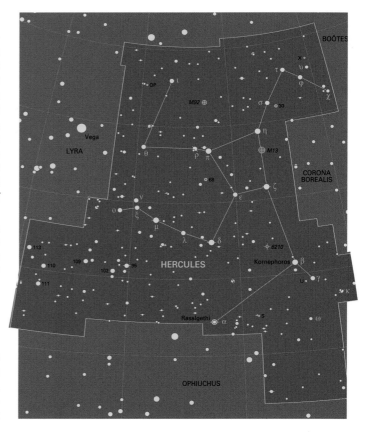

▲ **A close-up of M13** taken with a CCD imager on a 130 mm refractor in Dengie Marshes, Essex, UK.

M13 as they are only 9° apart. M92 is renowned for containing some of the most ancient stars of any globular cluster in our Galaxy. At around 14 billion years old, they are similar in age to the Universe as a whole. See also the image on page 111.

▼ **M13 (left) and M92** offer an interesting comparison. These telephoto lens shots give the same impression as you would get visually with a small telescope, though the brightness is greatly exaggerated.

Object	Type	M$_v$	Magnification	Distance
M13	Globular cluster	5.8	25×	25,100 light years
M92	Globular cluster	6.4	50×	26,700 light years

AQUILA

Despite being a prominent constellation in the Milky Way, Aquila is surprisingly devoid of bright and well-known deep-sky objects, though there are good starfields for binocular stargazing. Its brightest star, Altair, is easily identified as there are two less bright stars on either side of it. This trio lies at one side of a large diamond shape of stars, with an offshoot to a C-shape of stars at the northern end of Scutum.

Variable star Eta Aquilae

One of the brightest of the Cepheid type of variable star, Eta Aquilae is easy to find and to make estimates of because it is surrounded by convenient comparison stars. It varies between magnitudes 3.5 and 4.4 every 7.2 days. Members of the UK-based Society for Popular Astronomy have been observing this and other bright variable stars for many years.

Dark nebulae B142 and B143

For a constellation in the Milky Way, Aquila is surprisingly empty of bright deep-sky objects, with just a couple of minor clusters shown on the map. However, it does have a good number of stars, which makes it a good area for scanning with binoculars. In particularly good skies you may just notice an absence of stars just to the west of the star Tarazed, the most northerly of the two stars that flank Altair. The dark nebulae B142 and B143 are here, known collectively as "Barnard's E" from their shape, after the great American observer and astrophotographer E. E. Barnard who first noticed them on his photographs of the Milky Way.

Object	Type	M$_v$	Magnification	Distance
Eta Aql	Variable star	3.5–4.4	Naked eye	1400 light years
B142, B143	Dark nebula	–	7×	2000 light years

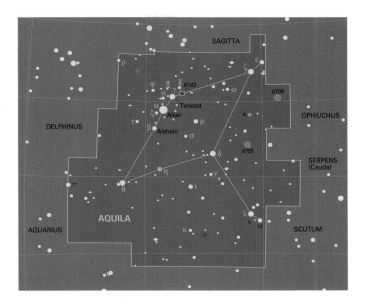

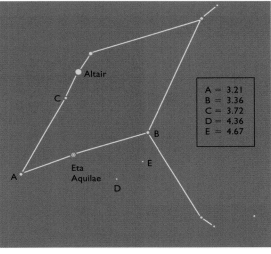

A =	3.21
B =	3.36
C =	3.72
D =	4.36
E =	4.67

◀ **Eta Aquilae,** *a variable star whose brightness is best estimated with the naked eye. Use these comparison stars to make your own estimates.*

▶ **The dark nebulae B142 and B143.** *Though this feature is more commonly photographed than observed visually, it is possible to make out the sideways dark E with binoculars fairly well in good country skies as long as the Milky Way itself is easily visible. It lies just to the west of Gamma Aquilae, Tarazed, and this dark nebula is often overlooked in favor of bright objects.*

SCUTUM

A distinctive C-shape of stars at the south end of Aquila directs you to Scutum, though most of the C-shape is actually in Aquila itself. The rest of this small constellation has no particular shape. However, Scutum contains a star cloud that is the brightest part of the Milky Way seen from northerly latitudes (from which the brighter star clouds of Sagittarius are too low in the sky to be prominent). Being just south of the celestial equator Scutum's star cloud is equally accessible from either hemisphere.

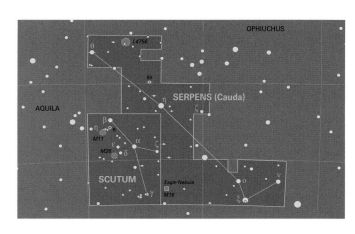

▶ *The Wild Duck Cluster,* M11. *The V-shaped pattern of stars is best seen using a telescope.*

▼ *The Eagle Nebula, M16, photographed with the 4 m Mayall telescope at Kitt Peak, Arizona. The central dark features resemble an eagle, though some observers see them as a "Star Queen."*

M11, Wild Duck Cluster

This beautiful cluster, visible in binoculars as a small misty patch, needs a telescope to do it justice as most of its stars are of magnitude 10 or 11. Collectively, however, they are sufficiently bright that in dark skies some people can see the cluster with the naked eye. Its name comes from the V-shape of its brighter stars, resembling wildfowl in flight. Only a few hundred of its nearly 3000 stars are within reach of amateur telescopes.

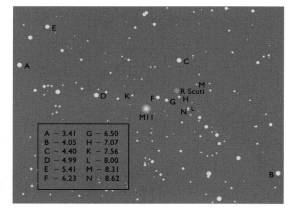

A — 3.41	G — 6.50
B — 4.05	H — 7.07
C — 4.40	K — 7.56
D — 4.99	L — 8.00
E — 5.41	M — 8.31
F — 6.23	N — 8.62

▲ *This chart of comparison stars for R Scuti is adapted from the one used by the British Astronomical Association's Variable Star Section.*

Variable star R Scuti

This is a favorite among variable star observers, because its location near the Wild Duck Cluster in a quadrilateral of stars makes it easy to find and to make estimates of its magnitude. It is the brightest of a type of pulsating variable known as RV Tauri stars (see A–Z section, page 293), and it spends much of its time at around magnitude 5. It may dim to magnitude 8 every four or five months, making the quadrilateral of stars look different, though for years on end its variations may be only a few tenths of a magnitude.

Object	Type	M$_v$	Magnification	Distance
R Sct	Variable star	4.2–8.6	Naked eye	3400 light years
M11	Open cluster	6.3	50×	6000 light years

SERPENS CAUDA

The stars of Serpens Cauda – the Serpent's Tail – are faint and undistinguished. They form a line to the west of Scutum, but one of the stars apparently in the line is actually in Ophiuchus. The brighter head of Serpens is on the other side of Ophiuchus. The constellation has little to offer the visual observer, as it is almost totally in the direction of a nearby dark cloud, part of the Great Rift (see page 29). Only at the extreme southern and northern ends are there many stars visible in binoculars. The cluster IC 4756 is an attractive sight with a low-power telescope.

M16, the Eagle Nebula

This is the location of one of the world's most famous astronomical images, the amazing "Pillars of Creation" from the Hubble Space Telescope. But visually it is rather a disappointment unless you have dark skies and the right instrument. In average skies it appears as a run-of-the-mill open cluster, with perhaps a hint of nebulosity. Binoculars show it clearly, though the individual stars are not obvious. Wide-field photographs pick up the red glow typical of a starbirth region, but close-up deep exposures show the darker lanes in the nebulosity to the south of the cluster. These "elephant trunks," as they are generally called, are pillars of denser gas within which stars are forming.

The Eagle Nebula gets its name from a fancied resemblance to an eagle with raised head and outstretched wing. These features can be glimpsed in large amateur telescopes but they are best enjoyed on photographs.

▼ *Hubble's image of the "Pillars of Creation"* in M16, in a 2014 reshoot. The left-hand trunk is the "head of the eagle."

Object	Type	M_v	Magnification	Distance
M16	Open cluster/Nebula	6.4	75×	7000 light years

OPHIUCHUS

Ask a stargazer to point to Ophiuchus and they will probably wave a hand at a barren area of sky somewhere above Scorpius and say, "It's around there." Though its brightest stars are of second magnitude, similar to those of, say, Cassiopeia, they are spread over such a large area of sky that there is no easily recognized pattern other than a misshapen pentagon. Possibly its most recognizable feature is a pair of stars, about 20° above the bright stars of Scorpius, named Yed Prior and Yed Posterior, meaning "hand in front" and "hand behind" in a mix of Arabic and Latin. Though these stars are unrelated to each other, their proximity in the sky makes an eye-catching pair that marks the western end of Ophiuchus. The constellation represents Aesculapius, a mythical healer, whose snake-entwined staff is still a symbol for the medical profession.

One peculiarity about Ophiuchus is that it is nowadays a major constellation of the zodiac – the band across the sky along which you find the Sun, Moon and planets. The Sun is actually within the borders of Ophiuchus between 30 November and 18 December, whereas it spends only the preceding six days in Scorpius. The dates given in the "Star Signs" that you read in the papers have nothing to do with the actual positions of the heavenly bodies in the sky, and the

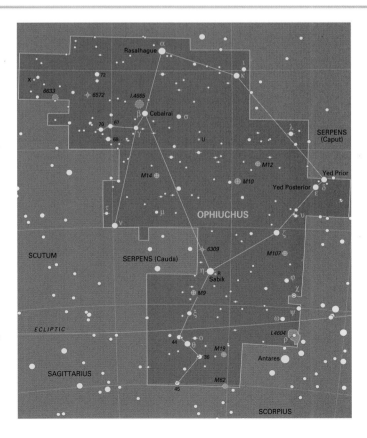

▶ **Nebulosity around Antares and Rho Ophiuchi** is a favorite target for astrophotographers fortunate to have good dark skies.

▲ **The center of the open cluster IC 4665** in Ophiuchus, photographed with a 130 mm reflecting telescope. Stars fainter than magnitude 14 are visible, but only the bright stars are cluster members.

definitions of the boundaries of the constellations have changed over the years. Ophiuchus extends both above and below the celestial equator, so it is visible from both hemispheres.

Open cluster IC 4665

Most objects in the *Index Catalogue* (the successor to the *New General Catalogue* or NGC) are rather faint objects. IC 4665 is an exception: a scattering of a dozen or so stars of around seventh magnitude, it is easily visible in binoculars. But the stars are so loosely clustered that in a telescope you may miss it.

Milky Way around Rho Ophiuchi

The Milky Way passes through Ophiuchus, and the area just above Antares in Scorpius is a favorite among photographers when there are dark skies, for in this area are beautiful bright and dark nebulae of differing colors. But in average skies little is visible. In the photo above, the bright area at the bottom is the star Antares in Scorpius, while to its right is the globular cluster M4. Rho Ophiuchi is at the top, surrounded by a blue nebula.

Globular clusters M10 and M12

These two clusters are interesting to compare because they are of similar brightness and are within the same field of view of many binoculars, which will show them looking like hazy stars in fairly dark skies. They are the brightest of many globular clusters in Ophiuchus. M10 is more condensed, while M12 appears larger. As with all globulars, the larger the telescope you use to view them, the better they appear.

▲ **Binoculars will show the globular clusters** M10 (lower left) and M12 (upper right) within the same field of view. This image shows stars to magnitude 13.5, which is fainter than the normal binocular limit.

Object	Type	M$_v$	Magnification	Distance
Rho Oph complex	Nebula	–	Photographic	400 light years
M10	Globular cluster	6.6	25×	14,300 light years
M12	Globular cluster	6.7	25×	16,000 light years
IC 4665	Open cluster	4.2	10×	1400 light years

SAGITTARIUS

A jewel among constellations, Sagittarius includes the richest areas of the Milky Way and contains more bright deep-sky objects than any other region of the sky. But to many northern observers, Sagittarius is hard to observe. From much of North America and Europe, it never rises more than 10° or 20° above the horizon. As a result, many of its wonders are lost in haze and are dismissed in favor of lesser objects higher in the sky.

The stars of Sagittarius are not particularly bright – the brightest, Kaus Australis, is only about 37th in the league table of bright stars. To many people, Sagittarius is most easily spotted by what is called "the Teapot," a pattern of eight stars that really does resemble a teapot. Most of the numerous deep-sky objects in the constellation are near this area, within the plane of the Milky Way. In a dark sky, with binoculars or a low-power telescope, the whole area is a superb sight with countless stars.

Sagittarius represents a centaur, not to be confused with Centaurus itself. He is an archer, aiming his arrow at the heart of neighboring Scorpius – the spout of the teapot is actually the point of the arrow.

The ecliptic runs through Sagittarius, this being its most southerly point.

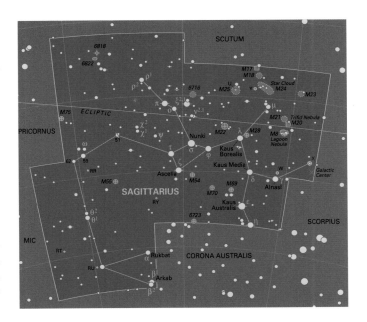

M8, the Lagoon Nebula

This is one of the brightest nebulae in the sky, following only the Eta Carinae Nebula and the Orion Nebula. In a good sky it is easily visible with the naked eye. With binoculars M8 looks like a cluster surrounded by an oval haze. Telescopically the dark central channel – romantically seen as a lagoon by 19th-century popularizer Agnes M. Clerke – is easy to see with low powers. The nebula is larger than the apparent size of the full Moon under good conditions.

The Lagoon Nebula lies at the southern end of a ragged line of nebulae and clusters, which includes most of the other diffuse nebulae in Messier's catalog. It makes an excellent starting point for a tour of deep-sky wonders. In Chapter 2, on page 21, you can compare a photograph of M8 taken with a large amateur telescope with a drawing made using a 300 mm telescope.

M20, the Trifid Nebula

This nebula is more difficult to see than the Lagoon, though it lies within the same binocular field of view. It is smaller and fainter than M8, and requires medium apertures for good views. The Trifid Nebula gets its name from its three dark lanes; the word is Latin for "cleft in three," reminiscent of John Wyndham's

classic sci-fi story *The Day of the Triffids*, about three-footed man-eating mobile plants – though there is no connection. The northern part of M20 is a blue reflection nebula. Though both are rather disappointing visually, they are easier to photograph, and the contrasting colors jump out at you.

Open clusters M23, M24 and M25

Lying across the north–south line of diffuse nebulae is this band of three clusters. M24, the central one, is sometimes known as the Little Sagittarius Star Cloud. It is really a large bright patch of the Milky Way, though there are a number of stars embedded in it which appear as a cluster in a small telescope. All three clusters are easily visible with binoculars and are best viewed at low power in a telescope.

▼ **M20, another famous poster object,** visible within a short distance of both M16 and M8. M20 is one of the brightest examples of a reflection nebula in the sky.

Object	Type	M$_v$	Magnification	Distance
M8	Nebula	4.6	25×	5200 light years
M17	Nebula	6.0	25×	5000 light years
M20	Nebula	9.0	25×	5200 light years
M22	Globular cluster	5.1	25×	10,400 light years
M23	Open cluster	6.9	25×	2150 light years
M24	Open cluster	4.6	10×	10,000 light years
M25	Open cluster	6.5	25×	2000 light years

▲ **From left to right,** *the open cluster M23, the star cloud M24 and the open cluster M25. Some books refer to a smaller cluster, NGC 6603, at the heart of the star cloud as being M24. At the top of the picture is the Swan or Omega Nebula, M17.*

▲ **One of the brightest diffuse nebulae,** *the Swan or Omega Nebula, M17, has a characteristic shape. This drawing was made using a 300 mm reflector.*

M17, the Swan or Omega Nebula

This nebula goes by several aliases. To our classically educated forebears it was a Greek capital omega, while to others it is the number 2, or a horse-shoe. But many people prefer to see it as a swan. It is bright and easy to spot in binoculars, while a low power on a 114 mm telescope readily reveals the swan shape. To its south lies a rather sparse cluster, M18, while to its north is M16 in Serpens.

Globular cluster M22

This is the brightest and largest of the globular clusters visible from the northern United States and northern Europe, though M13 in Hercules is often given that title. Being quite far south, however, M22 suffers, for many observers, from being low in the sky, which makes it appear less spectacular. M22 is the brightest of several globular clusters that can be found in Sagittarius.

▲ **If it were not for its southerly declination,** *globular cluster M22 would be better known among northern observers.*

JOURNEY TO THE CENTER OF THE GALAXY

We take it for granted that we live in a spiral galaxy, but this is not obvious just by looking at the Milky Way. The Sagittarius part of the Milky Way is the brightest, and indeed the center of the Galaxy lies within its boundaries. However, because the Galaxy spreads all around us, we are not in an ideal position to get an overview. The Solar System happens to be quite close to the plane of the Galaxy, so we are in the thick of it. The gas and dust clouds which fill the spiral arms obscure our view in many directions, and in the

diameter of about 100,000 light years. The Sun and its Solar System takes about 200 million years to complete one orbit of the Galaxy.

Looking toward the center of the Galaxy itself we find several circular features that are identified with supernova remnants, while at the very center lies an object known as Sagittarius A. This dates back to the early days of radio astronomy, when the brightest radio sources in the sky were given letters according to their brightness and the constellation in

▲ **The center of the Galaxy** *lies in the middle of this photo, which includes both Sagittarius and Scorpius. The brightest part of the Milky Way is the main Sagittarius Star Cloud —*

the fainter and smaller cloud known as M24 lies some way to its north. Dark dust lanes hide the direction in which the center itself lies. The dark nebula to the right of center is the Pipe Nebula.

direction of the galactic center in particular. Visible light is blocked completely, and only radio or infrared wavelengths reach us.

A wide-field photo of the Milky Way gives us a general impression, however, and with imagination we can picture the great bulge of the central regions. We know that Sagittarius is full of globular clusters, so we can picture these surrounding the bulge, a bit like bees round a hive. But to plot the spiral arms we must turn to radio and infrared astronomy and the techniques outlined in Chapter 1, of mapping the Doppler shifts of the gas clouds within the spiral arms. This produces a clear picture of the Milky Way with several well-defined spiral arms, with the Sun centrally placed in a minor spur. Our distance from the galactic center is about 26,500 light years and we are roughly half of the way out toward the edge of the main disk of stars, which has an overall

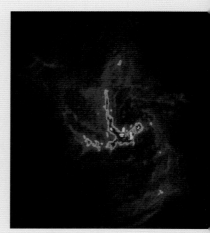

▲ **A radio image made with the Very Large Array radio telescope** *shows infalling gas at the heart of the Galaxy creating this mini-spiral. Sgr A* is the bright dot at the center of the spiral. It is surrounded by a ring of gas known as the Circumnuclear Disk. The area shown is the central 30 light years or so.*

which they were discovered. Since that time, however, more details have been revealed about the objects at the very center. In particular there are three features now known within Sagittarius A, known as Sagittarius A East, Sagittarius A West and the starlike Sagittarius A Star, the latter name usually being written Sgr A*.

Sagittarius A East is another supernova remnant, while Sgr A* marks the very center of the Galaxy. Doppler shifts show that material is orbiting this object very rapidly, at speeds of up to 1400 km/s, yet within a tiny area. This usually means just one thing – a supermassive black hole. The feature also shows up in X-rays, as would be expected from a black hole, though the intensity varies considerably, presumably as fresh supplies of gas fall into the accretion disk surrounding the black hole.

A radio map of the region shows a dramatic feature known as the mini-spiral, which is designated Sgr A West and has Sgr A* at its center. This looks at first sight as if it is at the heart of the spiral structure of the Galaxy as a whole, but in fact it is probably the

form taken by streamers of gas falling toward the black hole.

Stars also show up on infrared images. These can be observed moving over a period of just a few months, and their paths have been traced as they orbit the galactic center. Several stars can be picked out within the central light year of the black hole, which itself may have a diameter about a third of the size of Mercury's orbit around the Sun or even less. However, nothing can be seen at these wavelengths which corresponds with Sgr A*. The black hole is black indeed.

These stars are moving so quickly this close to the black hole that some have been tracked over most of their orbits, and animations of the scene are available on the Internet. One star has been watched racing within 17 light hours of the black hole – equivalent to three

times the distance of Pluto from the Sun. The fact that stars can orbit it shows that the popular notion that black holes suck in everything within their reach is a myth. However, somewhat closer to the hole itself the

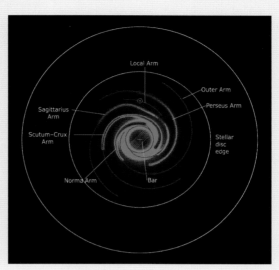

▲ **Observations taken from a wide range of wavelengths** *can be used to provide a plan view of the Milky Way. This composite view, produced by the Canadian Galactic Plane Survey, shows the Sun's location by the yellow circle, on a small spur known as the Local Arm or Orion Arm. The main arms are named after constellations that have prominent features in those directions.*

difference in gravitational pull from one side of the star to the other will be so great that the star will literally be pulled apart, though the star would have to get very much closer for this to be a real danger. Nevertheless, as the hole grows in mass these stars, too, will find its gravitational pull impossible to resist.

At the moment, the black hole's appetite for gas is tiny – some say as little as a fraction of an Earth mass a year, though others suggest a fraction

of the Sun's mass a year. It may be in a quiescent state at the moment, possibly because supernovae such as the nearby Sagittarius A East have temporarily blown away the supply of infalling gas on which it feeds.

However, intense X-rays from a hydrogen cloud some 350 light years away from the central region have been interpreted as being caused by a much greater X-ray brightness for the accretion disk in the recent past. As seen from Earth, the radiation is still flooding the gas cloud. Had Isaac Newton possessed an X-ray tele-scope, he might have observed a much brighter black hole than we do today.

There is no obvious jet from the Milky Way's black hole, though this is a characteristic of other supermassive black holes. The hole's mass is still being debated. The orbits of stars seen close to the area give fairly consistent values of 4.1 to 4.3 million Suns. The fact that some stars have been seen to survive passing within 6.5 light hours of the center show that whatever is there is no larger than the orbit of Uranus around the Sun. The only known way to squeeze that much matter into such a comparatively small area is to invoke a black hole. In 2014, a gas cloud was observed to pass close to the center and fireworks were expected, but in fact nothing unusual was observed. The area is still keeping its secrets!

▼ **The central two degrees** *of the plane of the Milky Way as seen by the Chandra X-Ray Observatory. Most of the bright spherical regions are supernova remnants, but the central bright object is Sagittarius A.*

▲ **Stars surrounding the galactic center** *(arrowed), seen by the Very Large Telescope at infrared wavelengths. These stars orbit the galactic center in a few years, compared with the Sun's orbital period around the same object of 200 million years. Beyond this area are several clusters of massive stars, such as the Arches Cluster and the Quintuplet Cluster. None of these features are visible optically.*

SCORPIUS

It is widely agreed that Scorpius is one of the few constellations in which you can join the dots and get a reasonable representation of the animal it is supposed to be. For many northern observers, however, the resemblance is lost because a crucial part of the constellation, the curve of the scorpion's sting, lies forever below the horizon. Antares marks the heart of the creature, with its claws extending westward, while its fearsome sting dips southward. Notice that the astronomical and astrological names for this constellation are subtly different – to astronomers it is Scorpius, not Scorpio. Capricornus is a similar trap for the unwary!

Scorpius is a very ancient constellation, and Greek mythology relates that it is the creature that killed the proud hunter Orion. To this day, Orion avoids the scorpion and his stars set as those of Scorpius rise.

The borders of the constellation include parts of both the Milky Way and the ecliptic, so occasionally the planet Mars can be seen very close to Scorpius' brightest star, Antares. The name Antares means "rival of Mars," and they do have similar ruddy colors. Antares has a fifth-magnitude companion star just 2.6 arc seconds away – so close that at least a 75 mm telescope is required to see it. It is a type B star, but is often described as green. In theory green stars should not occur, so this is probably a contrast effect.

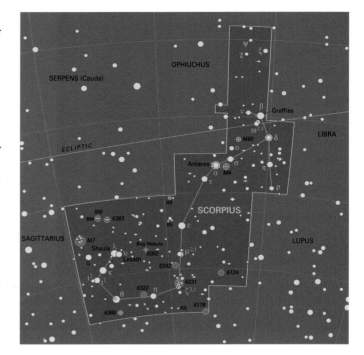

▼ **This Kitt Peak photo of open cluster M6** *suggests the butterfly-like arrangement of stars.*

Globular cluster M4

This is a bright object very close to Antares, but much larger and fainter than other globular clusters. Find it by centering Antares in the field of view, then moving just over 1° westward. M4 may not be easy to find in a hazy sky because it is not a particularly rich or condensed cluster. You will need a small telescope to begin to resolve it into stars.

Open clusters M6 and M7

Of several bright open clusters in Scorpius, these are the most prominent. Both are visible with the naked eye as misty patches and can be resolved into stars with binoculars. M7 has the brighter stars, while M6 is more condensed but richer in faint stars. In a low-power telescope of any size both clusters are impressive. Some people refer to M6 as the Butterfly Cluster because the pattern of the stars resembles a butterfly's open wings.

▼ **Globular cluster M4** *is large and bright enough to be visible with the naked eye in good skies, but a telescope shows it best.*

Object	Type	M_v	Magnification	Distance
M4	Globular cluster	5.6	25×	7200 light years
M6	Open cluster	4.2	10×	1600 light years
M7	Open cluster	3.3	7×	800 light years
NGC 6231	Open cluster	2.6	25×	5900 light years
NGC 6302	Planetary nebula	10	200×	4000 light years

Open cluster NGC 6231

This is a large and bright cluster at the bottom of the sting of the scorpion, at the point where it turns eastward. It is somewhat smaller than the Pleiades, and its stars are less bright, so it is not particularly obvious to the naked eye. But with binoculars it is a brilliant sight. Even its pattern of stars is reminiscent of the Pleiades, but on a smaller scale.

Planetary Nebula NGC 6302, the Bug

There seems to be no end to the wonderful variety of planetary nebulae. They delight visual observers by their unexpected shapes, which can make a refreshing change from galaxies which are often just elliptical. And they delight armchair observers too, particularly when they have been photographed by the Hubble Space Telescope, as in the image at right.

The Bug Nebula suits both. It is observable in medium-sized telescopes, but is less familiar to northern-hemisphere observers than some other equally bright planetary nebulae on account of its low declination. Its name comes from its antenna-like extensions, though these are not easy to observe visually. It has an irregular shape through a telescope, unlike any other object, with a fat body crossed by a dark band and wisps of gas at either end.

The Hubble view of this object shows a dramatic sight of the heart of the creature. The dark band that is visible through a telescope is seen as a donut surrounding the star, which is completely hidden. This star is said to be one of the hottest known, at 250,000°C, but the material it has thrown off has now condensed as ice, so it is actually surrounded by a blizzard of hailstones.

▶ **An Earth-based photo of the Bug Nebula,** *sometimes called the Butterfly Nebula, reveals its curious insect-like shape. The nebula is about a third of a light year across.*

▼ **The Hubble view of the same object** *reveals clouds of gas thrown off the dying star. This planetary nebula is only a few thousands years old, and will fade over the centuries as the nebula disperses and the star cools.*

◀ **M7 is in a richer part of the Milky Way than M6.** *It was mentioned by the ancient Greek astronomer Ptolemy, so it is sometimes referred to as Ptolemy's Cluster.*

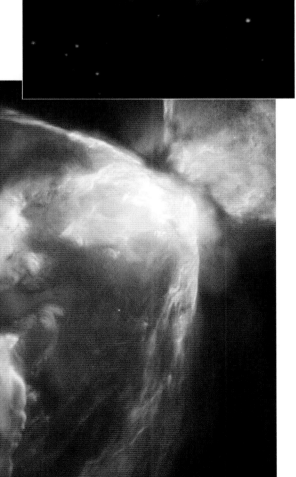

▲ **Scorpius carries a jewel in its sting** — *the open cluster NGC 6231. Many of its stars are type O, and are only about 3 million years old.*

LUPUS AND NORMA

The stars of Lupus lie between Antares and Alpha Centauri. None of them is brighter than second magnitude but the region is in a rich part of the Milky Way and offers good sweeping with binoculars. These constellations are mostly below the horizon for northern observers, except those in the southern United States or south of the Mediterranean.

Norma is undistinguished in terms of bright stars, but it contains some nice open clusters, as well as a bright patch of the Milky Way known as the Norma Star Cloud. A long rift in the Milky Way that runs through Lupus and Norma is referred to by Native Australians as "the Emu." The bird's head is the Coalsack in Crux, while the clouds in Lupus and Norma are its body. The emu's legs are hidden underneath its body as it crouches on its nest.

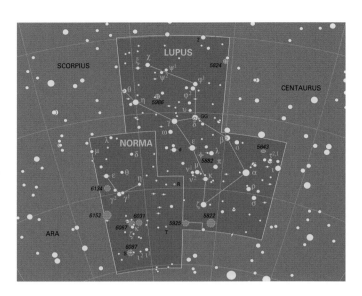

Open cluster NGC 5822

Some clusters, like this one, spring into view with binoculars when the sky is dark enough, but are otherwise difficult to see. It appears as a large diffuse haze with binoculars, while in a telescope you need a wide-field eyepiece because the cluster covers about 40 arc minutes and fills the view with stars. There are around 150 stars, all of around tenth to 11th magnitude. You will find the cluster about three-quarters of the way between Alpha Centauri and Zeta Lupi.

▼ *A large but faint cluster,* NGC 5822 *is best seen with a telescope.*

Open cluster NGC 6067

Having found NGC 5822, move exactly 10° eastward and you come to this rather brighter cluster. It is easily found in binoculars as a fairly condensed misty patch, and many eighth-magnitude and fainter stars are visible with a telescope.

Open cluster NGC 6087

Within the same binocular field as NGC 6067 lies NGC 6087, 4° farther south and 1° east. It is a more condensed cluster, with several eighth-magnitude stars and many fainter ones, some being yellow or orange when seen through a telescope. There is a double star near the center.

Object	Type	M$_V$	Magnification	Distance
NGC 5822	Open cluster	7	25×	3000 light years
NGC 6067	Open cluster	5.6	50×	4600 light years
NGC 6087	Open cluster	5.4	50×	2900 light years

▲ *The clusters NGC 6067* (top) *and NGC 6087 can both be seen within the same binocular field of view.*

ARA

To the south of the sting of Scorpius lies this small and easily identified constellation, consisting of two slightly curved lines of stars. It is on the fringes of the Milky Way and contains several open clusters that are visible with binoculars, NGC 6193 being the brightest. Despite being fairly far south, the constellation was named in ancient times and may represent the altar on which Greek gods swore an oath of allegiance.

Open cluster NGC 6193 and nebula NGC 6188

Though it is easily found in binoculars, the cluster NGC 6193 is not particularly noteworthy, as it has few bright stars. A couple of other open clusters – NGC 6200 and NGC 6204 – are visible within the same binocular field to the north. But this area comes into its own in long-exposure photographs, being surrounded by NGC 6188, with dark lanes and bays somewhat reminiscent of the Horsehead Nebula.

Globular cluster NGC 6397

Situated at the apex of a triangle made by Alpha and Beta Arae, this is one of the closest globular clusters to the Solar System, if not the closest. Being quite large, it is an easy binocular object, but a telescope and a little bit more power are needed to resolve it into stars. It is fairly loose and open, like M4 in Scorpius.

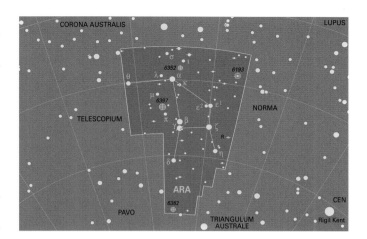

◀ *An amateur-taken image of NGC 6397.*

Because of its closeness to Earth, this cluster has been intensively studied using the Hubble Space Telescope.

Object	Type	M$_V$	Magnification	Distance
NGC 6188	Nebula	–	Photographic	3800 light years
NGC 6193	Open cluster	5.2	50×	3800 light years
NGC 6397	Globular cluster	5.9	25×	7200 light years

◀ *A small telescope shows* the sparse cluster NGC 6193 – the clump of bright stars – while the nebulosity NGC 6188 is a good photographic target. This view has north to the left.

SEPTEMBER–OCTOBER

Evening

1 September at 11.30 pm
15 September at 10.30 pm
30 September at 9.30 pm

Morning

15 June at 4.30 am
30 June at 3.30 am
15 July at 2.30 am

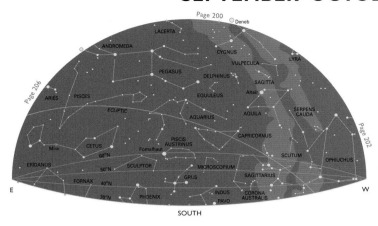

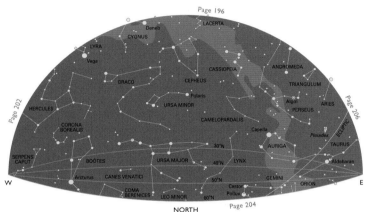

The northern-hemisphere sky

By September, the Milky Way is high in the sky. We are now looking more toward the center of the Galaxy, so it is much more obvious than it was in January. Deneb in Cygnus, the Swan, is at the top of a large cross-shape known as the Northern Cross. This extends along the Milky Way with Altair in Aquila, flanked by two fainter stars, to one side of the line. Follow the line southward to find Sagittarius, near the horizon, which has no very bright stars.

High in the south is the large but fairly faint Square of Pegasus, which you can use as a signpost to the even fainter constellation of Aquarius and the lone bright star Fomalhaut. From the upper left of Pegasus extends a widely spaced line of three stars in Andromeda. Any planets will be low in the west or in mid sky in the south. To the north, the Big Dipper or Plough is close to the horizon.

Evening

1 September at 11.30 pm
15 September at 10.30 pm
30 September at 9.30 pm

Morning

15 May at 6.30 am
1 June at 5.30 am
15 June at 4.30 am

The southern-hemisphere sky

Looking north there are few bright stars, but the Square of Pegasus, though not bright, is easy to pick out. The lower edge of the Square points down to the northeast and the line of three widely spaced stars that are the brightest in Andromeda. To the east of this lie a small group of three stars that identify Aries, the Ram. Take a diagonal line upward through the Square to find Aquarius, which is another constellation of the zodiac, yet with rather faint stars. Its most identifiable feature is an arrow shape known as the Water Jar. The ecliptic continues through faint Capricornus to the more obvious stars of Sagittarius.

Low in the north is the bright star Deneb in Cygnus, also known as the Northern Cross. The Milky Way forks here in what is called the Cygnus Rift, then runs past Altair to Sagittarius and Scorpius. This is the most brilliant part of the Milky Way, and the minor constellations of Vulpecula, Sagitta and Scutum are worth finding for their deep-sky riches.

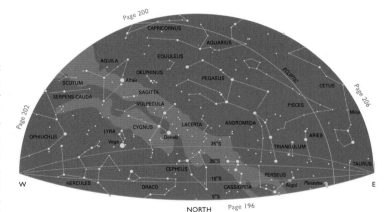

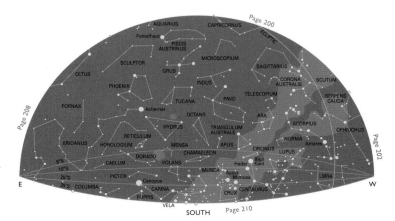

CYGNUS

This celestial swan flies forever down the Milky Way, its long neck and wings outstretched. It is also referred to as the Northern Cross, though it is much larger than the better known Southern Cross. The bright star Deneb marks its tail. This is a remarkable star, though telescopically it looks no different from any other. Deneb is a true searchlight, and its light dominates this part of the Galaxy. Though its distance is rather uncertain, a recent estimate puts it at over 1400 light years away. Compare that with Altair, a bright star some way south of Cygnus. Altair is a little brighter in the sky but is only 17 light years away – about 80 times closer.

In addition to containing the most distant bright star, Cygnus also contains one of the closest stars, fifth-magnitude 61 Cygni, just 11.4 light years away. This star was the first to have its distance measured. It is an easy double star, with components separated by 27 arc seconds, making it just visible as double with 10-power binoculars.

There are several interesting variable stars in Cygnus. Chi Cygni is a bright Mira-type star (see page 188). At its brightest, every 400 days or so, it is magnitude 3.3; when faint, it drops to magnitude 14.5 and is effectively invisible to all but quite powerful telescopes. Another now-you-see-it-now-you-don't star is W Cygni, which varies between magnitudes 5.0 and 7.6 in about 130 days. P Cygni is usually fifth magnitude, but it is massive and erratic and has been known to brighten to magnitude 3.

Cygnus is a happy hunting ground for most types of deep-sky object. For many people, just sweeping the area with binoculars is very enjoyable, as the Milky Way in this region is very rich. One feature of Cygnus is that its cross-shape makes it very easy to locate objects by star-hopping: there are easy routes to most objects within the constellation.

Double star Beta Cygni (Albireo)

Arguably the most popular double star in the sky, Albireo marks the head of the swan. It consists of two well-separated stars of which the brighter is yellowish and the fainter is bluish. This is no contrast effect (photographs show these colors clearly), though the proximity of the two makes the difference in color easier to see. The stars themselves are types K and B.

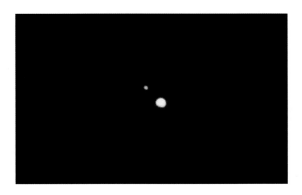

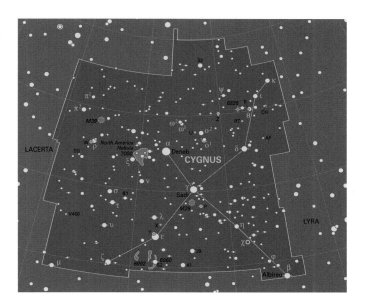

Open cluster M39

Not a spectacular open cluster, M39 is easy to find by following a line of stars northeast of Deneb. It is best seen with a low power.

◀ **Not far north of Deneb,** *M39 is in a rich part of the Milky Way.*

Planetary nebula NGC 6826, the Blinking Planetary

At eighth magnitude, this planetary is easy to find in binoculars or a finder at the end of a line of stars, though you may think it is just another star, as it is compact and round. It is about the same brightness as the more famous Ring Nebula in Lyra, but only half the diameter. Use considerable magnification – around 150 or more – and you will see that there is a central star within a circular bluish shell. This star is one of the brightest central stars of any planetary nebula, at about magnitude 11, and is visible with even a fairly small telescope. But as you stare at the star, you may notice an odd thing – the nebula surrounding it disappears!

◀ **Albireo's contrasting colors** *and wide separation make it one of the most easily observed and attractive double stars in the sky.*

Object	Type	M$_V$	Magnification	Distance
Albireo	Double star	3.1	10×	420 light years
M39	Open cluster	4.6	7×	825 light years
NGC 6826	Planetary nebula	8.8	250×	2200 light years
NGC 7000	Nebula	–	7×	1600 light years
Veil Nebula	Supernova remnant	5?	7×	2600 light years

175

▲ *This photo of the*
North America Nebula
was taken on film, but CCDs
and modified DSLR cameras
will also do the job. To its right
is the Pelican Nebula.

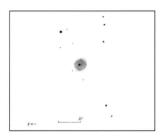

▲ *The Blinking Planetary*
is one of the smaller bright
planetaries. It requires a very
high power for the best views.
This drawing was made using
a 450 mm reflector with a
magnification of 508.

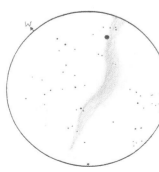

Milky Way star clouds

Although not the richest part of the Milky Way, the Cygnus region has the advantage for many northern-hemisphere observers that it is virtually overhead, and in light-polluted areas it is virtually the only part of the Milky Way that is visible, the brighter regions farther south being hidden in the haze and light pollution. Wide-field photography reveals countless stars, and also brings out the Cygnus Rift, the great dust lane which extends right down to Ophiuchus and Sagittarius.

You can now see that the North America Nebula is actually right at the tip of the dust lane, and is an illuminated part of the giant cloud that constitutes the Cygnus Rift. In fact there is more than one dust cloud making up the Rift. The eastern cloud, which includes the North America Nebula, is at a distance of about 2500 light years, while the western part is more distant at somewhere between 5000 and 7000 light years.

Apparently surrounding the star Gamma Cygni are patches of red nebulosity known as IC 1318, sometimes called the Butterfly Nebula. In fact they are not linked with the star, which is at a distance of about 750 light years, while the nebulae are at about 5000 light years. Again, they are just the glowing surface layer of a dense cloud. Within the cloud in this area lies an association of massive stars known as Cygnus OB2. They are only visible on infrared surveys.

The brightest nova of recent years occurred in Cygnus in 1975. It appeared to the east of Deneb, changing the familiar pattern of the Northern Cross when at its peak magnitude of 1.8. However, many people missed it because it was only bright for a few days and was visible to the naked eye for only about a week.

This is because when you look directly at the star it is bright enough to be seen using the cones in the center of your vision (see page 32), whereas with small telescopes the planetary surrounding it requires averted vision. If you look away, the nebula will reappear. This blinking effect depends on the aperture and the magnification.

NGC 7000, the North America Nebula

Some people say that they can easily see this nebula with the naked eye, but all that most can see is perhaps a slightly brighter area to the east of Deneb. Binoculars and an LPR filter might improve matters somewhat, but by far the best way to detect this impressive hydrogen gas cloud is to take a photograph of the area. This was much easier in the days of film, which was far more red-sensitive than most modern DSLR cameras. But in a dark sky, an exposure of a minute or two will reveal a pink glow with the unmistakable shape of North America. To its west is a fainter area known from its shape as the Pelican Nebula. In good conditions, the same setup may well show nebulosity around Gamma Cygni.

NGCs 6960 and 6992, Veil Nebula

This supernova remnant is a favorite target for deep-sky observers and photographers. It is not difficult to see in country skies, but light pollution soon wipes it out. The usual method of finding it is to locate the star 52 Cygni, which lies right in the middle of the western arc of the nebula, NGC 6960. A light-pollution filter will help. Again, the object can easily be photographed, though because it is thinner than the North America Nebula, it is worth using a modest telephoto lens (about 135 mm or longer).

◀ *An O III filter was used on a 222 mm reflector*
for this drawing of part of the Veil Nebula. The star at
the top is 52 Cygni. See also page 113.

▲ *Virtually the whole of Cygnus is shown in this very deep*
exposure. The North America Nebula is at top left, while Gamma
Cygni and IC 1318 are just above the center. Toward the lower left
is the Veil Nebula supernova remnant, also shown on page 113.

CEPHEUS

Most northern-hemisphere constellations, such as Cassiopeia and Cygnus, have clear patterns that make them easy to spot. But Cepheus is more like a southern constellation – a mass of stars of different brightnesses with no distinguishing features. Despite including a chunk of the Milky Way, it has no bright deep-sky objects. It is better known for its stars, one of which has made the constellation a household name, at least among astronomers. Cepheus himself was a king in the Perseus legends, the husband of Cassiopeia.

Variable star Delta Cephei

This star, at the apex of a little triangle at the southern end of the constellation, was one of the first variable stars to be recognized. It changes in brightness between magnitudes 3.5 and 4.4 over a regular period of 5 days 9 hours, with a slow decline and a more rapid increase. This is typical of a particular class of variable stars known as Cepheid variables, after this star.

The discovery of this star's variability makes an interesting story. It was first noticed in 1784 by a 20-year-old amateur astronomer named John Goodricke – a great feat, but even more so because Goodricke was deaf and dumb. In an age when such people were regarded as simpletons, his work on this and other variables such as Algol was a real triumph. He reported his work to the Royal Society and was awarded its prestigious Copley Medal. At age 21 he was elected a Fellow – though sadly he died of pneumonia within two weeks of his election.

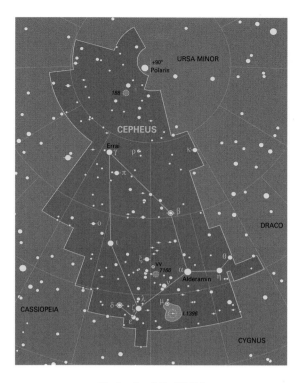

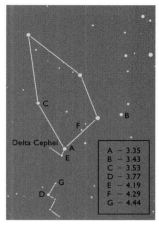

| A – 3.35 |
| B – 3.43 |
| C – 3.53 |
| D – 3.77 |
| E – 4.19 |
| F – 4.29 |
| G – 4.44 |

▲ **You can follow the brightness variations** of Delta by comparing its brightness with those of nearby stars. A light curve of the star is shown on page 109.

▼ **Spitzer's view of the Elephant Trunk Nebula** reverses the normal appearance of the object, shown in the inset in a photograph taken by the Canada–France–Hawaii Telescope on Hawaii.

Nebula IC 1396

A large and impressive nebula that is virtually impossible to see but which is a popular photographic target for CCD cameras, though at over 2° diameter it requires a large chip to get it all in. Mu Cephei is at one edge of the nebula. The distances of both Mu and IC 1396 are uncertain, so it may not be connected with the nebula.

▲ **Both Mu Cephei and IC 1396** feature in this view. Mu is the red star at top left. The time exposure also reveals the diffuse nebula IC 1396, which is over 2° across.

Mu Cephei, Garnet Star

Star colors are usually rather elusive. Diagrams in books may show red giant stars as the color of traffic lights, but in fact most are actually no redder than a 100-watt light bulb. Mu Cephei is known as the reddest star visible to the naked eye in the northern hemisphere, and in binoculars or a telescope its color is unmistakable. But it is not bright enough to appear particularly red to the naked eye. William Herschel famously described it "a very fine deep garnet color," so it is often called "the Garnet Star."

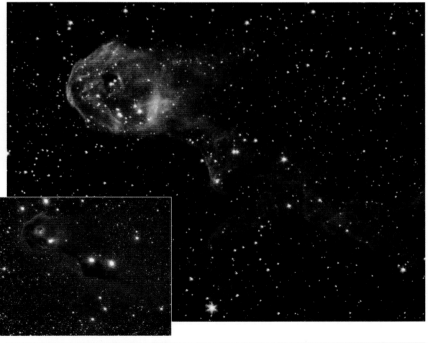

Object	Type	M$_v$	Magnification	Distance
Delta Cep	Variable star	3.5–4.4	Naked eye	863 light years
Mu Cep	Star	4.2	Naked eye	5260 light years
IC 1396	Nebula	–	Photographic	2450 light years

Elephant Trunk dark nebula in IC 1396

One feature of IC 1396 that is popular with astro-photographers is a small dark lane within the nebula. It can be seen on the main picture on page 177 midway between the star near the center of the nebula and its right-hand edge. This has been photographed in infrared by the Spitzer Space Telescope to great effect. Comparison with the visual view shows the great power of infrared astronomy to cut through the nebulosity to reveal the stars and other objects within.

The Elephant Trunk is normally seen as a dark nebula with a bright rim – turn the picture clockwise through 90° to see the resemblance to the head and trunk of an elephant. But the thin gas is transparent to infrared, which instead reveals the interior of the dark lane, now glowing. And instead of an elephant, we now see a dragon. In the "eye" of the nebula, two newborn stars have cleared a cavity with their stellar winds and ultraviolet light. Radiation from a nearby star is also eating away the gas from the outside.

CAPRICORNUS

This is a large rough triangle of stars, most of which are lost in the murk from more northerly and light-polluted skies as it is fairly far south. The best way into Capricornus is to follow the line of three stars in Aquila containing Altair, going as far to the southeast as Albireo (Beta Cygni) is to the northwest. You come to the constellation's brightest star, Alpha Capricorni or Algedi.

The traditional constellation figure of Capricornus is a sea goat – though this is not a creature that features in any myths, unlike centaurs and lions. It has been linked with the Greek god Pan, who had goat's feet and who developed a fish's tail to escape a sea monster. Being in the ecliptic, it is often home to bright planets, which then outshine any star in the constellation.

Object	Type	M$_V$	Magnification	Distance
Alpha Cap	Double star	3.7	Naked eye	106/568 light years
Beta Cap	Double star	3.0	Naked eye	340 light years
M30	Globular cluster	6.9	50×	26,000 light years

Double stars in Capricornus

Alpha Capricorni (Algedi) is a naked-eye double star, with the components separated by 6½ arc minutes, which most people can manage visually. The two stars are unrelated, however, with the fainter one being six times more distant than the brighter. The star is another candidate for the title "Double Double," like Epsilon Lyrae (see page 161), but unlike Epsilon Lyrae the secondary stars in each case are considerably fainter than the main stars.

Beta Capricorni is a double with stars of different colors – yellow and blue. The components are separated by 3½ arc minutes, which puts the star well into the binocular category, though some may manage it with the naked eye.

▶ **M30 is a smaller and fainter globular cluster** *than many, but is by far the brightest deep-sky object in this part of the sky. The star shown here is not 41 Capricorni but a much closer eighth-magnitude star.*

◀ **A close-up of naked-eye double stars** *Alpha (upper right) and Beta Capricorni. Stars to 11th magnitude are shown.*

Globular cluster M30

Capricornus is rather barren of bright deep-sky objects, which makes the soft glow of M30 all the more welcome. It lies close to the fifth-magnitude star 41 Capricorni, which is a good guide to its location for those northern observers for whom this is quite low in the sky. It is quite a distant object at 26,000 light years, so individual stars are not easy to resolve in smaller telescopes.

AQUARIUS

At the dawn of history, when people in the Middle East were telling the stories about the stars that have come down to us in constellation names, they noticed that when the Sun was in the regions of the sky we now call Aquarius and Capricornus, the rainy season was upon them. The constellations in this area still have a watery connection, Aquarius being the Water Carrier who constantly pours water from his water jar into the ocean. The Water Jar itself is a group of four fairly faint stars in the shape of a sideways Y, but as with many asterisms it is the shape that is more distinctive than the brightness. To modern eyes the Water Jar and Alpha together resemble a jet plane more than a water jar. But there is a ragged line of more faint stars, which represents the water tumbling from the jar into the sea.

The brightest star in Aquarius is only slightly brighter than third magnitude, and the whole area is rather barren of stars. One of the Messier objects shown on the map, M73, has frustrated astronomers for years as it is simply a group of four faint stars. Why it was included in the catalog remains a mystery. The constellation also contains two well-known planetaries and two globular clusters.

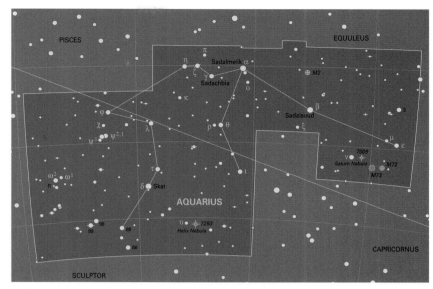

Object	Type	M_v	Magnification	Distance
NGC 7293	Planetary nebula	6.5	10×	450 light years
NGC 7009	Planetary nebula	8.0	200×	2400 light years.
M2	Globular cluster	6.5	50×	37,000 light years.
M72	Globular cluster	9.3	50×	55,400 light years

NGC 7293, Helix Nebula

You would expect the nearest planetary nebula to the Solar System, at magnitude 6.5, to be a fine sight, but in fact it is a great challenge to view, particularly for more northerly observers for whom it is rather low in the sky. The problem is that its brightness is spread over an area of sky more than a third of the diameter of the full Moon. As a result it has a very low surface brightness and is easily lost in haze or light pollution. A clear, dark country sky and binoculars or a low-power telescope are needed. Even photography is a challenge.

At first glance, a photograph of the Helix looks like the Ring Nebula in Lyra, but a closer look shows that it has two distinct rings of material arranged in a coil or helix seen end-on. This structure is hard to see visually, requiring large amateur telescopes as well as good skies and usually an O III filter.

NGC 7009, the Saturn Nebula

There is a great contrast between the Helix Nebula and the Saturn Nebula. The Saturn is also a planetary, and is less bright in total at magnitude 8. But it is small and its surface brightness is much higher than that of the Helix, so it is quite easily observable with small telescopes. The Saturn Nebula is about 1° due west of the fourth-magnitude star Nu Aquarii, making it simple to find. However, it is indistinguishable from a star unless you use a power of 50 or more.

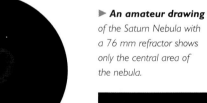

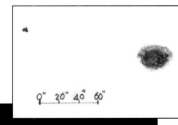

▶ **An amateur drawing** of the Saturn Nebula with a 76 mm refractor shows only the central area of the nebula.

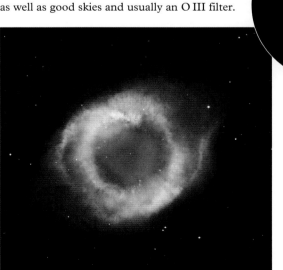

▲ **Viewed with a 200 mm reflector** at 72× from a suburban site near Brisbane, Australia, the Helix Nebula was only visible using an O III filter.

◀ **The Helix Nebula** as seen by the Hubble Space Telescope.

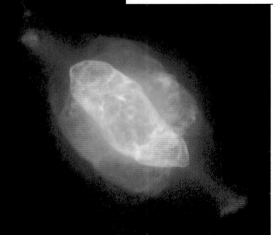

▶ **The extensions that give** the Saturn Nebula its name are easily seen on this Hubble Space Telescope image.

▲ **M2 as photographed** *using a 0.9 m telescope at Kitt Peak.*

▶ **The same instrument** *was used for this view of nearby M72.*

With a small telescope the Saturn Nebula appears as a misty oval patch about the same apparent size as the planet Saturn. Only with a large telescope are projections visible on either side, making the resemblance with that planet much greater.

Globular clusters M2 and M72

It is fortunate that M2 is a bright object, at magnitude 6.5, otherwise it would be quite hard to locate in a small telescope, not being near any bright stars. Find it by locating Beta Aquarii and moving about 5° north, as the two objects have almost exactly the same right ascension.

In a small telescope M2 is a misty ball, and at around 37,000 light years it is more distant than many other bright globular clusters. This means that its individual stars are quite faint, so you will need a medium-sized telescope to see many stars in it.

Not far from the Saturn Nebula lies M72, which is even more remote. It is notoriously hard to resolve into stars.

PEGASUS

This region of sky, well away from the Milky Way, is lacking in stars. So although the main stars of Pegasus are not particularly bright, they do stand out to form an easily recognizable square. You might think that the Square of Pegasus would be a compact box, but it is actually fairly large – getting on for about 20° across. Its simplicity makes it a classic signpost constellation for finding other stars – it will lead you to the stars of Andromeda, Pisces, Piscis Austrinus, Aquarius and Cetus very easily. Pegasus is, therefore, an excellent starting point for learning the evening skies toward the end of the year, though it is rather lacking in bright deep-sky objects. The Square itself is also rather barren; counting the number of stars within it is a useful means of comparing the eyesight of different observers, or of one observer on different occasions. Our chart shows 36 stars down to sixth magnitude, but keen-eyed observers in good skies may see more.

The mythological figure of Pegasus is a winged horse, though oddly the traditional depiction is upside down as seen from the northern hemisphere, and includes only the front half.

Globular cluster M15

The lone star Enif, well separated from the Square of Pegasus, is a good guide to finding M15 with binoculars or a finder. It is easily spotted with binoculars, and it makes an interesting comparison with the globular M2, which lies 13° to the south and slightly farther east. Both globulars are of similar total brightness, but M2 is more concentrated toward the center. However, unlike closer globulars, such as M13 or M22, M15 needs a medium-sized telescope to resolve it into stars.

▼ **Globular cluster M15,** *photographed using a CCD on a 355 mm Schmidt-Cassegrain telescope.*

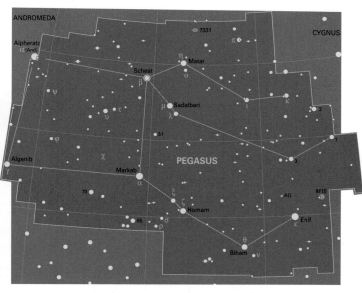

Object	Type	M$_v$	Magnification	Distance
M15	Globular cluster	6.2	50×	34,000 light years

THE MAGELLANIC CLOUDS

Many keen stargazers spend a lot time trying to locate distant galaxies, yet there are two galaxies that anyone far enough south can see at virtually any time – the Large Magellanic Cloud (LMC) and the Small Magellanic Cloud (SMC). Satellite galaxies to our own Galaxy, they are respectively a mere 163,000 and 200,000 light years away, compared with 2.5 million light years for the Andromeda Galaxy, M31. They are of the type known as irregular galaxies, meaning that they have no regular form, unlike spiral or elliptical galaxies. The LMC has a hint of organized structure, with an attempt at being a barred spiral.

These galaxies are named after the Portuguese navigator Ferdinand Magellan (1480?–1521), who voyaged south in 1519. They were certainly known to earlier European sailors, and were referred to as "Cape Clouds" after the Cape of Good Hope. Magellan's sailors used them as navigational aids. Though they lie within 20° of the south celestial pole, they can be seen at certain times of year, such as evenings from November to February, from Central America, West Africa, and southern Arabia and India. From much of the southern hemisphere they are visible all night. The LMC occupies part of the constellations of Dorado and Mensa, while the SMC is in Tucana.

They are often described as resembling fragments of the Milky Way that have become detached – they have a similar brightness to its brightest regions, and the LMC can be seen despite some light pollution or moonlight. The bright stars Canopus and Achernar act as rough guides – the LMC is about 15° south of Canopus, while the SMC is a similar distance south of Achernar. With the naked eye there is little detail visible, but binoculars hint at the amount of objects to be seen within.

Large Magellanic Cloud

The LMC has a diameter of 6° and is noticeably elongated along the line of the bar. A brighter patch within it, which to the naked eye resembles a hazy star just outside the main bar, becomes easily visible in binoculars. This is a massive diffuse nebula or H II region, known as the Tarantula Nebula. Its spidery extensions are visible in telescopes, particularly using O III filters. The Tarantula (NGC 2070) is huge: it is not very much smaller in the sky than the Orion Nebula, yet it is over 100 times more distant. By comparison, the Orion Nebula is insignificant, and would be virtually invisible if it were in the LMC. The Tarantula is a major star-forming region and a cluster of new stars, known as 30 Doradus, is visible in small telescopes at its center.

Numerous nebulae, star clusters and globular clusters are to be seen within the LMC, with the brighter individual stars being visible in some of the open clusters with medium-to-large telescopes.

In 1987, a star appeared to the naked eye within the LMC, in addition to the few foreground stars within our own Galaxy that are visible within its area. This was a supernova, the brightest in terms of apparent brightness to have been witnessed in modern times. It rose to magnitude 2.8 before taking a year or so to fade below naked-eye visibility. It is referred to as SN 1987A, and remains the subject of intense study by professional astronomers.

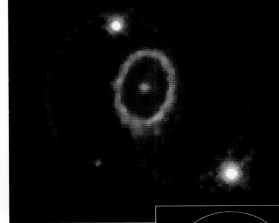

Object	Type	M_V	Magnification	Distance
LMC	Galaxy	0.1	Naked eye	160,000 light years
NGC 2070	Nebula	8	50×	163,000 light years
SMC	Galaxy	2.3	Naked eye	200,000 light years
47 Tuc	Globular cluster	4.0	25×	14,700 light years

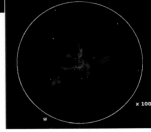

▲ **Rings surround the remains of Supernova 1987A** in the LMC, photographed with the Hubble Space Telescope in 1994.

◀ **The Large Magellanic Cloud** with the Tarantula Nebula clearly visible to its upper left.

▲ **A drawing of the Tarantula Nebula** made using a 250 mm Dobsonian.

The Small Magellanic Cloud

Smaller, fainter and slightly more distant than the LMC, the SMC requires a darker sky to be seen with the naked eye. It resembles a teardrop or tadpole. The tail of the tadpole contains several H II regions, visible with medium-to-large telescopes with an O III filter. A round nebulous object within the SMC is the globular cluster NGC 362, which is actually much closer than the Cloud and just happens to be in the same line of sight. Were it separate from the SMC, it would be ranked as one of the brighter globulars in the sky.

Anyone glancing at the SMC will undoubtedly be drawn to its near neighbor in the sky, the spectacular globular cluster 47 Tucanae. As its star name rather than NGC number implies, it appears to the eye as a star of about fourth magnitude. It ranks second only to Omega Centauri in splendor, and with a telescope of more than 100 mm aperture you can start to resolve it into stars. Some observers prefer the appearance of 47 Tucanae to Omega Centauri because, being smaller, it fits better into the low-power field of view of a telescope. The cluster contains upward of half a million stars.

▲ **The Small Magellanic Cloud** and the globular cluster 47 Tucanae to its right.

▲ **Globular cluster 47 Tucanae,** photographed with a 4 m telescope in Chile.

▼ **A wide-field view** of the southern skies shows Canopus (lower left), Achernar (upper right), the Large Magellanic Cloud (LMC, lower center) and the Small Magellanic Cloud (SMC, right).

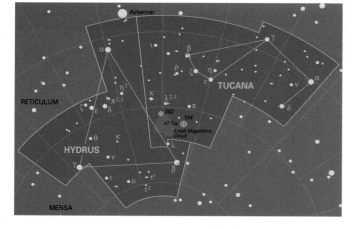

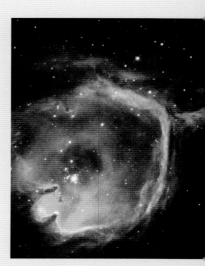

▲ **The bubble produced by a newborn star** is seen here in spectacular detail in an object named N44F. The massive star, on the left, has blown a bubble in the surrounding gas. By chance we can see inside this 35-light-year-diameter bubble, which Hubble scientists have likened to a geode, a rock that has a crystal-lined hollow inside, only apparent when it is split in two. The gas is flowing from the star at a rate 100 million times greater than the solar wind flows off our Sun. Where the flow meets the cold interstellar gas a shock wave is set up, often referred to as a stellar snowplow from the way the gas piles up to create the bubble. Only a small number of clear examples of such bubbles around individual stars are known, though clear zones are common around clusters of newborn stars.

▶ **The starforming region NGC 346** is the SMC's equivalent of the Tarantula Nebula. It is easily visible even in small telescopes, though it is by no means as prominent as the Tarantula. Here seen by the Hubble Space Telescope, it has a glittering array of newly formed stars surrounded by a ragged filament of gas and dust from which new stars will ultimately form. The strong radiation and outflow from the newly formed stars has created a clear bubble and the filament of gas has numerous fingers pointing inward. As in the Eagle Nebula in the Milky Way, these fingers are caused by denser blobs of gas or maybe protostars preventing the evaporation of a column behind it.

CURIOSITIES WITHIN THE CLOUDS

Both Magellanic Clouds are full of star-forming regions and have therefore been a prime target for the Hubble Space Telescope and the Very Large Telescope. These regions are visible in amateur telescopes as hazy patches within the Clouds, though even the brightest clusters are hard to resolve into individual stars. The best known and brightest object is the Tarantula Nebula, shown on page 181 and in the A–Z section on page 306, but the Clouds are full of other objects which illustrate different aspects of the life of a star. Most are unspectacular when viewed with amateur telescopes, if they are visible at all, but they take on a life of their own when photographed through a large instrument.

Quite often, it is possible to tell what an object is just looking at it. But sometimes it is necessary to take a spectrum, because there may be similarities in appearance between a newly formed star and a planetary nebula. One thing that astronomers have found is that stars in the

▲ **There are more than 35 supernova remnants** *in the Large Magellanic Cloud, and it is a happy hunting ground for such objects because of the ease with which we can spot them compared with our own Galaxy. The LMC is laid open for inspection, whereas much of our own Galaxy is hidden from view by the gas and dust clouds in which we*

Magellanic Clouds, particularly in the SMC, have fewer of what they call metals than stars in our own Galaxy. To astronomers, any element other than hydrogen or helium is regarded as a metal, though chemically this is clearly a falsehood. We just have to accept this foible of terminology when talking among astronomers. Even the youngest stars in the LMC have half the metal content of the Sun, while those in the SMC have only a quarter.

The reason for this difference is that, being much younger galaxies than our own, the stars of the Magellanic Clouds have gone through fewer generations of star formation than those in our own Galaxy. The cycle is that stars form and produce the heavier elements through the nuclear reactions that power them, then the most massive ones explode as supernovae which both scatters the metals throughout the neighborhood and triggers off more star formation. Over time, the proportion of metals to hydrogen increases and in our Galaxy about 5 billion years ago when our

Sun formed there were enough of the heavier elements to form terrestrial planets and eventually lifeforms.

It is possible that planetary systems with rocky planets and water are unlikely to have formed much before our own Earth. Maybe we are an early example of life, which may explain

are effectively immersed. This example is numbered N63A, and shows bright clouds of gas glowing as they plow into the gas between the stars. The brightest part of the visible remnant seen here is just a fraction of the whole shell, which is a strong radio and X-ray source. The remnant as we see it is between 2000 and 5000 years old.

▲ **An exquisite supernova remnant** *in the LMC, known as N49, which is about 50 light years across. This remnant is associated with a rare type of gamma-ray source known as a soft gamma-ray repeater, or SGR. Only a handful of these are known, so to find one in the same direction as a nearby supernova remnant is regarded as more*

why we do not see signs of other more advanced civilizations around us in the Universe.

The major star-forming episode in the LMC which is still under way seems to have begun about 3 billion years ago. It may have been a close encounter with our own Galaxy which triggered this off. There was an older bout of star-forming, but the process does not seem to have been continuous as in our own Galaxy. These differences are a boon to astronomers trying to understand how stars evolve, because they provide us with examples of stars and galaxies in an earlier stage of their development, right on our doorstep.

Studies of the stars in the Small Magellanic Cloud show that it is bigger than it appears. It is elongated in the direction of the Milky Way — maybe as a result of the gravitational pull of our larger Galaxy. It is about three times longer than it is wide.

Both Magellanic Clouds are enveloped in a long trail of neutral hydrogen known as the Magellanic Stream. Detectable by radio telescopes, this stretches for some 90° across the sky with the Clouds at one end, partially encircling our Galaxy. Its origin is still being argued over.

than a coincidence. SGRs are now thought to be magnetars, which are magnetic neutron stars. This particular SGR was spotted in 1979, and for just a few minutes sent out gamma rays which pulsed every eight seconds — a strong hint that a pulsar was involved. It lies within N49, and more work is needed to discover whether the two are linked.

NOVEMBER–DECEMBER

Evening

1 November at 11.30 pm
15 November at 10.30 pm
30 November at 9.30 pm

Morning

15 August at 4.30 am
30 August at 3.30 am
15 September at 2.30 am

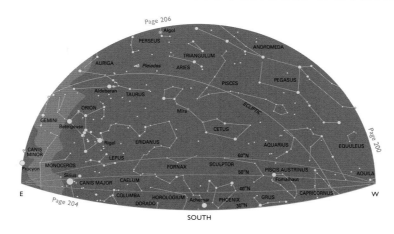

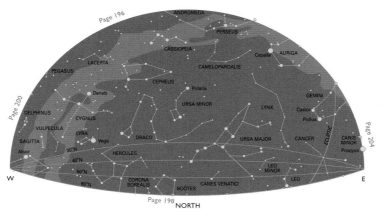

The northern-hemisphere sky

The late fall skies are empty of bright stars, but over to the west you can still see the large Summer Triangle of Vega, low down in the northwest, Altair, low in the west, and Deneb in mid sky. The signpost constellation is the Square of Pegasus, which stands out because there are few bright stars in this part of the sky. Follow the top of the Square upward and you find three bright stars of Andromeda, which lead to Perseus, almost overhead, then the W-shape of Cassiopeia.

Below the Square lies faint Cetus, and to its east lies a small group of three stars marking Aries. Beyond this are stars that will grace the winter skies – bright, reddish Aldebaran within the V-shaped Hyades cluster, the glittering Pleiades cluster, and, rising in the east, the brilliant stars of Orion, with the three stars of Orion's belt pointing almost vertically upward.

Evening

1 November at 11.30 pm
15 November at 10.30 pm
30 November at 9.30 pm

Morning

15 July at 6.30 am
1 August at 5.30 am
15 August at 4.30 am

The southern-hemisphere sky

The familiar Southern Cross is low down in the south during the early summer, and may even be partially below the horizon for some people. But higher up in the sky are the three well-separated bright stars Achernar, in mid sky to the south, Canopus, in the southeast, and Sirius, lower down in the east, with the bright stars of Orion and Aldebaran in Taurus beyond that as a taste of things to come later in the summer.

Looking north you can see the M-shape of Cassiopeia fairly low down, with a compact line of stars marking Perseus to its east. Above that is a pattern of three stars in Aries, and more or less overhead lies a long wriggling line of fairly faint stars in Cetus. The Square of Pegasus is now setting to the west. Use it to find the faint stars of Aquarius, then the bright star Fomalhaut, and a pair of fairly bright stars marking Grus, the Crane.

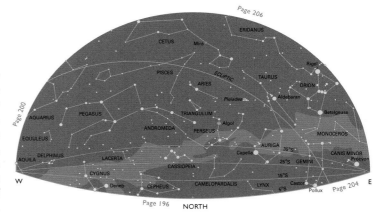

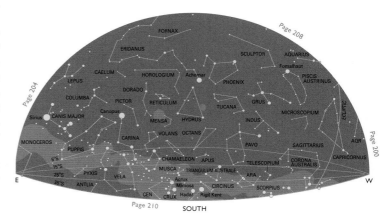

PISCES

The sign of the fish continues the watery theme of this part of the sky (see page 179). The constellation itself is particularly faint, however, and none of the stars is brighter than magnitude 3.7, which places virtually the whole constellation out of reach to the naked eye in light-polluted areas. In a good country sky, however, you can see that there are two lines of stars starting from Alpha Piscium (Alrescha) – one running between Andromeda and Aries, and the other below the Square of Pegasus, where it ends in a circlet of stars that marks the head of one of the fish. Being close to the celestial equator Pisces is equally visible from both hemispheres.

Alpha Piscium (Alrescha) is a double star of roughly equal components separated by 1.9 arc seconds, making it a good test for a 60 mm refractor and more easily visible in a 100 mm or larger telescope.

The Circlet asterism and TX Piscium

This evenly spaced circlet of stars just below the Square of Pegasus is easy to spot despite its faintness. Its brightest member, Gamma Piscium, is magnitude 3.9. But its greatest claim to fame is the irregular variable star 19 Piscium, also known as TX Piscium. This alias results from its being a variable star – variables in a constellation are given letters in a complicated sequence which need not concern us here. The star itself is famous for its strong color, which is easily visible in binoculars or a small telescope. It varies erratically between magnitudes 4.5 and 6.2, though sources differ as to its normal range and there is a lot of scatter in the estimates. Old books will tell you that it is a type N star, but these days it is known as type C, for carbon. TX Piscium is one of the brightest carbon stars, and it is the carbon in its atmosphere that filters out much of its blue light, resulting in a red star.

Galaxy M74

Pisces is barren not just of stars but also of deep-sky objects. There is just one of any note, the galaxy M74. Like many other face-on spirals, this is notoriously difficult to see, as it is ninth magnitude and quite large. Some people can see it better in a good finder than the

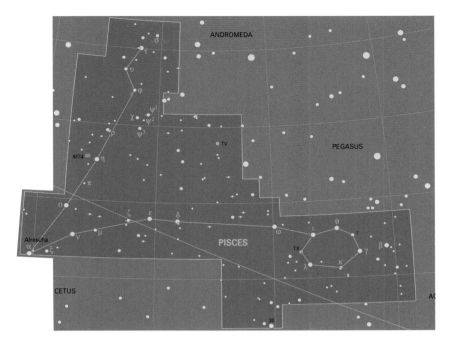

main instrument. Fortunately, it is close to the third-magnitude star Eta Piscium, which itself is easy to locate using the two brightest stars of Aries as pointers. But many observers can miss the arms of the galaxy even though they may see the nucleus, as it appears almost starlike.

Its large size and elusive nature have earned for M74 the nickname "Phantom Galaxy." One oddity is that three supernovae have already been seen in this galaxy in the 21st century. Just 2½° southeast of the galaxy lies another member of the galaxy cluster of which it is the brightest member, NGC 660, which is magnitude 10.7.

Phantom it may be, but M74 is a very elegant galaxy. It is what is known as a *grand design* spiral, which means simply that it has clearly defined and well-organized spiral arms, characteristically just two of them wound round the nucleus. Another common type of spiral galaxy is known as *flocculent*, in which there is an obvious spiral structure but the arms are by no means so continuous as those of M74, an excellent example being NGC 4414 on page 115.

▲ *The Phantom Galaxy, M74, photographed with the 2.5 m Isaac Newton Telescope. It is some 30 million light years away, and its diameter is about the same as our own Milky Way Galaxy.*

▲ *A photograph of the Circlet brings out the different star colors, notably that of TX Piscium at the left.*

Object	Type	M$_v$	Magnification	Distance
TX Psc	Variable star	4.5–6.2	Naked eye	700 light years
M74	Galaxy	9.4	75×	30 million light years

TAURUS

Although Taurus is a well-known and picturesque constellation, most of its stars are below third magnitude. Yet it is one of the best-known patterns in the northern sky, and it bears some resemblance to a bull on a join-the-dots basis. Red giant star Aldebaran represents the eye of the bull, set within a V-shape of stars, the Hyades, which mark its face. Two outlying stars indicate the tips of its horns, while on its back sits the Pleiades star cluster. There is a suggestion that dots on a famous prehistoric cave painting at Lascaux, France, represent the Pleiades in exactly the same relationship to the head of a long-horned bull, and Taurus may be an even more ancient constellation than we realize. It is intriguing that the ancient cave painting shows just the front half of the bull, in just the same way as the more recent constellation depictions.

The most northerly extent of the ecliptic lies on the border between Taurus and neighboring Gemini, so the Moon and planets may be seen in Taurus and indeed can pass close to or even in front of Aldebaran and the stars of the Hyades and Pleiades. The constellation can be seen from both hemispheres.

Hyades star cluster

The stars of the Hyades are more widely spread than those of any other cluster, as it is the closest cluster to the Solar System, at a distance of just 150 light years. But Aldebaran, which seems embedded within it, is less than half that distance away. The individual stars are easy to count with the naked eye, with several at between magnitudes 3 and 5; in binoculars many more are visible.

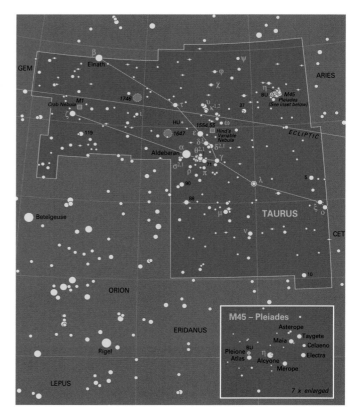

▲ **A long-exposure photograph of the Pleiades** shows the Merope Nebula. Its blue color indicates that it is interstellar dust rather than gas. A short-exposure image of the Pleiades appears on page 110.

▼ **Aldebaran and the Hyades,** with stars shown down to about tenth magnitude. The field of view here is 15°.

Object	Type	M_v	Magnification	Distance
Aldebaran	Star	0.9	Naked eye	65 light years
Hyades	Open cluster	–	Naked eye	150 light years
M45	Open cluster	1	Naked eye	445 light years.
M1	Supernova remnant	8.4	50×	6300 light years

M45, Pleiades star cluster

The Pleiades attract the eye like nothing else in the sky. The individual members of this brilliant cluster range from Alcyone at magnitude 2.9 down to very faint stars, and there are probably 500 or more in all. They have a total brightness of about first magnitude, and by averted vision the cluster is just as easily visible as a first-magnitude star, though appearing as a haze rather than a point of light. Much of the cluster is enveloped in a blue reflection nebula, known as the Merope Nebula after the star within its brightest part, though this requires binoculars or a low-power telescope in a good sky to be seen properly.

A popular name for the cluster is the Seven Sisters, and opinions vary as to how many stars are visible with the naked eye. Most people can certainly see six, and seven or eight are not hard, though some people can manage 14 or more. But the number seven was probably chosen for symbolic rather than numerical reasons. Although nine of the stars have names, the Pleiades were the seven mythological daughters of Atlas and Pleione, and their parents are represented by the two easternmost bright stars.

The cluster is a beautiful sight in any telescope, though a low power is needed for the best views. In a large telescope equipped with an ultra-wide-angle eyepiece in a dark sky the view is stunning, but even a city dweller will get a pretty sight with binoculars.

M1, Crab Nebula

Long-exposure photographs of the Crab Nebula have made it one of the most famous astronomical objects, and every owner of a telescope will want to see it. But the reality is less dramatic than the photographs of an exploding star: a hazy oval is all that most people see. At eighth magnitude, the Crab Nebula is within reach of binoculars on a clear, dark night, but it has a low surface brightness and is easily wiped out by any light pollution or haze.

Fortunately, it is located close to a bright star, Zeta Tauri. The Crab Nebula lies just over a degree to the northwest of this star, about halfway toward a sixth-magnitude star that should be visible in a finder. M1 is about 8 arc minutes in width, which makes it larger than most planetary nebulae and galaxies, so a fairly low power is all that is needed. Few people see the claw-like filaments that gave the Crab its name, however, unless they are using large telescopes in very transparent skies.

The nebula itself is unique. It is a supernova remnant like few others in the sky, consisting of a shell of material kept glowing by radiation from a tiny central star. The supernova that created it was seen to explode in the year 1054 by Chinese astronomers and presumably by most others of the world's population, but only the Chinese definitely recorded it. It shone as bright as Venus, and was one of only four supernovae to have been observed in our Galaxy during the last millennium.

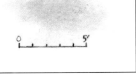

▲ *A drawing of the Crab*
made with a 76 mm refractor.
It shows just the oval outline
of the nebula.

◄ *The Crab Nebula*
photographed using a 3.5 m
telescope. It shows the dramatic
filaments from the star that
was seen to explode there
over 950 years ago.

CETUS

Lurking below Pisces is the sea monster or whale, Cetus. Its head is a group of stars below Aries, and Alpha Ceti (Menkar) forms an equilateral triangle with Hamal in Aries and the Pleiades. Beta Ceti, Deneb Kaitos, is a long way south of the eastern edge of the Square of Pegasus, and is the only bright star in this part of the sky. The constellation itself straggles between Alpha and Beta, and is well placed for viewing from both hemispheres.

Cetus is not rich in bright deep-sky objects – the most notable is the galaxy M77, which is conveniently located near to Delta Ceti and forms a neat triangle with Delta and fifth-magnitude 84 Ceti. M77 is a Seyfert galaxy, which is a type of spiral galaxy with a bright nucleus. The active nuclei of Seyfert galaxies are thought to be low-power versions of quasars – highly luminous galactic nuclei that are powered by massive black holes. All quasars, however, are very remote, so Seyfert galaxies such as M77 allow us to glimpse some of the most energetic processes taking place in the Universe. The galaxy itself is eighth magnitude, and its condensed nucleus makes it easier to observe with small telescopes than many other Messier galaxies.

▶ **The disk of the giant star Mira** *in visible light, as seen with the Hubble Space Telescope. The disk appears distorted, possibly as a result of starspots on its surface.*

Variable star Omicron Ceti (Mira)

As you follow your way down the line of stars from Alpha Ceti toward the tail of the monster, you may or may not come across one particular star – Mira. At times it can be easily seen with the naked eye, but at others it requires binoculars and a detailed star chart because it has faded to about ninth magnitude. Such behavior brought it to the attention of astronomers as early as the 17th century, at a time when changes among the stars were largely unknown. The name Mira means "the wonderful," in honor of what were then thought to be amazing properties.

Mira has given its name to a whole class of stars that behave in the same sort of way, with considerable variations in brightness over a matter of months. The rise is fairly steep, so the star can appear almost nova-like by suddenly appearing in the sky over just a few weeks, reaching a maximum of about third magnitude. Mira itself has a period of variability of about 330 days, so its maximum is around a month earlier each year. Its closeness to the ecliptic means that a maximum occurring between April and June will be unobservable because it will be lost in the twilight at the crucial time; for several years at a time Mira may always be below naked-eye visibility in the night sky.

Mira is one of only two stars whose actual disk has been imaged by the Hubble Space Telescope. Although it is about 300 light years away, it swells to such a size when at maximum that it is just within the HST's resolving power.

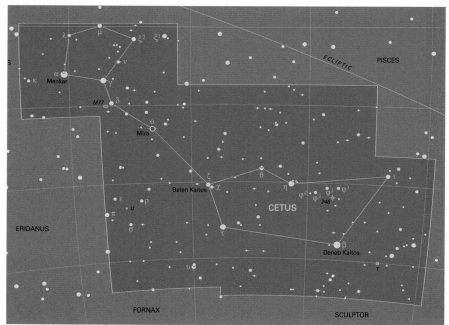

Object	Type	M_v	Magnification	Distance
Mira	Variable star	3–9	Naked eye	300 light years

TRIANGULUM

This is a constellation that no one can argue does not resemble its namesake – a triangle. It dates from Greek times, but consists of stars of only third and fourth magnitude. Its modern claim to fame is that it is home to the galaxy M33, the Pinwheel Galaxy. This is a member of our Local Group of galaxies, of which the other leading members are our own Galaxy and M31. The Pinwheel is about 2.8 million light years away, so is slightly more distant than M31 and its companion galaxies.

Galaxy M33, the Pinwheel Galaxy

For those few observers able to see this galaxy with the naked eye, it is the most distant object visible. But the majority of us need binoculars. If you are used to galaxies in binoculars being tiny blurs among the stars, you may overlook M33 as it is very large. Once you are aware of this, however, it is not too hard to find in reasonably dark skies. The map shows its location not far from Alpha Tri, and it lies about a binocular field away from that star. Though the nucleus appears to be surrounded by a fainter area, the spiral structure that is seen in photographs is not obvious. The total brightness is magnitude 5.3, but this is spread over an area twice the diameter of the full Moon so the surface brightness is low.

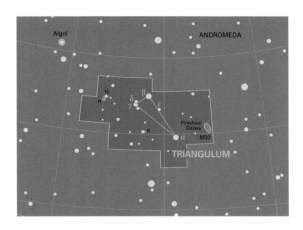

Object	Type	M_v	Magnification	Distance
M33	Galaxy	5.7	10×	2.8 million light years

▶ **This telephoto lens shot of the region around M33** *has a field of view of 5.5° and shows stars down to magnitude 11. It closely resembles the galaxy's appearance in binoculars.*

◀ **The galaxy M33** *photographed with the Mayall 4 m telescope at Kitt Peak.*

ANDROMEDA

The region around Andromeda can be likened to a mythological cartoon strip. Andromeda, the daughter of Queen Cassiopeia, is chained to a rock as a sacrifice to the sea monster Cetus. In adjacent frames we see the hero Perseus who rescues her, and her worried mother Cassiopeia. The whole story remains in the sky to be told and retold around camp fires.

The four main stars of Andromeda are in a rough line, and hardly represent a beautiful maiden, but the constellation does include one of the sights of the sky, the Andromeda Galaxy. At one end of the line is Perseus – in fact, you can reach Alpha Persei by continuing the line of stars to the east. At the other end is the Square of Pegasus, and the star Alpha Andromedae doubles up as one corner of the Square. Andromeda is a northern constellation, but most of its stars can be seen from the southern hemisphere, though very low down for the more southerly observers.

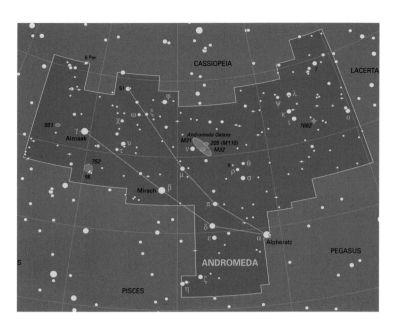

Double star Gamma Andromedae (Almaak)

Contrasting colors are always popular in double stars, and Gamma Andromedae is a good example. Its stars are yellow and blue, separated by just under 10 arc seconds, so even a small telescope will split them. The sixth-magnitude secondary star is itself a double, though the two stars are very close together.

M31, Andromeda Galaxy, M32 and NGC 205 (M110)

It comes as a surprise to many people that the Andromeda Galaxy is not only easy to find, but also visible with the naked eye even in average skies. With binoculars it can be viewed even from city centers on clear nights. Although it appears as only an oval smudge under such circumstances, it is something just to be able to observe the heart of another galaxy over 2.5 million light years away armed with nothing more than binoculars or the naked eye. The Andromeda Galaxy is generally regarded as the most distant object that can be seen with the naked eye, though there are always people who can see more distant objects under the right conditions.

To locate it, find the line of stars Alpha, Delta, Beta, Gamma. From Alpha, go two stars along to Beta and from there take a right angle to the north and count along two fainter stars, Mu and Nu. The Andromeda Galaxy is alongside Nu Andromedae. If you don't see it by direct vision, look away from it slightly and you should pick up a misty ellipse. In light-polluted skies you are seeing just the nucleus of the galaxy, but from a good country site and with good eyesight you can trace it much farther out – its outline on the map shows that it covers more than 3° of sky.

In a telescope, the smudge becomes larger and more extensive, and you should be able to pick up M31's two companion galaxies, M32 and NGC 205. M32 is a circular elliptical galaxy; on photographs it appears on the edge of M31, but visually it is usually separate. NGC 205, now generally added to Messier's catalog as M110 on the grounds that he did actually observe the object, is an ellipse on the other side of M31 from

▲ **The closeness and contrasting** colors of its two stars make Gamma Andromedae particularly beautiful.

▲ **Locating the Andromeda Galaxy.** The bright star at lower left of center is Beta And; from there move up past Mu and Nu to find M31 itself.

▲ *A close-up of galaxy* **M32,** *taken at Kitt Peak in Arizona. For an amateur drawing of M31 and M32, see page 120.*

magnitude star, and is of similar brightness but noticeably larger, even with a low power. The blue color is not as obvious to every observer as its name would imply. Higher magnifications will help to reveal some internal detail, particularly with larger instruments.

Galaxy NGC 891

Andromeda is host to a few galaxies in addition to the M31 group, and NGC 891 is the brightest example. It requires a medium aperture, but is quite easily found by sweeping 3½° eastward from Gamma Andromedae. It lies midway between that star and the cluster M34 in Perseus. It is an edge-on spiral, and the dark lane that bisects it can be seen under good conditions.

M32. If it were not for M31, both these companion galaxies would still be among the brightest in the sky, but we tend to relegate them to minor objects because of their proximity to the great galaxy itself.

Large telescopes will show a surprising amount of detail in M31. The dark lane shown on photographs is visible in good skies with medium apertures, but individual star clouds and globular clusters can be made out under good conditions.

Open cluster NGC 752

Some clusters, like this one, are comparatively faint but quite rich in stars. This scattering of stars of eighth to tenth magnitude is particularly attractive in a telescope, though it is also visible in binoculars. The unrelated star 56 Andromedae is a good guide to its location.

Planetary nebula NGC 7662, the Blue Snowball

With a nickname like "the Blue Snowball," the popularity of this bright planetary nebula is assured. Despite the lack of nearby bright stars it is fairly easy to find without a Go To telescope, as it lies just under 5° to the east of the pair of stars Omicron Andromedae and 2 Andromedae. With these in the field of view of the telescope, either sweep eastward or leave the telescope stationary for 23 minutes and the planetary will drift into the center of the field of view. It lies near an eighth-

▶ **NGC 752 is the** **scattering of stars** *below the center. The double star just below it is 56 And.*

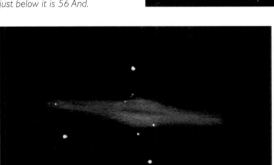

◀ *A 250 mm reflector was used for this sketch of the galaxy NGC 891, with powers of 113 and 184.*

▼ **This annotated sketch** **of the Blue Snowball** *was also made with a 250 mm reflector.*

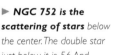

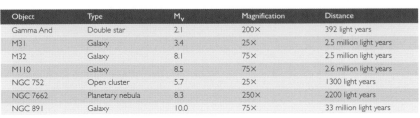

Object	Type	M$_v$	Magnification	Distance
Gamma And	Double star	2.1	200×	392 light years
M31	Galaxy	3.4	25×	2.5 million light years
M32	Galaxy	8.1	75×	2.5 million light years
M110	Galaxy	8.5	75×	2.6 million light years
NGC 752	Open cluster	5.7	25×	1300 light years
NGC 7662	Planetary nebula	8.3	250×	2200 light years
NGC 891	Galaxy	10.0	75×	33 million light years

THE PERSEUS CLUSTER OF GALAXIES AND NGC 1275

To the amateur observer, Perseus is well known for its star clusters, but many professional astronomers are more interested in another type of cluster – the Perseus Cluster of galaxies. The brightest members of this cluster are about 12th magnitude and are faintly visible to keen observers with dark skies, just over 2° east of Algol. The cluster itself spreads over a degree or so of sky – a tiny area compared with the better known Virgo Cluster, which is much nearer. While the Virgo galaxies are about 55 million light years away, those of the Perseus Cluster are over four times more distant.

Most clusters of galaxies have a dominant member, and in the case of Perseus this is NGC 1275. Through a medium-sized telescope in good skies it appears to have a starlike nucleus rather than being just a fuzzy blob, a characteristic that sets it aside from many other galaxies. It is of a type known as a Seyfert galaxy, and the brilliant nuclei of such galaxies can emit 10,000 times as much light as those of normal galaxies. Active

▼ **The Perseus Cluster of galaxies,** with NGC 1275 at its heart. This view from the Digitized Sky Survey covers an area of just a quarter of a degree. As this area is in the plane of the Milky Way, there are many foreground stars in our own Galaxy visible.

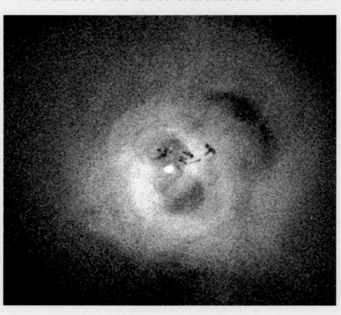

▲ **The central area of the Perseus Cluster** as revealed by the Chandra X-ray Observatory. The bright spot and yellow glow at the center coincide with NGC 1275 itself, while the dark fragments may be the remains of a galaxy that has fallen into NGC 1275, adding to its mass. Large dominant galaxies in galaxy clusters are cannibals,

dragging in smaller galaxies in the group that orbit too close. X-rays surround the area, stretching out beyond the galaxies nearest to NGC 1275. The large dark blobs on either side of NGC 1275 are strong in radio waves and are regions where particles and magnetic fields produced by the central black hole in NGC 1275 have swept aside the hot X-rays.

galaxies such as this are of great interest to professional astronomers, as they probably contain super-massive black holes which are far

more powerful than the quiescent one at the center of our own Galaxy. The activity is created as gas spirals on to the accretion disk surrounding the black hole.

Not only does this gas produce light, it also produces more energetic forms of radiation, notably X-rays. To astronomers who study the sky by the X-rays detected by satellite observatories, this is known as Perseus A, an intense source of X-rays, while radio astronomers call it 3C84. The whole area around NGC 1275 is surrounded by thin but very hot gas – at temperatures greater than 30 million degrees C, which is flowing down on to the galaxy. But NGC 1275 is also in the throes of consuming another galaxy, which presumably also helps to feed its voracious black hole.

▼ **NGC 1275 caught in the act** by the Hubble Space Telescope of swallowing another galaxy. Dark dust lanes are rimmed with bright star formation triggered off by the event, while the bright nucleus of NGC 1275 itself is in the background. After several such galaxy mergers, the cannibal spiral galaxy begins to resemble a giant elliptical galaxy.

PERSEUS

Few constellations have as much to offer as Perseus. This section of the northern Milky Way contains a range of interesting objects to suit all tastes.

In Greek mythology, Perseus was the hero of many adventures. But there is one particular story told in the stars – his rescue of the beautiful princess Andromeda from the sea monster Cetus, far to the south. In a previous episode he had slain the Gorgon Medusa, whose mere glance would turn a man into stone. He still carries her head hanging from his belt.

Surrounding Alpha Persei (Mirphak) is a glittering loose cluster of stars, known simply as the Alpha Persei Cluster. It is best viewed in binoculars, where it appears as a prominent S-shape of stars resembling a rollercoaster. Among other open clusters in Perseus is M34, which is easily resolved into stars with binoculars.

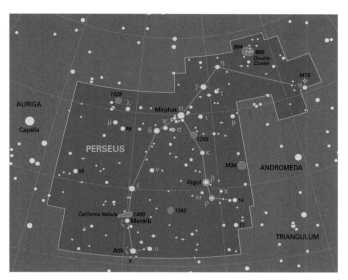

▲ *The Alpha Persei Cluster,* also known as Melotte 20. Mirphak itself is at one end of a long S-shape of stars.

Variable star Beta Persei (Algol)

The Gorgon's head is marked by a particularly famous star, Algol, whose name means "the ghoul." Every 2 days 21 hours, this star winks – dimming from magnitude 2.1 to 3.4, and taking about ten hours to do so. Algol was the first eclipsing binary to be identified, and stars of this sort continue to be called Algol-type stars.

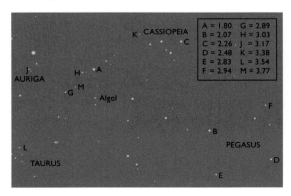

▲ *A comparison star chart,* which can be used to follow Algol's variations. They are predicted in sky handbooks and are available online.

Double Cluster, NGCs 869 and 884

To many observers this pair of clusters is as great a jewel of the sky as the Pleiades. Though they are by no means bright they are visible with the naked eye, by averted vision at least, in a good sky midway between Perseus and Cassiopeia. In binoculars they are a delight, and in a telescope under good conditions they can be breathtaking. One of the clusters by itself would give most others a run for its money; but side by side they are unbeatable. Yet these are not nearby clusters – they are both over 7000 light years away, getting on for twice the distance of M37, say. It is the sheer number of faint stars that makes them impressive, and if they were much closer they would be an amazing sight. There are several red stars in NGC 884, which add to the spectacle.

▼ *The Double Cluster, NGC 884* (left), also known as Chi Persei, and NGC 869 (right), also known as h Persei.

Object	Type	M$_v$	Magnification	Distance
Alpha Per	Star	1.8	Naked eye	500 light years
Beta Per	Variable star	2.1–3.4	Naked eye	90 light years
NGC 869/884	Open cluster	4.3/4.4	10×	7100/7400 light years
M76	Planetary nebula	10.1	100×	3400 light years
NGC 1499	Nebula	–	Photographic	1000 light years
GK Persei	Former nova	13.5	100×	1500 light years

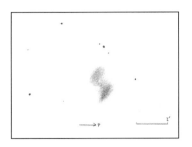

▲ **A drawing of M76** *made using a 460 mm Dobsonian telescope.*

▶ **This close-up of the California Nebula** *has east at the top. Xi Persei is the blue star to the right of the nebula.*

Planetary nebula M76, the Little Dumbbell

In theory, the Little Dumbbell should be a great challenge to owners of small telescopes because it is claimed to be the faintest object in the *Messier Catalogue*. In practice, however, this tenth-magnitude planetary has a fairly high surface brightness so it is a good deal easier than some objects that are supposedly much brighter. M76 is simple to locate once you have found fourth-magnitude Phi Persei, which is easier to find by star-hopping from Gamma Andromedae than from the stars of Perseus. M76 is just to the west of a sixth-magnitude star a degree north of Phi.

NGC 1499, the California Nebula

One for the photographers, this patch of pink nebulosity near Xi Persei looks at first sight like a rather fat pink banana. But turn it with east at the top, and it becomes a fair representation of the state of California. It is surprisingly easy to capture using a CCD and shows through moderate light pollution; visually, however, it requires dark skies and an O III or hydrogen-beta filter. The nebula is a hydrogen cloud that just happens to be illuminated by Xi Persei.

Nova GK Persei

Some of the great discoveries in astronomy are made not by professionals but by interested amateurs who spend a lot longer actually staring at the stars than do their observatory-bound counterparts. So it happened in February 1901 to Dr T. D. Anderson of Edinburgh,

Scotland, a clergyman who was just walking home when he glanced upward, as astronomers often do, before entering his house. He spotted a third-magnitude star in Perseus, which he knew should not be there, making an unfamiliar triangle with Alpha Persei and Algol, and his first reaction was irritation that during his walk he had not spotted it sooner.

This object was Nova Persei 1901, and Anderson had found it just as it was increasing in brightness. Two days previously, a photograph of the area showed it ten magnitudes fainter. But within two days of Anderson's discovery it had increased in brilliance to become one of the brightest stars in the sky. Thereafter it faded rather more slowly.

Today, the star is known as GK Persei, and a shell of material surrounding it is visible on photographs taken with long exposure times. Research tells us that these novae or new stars are actually otherwise unremarkable stars that have undergone a tremendous outburst. They are generally known as cataclysmic variables, and are members of close double-star systems of which one is a white dwarf that is slowly accumulating material from its companion. Eventually this material flares up in an explosion and we see a great increase in brightness. Then the episode is over for maybe 10,000 years until the material builds up again.

Novae as bright as Nova Persei are rather rare. The last one was in Cygnus in 1975, but another could happen at any time. Fainter novae, however, occur a few times a year, and are often spotted by amateurs. Typically they may reach magnitude 8 or 9. It is likely that many go unnoticed because they usually occur in the Milky Way and one additional ninth-magnitude star for a few days among thousands of others can easily be missed.

▲ **The shell of gas surrounding GK Persei** *is expanding year by year, which allows astronomers to compare its change in size with its Doppler shift, yielding an accurate value for its distance of 1533 light years. The shell shown here is about 0.7 light years across, but material presumably from a previous outburst can be detected 20 times farther out.*

CASSIOPEIA

To those viewing in the northern hemisphere the W-shape of the main stars of Cassiopeia (or M-shape, depending on which way up it is) is a familiar sight – for it is always in the sky for most observers. As usual, its star pattern has very little to do with its representation of Queen Cassiopeia on her throne. It is in a rich area of the northern Milky Way, and although it lacks the very brightest of splendors, there are many fainter deep-sky objects within it. It was also the location of one of the few supernovae to be seen in our Galaxy in the past millennium, in 1572, though no visible trace of this event remains.

The star Gamma Cassiopeiae, the middle star of the W, is an erratic variable star. For most of the time it remains at about magnitude 2.2, but it can fade to magnitude 3 and has been known to brighten to 1.6. A blue giant star, it from time to time throws off shells of material and could undergo another shell episode at any time. Unusually for such a prominent star, it has no accepted popular name and is usually known by the abbreviation Gamma Cas.

Long-exposure photographs show several pink wisps of nebulosity in Cassiopeia, notably IC 1805 and IC 1848.

Open cluster M52

Follow a line from Alpha through Beta Cas an equal distance and you come to the open cluster M52. It is visible with binoculars as a misty blur, and a small telescope will show numerous stars of about tenth magnitude. It is quite compact and distant, and requires more magnification than many other clusters. Nearby lies a nebula, NGC 7635, known as the Bubble Nebula. In medium and large telescopes, under good conditions, it is just visible as a crescent-shaped nebula surrounding a star. Photographs reveal a curious circular bubble which results from hot gas expelled from a central star.

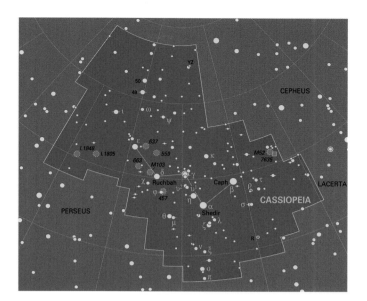

NGC 457, Owl Cluster

Many clusters have nicknames, but this one has at least two aliases – the Owl Cluster and the E.T. Cluster. At one edge of it is fifth-magnitude Phi Cas, just south of Delta Cas. Like the Jewel Box in Crux, NGC 457 is a cluster to which binoculars do not really do justice because the bright stars overpower the underlying fainter stars. The alternative names derive from two long streams of stars at an angle to the main cluster, which some see as the wings of an owl and others as the arms of Spielberg's famous extra-terrestrial. In a telescope the arms or wings become very obvious, and Phi Cas and a neighboring star look like the huge eyes of an owl or of E.T.

Open cluster M103

Smaller and less impressive than NGC 457, M103 can be found by returning to Delta Cas and moving a degree east and slightly north. Although visible in binoculars, a higher magnification is a great help. M103 is noticeably fan-shaped. Beyond it is the cluster NGC 663, which is more obvious in binoculars.

▲ **The two bright stars in NGC 457** have the appearance of huge eyes, an effect that is even more vivid in a telescope than on a photograph.

▶ **M103 requires a telescope** for the best views and is not to be confused with the nearby and larger NGC 663.

◀ **A small telescope will show cluster M52** (top left), but a larger telescope or a long-exposure photograph is needed to reveal the Bubble Nebula, bottom right.

Object	Type	M_v	Magnification	Distance
M52	Open cluster	7.3	50×	5000 light years
NGC 7635	Nebula	–	Photographic	7100 light years
NGC 457	Open cluster	6.4	25×	2400 light years
M103	Open cluster	7.4	50×	8500 light years
NGC 663	Open cluster	7.1	25×	2000 light years

8

KEY TO STAR MAGNITUDES

- 0.0 and brighter
- 0.1 – 0.5
- 0.6 – 1.0
- 1.1 – 1.5
- 1.6 – 2.0
- 2.1 – 2.5
- 2.6 – 3.0
- 3.1 – 3.5
- 3.6 – 4.0
- 4.1 – 4.5
- 4.6 – 5.0
- 5.1 – 5.5

KEY TO MAP SYMBOLS

- Double stars
- Variable stars
- Open clusters
- Globular clusters
- Planetary nebulae
- Diffuse nebulae
- Galaxies

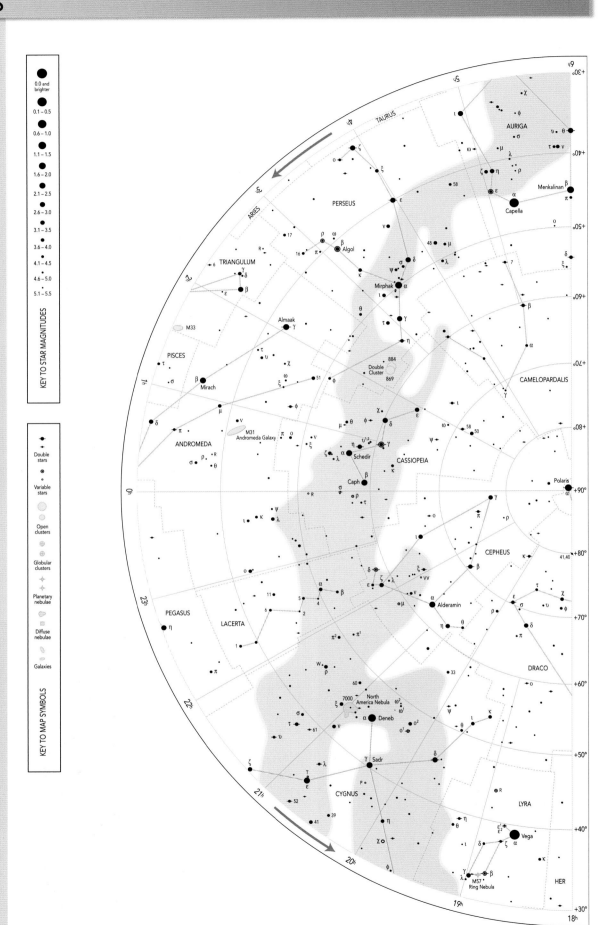

The star maps in this chapter cover the whole of the sky, with some overlap. Maps 1 and 2 represent the northern stars down to declination +30°. Maps 3 to 6 cover a broad strip of the sky extending 60° either side of the celestial equator. The maps are suitably oriented for northern-hemisphere observers; southern-hemisphere observers should simply turn the book through 180°. Maps 7 and 8 show the southern stars, to declination −30°.

A photo-realistic map appears opposite each of the more conventional maps. It shows exactly the same region of sky, but in a manner that closely resembles what you will actually see. You can match this map with your view, and then use the conventional map to find the names of the stars and constellations.

All 88 constellations are shown on these maps. An index listing on which map(s) each constellation appears can be found opposite. The constellation names are in capital letters on the maps, while star names are in lower case with an initial capital. A few prominent asterisms, such as the Square of Pegasus, are also named. The Milky Way is shown in light blue. The ecliptic is shown as a dashed red line. The borders between the constellations are represented by a dashed pink line.

Stars to magnitude +5.5 are shown, which is roughly the naked-eye limit in semirural locations, together with major deep-sky objects. A key to star magnitudes and a key to map symbols accompany each chart.

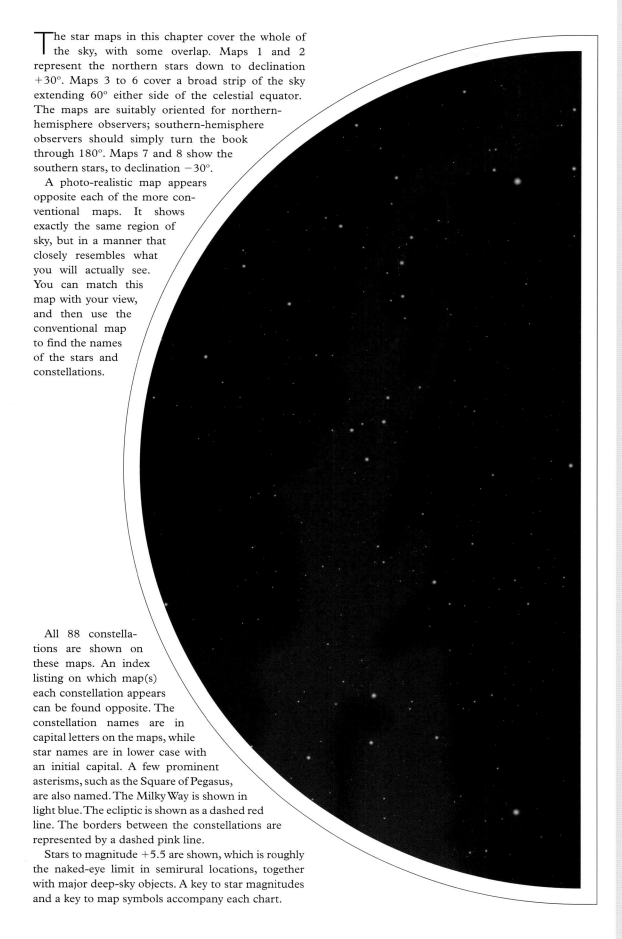

CONSTELLATION	STAR MAP
Andromeda	1, 3, 6
Antlia	5
Apus	8
Aquarius	3
Aquila	3
Ara	3, 4, 8
Aries	6
Auriga	1, 2, 5, 6
Boötes	4
Caelum	6
Camelopardalis	1, 2
Cancer	5
Canes Venatici	4
Canis Major	5
Canis Minor	5
Capricornus	3
Carina	8
Cassiopeia	1
Centaurus	4, 5, 8
Cepheus	1
Cetus	6
Chamaeleon	8
Circinus	4, 8
Columba	5, 6
Coma Berenices	4
Corona Australis	3, 7
Corona Borealis	4
Corvus	4
Crater	5
Crux	8
Cygnus	3
Delphinus	3
Dorado	6, 7
Draco	3
Equuleus	3
Eridanus	6
Fornax	6
Gemini	5
Grus	3
Hercules	3, 4
Horologium	6, 7
Hydra	4, 5
Hydrus	7
Indus	7
Lacerta	1, 3
Leo	5
Leo Minor	5
Lepus	5, 6
Libra	4
Lupus	4, 8
Lynx	2, 5
Lyra	3
Mensa	7, 8
Microscopium	3, 7
Monoceros	5
Musca	8
Norma	8
Octans	7, 8
Ophiuchus	3, 4
Orion	5, 6
Pavo	3, 7, 8
Pegasus	3
Perseus	1, 6
Phoenix	3, 6
Pictor	5, 6, 7, 8
Pisces	3, 6
Piscis Austrinus	3
Puppis	5, 8
Pyxis	5
Reticulum	6, 7
Sagitta	3
Sagittarius	3, 4
Scorpius	4
Sculptor	3, 6
Scutum	3
Serpens	3, 4
Sextans	5
Taurus	6
Telescopium	3, 7
Triangulum	6
Triangulum Australe	8
Tucana	3, 7
Ursa Major	2, 4, 5
Ursa Minor	1, 2
Vela	5, 8
Virgo	4
Volans	8
Vulpecula	3

KEY TO STAR MAGNITUDES

- 0.0 and brighter
- 0.1 – 0.5
- 0.6 – 1.0
- 1.1 – 1.5
- 1.6 – 2.0
- 2.1 – 2.5
- 2.6 – 3.0
- 3.1 – 3.5
- 3.6 – 4.0
- 4.1 – 4.5
- 4.6 – 5.0
- 5.1 – 5.5

KEY TO MAP SYMBOLS

- Double stars
- Variable stars
- Open clusters
- Globular clusters
- Planetary nebulae
- Diffuse nebulae
- Galaxies

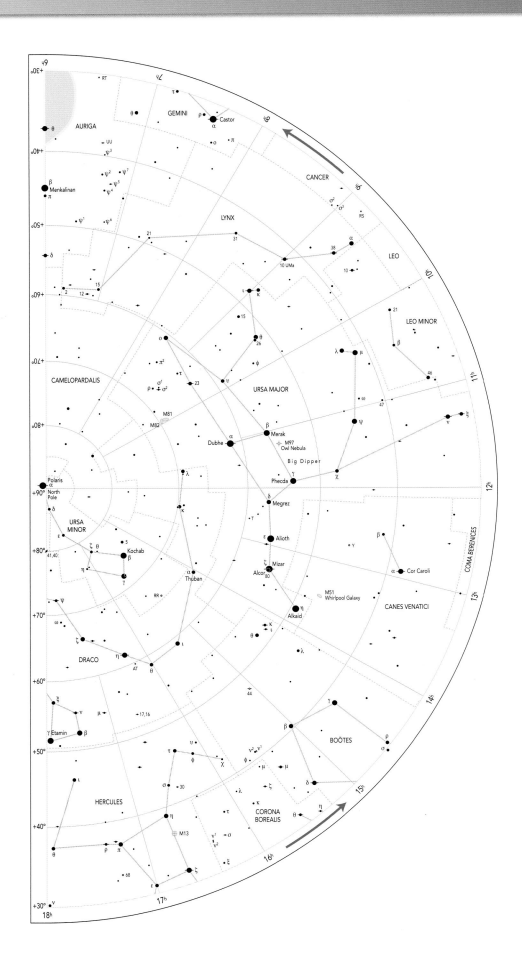

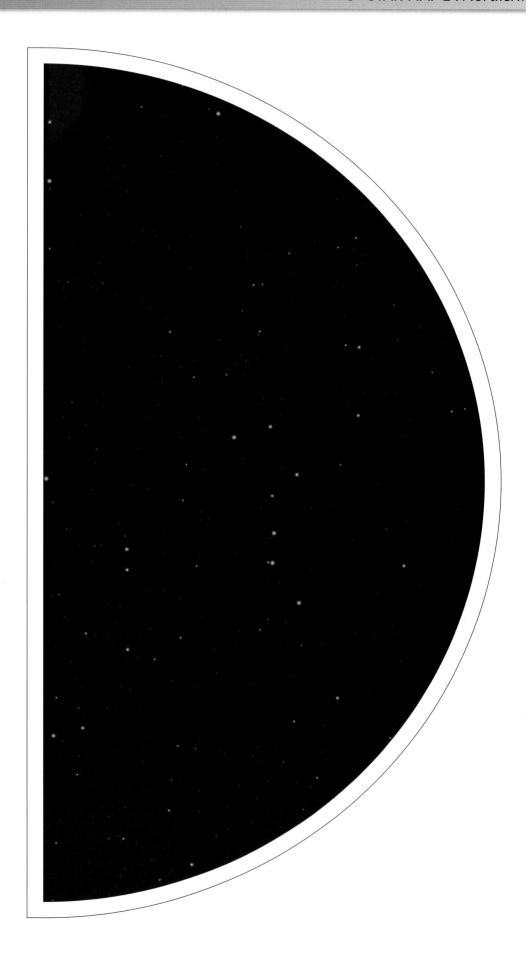

KEY TO STAR MAGNITUDES

0.0 and brighter
0.1 – 0.5
0.6 – 1.0
1.1 – 1.5
1.6 – 2.0
2.1 – 2.5
2.6 – 3.0
3.1 – 3.5
3.6 – 4.0
4.1 – 4.5
4.6 – 5.0
5.1 – 5.5

KEY TO MAP SYMBOLS

Double stars
Variable stars
Open clusters
Globular clusters
Planetary nebulae
Diffuse nebulae
Galaxies

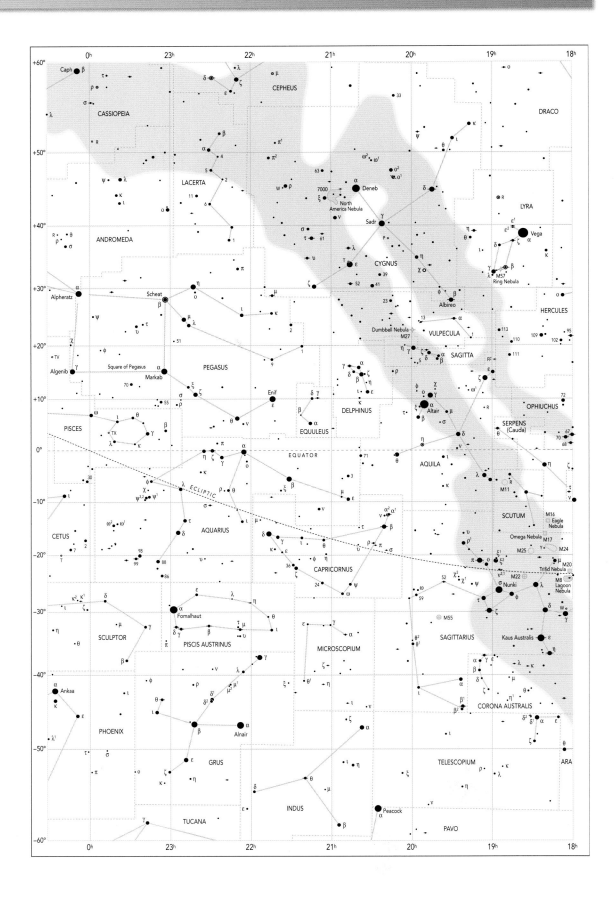

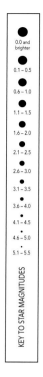

KEY TO STAR MAGNITUDES

0.0 and brighter

0.1 – 0.5

0.6 – 1.0

1.1 – 1.5

1.6 – 2.0

2.1 – 2.5

2.6 – 3.0

3.1 – 3.5

3.6 – 4.0

4.1 – 4.5

4.6 – 5.0

5.1 – 5.5

KEY TO MAP SYMBOLS

Double stars

Variable stars

Open clusters

Globular clusters

Planetary nebulae

Diffuse nebulae

Galaxies

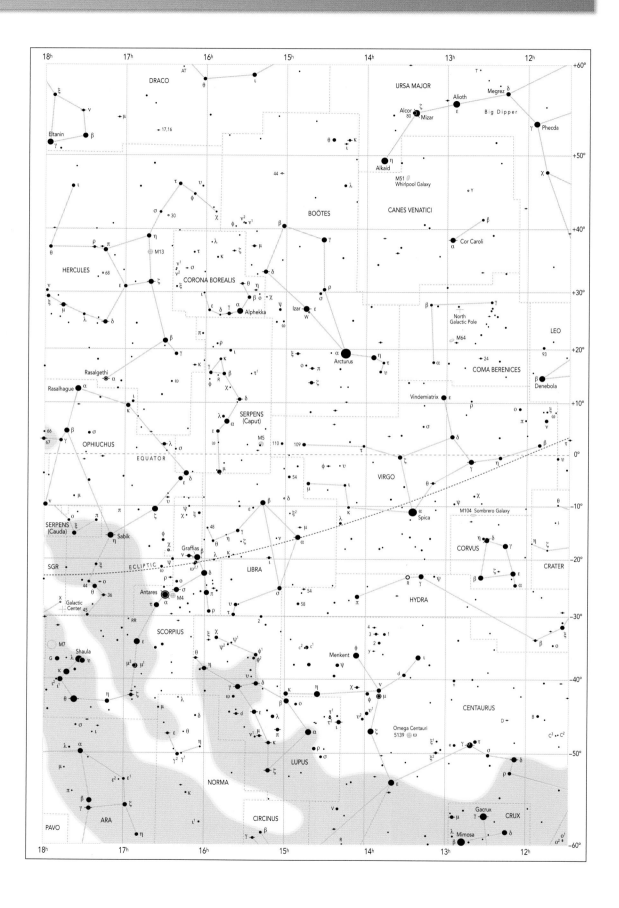

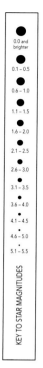

KEY TO STAR MAGNITUDES

0.0 and brighter
0.1 – 0.5
0.6 – 1.0
1.1 – 1.5
1.6 – 2.0
2.1 – 2.5
2.6 – 3.0
3.1 – 3.5
3.6 – 4.0
4.1 – 4.5
4.6 – 5.0
5.1 – 5.5

KEY TO MAP SYMBOLS

Double stars

Variable stars

Open clusters

Globular clusters

Planetary nebulae

Diffuse nebulae

Galaxies

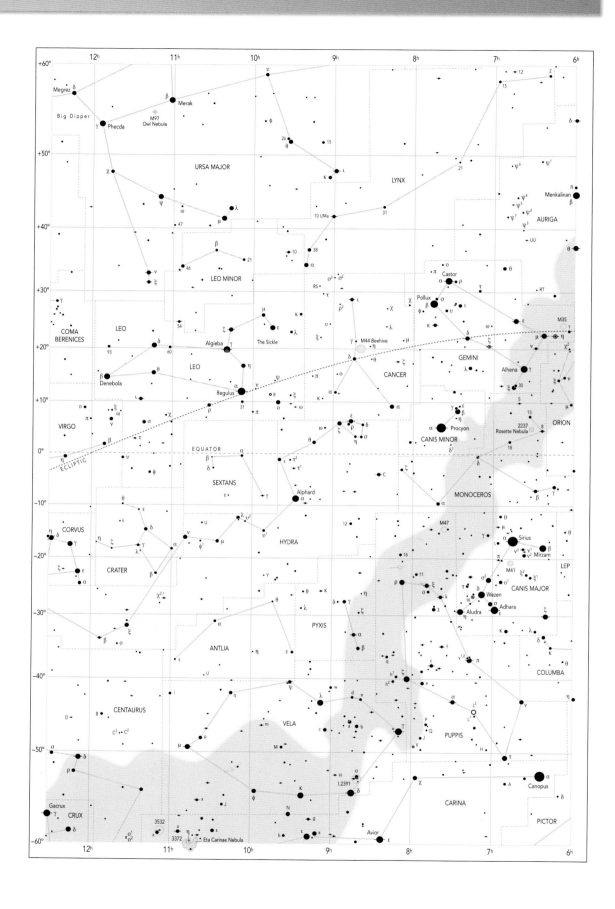

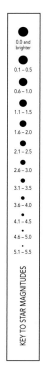

KEY TO STAR MAGNITUDES

- 0.0 and brighter
- 0.1 – 0.5
- 0.6 – 1.0
- 1.1 – 1.5
- 1.6 – 2.0
- 2.1 – 2.5
- 2.6 – 3.0
- 3.1 – 3.5
- 3.6 – 4.0
- 4.1 – 4.5
- 4.6 – 5.0
- 5.1 – 5.5

KEY TO MAP SYMBOLS

- Double stars
- Variable stars
- Open clusters
- Globular clusters
- Planetary nebulae
- Diffuse nebulae
- Galaxies

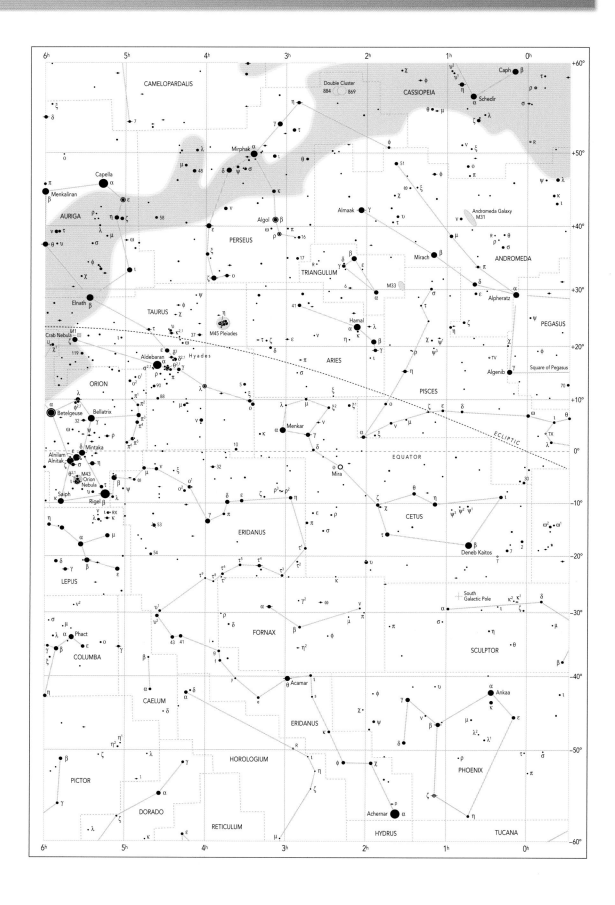

KEY TO STAR MAGNITUDES

0.0 and brighter

0.1 – 0.5

0.6 – 1.0

1.1 – 1.5

1.6 – 2.0

2.1 – 2.5

2.6 – 3.0

3.1 – 3.5

3.6 – 4.0

4.1 – 4.5

4.6 – 5.0

5.1 – 5.5

KEY TO MAP SYMBOLS

Double stars

Variable stars

Open clusters

Globular clusters

Planetary nebulae

Diffuse nebulae

Galaxies

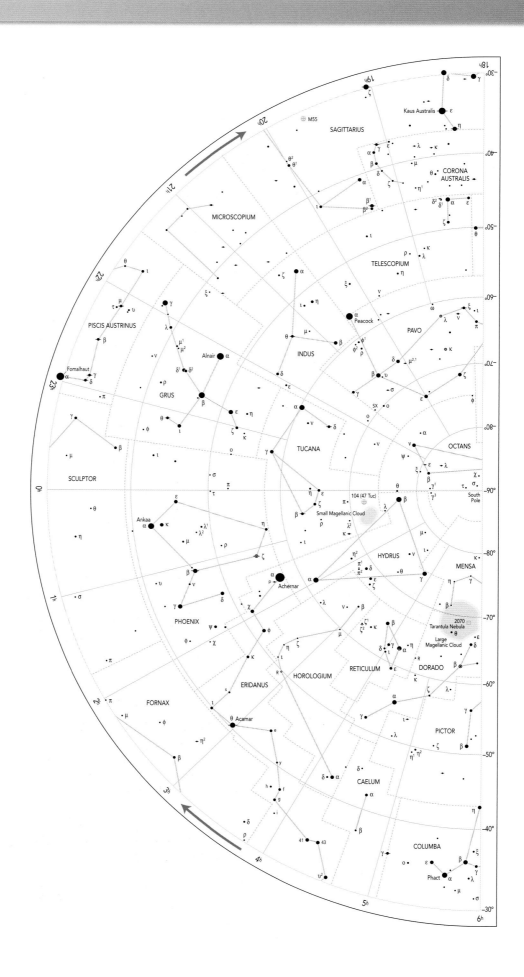

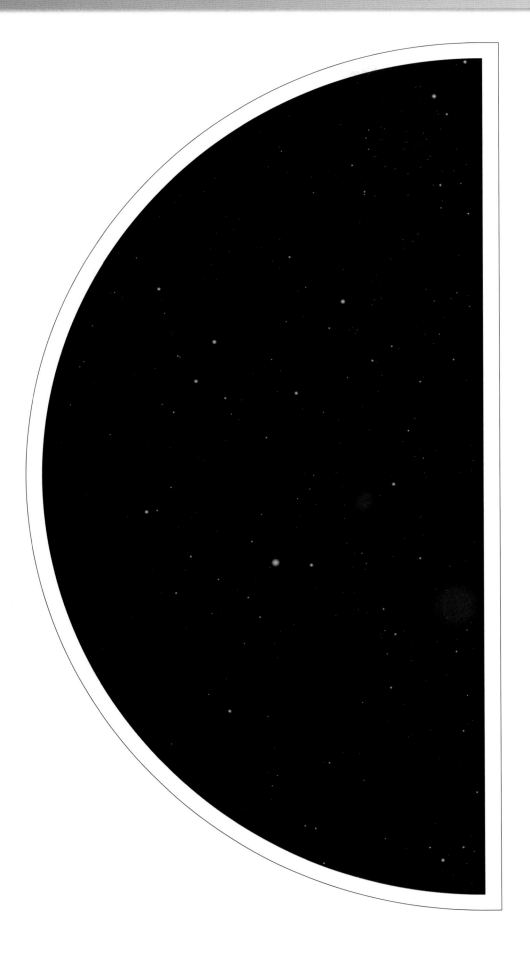

KEY TO STAR MAGNITUDES

- 0.0 and brighter
- 0.1 – 0.5
- 0.6 – 1.0
- 1.1 – 1.5
- 1.6 – 2.0
- 2.1 – 2.5
- 2.6 – 3.0
- 3.1 – 3.5
- 3.6 – 4.0
- 4.1 – 4.5
- 4.6 – 5.0
- 5.1 – 5.5

KEY TO MAP SYMBOLS

- Double stars
- Variable stars
- Open clusters
- Globular clusters
- Planetary nebulae
- Diffuse nebulae
- Galaxies

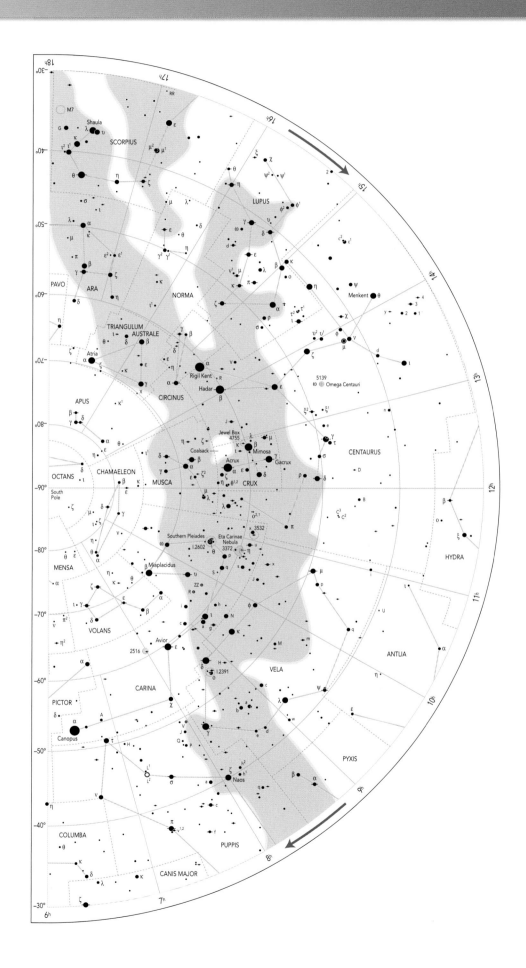

USING THE A–Z SECTION

The alphabetical order of headwords in this A–Z section is letter by letter, ignoring the spaces between words. So, for example, "**G star**" comes between "**Grus**" and "**Gum Nebula**." The exception is that "**H α**," "**H I region**," and "**H II region**" appear at the beginning of the "H" section. Terms beginning with a number are alphabetized as if the number were spelled out; an example is "**Sixty-one Cygni**."

Cross-references are used extensively, and appear in **bold type**. A cross-reference usually indicates that the entry referred to provides secondary information on the topic under discussion. Sometimes it means that the term in **bold type** is defined or discussed at greater length in its own entry. An entry which is just a cross-reference means that the term

is defined in another entry, where it will be found in italic type. A heading or cross-reference followed by a number, for example "**aberration (2)**," is used whenever the same term has more than one entry.

Most entries are fairly concise, but a number of subjects are treated at greater length, for example "**spectral classification**" and the entries for the major planets. The term "planetary body" is used in this dictionary to refer to large satellites and asteroids in addition to the major planets. For the purposes of the values given for the diameters of the giant planets, their "surfaces" have been taken to be the levels in their atmospheres at which the pressure is 1 bar, i.e. the average pressure at the surface of the Earth.

In the past decade numerous small satellites of the outer planets have

been discovered, and many have been named. To avoid overburdening this A–Z section with the names of minor objects, only named satellites larger than a certain diameter have been added for this edition (see the note to the table on page 294).

Particularly in tables, large numbers are sometimes given in exponential form. Thus 10^3 is a thousand, 2×10^6 is two million, and so on. "Billion" means a thousand million, or 10^9.

Dates in this section have been expressed in the order day, month, year, although it is customary in astronomy to show dates in the order year, month, day.

In biographical entries, if the first forename of a particular person is not the one by which they were most commonly known, it is enclosed in parentheses.

aberration (1) The small apparent displacement of a star's position produced by the combination of the motion of the Earth and the finite velocity of light.

Annual aberration is the displacement resulting from the motion of the Earth in its orbit around the Sun. This is seen by an observer as an apparent motion of a star in an ellipse over the course of the year. The maximum displacement of the star, given by the **semimajor axis** of this ellipse, is just over 20 arc seconds. This angle (α) is the *constant of annual aberration* and is determined from the relationship

$$\tan \alpha = v/c,$$

where v is the Earth's orbital velocity and c is the speed of light. For a star at the pole of the ecliptic the ellipse becomes a circle; for a star on the ecliptic the ellipse becomes a straight line.

Diurnal aberration is a very much smaller displacement of a star, to the east, resulting from the rotation of the Earth on its axis. It reaches a maximum of 0.32 arc seconds, for an observer at the equator, and is zero at the poles. The Earth's rotation carries an observer in a circle with a speed that is greatest at the equator and zero at the poles.

The aberration of starlight was discovered by James **Bradley**, who published his findings in 1729. In demonstrating that the Earth is not fixed in space – thus furnishing the first observational evidence to support **Copernicus**'s theory – and that light travels at a finite velocity, the discovery is important in the development of cosmological ideas.

aberration (2) An optical defect, such as blurring, distortion or false coloration that can occur in the image produced by a lens or curved mirror. The major aberrations are **chromatic aberration**, **spherical aberration**, **coma (1)** and **astigmatism**.

*▶ **absorption line** The visible light spectrum of the cool giant star Arcturus (α Boötis) is shown here. The dark vertical lines in the spectrum are caused by atoms in the star's atmosphere absorbing radiation. Because each element absorbs radiation at characteristic wavelengths, the spectrum of a star can be used to determine which elements are present.*

ablation Wearing away of the surface of an object. An example is the erosion of a meteoroid during its high-speed passage through the Earth's atmosphere: atmospheric molecules striking the meteoroid vaporize its surface layers.

absolute magnitude (symbol M) The apparent magnitude a star would have if it were at a standard distance of 10 parsecs from the Earth. The absolute magnitude may be derived from the **apparent magnitude** (m) and the **parallax** (π) by the formula

$$M = m + 5 + 5 \log \pi,$$

where π is in arc seconds. Most stars have an absolute magnitude between -5 and $+15$. The Sun has an absolute magnitude of $+4.8$. SEE ALSO **magnitude**.

absolute temperature Temperature as measured on a scale whose zero point is *absolute zero*, the point at which almost all motion at the molecular level ceases. The unit of absolute temperature is the degree kelvin; an interval of one kelvin (K) is equivalent to an interval of one degree Celsius (°C). The relationship between an absolute temperature (T) and the same temperature as measured on the Celsius scale (t) is

$$T = t + 273.15,$$

absolute zero being -273.15°C. Absolute temperatures are widely used in astronomy. (SEE ALSO the table of conversion factors on page 317.)

absorption line A dark line or band in a **continuous spectrum**. Absorption lines are produced when particular wavelengths of radiation from a hot source are absorbed by a cooler, intervening medium. They are typical of the spectra of stars – radiation from the hot interior is absorbed by a star's cooler outer layers, producing absorption lines. The absorption lines in the Sun's spectrum are called **Fraunhofer lines**. The wavelengths at which absorption lines occur are characteristic of the chemical elements present in the radiating source. This enables a star's chemical composition to be determined from its spectrum.

accelerating Universe An increase in the speed of expansion of the Universe. It had been expected that the expansion of the Universe imparted to it by the **big bang** would be slowed by the effects of gravity, but measurements of the distances of Type Ia **supernovae** and of X-ray emissions from clusters of galaxies have shown that about 5 to 6 billion years ago the rate of expansion started to increase. This acceleration has been attributed to the effects of **dark energy**. SEE ALSO **Hubble constant**.

accretion The coalescence of small particles in space as a result of collisions, and the gradual building up of larger bodies from smaller ones by gravitational attraction; also, the accumulation of matter by a star or other celestial object. Matter can be transferred from one component of a close binary system to another by an accretion process, via an **accretion disk**. Accretion is an important factor in the evolution of stars, planets and comets.

accretion disk A disk of gas in orbit around an object, formed by inflowing matter. In a binary star system in which the two component stars are very close, gas can be transferred from the larger to the smaller star, creating an accretion disk. If the accreting star has a strong gravitational field, as it will have if it is a neutron star or black hole, friction will heat the orbiting gas to millions of degrees, hot enough to emit X-rays. Accretion disks also occur at the center of galaxies.

Achernar The star Alpha Eridani, magnitude 0.46, distance 144 light years, luminosity 1150 times that of the Sun. It is a main-sequence star of spectral type B3 and the ninth brightest star in the sky.

Achilles Asteroid no. 588, one of the **Trojan asteroids** sharing Jupiter's orbit, and the first to be discovered, by Max Wolf in 1906. It has a diameter of 147 km (91 miles).

achondrite A type of **stony meteorite**, consisting of rock that was once molten, and which therefore was once part of a planetary body that underwent **differentiation**. Like all stony meteorites, achondrites contain no uncombined metallic elements, and they also lack chondrules – millimeter-sized globules of silicate rock. The principal types are *howardites*, *eucrites* and *diogenites* (collectively known as *HED achondrites*), differing in chemical makeup and structure but all with properties, such as radiation damage, that suggest they were once part of a planetary surface. A main candidate for the HED achondrite parent is the asteroid **Vesta**. Achondrites include a handful of meteorites known to have originated on the Moon, and the **SNC meteorites**, believed to have originated on Mars. SEE ALSO **chondrite**.

achromatic lens (achromat) A lens designed to reduce **chromatic aberration**, cutting down the amount of false color in the image. This is usually achieved by using two lens elements made from different types of glass, for example one element of crown glass and the other of flint glass. The combination is chosen so that two selected wavelengths of light come to the same focus, and at the same time the residual aberration is reduced to a minimum. SEE ALSO **apochromat**.

active galactic nucleus (AGN) The region, at the heart of an *active galaxy*, in which huge amounts of energy are generated by processes other than those that power stars. The energy is believed to be released as matter is accreted by a supermassive **black hole**, of up to a few billion solar masses, at the galaxy's center. **Jets** are often emitted by AGNs. Active galaxies show a range of behavior, depending on the amount of energy generated in their nuclei: SEE **Seyfert galaxy**, **BL Lac object**, **blazar**, **quasar**.

▲ *active optics* The active optics actuators on the reverse of the primary mirror of the WIYN telescope at Kitt Peak National Observatory allow the mirror to be flexed continually to compensate for the effects of gravity as the telescope moves. This system means that far thinner mirrors can be used without risking distortion of the images and data obtained.

▲ *adaptive optics* The Very Large Telescope (VLT) uses adaptive optics. Here the light from a close binary pair has been reflected from the primary mirror on to a subsidiary mirror, which is continually adjusted to compensate for variations in the Earth's atmosphere; it is then computer processed.

active optics A technique for maintaining the primary mirror of a large telescope in accurate shape and alignment, countering the distortions caused by temperature changes or tilting of the telescope. The mirror rests on numerous supports which are automatically adjusted by computer to keep the mirror's figure smooth under all conditions. The technique can also compensate for any slight imperfections that remain after the manufacturing of the mirror. Mirrors with active supports can be made thinner, and thus lighter and cheaper, than ordinary telescope mirrors.

Adams, John Couch (1819–92) English astronomer. While still an undergraduate at Cambridge he set out to analyze the motion of Uranus. The planet's observed path was not in agreement with its calculated orbit, and Adams believed that the discrepancies could be accounted for by the gravitational influence of an undiscovered planet. He calculated an orbit for the new planet, but no search was mounted from England. It was Johann **Galle** who located Neptune, as it was subsequently called, near a position predicted by the French astronomer Urbain **Le Verrier**. After a lengthy priority dispute, the honours of predicting the existence of Neptune came to be shared between Adams and Le Verrier, although it has since emerged that Adams' calculations were not as accurate or complete as those of Le Verrier.

adaptive optics A technique to counteract the blurring of images caused by **seeing** conditions in the Earth's atmosphere. Sensors measure the distortions in a telescope's image caused by atmospheric effects. The image falls on to a small, wafer-thin subsidiary mirror, which can be flexed in a controlled fashion hundreds of times a second to cancel out the imperfections in the image. An artificial guide star created by shooting a laser beam high into the atmosphere is often used as a reference. The resulting image is almost as sharp as if the telescope were in space.

Adhara The star Epsilon Canis Majoris, magnitude 1.50, 431 light years away, and 3700 times as luminous as the Sun. It is a blue giant of spectral type B2.

Adrastea One of the small inner **satellites** of Jupiter discovered in 1979 during the Voyager missions. It is irregular in shape, averaging 16 km (10 miles) in diameter, and orbits near the outer edge of Jupiter's main ring.

aeon (US **eon**) A period of a billion (10^9) years.

Agena Another name for **Beta Centauri**.

age of the Universe The length of time for which the Universe has been in existence. Assuming, as seems likely, that we inhabit a **flat universe**, then the results from the **Planck** mission published in 2013 give the time from the **big bang** as 13,800 million years. The oldest stars are about 12,100 million years old, so the "age problem," whereby some earlier measurements appeared to imply that the Universe is younger than some of its stars, has been resolved. SEE ALSO **Hubble constant**.

AGN ABBREVIATION FOR **active galactic nucleus**.

airglow A faint glow in the Earth's atmosphere. Molecules in the tenuous upper atmosphere, where the solar radiation is strong, are ionized (split up into charged particles) by ultraviolet radiation or by collision with other charged particles. These particles are short-lived and soon recombine to form the original molecules, emitting a faint light as they do so. The recombination of oxygen

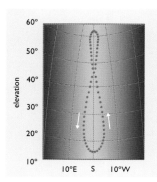

▶ **albedo** This albedo map of Mars was produced by NASA's Mars Global Surveyor. Red areas are bright and show where there is dust. Blue areas show where the underlying, darker rocks have been exposed.

▼ **altazimuth** This simple form of telescope mount allows free movement in both horizontal and vertical axes. An altazimuth mounting is not suitable for use with motor drives, however, unless they are computer controlled.

horizontal axis

vertical axis

▲ **analemma** A plot of the Sun's apparent position from 52°N, looking south at midday, at 5-day intervals throughout the course of a year. The Sun is at the top of the "figure 8" at the summer solstice and at the bottom at the winter solstice.

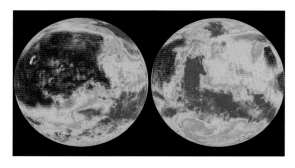

molecules is a principal source of emission, which continues long after sunset. From a dark site on a clear, moonless night the airglow is visible as a gray luminous background to the stars, but for many observers **light pollution** makes the airglow impossible to see. The **Ashen Light** observed on Venus's night side may be a similar phenomenon.

Airy, George Biddell (1801–92) English astronomer and geophysicist. He became Astronomer Royal and Director of the Royal Greenwich Observatory, and re-established the Observatory's standards and authority, equipping it with new instruments. Airy determined the mass of the Earth by taking gravity measurements at the top and bottom of a mineshaft.

Airy disk The central spot in the image of a star formed in a telescope. The effects of **diffraction** make it impossible for the image to be formed as a point. Instead, it is spread out into an Airy disk, surrounded by diffraction rings and spikes. (After George Biddell **Airy**.)

al-Battānī, Muhammad ibn Jābir (Latinized name **Albategnius**) (858–929) Arab astronomer. His 40 years of observations yielded a catalog of star positions more accurate than those given in Ptolemy's **Almagest**, and he introduced trigonometry into Arab astronomy. He made accurate determinations of the precession of the equinoxes and the obliquity of the ecliptic, and showed that the Earth's distance from the Sun varies over the course of the year.

albedo A measure of the reflectivity, or brightness, of a material or body. The albedo scale goes from 1, for a perfectly reflecting (white) surface, to 0, for a perfectly absorbing (black) surface. In astronomy, albedo is used to indicate the fraction of sunlight reflected by Solar System bodies. Rocky bodies such as Mercury and the Moon have low albedos, while cloud-covered bodies like Saturn, or those consisting of a high proportion of water ice – like many of the satellites of the outer Solar System – have high albedos. Albedo can be defined in various ways. The one most used now is *geometric albedo*, defined as the ratio of the reflectance of the body to that of a flat white surface of the same surface area at the same position.

Albiorix A small outer **satellite** of Saturn, discovered in 2000.

Aldebaran The star Alpha Tauri, magnitude 0.87 (but slightly variable, of irregular type), 65 light years distant, luminosity about 150 times that of the Sun. Aldebaran is a giant of type K5. It appears to lie in the Hyades cluster, but is actually a foreground object at about half the cluster's distance.

Algol The star Beta Persei, the prototype of a class of eclipsing binaries known as Algol-type variables. The first recorded observation of Algol's variability was made by Geminiano Montanari of Bologna in 1669. John **Goodricke** in 1782 was the first to suggest that the variability resulted from two stars in binary motion, the fainter and larger one periodically passing in front of the brighter one. There are two stars, of spectral types B8 and G, orbiting each other in 2.867 days. The B8 star is a dwarf and is the visible component; the fainter star is a subgiant. During eclipses, Algol's brightness drops from magnitude 2.1 to 3.4. There is also a third star in the system, but it does not take part in the eclipses. Algol lies 93 light years away.

Algonquin Radio Observatory A radio astronomy observatory at Lake Traverse, Ontario, run by Thoth Technology Inc. Its

principal instrument is a 46 m (150 ft) fully steerable radio dish which has been in operation since 1966.

ALMA ABBREVIATION FOR **Atacama Large Millimeter Array**.

Almagest An encyclopedia of astronomy compiled by **Ptolemy** in about AD 140. It is thought to be based mainly on the work of **Hipparchus**, whose star catalog it incorporates. It was the ancient world's standard accurate description of the heavens and remained the recognized authority until the middle of the 16th century.

almanac A yearbook containing predicted positions of celestial objects, and other astronomical and calendrical data. For astronomical and navigational purposes the leading publication is *The Astronomical Almanac*, produced each year jointly by Her Majesty's Nautical Almanac Office and the US Naval Observatory.

Alpha Centauri The brightest star in the constellation Centaurus, and the third brightest star in the sky, magnitude −0.28; also known as Rigil Kentaurus. It is a visual binary, with G2 and K1 main-sequence components of magnitudes −0.01 and 1.35, orbital period 80 years, distance 4.31 light years. SEE ALSO **Proxima Centauri**.

al-Sūfi, Abu'l-Husain (903–86) Arab astronomer. In his *The Book of the Fixed Stars* he presented a detailed revision of the star positions in Ptolemy's **Almagest**, based on his own observations. It contains the first recorded reference to the Andromeda Galaxy. He measured the length of the year, and attempted to determine the length of a degree of the meridian.

Altair The star Alpha Aquilae, magnitude 0.76, distance 17 light years, luminosity 11 times that of the Sun. It is a main-sequence star of type A7.

altazimuth A mounting for an instrument such as a telescope that allows the instrument to be moved in both altitude and azimuth – that is, about a horizontal and a vertical axis. The disadvantage of this mounting for an amateur telescope is the problem of moving it simultaneously in azimuth and altitude in order to counteract the Earth's diurnal motion, and so keep the object being observed in the field of view. This is made easier in the **Dobsonian telescope**, and circumvented in the **equatorial telescope**. Since the introduction of computer-controlled drive mechanisms, the altazimuth has been the preferred mounting for all large professional instruments.

altitude The angular distance of a celestial body above the observer's horizon. It is measured in degrees from 0 (on the horizon) to 90 (at the zenith) along the **great circle** passing through the body and the zenith. If the object is below the horizon, the altitude is negative. SEE ALSO **coordinates**.

Amalthea The largest of Jupiter's small inner **satellites**, discovered in 1892 by E. E. Barnard. It is an irregular body, averaging 167 km (104 miles) in diameter. Amalthea is dark red, which may indicate that its surface is covered with sulfur from the volcanoes of Io, the innermost of Jupiter's large satellites.

Amor asteroid One of a class of small **asteroids** whose **perihelion** distances lie inside the orbit of Mars but outside that of Earth (specifically, those with perihelion between 1.017 and 1.3 au). They are named after the 1 km (0.6-mile) diameter asteroid 1221 Amor. The first to be discovered, and the most studied is 433 **Eros**. Over 5000 Amors were known as of early 2015.

Some of these small bodies may have been perturbed out of the main asteroid belt by Mars, while others may be the nuclei of extinct comets. Close encounters with Mars or Earth can further perturb Amor asteroids into Earth-crossing **Apollo asteroid** orbits; the two groups are often classified together as *Apollo–Amor objects*. Like other small bodies in the inner Solar System, Amor asteroids have limited lifetimes, typically about 10 million years, and many of them will eventually impact with the Earth or Mars. SEE ALSO **Aten asteroid**.

analemma A long, thin figure-of-eight formed by plotting the position of the Sun in the sky at the same time of day throughout the year. The variation in the Sun's position along the long axis of the "8" – highest at the summer **solstice**, lowest at the winter solstice – is caused by the Earth's axial inclination. The variation across the short axis is a result of the eccentricity of the Earth's orbit.

Ananke An outer **satellite** of Jupiter discovered in 1951 by Seth Nicholson. It is in a **retrograde** (1) orbit and may be a captured asteroid.

anastigmatic lens (anastigmat) A type of lens designed to compensate for **astigmatism**.

Andromeda A large constellation of the northern hemisphere, adjoining the Square of Pegasus. The main stars lie in a line leading away from Pegasus, and the star Alpha Andromedae (Sirrah or Alpheratz, magnitude 2.1) actually forms one corner of the Square. In mythology, Andromeda was the princess rescued by Perseus. Gamma (Almach or Alamak) is a fine double of magnitudes 2.3 and 4.8, spectral types K3 and B9. The open cluster NGC 752 is an easy binocular object, and NGC 7662 is a planetary nebula visible in small telescopes. The most famous object in the constellation is the **Andromeda Galaxy**.

Andromeda Galaxy A spiral galaxy 2.5 million light years away in the constellation Andromeda, the most distant object clearly visible to the naked eye, also known as M31 or NGC 224. The Andromeda Galaxy has a mass of over 300,000 million Suns. Cepheid and long-period variable stars, novae and a supernova have been observed in it. To the naked eye it appears as a smudge of magnitude 3.4; through binoculars and small telescopes it shows an elongated structure some 1° × 3° in extent, and its plane is tilted at 13° to our line of sight. Its true diameter is about 150,000 light years, somewhat larger than our own Galaxy. The Andromeda Galaxy is accompanied by two small satellite elliptical galaxies, M32 (NGC 221) and M110 (NGC 205).

Andromedids A meteor shower which gave spectacular displays on 27 November in 1872 and 1885, following the breakup in the 1840s of its parent comet, **Biela's Comet**. The shower, also known as the *Bielids*, had its radiant in the constellation Andromeda, near the star Gamma. After more than a century of quiescence, the shower returned in 2011 and 2013, when modest displays were seen.

Anglo-Australian Observatory Former name of the **Australian Astronomical Observatory**.

angstrom (angstrom unit) (symbol Å) A unit of length, equal to 10^{-10} m, formerly widely used in spectroscopy for wavelengths of light. It is no longer recommended for use in astronomy, having largely been superseded by the nanometer, symbol nm (1 nm = 10^{-9} m, 1 Å = 0.1 nm). It is named after the Swedish physicist and astronomer Anders Ångström (1814–74), who introduced it in his atlas of the solar spectrum, published in 1868.

angular diameter The apparent diameter of a celestial body expressed in angular measure (usually in **arc minutes** and **arc seconds**). It is the angle subtended at the observer by the actual diameter of the body under observation. If the distance of a celestial body from the observer is known, its true diameter can be calculated.

annual parallax (heliocentric parallax) The angle subtended at a celestial object by the radius of the Earth's orbit, which is 1 astronomical unit. It is measured by determining the **semimajor axis** of the parallactic ellipse traced by the star on the celestial sphere. The reciprocal of the annual parallax in arc seconds is the distance of the object in **parsecs**. SEE ALSO **parallax**.

annular eclipse A type of **solar eclipse** in which the Sun's disk is not completely hidden by the Moon, a thin ring of light (*annulus* is Latin for "ring") remaining visible around the dark body of the Moon. A solar eclipse is annular if it occurs when the Moon is close to **apogee**, at its farthest point from the Earth in its elliptical orbit. The tip of its shadow-cone does not reach down to the Earth's surface, and its apparent diameter is a little less than the Sun's.

anomalistic month The time taken for one revolution of the Moon around the Earth relative to its **perigee**. It is equal to 27.55455 days of **mean solar time**.

anomalistic year The time taken for one revolution of the Earth around the Sun relative to the orbit's **perihelion**. It is equal to 365.25964 days of **mean solar time**. This is about 4m 43.5s longer than a **sidereal year**, because the perihelion point moves eastward.

anomaly An angular measurement used in determining the position of a body in an elliptical orbit. It is reckoned from periapsis (SEE **apsides**), in the direction of the object's motion. There are three types of anomaly. *True anomaly*, in the case of a planet, is the angle between the planet's radius vector and its perihelion. In the diagram it is the angle PSX. *Mean anomaly* is the same angle, but

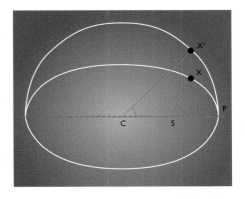

▶ **anomaly** The true anomaly is described by the angle PSX in the diagram here. The eccentric anomaly, PCX', is given by projecting the body's position on to a circumscribing circle, and can be used as a matter of convenience in orbital calculations. C, center of ellipse; S, Sun; P, perihelion; X, position of planet.

for an imaginary planet moving along its orbit at constant angular speed, rather than at non-uniform angular speed as in reality. *Eccentric anomaly* relates to the motion of a point along a circle whose diameter is the long axis of the planet's orbit. In the diagram, the eccentric anomaly is the angle PCX'. The point X' is where a perpendicular from the long axis through the planet crosses the circle that circumscribes the ellipse.

ansae (singular **ansa**) The outer parts of Saturn's rings, which through a telescope can look like handles projecting from the planet's disk. The word is Latin for "handles."

antapex SEE **apex**.

Antares The star Alpha Scorpii, a supergiant of type M1 that varies irregularly from magnitude 0.9 to 1.2. Its distance is 604 light years, and it is more than 10,000 times as luminous as the Sun. Antares has a main-sequence companion star of magnitude 5.4, type B2, orbiting it with a period of about 900 years.

anthropic principle The argument that life in the Universe is somehow related to the physical properties of the Universe. Its simplest form is the *weak anthropic principle*, which states that the Universe we inhabit has properties that have made it possible for life to develop. This is not controversial, but the *strong anthropic principle* goes further. It argues that the laws of physics that have shaped the Universe are such that the emergence of intelligent life is an inevitable outcome.

Antlia A small southern constellation representing an air pump, added to the sky by **Lacaille**. It has no star brighter than magnitude 4.3.

Antoniadi, Eugène Michael (1870–1944) French astronomer, born in Turkey of Greek parents. He was a highly skilled observer, particularly of the inner planets. His maps of the surface features of Mars and Mercury were not bettered until those planets were visited by spacecraft. The **canals** of Mars were shown by Antoniadi to be illusory. He devised a system known as the **Antoniadi scale**, used to rate **seeing**.

Antoniadi scale The scale devised by Eugène **Antoniadi**, widely used by amateur astronomers to rate the **seeing** conditions (image quality as affected by air movements) under which lunar and planetary observations are made. It distinguishes five gradations of seeing:

I   Perfect seeing, without a quiver.
II   Slight undulations, with periods of calm lasting several seconds.
III   Moderate seeing, with larger air tremors.
IV   Poor seeing, with constant troublesome undulations.
V   Very bad seeing, scarcely allowing a rough sketch to be made.

ap-, apo- Prefixes referring to the farthest point in an object's orbit around another body, as in **aphelion** and **apogee**.

Apache Point Observatory An observatory in the Sacramento Mountains of New Mexico. Its main instrument is a 3.5 m (138-inch) reflector, and the **Sloan Digital Sky Survey**'s 2.5 m (100-inch) reflector is also sited there. The observatory is operated by a group of US universities called the Astrophysical Research Consortium.

apastron The point in the orbit of one star in a binary system at which it is farthest from the other star.

aperture The clear diameter of an optical telescope's objective lens or primary mirror. More generally, for other types of telescope, the aperture is the size of the principal radiation-collecting element – for example, the antenna of a radio telescope. Aperture is the most significant parameter of any telescope, for the amount of radiation collected from distant objects is all-important.

▲ **Apollo program** *The Lunar Roving Vehicles allowed the astronauts on the final three Apollo lunar missions to range farther from the landers and so collect more varied samples from the Moon's surface. Here, James B. Irwin works next to the Vehicle during the Apollo 15 mission. The photograph was taken by David R. Scott.*

aperture synthesis A technique in radio astronomy in which the signals from a number of radio dishes are used to synthesize the resolution of a single, much larger telescope. The output from numerous small antennas, which are often arranged in a linear or two-dimensional array many kilometers long, are combined electronically. As the Earth rotates, the array sweeps over an area equivalent to a single large dish with a diameter equal to the maximum dimension of the array. An example is the **Very Large Array**. SEE ALSO **radio interferometry**.

apex The point on the celestial sphere toward which the Sun and the entire Solar System appear to be moving, at a velocity of 19–20 km/s (12 miles/s) relative to the nearby stars; it is also known as the *solar apex*. The apex lies in the constellation Hercules at about RA 18h, dec. +30°. The point on the celestial sphere opposite the apex is the *antapex*, in the constellation Columba.

aphelion The point in the orbit of a planet or a comet at which it is at its farthest from the Sun.

apochromat A lens especially corrected to a high degree to eliminate **chromatic aberration**. By using three elements, all of different types of optical glass, a greater amount of correction is possible than with the two-element **achromat**.

apogee The point in the orbit of a body around the Earth at which it is at its farthest from the Earth.

Apollo asteroid One of a class of small **asteroids** whose **perihelion** distances lie inside the Earth's orbit (specifically, those with perihelion less than 1.017 au and **semimajor axis** greater than 1 au). They are named after the 2 km (1.25-mile) diameter object 1862 Apollo. Over 6000 were known as of early 2015.
  Like the **Amor asteroids**, with which they are often grouped as *Apollo–Amor objects*, some may be extinct cometary nuclei, but others may have been perturbed inward by Jupiter from one of the **Kirkwood gaps** in the main asteroid belt. Many Apollo asteroids could impact the Earth within a few million years. SEE ALSO **Aten asteroid**.

Apollo program A US space project that landed astronauts on the Moon. The Apollo spacecraft had three main parts: a pressurized *command module* for the three crew members; a *service module* containing oxygen and electrical supplies and the main manoeuvring engine; and the *lunar module*, in which two astronauts made the actual Moon landing. The craft was launched by a Saturn V rocket. After four preparatory crewed flights, Apollo 11 made the first Moon landing in July 1969. Neil Armstrong and Edwin "Buzz" Aldrin became the first men to step on the Moon, remaining on the surface for nearly a day and bringing back 22 kg (49 lb) of lunar rock and dust. Subsequent Apollos extended the exploration of the Moon's surface, bringing back further samples and leaving behind instruments which transmitted data about moonquakes. The last mission of the series, Apollo 17, was in December 1972.

apparent magnitude (symbol *m*) The brightness of an object as seen from Earth. The apparent magnitude differs from the **absolute magnitude** because of the object's distance. SEE ALSO **magnitude**.

apparent solar time Local timescale, based on the apparent daily movement of the Sun. It is the time shown on a sundial. Noon occurs when the Sun reaches its maximum altitude, i.e. when it crosses the observer's meridian. Time reckoned in this way is subject to considerable variations because of the Sun's non-uniform motion, a result of both the elliptical shape of the Earth's orbit and the fact that the Sun moves along the ecliptic, not the celestial equator. The time kept by clocks is **mean solar time**, based on the motion of a hypothetical mean Sun.

apparition The appearance in the sky of a comet, or a period during which a planet or satellite is best observed.

appulse The apparent close approach of two celestial bodies, such as a planet and a star, whose directions of motion as seen on the celestial sphere cause them to converge (though in reality they are remote from each other). SEE ALSO **occultation**.

apsides (singular **apsis** or **apse**) The two points in an orbit that are respectively closest to (*periapsis*) and farthest from (*apapsis*) the primary body. For example, the apsides of the Earth's orbit are its perihelion and aphelion; in the Moon's orbit the apsides are its perigee and apogee. The *line of apsides* is the line connecting these two points.

Apus A far-southern constellation representing a bird of paradise, first shown in Johann **Bayer**'s atlas of 1603. Its brightest star is of magnitude 3.8.

Aquarids Either of two **meteor showers**, the **Delta Aquarids** and the **Eta Aquarids**.

Aquarius A constellation of the **zodiac**, representing a figure pouring water from a jar. Alpha (Sadalmelik, magnitude 3.0) and Beta (Sadalsuud, magnitude 2.9) are its brightest stars. Zeta is a fine binary with a period of 590 years, and components of magnitudes 4.4 and 4.6. R Aquarii is a **symbiotic variable** with a range from magnitude 5.8 to 12.4 and a mean period of 387 days. There are two important planetary nebulae: the **Saturn Nebula** and the **Helix Nebula**. Aquarius also contains two globular clusters, M2 and M72.

Aquila A distinctive constellation, led by first-magnitude **Altair**; most of it lies in the northern hemisphere of the sky, although the celestial equator crosses it. Mythologically it represents the eagle sent by Zeus to collect the shepherd-boy Ganymede, who was to be cup-bearer to the gods. Gamma (Tarazed), magnitude 2.7, and Beta (Alshain), magnitude 3.7, stand either side of Altair. Eta, a Cepheid variable, has a period of 7.18 days and a range from magnitude 3.5 to 4.4; it lies between Theta (magnitude 3.2) and Delta (magnitude 3.4), which make good comparison stars. The Mira variable R Aquilae has a range from magnitude 5.5 to 12.0 and a period of 284 days. The Milky Way runs through the constellation and is very rich in this region.

Ara A southern constellation between Scorpius and Triangulum Australe, representing the altar of the gods. The brightest stars are Alpha and Beta, both magnitude 2.8. The brightest variable in the constellation is the eclipsing binary R Arae (magnitude 6.0 to 6.9; period 4.43 days, Algol type). There is also a prominent open cluster, NGC 6193, and a scattered globular, NGC 6397.

arc minute, second (symbols ′, ″) Units of angular measure, equal to 1/60 of a degree and 1/3600 of a degree, respectively. These units are used to express the apparent diameter or separation of

▲ **Arecibo Observatory** *As well as its more famous role in the SETI project, the radio telescope is a powerful radar transmitter and receiver, and is used to map small bodies such as asteroids and to examine the Earth's upper atmosphere.*

astronomical objects and the **resolving power** of a telescope. Often abbreviated to arcmin, arcsec.

Arcturus The star Alpha Boötis, the fourth brightest in the sky at magnitude −0.05, distance 37 light years, luminosity 100 times that of the Sun. It is a giant of type K2.

Arecibo Observatory A radio and radar telescope near Arecibo in Puerto Rico, diameter 305 m (1000 ft), the largest single-bowl type of instrument in operation. It is built into a natural hollow in the limestone terrain and cannot be steered. Instead, an ingenious feed system enables sources as far as 20° from the zenith to be tracked. It was opened in 1963 and is operated by Cornell University. An upgrade completed in 1998 corrected the telescope's spherical aberration.

areo- Prefix referring to the planet Mars, as in *areography* – the description and mapping of the surface features of Mars. (From *Areos*, Greek for "Mars.")

Argelander, Friedrich Wilhelm August (1799–1875) German astronomer. In 1852 he began the vast undertaking of preparing an atlas and catalog of all stars down to magnitude 9.5 in the northern hemisphere. This immense work, covering over 300,000 stars, was published as the ***Bonner Durchmusterung***. Argelander introduced the system of nomenclature for variable stars in which they are assigned one or two capital Roman letters, beginning with R and RR, and also the subdivision of magnitudes into tenths.

Ariel One of the five main **satellites** of Uranus, diameter 1158 km (720 miles), discovered in 1851 by William Lassell. It is composed largely of a mixture of rock and water ice, with a density of 1.6 g/cm³. Ariel has been geologically active in the past. Its characteristic features are wide, steep-sided troughs that wind across a landscape of cratered terrain, ridged terrain and plains. **Tectonics** accounts for the ridges, and volcanic flooding for the plains; the troughs formed when the water interior froze and the satellite expanded, splitting the crust.

Aries The first constellation of the **zodiac** – although, since **precession** has now shifted the **vernal equinox** into Pisces, Aries has technically lost this distinction. In mythology it represents the lamb with the golden fleece. Its brightest stars are Alpha (Hamal), magnitude 2.0, and Beta (Sheratan), magnitude 2.6.

Aristarchus of Samos (3rd century BC) Greek mathematician and astronomer. He tried to calculate the distances of the Sun and Moon from the Earth, as well as their sizes. His method was sound but the results were inaccurate. He explained the immobility of the fixed stars by their great distance from the Earth's orbit. He put the Sun at the center of the Universe, making him the first to propose a heliocentric theory, but the idea was not taken up because it did not seem to make the calculation of planetary positions any easier.

Aristotle (384–322 BC) Greek philosopher and encyclopedist. He developed a view of the Universe based on the four "elements" (earth, air, fire and water) and the system of concentric spheres proposed by Eudoxus of Cnidus upon which various celestial bodies were carried. Aristotle added more spheres to account for the motion of all celestial bodies. The celestial bodies were, he said, made of a substance called aether, and were perfect and incorruptible, unlike the Earth. He attempted to estimate the size of the Earth, which he maintained was spherical but unmoving – the center of the Universe. This "Aristotelian" view of the world remained almost unchallenged until Copernicus.

Arizona Meteor Crater SEE **Meteor Crater**.

array An arrangement of antennas used in radio astronomy. SEE **radio telescope**.

artificial satellite A spacecraft, in particular one that is uncrewed, in orbit around the Earth or some other body, as opposed to a natural satellite. The first artificial satellite, Sputnik 1, was successfully put into orbit by the USSR on 4 October 1957.

ascending node The point at which an orbit crosses from south to north of, for example, the celestial equator or the ecliptic. SEE **node**.

Ashen Light A faint luminosity on the night side of Venus when the planet appears as a thin crescent, close to inferior conjunction.

The Ashen Light is occasionally reported by observers, but has never been imaged. Its appearance is similar to **earthshine** on the Moon, although that is produced in a different way. If it is a real phenomenon, and not illusory, it is probably the result of electrical disturbances in the planet's ionosphere, and may therefore be similar in nature to the **airglow** in the Earth's atmosphere.

aspect The position of a planet or the Moon relative to the Sun as seen from the Earth. SEE **conjunction, elongation, opposition, quadrature**.

association, stellar A group of stars that have formed together, but are more loosely linked than the members of a star cluster. Many are young and define portions of the spiral arms of our Galaxy. SEE **OB association, T association**.

A star A star of spectral type A, characterized by extremely strong hydrogen absorption lines in its spectrum. On the main sequence, A stars range in surface temperature from 7500 K at type A9 up to 9900 K at type A0. Their corresponding masses and radii are, respectively, 1.8 and 1.4 times the Sun's at A9, increasing to 3.2 and 2.5 at A0. Examples are Sirius, Vega, Altair and Deneb. Roughly 10% of A stars have strong magnetic fields. In these, certain chemical elements become concentrated in their atmospheres and produce unusually strong spectral lines. These are classified as Ap ("p" for peculiar) stars. Another kind of spectral anomaly produces the metallic-line (Am) stars, of which Sirius is an example.

asterism A group of easily recognizable stars, part of a constellation. Examples are the seven stars that form the shape of the Big Dipper or Plough in Ursa Major, and the "Teapot" in Sagittarius. The term is also used for distinctive groupings of stars visible in binoculars or small telescopes, some of which have been given descriptive names by observers, such as the Coathanger (in Vulpecula) and Kemble's Cascade (in Camelopardalis).

asteroid (minor planet) A small Solar System body in an independent orbit around the Sun. The majority move between the orbits of Mars and Jupiter. The largest asteroid, and the first to be discovered, on 1 January 1801 by Giuseppe **Piazzi**, is **Ceres**, which has a diameter of 934 km (580 miles). The smallest that have been

▲ **Ariel** *Voyager 2 captured this image of Uranus's satellite Ariel in 1986. Ancient faults and valleys are overlain by abundant impact craters, indicating that the moon has not been geologically active for billions of years.*

ASTEROIDS						
Name	Number	Discovery	Diameter (km)	Average distance from Sun (million km)	Orbital period (years)	Rotation period (hours)
The first discovered						
Ceres	1	1801, Piazzi	934	413.6	4.60	9.08
Pallas	2	1802, Olbers	525	414.0	4.61	7.81
Juno	3	1804, Harding	268	399.3	4.36	7.21
Vesta	4	1807, Olbers	510	353.2	3.63	5.34
Astraea	5	1845, Hencke	120	385.2	4.13	16.81
Other large asteroids						
Hygeia	10	1849, De Gasparis	408	469.2	5.55	27.6
Davida	511	1903, Dugan	326	475.4	5.67	5.13
Interamnia	704	1910, Cerulli	316	458.1	5.36	8.72
Europa	52	1858, Goldschmidt	302	463.3	5.46	5.63
Sylvia	87	1866, Pogson	260	521.5	6.52	5.19
Some Trojan asteroids						
Hektor	624	1907, Kopff	225	775.3	11.80	6.92
Patroclus	617	1906, Kopff	149	782.7	11.97	—
Achilles	588	1906, Max Wolf	147	774.4	11.78	—
Some Amor asteroids						
Ganymed	1036	1924, Baade	41	398.7	4.35	—
Eros	433	1898, Witt	21	218.6	1.76	5.27
Amor	1221	1932, Delporte	1	287.2	2.66	
Some Apollo asteroids						
Phaethon	3200	1983, Green	5.0	190.2	1.43	4
Daedalus	1864	1971, Gehrels	3.3	218.5	1.77	8.57
Apollo	1862	1932, Reinmuth	2.0	220.1	1.78	3.07
Some Aten asteroids						
Ra-Shalom	2100	1978, Helin	3.5	124.5	0.76	19.79
Aten	2062	1976, Helin	1	144.6	0.95	—
Hathor	2340	1976, Kowal	0.5	126.2	0.78	—

Diameters and rotation periods of asteroids are often uncertain, and different measurement techniques give different results. For irregular asteroids, the diameter given is an average.

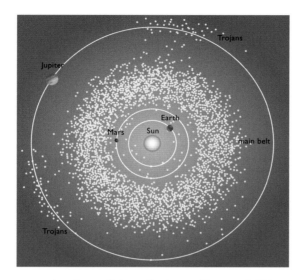

▶ **asteroid** *Plots of the larger known asteroids: most of them lie in the main belt between the orbits of Mars and Jupiter, while the Trojans move in the same orbit as Jupiter, 60° in front of or behind it.*

measured are the so-called **near-Earth asteroids**, some of which are just a few meters across. Some of the larger asteroids are spherical, but most are irregularly shaped. Since the 1970s, polarization, infrared and radar techniques have allowed improved measurements to be made of asteroids' albedos, sizes and shapes. There are thought to be over a million asteroids with a diameter greater than 1 km (0.6 mile); below this, they extend in size down to dust particles. Objects less than about 10 m (30 ft) in diameter are best classed as **meteoroids**. Some of these very small objects have found their way to Earth as **meteorites**.

The second asteroid to be found was **Pallas**, in 1802, followed by **Juno** in 1804 and **Vesta** in 1807. It was not until 1845 that the fifth asteroid, Astraea, was discovered, but from then on the pace of discovery quickened. In 1891 Max **Wolf** made the first asteroid discovery by photography, and went on to find hundreds more. As of early 2015, over 650,000 asteroids had been cataloged and had their orbits calculated, and this figure is now increasing by thousands each year. Tens of thousands more have been observed, but not often enough for an accurate orbit to be calculated.

Most asteroids orbit within the main *asteroid belt*, a torus-shaped region of space between about 2.2 and 3.3 au from the Sun. Within the main belt are the **Kirkwood gaps** at distances from the Sun at which **resonances** rule out stable orbits. The typical asteroid orbit is elliptical but more eccentric than those of the major planets, and with a greater inclination to the ecliptic. Some asteroids have orbits that bring them into the inner Solar System (SEE **Amor asteroid**, **Apollo asteroid** and **Aten asteroid**), while others, such as **Hidalgo** and **Chiron**, have orbits that lie partially or entirely in the outer Solar System.

The great majority of asteroids almost certainly originate from small **protoplanets** left over from the formation of the Solar System, prevented by Jupiter's gravitational influence from forming into a major planet. A few may be the extinct nuclei of **comets**. The present asteroid population does not represent the remnants of a single large planet that disintegrated, as was once thought: put together, they would make up a body only about 1500 km (950 miles) in diameter. New asteroid discoveries are designated by the year of discovery, and a pair of letters and subscripted numbers to denote the half-month and order of discovery (e.g. 2004NB denotes the second asteroid discovered in the first half of July in that year; the letter I is not used). A permanent number is given when the orbit has been accurately calculated, and the discoverer may then add a name to the number (as with, for example, 1707 Chantal), subject to the approval of the International Astronomical Union. SEE ALSO **Centaur**, **Hirayama family**, **Kuiper Belt**, **Trojan asteroid**.

astigmatism An **aberration** (2) in lenses and mirrors that affects light from objects away from the center of the observer's field of view. Instead of being brought to a point focus, these oblique rays of light form an image which is either an ellipse or a straight line, depending on the distance from the lens at which it is viewed. A lens designed to minimize astigmatism is called an *anastigmat*.

ASTRONOMERS ROYAL	
Name	Held office
John **Flamsteed** (1646–1719)	1675–1719
Edmond **Halley** (1656–1742)	1720–42
James **Bradley** (1693–1762)	1742–62
Nathaniel Bliss (1700–64)	1762–64
Nevil **Maskelyne** (1732–1811)	1765–1811
John Pond (1767–1836)	1811–35
George Biddell **Airy** (1801–92)	1835–81
William Christie (1845–1923)	1881–1910
Frank Watson Dyson (1868–1939)	1910–33
Harold Spencer **Jones** (1890–1960)	1933–55
Richard Woolley (1906–86)	1956–71
Martin **Ryle** (1918–84)	1972–82
Francis Graham-Smith (1923–)	1982–90
Arnold Wolfendale (1927–)	1991–94
Martin Rees (1942–)	1995–

astrobiology (exobiology) The study of the possible existence of life beyond the Earth. This includes looking for life forms on other planetary bodies in the Solar System, such as Mars and Europa; the study of organic molecules in nebulae, meteorites and comets; the study of planetary surfaces (and subsurfaces) and atmospheres; questions relating to the origin of life and the range of conditions under which it can survive; and the search for signs of life in space, such as incoming messages at radio or other wavelengths and investigating **extrasolar planets**. SEE ALSO **SETI**.

astrolabe An early astronomical instrument for showing the appearance of the celestial sphere at a given moment, and for determining the altitude of celestial bodies. The basic form consisted of two concentric disks – one with a star map and one with a scale of angles around its rim – joined and pivoted at their centers (rather like a modern **planisphere**), with a sighting device attached. Astrolabes were used from the time of the ancient Greeks through to the 17th century for navigation, measuring time, and terrestrial measurement of height and angles.

astrology A pseudoscience professing to assess people's personality traits and to predict events in their lives and future trends in general from aspects of the heavens, in particular the positions of the planets. Astrology is based on ideas that are scientifically unsound and dismissed by the majority of rational people, but horoscopes are still published in the popular press, though mainly for their entertainment value. In many ancient cultures it was for astrological reasons that observational records first came to be kept. Such records from ancient China have proved of great historical importance in research into past eclipses, novae and comets. In later times, casting horoscopes provided astronomers such as Johann **Kepler** with the livelihood to sustain them in their astronomical work.

astrometric binary A star that is recognized as a **binary star** from its wave-like **proper motion**, caused by its orbital motion with an unseen companion.

astrometry The branch of astronomy concerned with the measurement of precise positions of celestial objects, usually determined from images obtained specially for the purpose. **Parallaxes** and **proper motions** of stars are derived from positions obtained at widely differing times. The most accurate positional measurements available were made by the astrometry satellite **Hipparcos**.

Astronomer Royal An honorary title bestowed upon a prominent British astronomer (until 1972, an additional title of the Director of the Royal Greenwich Observatory). The first, John Flamsteed, was appointed by Charles II when he founded the Greenwich Observatory. SEE table bottom left.

astronomical twilight The dawn and dusk periods when the Sun is less than 18° below the horizon. SEE ALSO **twilight**.

astronomical unit (symbol au) The mean distance between the Earth and the Sun, used as a fundamental unit of distance, particularly for distances in the Solar System. It is equal to 149,597,870.7 km (92,955,807.3 miles).

astrophotography The application of photography to astronomy. Astrophotography makes it possible to obtain a permanent record of the positions and appearance of objects, while long exposures show faint objects such as remote galaxies invisible to the naked eye. It differs from normal photography in that exposures can be very long, often an hour or more, and the camera or telescope must accurately track the object as the Earth rotates during the exposure.

In 1850 the first star images, of Vega and Capella, were obtained by W. C. Bond at Harvard College Observatory. Henry **Draper** obtained a spectrogram of Vega in 1872 and photographed the Orion Nebula in 1880. The 1880s marked photography's breakthrough in astronomy, as techniques improved and plates became more sensitive. The first photographic atlases of the sky were made, and libraries of stellar spectra were built up. In the 20th century astrophotography rapidly supplanted visual observation for many purposes, but has itself now been superseded by electronic detectors particularly **CCD**-based systems. Film has now been replaced almost entirely by digital systems, either specialist CCDs, digital cameras or webcam-type cameras which are used for planetary imaging.

astrophysics The study of the physical properties of celestial bodies. It is based mainly on the study of radiation from these bodies, and has developed since the 19th century through the application of photography, photometry and spectroscopy. New observing techniques at wavelengths from radio and infrared to X-rays and gamma rays enormously increased the scope of astrophysics in the second half of the 20th century. Theoretical astrophysics applies mathematical methods and physical laws to explain or predict the behavior of the Universe and objects within it.

Atacama Large Millimeter Array (ALMA) A millimeter-wavelength telescope built and operated by an international consortium (Europe, USA, Canada, Japan and Taiwan) at Llano de Chajnantor, in the Chilean Andes. The instrument, at an elevation of 5000 m (16,400 ft), consists of 66 antennas, 54 with a diameter of 12 m (39 ft) and 12 with a diameter of 7 m (23 ft). The high altitude will give the telescope access to wavebands between 350 μm and 10 mm. ALMA began observations in 2011 and the last of the antennas was installed at the end of 2013.

Aten asteroid One of a class of small **asteroids** whose orbits lie largely within the Earth's (specifically, those with **semimajor axis** less than 1 au). They are named after the 1 km (0.6-mile) diameter object 2062 Aten. Over 900 were known by early 2015. SEE ALSO **Amor asteroid**, **Apollo asteroid**.

Atlas One of the small inner **satellites** of Saturn, discovered in 1980 during the Voyager missions. Atlas orbits just outside Saturn's A Ring, and seems to act as the **shepherd moon** to the whole bright ring system.

atmosphere The envelope of gases around a star, planet or satellite. The ability of a planet or satellite to retain an atmosphere depends on its outer temperature, determined by its distance from the Sun, and its escape velocity, determined by its mass. Smaller bodies in the inner Solar System, such as Mercury and the Moon, have almost no atmosphere or none at all, while the giant planets have deep, massive atmospheres. Far from the Sun, even small worlds like Pluto and Triton have clung on to a tenuous atmosphere. Venus, Earth and Mars were too small to hang on to the lighter gases, such as hydrogen, from their original atmospheres, which had the composition of the solar nebula from which the Solar System formed. Subsequent geological processes such as **volcanism** have modified their atmospheres considerably, and they now contain carbon dioxide, oxygen and nitrogen (although in very different proportions). The giant planets have their original, largely hydrogen, atmospheres. Saturn's largest moon, **Titan**, is the only satellite with a substantial atmosphere, mainly of nitrogen.

The outer layers of a star, which is where the lines in the star's spectrum originate, are referred to as its atmosphere.

atmospheric extinction The reduction in the brightness of light from celestial bodies when it passes through the Earth's atmosphere. The effect is more pronounced for objects at low altitude, for their light has then had to pass through a greater volume of atmosphere. Observations of, for example, a variable star have to be corrected for extinction, which can dim the star's light by half a magnitude when viewed at an altitude of 20°, and by a whole magnitude at 10°. The main cause of atmospheric extinction is the **scattering** of light by molecules in the atmosphere. As scattering affects wavelengths at the blue end of the visible spectrum more than it does the red, the light becomes reddened, as is most obvious at sunrise or sunset.

atmospheric refraction The small apparent change in altitude of a celestial object, seen by an observer at the Earth's surface, caused by its light passing through the Earth's atmosphere. Just as light is refracted when it passes from air into glass, so it is refracted when it passes from the vacuum of space into the atmosphere. Atmospheric refraction makes a body appear to be at a higher altitude than it actually is. The effect is greater the closer the body is to the horizon, and a body can even become visible before it has risen. All astronomical observations have to be corrected for atmospheric refraction in order to obtain the true position from the observed apparent position.

au ABBREVIATION FOR **astronomical unit**.

Auriga A large northern constellation, containing the first-magnitude star **Capella**. Mythologically it represents Erichthonius, a king of Athens said to have invented the four-horse chariot. Beta (Menkalinan), magnitude 1.9, is an eclipsing variable of very small range. The stars Eta and Zeta are popularly called the Haedi or Kids. **Epsilon Aurigae** is a remarkable eclipsing binary. Zeta is also an eclipsing binary, consisting of a K5 giant orbited by a B7

main-sequence star with a period of 972 days; the visual range is magnitude 3.7 to 4.0. There are three prominent open clusters: M36, M37 and M38.

aurora (plural **aurorae**) An illumination of the night sky, also known as the *aurora borealis* in the northern hemisphere and as the *aurora australis* in the southern. Aurorae can take many forms, including *arcs* and *bands*, which may be *homogeneous* or show vertical *ray* structure; folded rayed bands resemble rippling curtains. Green and red colors predominate, resulting from excitation of atmospheric oxygen and nitrogen by electrons accelerated in Earth's **magnetosphere**. Emissions are seen at altitudes typically between 100 and 1000 km (60 and 600 miles), sometimes reaching as low as 80 km (50 miles) during strong displays. Auroral activity is present more or less continuously in oval regions surrounding either geomagnetic pole at high latitudes. Aurorae are commonest at lower latitudes at times of high solar activity, when solar **flares** and **coronal mass ejections** deliver powerful bursts of charged particles via the **solar wind**. Aurorae have been observed on all the giant planets, which have strong magnetic fields.

Australian Astronomical Observatory An observatory on Siding Spring Mountain in New South Wales, founded in 1973 as the Anglo-Australian Observatory. It contains the 3.9 m (150-inch) Anglo-Australian Telescope and the 1.2 m (48-inch) UK Schmidt Telescope. The observatory changed to its current name in 2010 when UK funding was withdrawn.

Australia Telescope National Facility A large **aperture synthesis** telescope in New South Wales, the most powerful radio astronomy instrument in the southern hemisphere. With three separate parts, it is an example of the use of **very long baseline interferometry**. One part consists of a 6 km line of six dish aerials, each 22 m (72 ft) in diameter, at Culgoora. These can be linked to another 22 m dish, 100 km to the south at Mopra, and also the large radio dish 200 km farther south at **Parkes Observatory**. The Australia Telescope came into operation in 1988.

autumnal equinox SEE equinox.

averted vision Looking slightly to one side of a faint object in order to see it better. The part of the retina in the human eye most sensitive to low light levels is not right at the center, but in a small region surrounding the center. Using averted vision when observing causes the light from the object being observed to fall in this region, making it easier to see.

axis (plural **axes**) The imaginary line, joining the north and south poles of a celestial body, about which it rotates. The *magnetic axis* of a body possessing a magnetic field is the line joining the north and south poles of the field. A body's rotational and magnetic axes do not necessarily coincide; with **Uranus** and **Neptune**, for example, they are markedly different.

azimuth The angle measured westward from north in a horizontal plane to the vertical circle (the meridian) that runs through a celestial object. SEE ALSO **coordinates**.

▲ **Atacama Large Millimeter Array** ALMA antennas on the Chajnantor Plateau, 5000 m above sea level. The ALMA array consists of a mixture of antennas of 12 m and 7 m diameter.

Baade, (Wilhelm Heinrich) Walter (1893–1960) German–American astronomer. From Mount Wilson Observatory in the 1943 wartime blackout he was able to observe individual stars in the Andromeda Galaxy and distinguish the younger, bluer Population I stars from the older, redder Population II stars (SEE **stellar populations**). He went on to improve the use of **Cepheid variable** stars as distance indicators, and showed that the Universe was older and larger than had been thought.

background radiation SEE **cosmic microwave background**.

Baily's beads A very short-lived phenomenon seen during a solar eclipse, just before or after totality, in which the extreme edge of the Sun's disk appears to break into a string of bright lights. It is caused by rays of light from the Sun shining through the valleys on the Moon's limb, while other rays are blocked off by mountains on the limb. It was first described by English astronomer Francis Baily (1774–1844), who observed it in 1836. SEE ALSO **diamond-ring effect**.

Balmer series The distinctive series of lines in the **hydrogen spectrum**. The Balmer series is a dominant characteristic of the spectra of A stars. The first Balmer line is called the **hydrogen alpha line** (Hα), and is at 656.3 nm; the second line is Hβ, at 486.1 nm; the third is Hγ, at 434.2 nm; and so on. The formula governing the wavelengths of lines in the spectrum of hydrogen was worked out in 1885 by Swiss mathematics teacher Johann Balmer (1825–98).

Barlow lens An extra lens used in conjunction with a telescope's eyepiece in order to increase the magnification, usually by a factor of two. It can also improve definition. (After English physicist Peter Barlow, 1776–1872.)

Barnard, Edward Emerson (1857–1923) American astronomer. He was a skilled visual observer, and also a pioneer of photographic astronomy. Barnard discovered many comets, the fifth satellite of Jupiter (Amalthea) and **Barnard's Star**. He speculated (correctly) that what appear to be regions containing few stars are dark nebulae, and published a catalog of them.

Barnard's Loop A large but faint emission nebula, spanning 14°, in the form of an arc centered on the Orion **OB association**. It appears to be the result of gas and dust blown away from the association by radiation pressure. It was discovered by E. E. **Barnard**.

Barnard's Star A red dwarf star 6 light years away in the constellation Ophiuchus. It has the largest known **proper motion**, at 10.4 arc seconds per year, and is the closest star to the Sun after the Alpha Centauri system. It was discovered in 1916 by E. E. **Barnard**. Barnard's Star is magnitude 9.5, with a luminosity 0.0004 times the Sun's, and type M4.

barred spiral galaxy A spiral **galaxy** in which the arms extend from the ends of a bar, an elongated distribution of stars extending either side of the center of the galaxy's disk. Most galaxies – including our own – are thought to possess a bar, which plays an important role in their evolution and in triggering star formation.

Barringer Crater SEE **Meteor Crater**.

barycenter The **center of mass** of two or more celestial bodies and the point around which they revolve. The term is used in particular for the Earth–Moon system. As the Earth is so much more massive than the Moon, the barycenter lies within the Earth.

Bayer, Johann (1572–1625) German astronomer. In 1603 he published a star atlas, *Uranometria*, in which he introduced 12 new constellations. In it he allocated letters of the Greek alphabet (SEE the table on page 317) to the stars in each constellation: alpha (α), beta (β), gamma (γ), and so on, usually in order of brightness. This system of what came to be known as *Bayer letters* is still in use.

▲ **barred spiral galaxy** NGC 1300 in Eridanus is a typical barred spiral galaxy, with arms connected to the ends of a bar, which has the nucleus at its center.

BD ABBREVIATION FOR *Bonner Durchmusterung*.

Beijing Astronomical Observatory An observatory with its main optical/infrared observing site at Xinglong, about 150 km (95 miles) northeast of Beijing. It houses China's largest telescope, the Large Sky Area Multi-Object Fiber Spectroscopic Telescope (LAMOST), completed in 2008. LAMOST is an unusual variant of the Schmidt telescope design. The main mirror, 6.7 × 6.0 m (265 × 236 inches), is fed light by a reflective corrector plate which moves to track objects and transmits light via fiber-optic tubes to stationary spectrographs.

Belinda One of the small inner **satellites** of Uranus discovered in 1986 during the Voyager 2 mission.

Bell Burnell, (Susan) Jocelyn (1943–) English astronomer. While a graduate student she ran a survey of radio galaxies which led to the discovery of the first **pulsar**. She is now generally credited, with her then supervisor, Antony **Hewish**, as the co-discoverer of pulsars, but she received no share of the Nobel prize subsequently won by Hewish.

Bennett, Comet (C/1969 Y1) One of the brightest comets of the 20th century, discovered by South African amateur astronomer John Caister Bennett (1914–90) in 1969, and a first-magnitude object in April 1970. Its tail extended for 20°.

BepiColombo A European Space Agency spacecraft scheduled to enter orbit around Mercury in 2024. After launch in 2017 the craft will obtain eight gravity assists: one from the Earth, two from Venus and five from Mercury itself. It will consist of two separate spacecraft: ESA's Mercury Planetary Orbiter (MPO), which will study the surface and internal composition of the planet, and a Japanese-built Mercury Magnetospheric Orbiter (MMO), which will be ejected from the MPO to study Mercury's **magnetosphere**, the region of space around the planet that is dominated by its magnetic field.

Bessel, Friedrich Wilhelm (1784–1846) German mathematician and astronomer. His catalog of 63,000 accurate star positions, based on refinements of previous observations as well as his own, marks the beginning of modern astrometry. In 1838 Bessel became the first to announce that he had determined the parallax of a star (61 Cygni). From the "wobbles" in the motions of Sirius and of Procyon he deduced the existence of their companion stars, which were not detected until many years after his death.

Beta Centauri The second brightest star in the constellation Centaurus, magnitude 0.61 (slightly variable), also known as Hadar or Agena. It is a giant of spectral type B1, and lies 525 light years away.

Beta Lyrae star A type of **eclipsing binary** in which gas escapes from a bloated primary star and falls on to an **accretion disk** that surrounds the secondary star. Beta Lyrae stars are members of a broad class of double stars known as *semidetached binaries*, so called because only one of the two stars fills its **Roche lobe**. In Beta Lyrae itself the primary star is overflowing its Roche lobe, and gases stream on to the disk at the rate of 0.00001 solar masses per year; some gas

▼ **Beta Lyrae star** The continuous light curve of Beta Lyrae stars is believed to be caused by an accretion disk of material which surrounds the secondary star and eclipses the primary. This theory also explains why the secondary star appears to be dimmer than the primary, despite being far more massive.

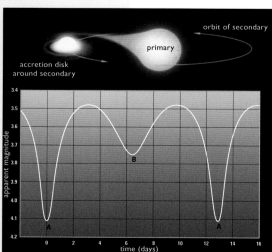

▲ **Big Bear Solar Observatory** *The suite of telescopes in the dome of the Big Bear Solar Observatory observe the Sun in visible, near-infrared and ultraviolet light, thus giving astronomers detailed information about different parts of the solar surface and atmosphere. This information helps them to monitor and understand the way in which features on the Sun's surface form and evolve.*

escapes from the system altogether. In contrast to **Algol**-type stars, Beta Lyrae variables do not show intervals of constant (or near-constant) light over part of their orbital period. The lower half of the diagram shows the light curve of Beta Lyrae; the primary minimum (A) is when the secondary star passes in front of the brighter primary, and the shallower minimum (B) is when the opposite occurs.

Beta Pictoris SEE **Pictor**.

Betelgeuse The star Alpha Orionis, the second brightest star in Orion. It is an orange-red supergiant of type M2 and a semiregular variable that ranges from magnitude 0 to 1.3 with a period of about 6½ years. Betelgeuse lies 500 light years away. It has a diameter about 500 times the Sun's (about 700 million km/400 million miles) and is about 10,000 times as luminous.

Bethe, Hans Albrecht (1906–2005) German–American physicist. In 1938 he and Carl von **Weizsäcker** independently proposed a detailed theory for the production of energy in the Sun and other stars (SEE **carbon–nitrogen cycle**) in which hydrogen is converted into helium by nuclear fusion; this is essentially still accepted today. In recognition, he was awarded the 1967 Nobel Prize for Physics. SEE ALSO George **Gamow**.

Bianca One of the small inner **satellites** of Uranus discovered in 1986 during the Voyager 2 mission.

Biela's Comet (3D/Biela) A now-disintegrated comet of orbital period 6.62 years, discovered by Wilhelm von Biela (1782–1856) in 1826. At its return in 1846 it was observed to have split into two parts. At the next return in 1852 the two parts were about 2.3 million km (3.7 million miles) apart. The comet has not been seen since, but in 1872 there was a magnificent display of the **Andromedid** meteor shower, at precisely the time when the Earth crossed the orbital path of the vanished comet. This was a dramatic demonstration that the particles that produce meteors are the debris of disintegrated comets.

big bang The theory that a giant explosion about 14 billion years ago created the Universe and started its expansion. It was developed in the 1940s by George **Gamow** from the ideas of Georges **Lemaître**.

According to the big bang theory, everything in the Universe once constituted an exceedingly hot and compressed gas with a temperature exceeding 10,000 million degrees K. When the Universe was only a few minutes old, its temperature would have been 1000 million degrees K. As it cooled, nuclear reactions took place that led to the material emerging from the fireball consisting of about 75% hydrogen and 25% helium by mass, the composition of the Universe as we observe it today. Small quantities of deuterium, helium-3 and lithium were also produced, and their abundances, together with the lack of heavier elements, enable a limit to be placed on the amount of ordinary matter in the Universe (SEE **critical density**). There were fluctuations from place to place in the density or expansion rate. Slightly denser regions of gas whose expansion rate lagged behind the mean value collapsed to form galaxies when the Universe was perhaps 10% of its present age.

The **cosmic microwave background** radiation, first detected in 1965, is considered to be the residual radiation of the big bang explosion. The *ripples* in this background radiation discovered in 1992 by the **Cosmic Background Explorer** and later confirmed by the **Wilkinson Microwave Anisotropy Probe** and **Planck** are evidence of the initial density variations from which galaxies formed.

The expansion of the Universe, the abundance of the light elements and the **cosmic microwave background** radiation are the three strongest observational supports for the big bang theory. However, there remain some problems that the basic or standard version of the big bang does not explain.

We can trace the Universe back at least to an age of 10^{-5} seconds after the big bang, when it contained matter, antimatter, and radiation at a temperature of around 10^{12} K. In the standard big bang theory we cannot extrapolate back to the moment of creation itself because the theory predicts the existence of a singularity, where the laws of physics break down. Suggestions for an origin have arisen from progress in the unification of the fundamental forces of physics, known as **grand unified theories**. This, though, leads back to one of the three main problems with the big bang theory: the so-called causality problem, of why the Universe should have expanded from a singularity (SEE ALSO **oscillating Universe**).

Other observations that the standard big bang theory fails to explain have led some cosmologists to postulate that there was a period of **inflation** very soon after the big bang occurred. The observed amount of ordinary matter in the Universe is only 5% of that needed for the **critical density**, leading to the **missing mass** problem and the possible existence of **dark energy** and **dark matter**. Finally, there is the antimatter problem: the question of why there are not equal amounts of matter and antimatter in the Universe, the solution to which may lie with particle physics.

Big Bear Solar Observatory A solar observatory built on an artificial island in Big Bear Lake, in the San Bernardino Mountains, California. The surrounding expanse of water ensures that there is none of the bad seeing caused by heat haze over land. The main instrument is a 1.6 m (63-inch) reflector, opened in 2009.

Big Dipper Popular US name for the group of stars in Ursa Major known in the UK as the **Plough**.

binary pulsar A **pulsar** orbiting another star, forming a binary. The first to be discovered was PSR 1913+16 in 1974; its companion is also thought to be a neutron star, but one that does not emit pulses. The orbital period is nearly 8 hours. Observations show that the pulsar's orbit is gradually contracting, caused by energy being emitted in the form of **gravitational waves**, as predicted by Einstein's general theory of relativity. American radio astronomers Joseph Taylor and Russell Hulse were awarded the 1993 Nobel Prize for Physics for the discovery of PSR 1913+16. In 2003, a binary system was discovered in which both stars are pulsars.

binary star Two stars in orbit around a common **center of mass**. They are usually classified as visual, spectroscopic or eclipsing binaries, according to the means by which they are observed. Their orbital periods vary enormously, from hours to years for spectroscopic and eclipsing binaries, and from a decade to many centuries for visual binaries. A star with more than two components is called a *multiple star*, **Castor** being an example.

In a *visual binary*, of which about 50,000 are known, the components are sufficiently far apart, with an angular separation greater than about 0.1 arc seconds, for the two to be seen separately through a telescope. The first visual binary discovered was **Mizar**, in 1650 by Giovanni Battista Riccioli, who saw that it had a close companion (as distinct from its wider companion, Alcor).

In an **eclipsing binary**, one star periodically passes in front of the other so that the total light output appears to fluctuate. Most eclipsing binaries are also spectroscopic binaries. **Algol** was the first eclipsing binary to be identified, by John **Goodricke** in 1782.

A **spectroscopic binary** is a system whose components are too close for their separation to be measured visually. Instead, the binary nature is revealed by a Doppler shift in the absorption lines of the stars' spectra as they move around their orbits. The brighter component of Mizar was the first star shown to be a spectroscopic binary, by Edward **Pickering** in 1889.

The study of irregularities in the **proper motions** of stars has in some cases revealed the existence of invisible companions; such a star is known as an *astrometric binary*.

binoculars Two small telescopes joined by a hinge, giving the observer a low-magnification view with both eyes simultaneously. The hinge allows the binoculars to be adjusted for the distance

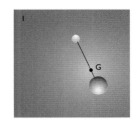

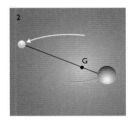

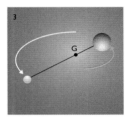

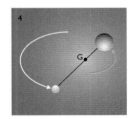

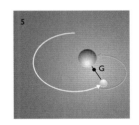

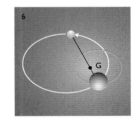

▲ **binary star** *The two components of a binary system move in elliptical orbits around their common center of gravity (G), which is not halfway between them but nearer to the more massive component. In accordance with Kepler's second law of motion, they do not travel at uniform rates. In (2) they are greatly separated and moving slowly. In (6) they are closer and moving more rapidly.*

between the eyes, and there is usually a means of focusing the eyepieces independently. Compactness is achieved by using prisms to fold the light-paths within the instrument. Binoculars have designations such as 8 × 40, where 8 is the magnification and 40 is the diameter of each object lens in millimeters.

bipolar flow The flow of material from a star in two streams in opposite directions. In young stars this can produce **Herbig–Haro objects**; in old stars, bipolar planetary nebulae. A star that generates bipolar flow is usually surrounded by a dusty envelope that is oriented perpendicularly to its rotation axis. The interplay of the star's rotation and mass outflow, and its magnetic field, is thought to create bipolar flows.

black-body radiation The radiation that would be emitted by a *black body*, a hypothetical object that absorbs all thermal radiation (heat) falling on it, and is a perfect emitter of thermal radiation. The spectrum of the radiation emitted by a black body is a **continuous spectrum**, and the intensity of the radiation is greatest at a wavelength that depends only on the body's temperature. Although the black body is only hypothetical, its importance in astronomy is that the spectra of stars can be interpreted by assuming that they approximate to black bodies. SEE ALSO **effective temperature**.

black drop An optical effect, visible at **second contact** or **third contact** during a transit of Mercury or Venus across the Sun's disk, in which a dark ligament – the "drop" – appears to connect the limb of the planet with the limb of the Sun. The phenomenon limits the accuracy of the timing of transits.

Black Eye Galaxy A spiral galaxy (M64, NGC 4826) in the constellation Coma Berenices which contains a dark cloud of dust near the nucleus, creating the "black eye" effect. The Black Eye Galaxy lies 24 million light years away.

black hole A localized region of space from which neither matter nor radiation can escape – in other words, where the escape velocity exceeds the velocity of light. The boundary of this region is called the *event horizon*. Its radius, the **Schwarzschild radius**, depends on the amount of matter that has fallen into the region and it increases linearly as the mass increases.

A black hole of stellar mass is thought to form when a massive star undergoes total gravitational collapse. For stars up to about 1.4 solar masses, gravitational collapse can be physically halted to produce a **white dwarf**. A slightly more massive object will collapse to form a **neutron star**. If, however, the mass exceeds about 3 solar masses, even after a supernova explosion has blasted away the outer layers of the star, the collapse continues beyond even the neutron star stage. As it contracts below its Schwarzschild radius, the object becomes a black hole and effectively disappears. A star of 3 solar masses has a Schwarzschild radius of about 9 km. Inside the event horizon of the black hole, space and time are highly distorted and the stellar matter is increasingly compressed until it forms an infinitely dense **singularity**.

Black holes can have an immense range of sizes. *Supermassive black holes* with up to a billion solar masses are believed to be the source of energy in **quasars** and **active galactic nuclei**. At the other end of the scale, some primordial black holes (SEE below) could be truly microscopic. Because no light or other radiation can escape from black holes, they are extremely difficult to detect. Any matter encountered by a black hole would probably first go into orbit rather than being drawn directly into it. A rapidly spinning disk of matter, known as an **accretion disk**, forms around the object and heats up through friction to such high temperatures that it emits X-rays. Black holes may therefore appear as X-ray sources in binary stars. An example is **Cygnus X-1**.

Not all black holes result from stellar collapse. During the big bang, some regions of space might have become so compressed that they formed so-called *primordial black holes*. Such black holes would not be completely black, because radiation could still "tunnel out" of the event horizon at a steady rate, leading to the evaporation of the hole. (Such small-scale effects are not important for larger black holes.) Primordial black holes could thus be very hot.

blazar A term, compounded from **BL Lacertae object** and **quasar**, for a galaxy with an exceptionally **active galactic nucleus**. Blazars show variable optical brightness, strong and variable optical polarization, and strong radio emission. The optical variations may be on timescales as short as days. The explanation for their exceptional activity is that in these galaxies we are viewing relativistic **jets** end-on.

BL Lacertae object (BL Lac object) A highly luminous **active galactic nucleus**, the prototype of which, BL Lacertae, was originally classified as a variable star. BL Lacertae objects are variable at all wavelengths from radio to X-rays, sometimes on a timescale of just a few hours, and their optical spectra are unusual in being completely featureless, containing neither absorption nor emission lines. This is consistent with them being jets from **quasars** or **radio galaxies** emitted along our line of sight.

blue giant A massive star that has exhausted the hydrogen at its core and initiated helium burning, causing it to expand and leave the **main sequence**. Blue giants have high surface temperatures, of about 30,000 K, and luminosities about 10,000 times that of the Sun. At a later stage of evolution, they expand still further and cool, becoming red **supergiants**.

blue moon An occasional blue coloration of the Moon caused by particles injected into the upper atmosphere by, for example, a large volcanic eruption or an extensive forest fire. The particles scatter light, making it appear bluer. The term has also come to mean a second full Moon in the same calendar month, which happens about seven times in every 19 years, and is used in everyday speech to denote a rare event.

blue straggler A bright, blue star in a globular cluster that remains on the **main sequence** long after stars of similar or lower mass have evolved into red giants. It is thought that they "straggle behind" because they are close binaries in which mass has been transferred from one to the other, or, in some cases, the two stars have merged.

Bode, Johann Elert (1747–1826) German astronomer. He published the theoretical relationship between the distances of the planets that had previously been pointed out by Johann Titius (1729–96). On the basis of this relationship (SEE **Bode's law**), Bode suggested the existence of an undiscovered planet between Mars and Jupiter. In 1801 he published the beautifully illustrated star atlas *Uranographia*, and a catalog of 17,000 stars and nebulae.

Bode's law More properly called the *Titius–Bode law* (SEE **Bode**), a numerical sequence announced by Bode in 1772 that matches the distances from the Sun of the then known planets. It is formed by adding 4 to each number in the sequence 0, 3, 6, 12, 24, 48, 96 and 192, giving 4, 7, 10, 16, 28, 52, 100 and 196. Taking the Earth's distance from the Sun as 10, Mercury, Venus and Mars fell into place quite well (at 3.9, 7.2 and 15.2), as did Jupiter and Saturn (52 and 95). The discovery in 1781 of Uranus at a distance corresponding to 192 prompted a search for a major planet to fill the gap at 28, but led instead to the discovery of the asteroids. The next number in the sequence is 388, but although Neptune (301) does not fit the bill, Pluto (395) does.

Although, in the form in which it was originally proposed, Bode's law has no real basis, it does give an indication of how **resonances** and tidal interactions (SEE **tides**) between bodies in the Solar System can result in them having orbits which are **commensurable**.

Bok, Bartholomeus Jan ("Bart") (1906–83) Dutch–American astronomer. He made extensive studies of the Milky Way, particularly of its star-forming regions (SEE **Bok globule**). Bok showed that stellar **associations** consist of young stars.

Bok globule A round, dark cloud of gas and dust which is a likely precursor of a **protostar**; named after Bart **Bok**. Small globules have diameters of 0.1 light year (6000 au) and can be seen in front of bright nebulae. Larger globules, up to 3 light years across, are seen as dark patches against the stellar background of the Milky Way. Estimated globule masses range from about 0.1 solar masses for the smallest to about 2000 solar masses for those in the Rho Ophiuchi dark cloud.

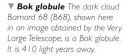
▼ **Bok globule** *The dark cloud Barnard 68 (B68), shown here in an image obtained by the Very Large Telescope, is a Bok globule. It is 410 light years away.*

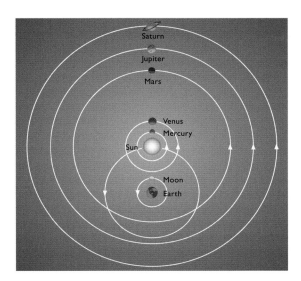

▲ **Brahe, Tycho** *In the Tychonian model of the Solar System, there is a static Earth at the center. The Moon and Sun orbit the Earth, and the other planets move around the Sun.*

bolide An exceptionally bright meteor – a major **fireball** – that is accompanied by a sonic boom. Objects whose passage through the atmosphere produce bolides may survive to reach the surface as meteorites.

bolometer An instrument for measuring all the electromagnetic radiation from a source, used in astronomy to measure the total radiation from a star, by registering the change in electrical resistance that results when radiation falls on a thin conductor.

bolometric magnitude (symbol m_{bol}) A measure of the total amount of energy radiated by a body at all wavelengths and expressed on the visual magnitude scale.

Bondi, Hermann (1919–2005) British cosmologist and mathematician, born in Austria. With Thomas **Gold** and Fred **Hoyle** he developed the **steady-state theory**, in which the continuous creation of matter drives the expansion of the Universe. A byproduct was Bondi and Hoyle's elucidation of the production of heavy elements inside stars. Bondi later demonstrated the reality of gravitational waves.

Bonner Durchmusterung (BD) A catalog containing data on over 300,000 stars, published in 1859–62 by F. W. A. **Argelander** and extended in 1886 to over 450,000 stars by Eduard Schönfeld. The stars are numbered in declination zones from +90° to −23°, and are cited in the form "BD +52° 1638".

Boötes A prominent northern constellation containing **Arcturus**, the brightest star north of the celestial equator. In mythology Boötes represents a herdsman. Epsilon (Izar or Pulcherrima) is a beautiful double star, with K0 and A0 components, magnitudes 2.6 and 4.8, separation 2.9 arc seconds. Mu (Alkalurops) is a wide double, magnitudes 4.3 and 6.5.

Borrelly, Comet 19P/ A periodic comet discovered by Alphonse Borrelly (1842–1926) in 1904. It has an orbital period of 6.9 years, and became the fourth comet to be visited by a spacecraft when the **Deep Space 1** probe flew past at a distance of 2200 km (1380 miles) in 2001. During the flyby, the nucleus was imaged as a dark (albedo 0.03) elongated body with a long axis of 6 km (4 miles) from which jets of gas were emerging.

bow shock A curved region ahead ("upwind") of a planet's **magnetosphere** where the solar wind is decelerated and deflected, rather like the effect set up in front of a ship's bow as it moves through the water. SEE ALSO **heliosphere**.

Bradley, James (1693–1762) English astronomer. In 1742 he succeeded Edmond **Halley** as **Astronomer Royal**. He discovered the **aberration** (1) of light – the first observational evidence for **Copernicus**'s heliocentric theory – and the **nutation** of the Earth's axis. His accurate catalog of 3000 stars was published posthumously, and formed the basis of Friedrich **Bessel**'s catalog.

Brahe, Tycho (1546–1601) Danish astronomer. In 1572 he observed the supernova in Cassiopeia, and his report of it, *De nova stella*, soon made him famous. Under the patronage of King Frederick II of Denmark he built and equipped two observatories – Uraniborg and Stjerneborg – on the island of Hven in the Baltic. He became the most skilled observer of the pre-telescopic era, expert in making accurate naked-eye measurements of the positions of stars and planets. He calculated the orbit of the comet seen in 1577, and this, together with his study of the supernova, showed that **Aristotle**'s picture of the unchanging heavens was wrong. He could not, however, accept the world system put forward by **Copernicus**. In his own planetary theory (the *Tychonian system*), the planets move round the Sun, and the Sun itself, like the Moon, moves round the stationary Earth. In 1597 he left Denmark and settled in Prague, where Johann **Kepler** became his assistant. The accurate series of planetary observations made by Tycho Brahe were used by Kepler in deriving his laws of planetary motion.

breccia A type of rock made up of coarse, angular fragments set in a finer-grained material. Impacts produced the fragments, which were later incorporated in a younger rock formation. Most of the lunar rock samples returned by the Apollo missions were breccias, and the surface rocks of many other cratered, geologically inactive bodies in the Solar System are expected to be made up largely of breccias.

bremsstrahlung (German, literally "braking radiation") Electromagnetic radiation produced when energetic electrons are decelerated, for example when approaching the nucleus of an atom. Also known as *free–free radiation*, it is generated in ionized gas clouds (SEE **H II region**).

Bright Star Catalog A catalog of naked-eye stars (above magnitude 6.5) first published by Yale University in 1930. The current (fifth) edition is in electronic form.

brown dwarf An object with a mass less than 0.08 solar masses, in which the core temperature does not rise high enough to initiate thermonuclear reactions. Such a *substellar object* is, however, luminous, because as it slowly shrinks in size it radiates away its gravitational energy. The surface temperature of a brown dwarf is below the 2500 K lower limit of a **red dwarf**. The very coolest exhibit specific spectral features, and three new spectral classes – **L stars**, **T stars** and Y stars – have been introduced to describe these stars. About 2000 brown dwarfs had been found by early 2015, both singly and in orbit around larger stars.

B star A star of spectral type B, characterized by strong neutral helium absorption lines and the presence of hydrogen lines (although not as strong as in the **A stars**). On the **main sequence**, the temperatures of B stars range from 10,500 K at B9 to 31,000 K at B0. Their masses and sizes range correspondingly from 2.3 to 14 solar masses, and from 2.5 to 10 solar radii. At their hottest, B stars have 25,000 times the luminosity of the Sun. Examples of B stars are Achernar, Regulus, Rigel and Spica. A B9 star stays on the main sequence for about 800 million years, but a B0 star remains there for only 11 million years. The hottest B stars (and the even hotter **O stars**) form loose groupings known as **OB associations**. O and B stars evolve to become supergiants with luminosities up to 100,000 times that of the Sun.

butterfly diagram A diagram illustrating the changing distribution of sunspots in solar latitude over the course of the 11-year cycle. At the start of a cycle there are spots at latitudes of up to around 30° north and south, but very few near the equator. As the cycle progresses, spots appear nearer the equator. When all the spots in a cycle are plotted on the diagram, a characteristic pattern in the shape of a butterfly appears. The diagram was devised by Edward Maunder in 1904.

▼ **butterfly diagram** *During the solar cycle, the latitude at which sunspots form changes. This is related to changes in the Sun's magnetic field during its cycle. Cycles usually overlap, with the first spots of a new cycle forming concurrently with the last spots of the old cycle.*

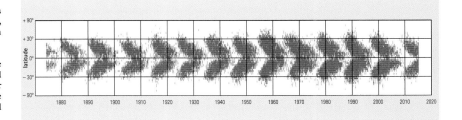

223

Caelum A constellation of the southern sky, adjoining Columba, introduced by **Lacaille**; it represents an engraving tool. It has no star above magnitude 4.4. One variable, R Caeli, of Mira type, ranges from magnitude 6.7 to 13.7 with a period of 391 days.

Calar Alto Observatory A German–Spanish observatory near Almeria in southern Spain. Its main telescope is a 3.5 m (138-inch) reflector, opened in 1984.

calendar A list of the days of the year, grouped into months, and based on the apparent motions of the Sun or Moon, or both. The ancient Egyptians used a calendar based on a solar year. The Babylonians (and the modern Hebrews and Muslims) used a lunar year of 12 months, which is 11 days shorter than a solar year, so an extra month is added every third year.

The present *Gregorian calendar* was introduced by Pope Gregory in 1582 to correct a discrepancy of 10 days that had accumulated in the earlier *Julian calendar*, which had a year of 365 days and added a leap day every four years. This gave an average calendar year of 365.25 days, longer than a **tropical year** (the interval between two successive passages of the Sun through the vernal equinox) by over 11 minutes. The Gregorian calendar reduces the average length of the calendar year by not inserting a leap day in century years that are not divisible by 400. Thus the years 1600 and 2000 are leap years, but 1700, 1800 and 1900 are not. This gives a calendar year of 365.2425 days, which is accurate to one day in 3300 years. For religious and political reasons many countries did not adopt the Gregorian calendar for many years or even centuries. It was adopted in Great Britain and its territories in 1752, when 2 September was followed immediately by 14 September. Dates before and after this "gap" are sometimes referred to as *Old Style* and *New Style*, respectively. SEE ALSO **Julian Day**.

Caliban An outer **satellite** of Uranus in a **retrograde** (1) orbit, discovered by Brett J. Gladman and Philip D. Nicholson in 1997.

California Nebula An emission nebula, designated NGC 1499, in the constellation Perseus. Discovered by E. E. **Barnard** in 1884–5, it takes its name from the similarity between its outline in long-exposure images to that of the American state. The nebula is part of a complex of gas and dust, 1000–1600 light years distant, from which stars are forming in an adjacent spiral arm of our Galaxy.

Callisto The second largest of Jupiter's **Galilean satellites**, with a diameter of 4821 km (2996 miles), and the outermost of them. It has the lowest average density, at 1.86 g/cm^3, indicating that it contains a high proportion of water ice. The surface of the satellite is dark and very heavily cratered – in fact it is the most densely cratered object known. Callisto's ancient surface bears scars from the heavy bombardment phase of the Solar System's formation, ending 3.8 billion years ago. As well as the dense craters, there are several large, multiringed **impact features**, the largest being **Valhalla**, with an overall diameter of 4000 km (2500 miles) and Asgard (1600 km; 1000 miles). The **Galileo** probe found evidence for a 10 km (6-mile) deep subsurface ocean 100 km (60 miles) below Callisto's surface, and detected a weak magnetic field and a thin carbon dioxide atmosphere. Callisto orbits just outside Jupiter's main **radiation belt**. SEE ALSO **satellite**.

Caloris Basin (Caloris Planitia) The largest **impact feature** on Mercury. About half of it was photographed by the Mariner 10 probe in 1974–5. It is about 1500 km (950 miles) in diameter and, like a number of similar features elsewhere in the Solar System, has a multi-ringed structure.

Calypso A small **satellite** of Saturn, discovered by Bradford Smith and others in 1980 between the two Voyager encounters. Calypso is irregular in shape, with an average diameter of 19 km (12 miles). It is **co-orbital** with Tethys and Telesto; Calypso and Telesto orbit near the L$_5$ and L$_4$ **Lagrangian points**, respectively, of Tethys' orbit around Saturn.

Camelopardalis A barren northern constellation representing a giraffe. Its brightest star, Beta, is of magnitude 4.0 and has a wide magnitude 8.6 companion. It contains the open cluster NGC 1502.

canals A network of dark linear markings on the surface of Mars, reported to exist by observers from the 1870s to the early 20th century. They were first described by Angelo **Secchi** and Giovanni **Schiaparelli** as *canali*, the Italian word for "channels." In some English reports *canali* was mistranslated as "canals" – with the inference that they were artificial constructions. This idea was seized upon by, in particular, Percival **Lowell**, who was convinced that the canals were artificial waterways built by intelligent beings to irrigate

a dying planet. Despite the demonstration in 1909 by Eugène **Antoniadi** that the canals were illusory, this romantic notion of Mars persisted in many minds until spacecraft visited Mars and sent back pictures that showed no sign of a canal network.

Cancer A dim zodiacal constellation, between Gemini and Leo. In legend it represents the crab sent by Juno, queen of the gods, and trodden on by Hercules during his fight with the multi-headed Hydra. Beta, magnitude 3.5, is its brightest star. Zeta is a multiple system; the main components are of magnitudes 5.0 and 6.2, and the brighter star is itself a close double. R Cancri is a Mira variable, magnitude range 6.1 to 11.8, period 362 days. RS Cancri is a semiregular variable (magnitude 5.1 to 7.0, period about 120 days), as is X Cancri (magnitude 5.6 to 7.5, period about 195 days). The large open cluster **Praesepe**, M44, is flanked by Gamma, magnitude 4.7, and Delta, magnitude 3.9, known as the Asses. Also in Cancer is M67, a very old open cluster visible in binoculars.

Canes Venatici A constellation introduced by **Hevelius** on his map of 1690, representing the dogs Asterion and Chara, held by the herdsman Boötes. The only star above fourth magnitude is Cor Caroli ("Charles' Heart"), magnitude 2.9. This name was given to it by Edmond **Halley**, in honor of King Charles I of England. The star is the prototype of the class of *magnetic variable* stars whose spectra contain lines that vary in intensity. It is a double, with a companion of magnitude 5.6, separation 20 arc seconds, excellent for small telescopes. Y is a semiregular variable, ranging in magnitude from 5.0 to 6.5 in a period of about 160 days. The famous **Whirlpool Galaxy**, M51, lies in Canes Venatici. Other spiral galaxies are M63, M94 and M106. There is also the sixth-magnitude globular cluster M3, which is distinct in binoculars.

Canis Major A constellation representing one of Orion's hunting dogs. It is distinguished by the presence of **Sirius**, the brightest star in the sky. Other bright stars are Epsilon (Adhara), magnitude 1.4; Delta (Wezen), magnitude 1.8; and Beta (Mirzam), magnitude 2.0. Beta is variable over a very small range and is a prototype of the class of variables known either as *Beta Canis Majoris stars* or as *Beta Cephei stars*. Other variables are R (magnitude 5.7 to 6.3, period 1.14 days, Algol type), W (magnitude 6.4 to 7.9, irregular) and UW (magnitude 4.8 to 5.3, period 4.39 days, Beta Lyrae type). There is one prominent open cluster, M41, of fifth magnitude, south of Sirius.

Canis Minor A constellation representing the smaller of Orion's two dogs. The first-magnitude star **Procyon** makes it easy to locate. Beta (Gomeisa), magnitude 2.9, is a slightly variable **shell star**. There are no other objects of note, although three Mira variables (V, R and S) can rise to above magnitude 8 at maximum.

Cannon, Annie Jump (1863–1941) American astronomer. Working at the Harvard College Observatory, initially as an assistant to E. C. **Pickering**, she reorganized the classification of stars into spectral types, and classified the 225,000 stars in the *Henry Draper Catalog* of stellar spectra.

Canopus The star Alpha Carinae, the second brightest star in the sky, magnitude −0.62. It is a bright giant of type F0, more than 10,000 times as luminous as the Sun, and lies 313 light years away.

Capella The star Alpha Aurigae, magnitude 0.08 (making it the sixth brightest star in the sky), distance 42 light years. It is a spectroscopic binary, consisting of two giant stars of types G6 and G2 with an orbital period of 104 days.

Capricornus One of the less prominent of the zodiacal constellations; it has been identified with the Greek god Pan. Its brightest star is Delta (Deneb Algedi), magnitude 2.9. Alpha (Algedi) is a wide optical double, magnitudes 3.6 and 4.2. The semiregular variable RT has a range from sixth to ninth magnitude and a period of 393 days. The most important cluster is M30, a globular of magnitude 7.5.

captured rotation SEE **synchronous rotation**.

carbonaceous chondrite A rare type of chondritic meteorite (SEE **chondrite**) with a higher-than-average content of carbon, in the form of organic compounds. Not counting the **volatiles**, the proportions of chemical elements in a carbonaceous chondrite are very similar to the composition of the Sun. These meteorites are therefore believed to have come from bodies that formed very early in the Solar System's history, which makes them important in the study of its origin.

▲ *Cassini–Huygens* Saturn and its largest moon, Titan, seen by the Cassini orbiter in May 2012. The rings are seen almost edge-on, and their broad shadow falls on the planet's cloud tops.

carbon–nitrogen cycle A sequence of nuclear reactions that accounts for the production of energy inside main-sequence stars that are more massive and hence hotter than the Sun. It was first described in 1938 by Hans **Bethe**. The reactions involve the fusion of four hydrogen nuclei into one helium nucleus with the release of a huge amount of energy. Nitrogen and oxygen are formed as intermediate products. The presence of carbon is essential, but it behaves as a catalyst and remains unchanged at the end of the cycle. The process takes place at temperatures above 15 million degrees K. It is also known as the *carbon cycle* and the *carbon–nitrogen–oxygen cycle*.

carbon star (C star) A cool red giant star whose surface composition shows carbon-based molecules such as carbon monoxide (CO), cyanogen (CN), molecular carbon (C_2) and other compounds of carbon, and also the presence of much lithium. A star of about the Sun's mass in the red-giant phase has carbon, produced by nuclear fusion at its core, brought to the surface by convection currents. The carbon reacts to form the molecules visible in its spectrum. Carbon stars are allocated spectral type C in the Morgan–Keenan system of **spectral classification**.

Carina A prominent southern constellation, once part of the larger grouping of Argo Navis, the ship Argo, and representing the ship's keel. It contains **Canopus**, the second brightest star in the sky. Other bright stars are Beta (Miaplacidus), magnitude 1.7, and Epsilon (Avior), magnitude 1.9. Epsilon and Iota (magnitude 2.2) form part of the **False Cross**. Upsilon is a double, magnitudes 3.0 and 6.3. Carina's variable stars include the unique **Eta Carinae**, which lies within the nebula NGC 3372. Other variables with maximum above magnitude 6.5 are the Cepheids U (magnitude range 5.7 to 7.0, period 38.77 days) and l ("ell," magnitude 3.3 to 4.2, period 35.54 days), as well as the Mira-type stars R (magnitude 3.9 to 10.5, period 309 days) and S (magnitude 4.5 to 9.9, period 150 days). There are several open clusters visible to the naked eye, principally NGC 2516, NGC 3114 and IC 2581. The open cluster IC 2602 around Theta is spectacular when viewed with binoculars. There is also a sixth-magnitude globular cluster, NGC 2808.

Carme An outer **satellite** of Jupiter discovered in 1938 by Seth Nicholson. It is in a **retrograde** (1) orbit, and may be a captured asteroid.

Cassegrain telescope A **reflecting telescope** in which the light received by the primary mirror is reflected back from the secondary mirror through a hole in the primary. The light is brought to a focus – the *Cassegrain focus* – on the axis of the telescope, where the image can be observed by the eye or by instruments. The combination of a concave, **paraboloidal** primary and a convex (usually **hyperboloidal**) secondary tends to cancel **aberrations** (2), such as coma, of the separate mirrors. The arrangement is used for both amateur and professional telescopes. It was devised in 1672 by the French priest Laurent Cassegrain (1629–93). SEE ALSO **Schmidt-Cassegrain telescope**.

Cassini French family of Italian origin that produced four generations of astronomers and cartographers, all of whom ran the Paris Observatory. The two most prominent were:

Cassini, Giovanni Domenico, known as **Jean Dominique** (1625–1712) after he moved to Paris in 1669. He was the first to accurately measure the dimensions of the Solar System. He discovered the division in Saturn's rings that now bears his name, and also four of that planet's satellites. He improved the **ephemerides** of Jupiter's satellites, which helped Ole **Römer** in his determination of the velocity of light.

Cassini, Jacques (1677–1756), son of Giovanni. He accurately measured an arc of a meridian running through France. The results helped to show that the Earth is flattened at the poles, although he and his father had believed there to be a polar elongation. He determined the proper motion of the star Arcturus.

Cassini–Huygens A joint space mission between NASA and the European Space Agency to explore Saturn and its satellites, in particular Titan. It was launched on 15 October 1997 and reached Saturn in July 2004, having flown past Jupiter on the way. The main Cassini craft entered orbit around Saturn, while the Huygens subprobe landed on **Titan** in January 2005 and returned data and images from the surface.

Cassini Division The gap between Rings A and B in the ring system of **Saturn**, discovered by Giovanni Cassini in 1675. It is about 4700 km (2900 miles) wide. The Voyager spacecraft showed that it is not a true gap, but contains several narrow rings.

Cassiopeia A distinctive northern constellation, representing the mother of Andromeda. The five leading stars make up a "W" or "M" pattern, on the opposite side of Polaris from Ursa Major. Alpha (Schedar) is its brightest star, magnitude 2.2. Gamma, magnitude range 1.6 to 3.0, is the prototype of the variables known as **shell stars**. Rho is a highly luminous semiregular variable, usually of around magnitude 5 but which on rare occasions falls to below 6. Other variables are R (magnitude 4.7 to 13.5, period 430 days, Mira type), RZ (magnitude 6.2 to 7.7, period 1.2 days, Algol type) and SU (magnitude 5.7 to 6.2, period 1.95 days, Cepheid). Eta is a fine colored binary, magnitudes 3.5 and 7.5; the orbital period is just under 500 years. Cassiopeia contains several open clusters resolvable in binoculars: M52, NGC 457, NGC 559, NGC 663 and M103.

Cassiopeia A The strongest radio source in the sky apart from the Sun, and the remnant of a supernova that exploded around 1660 but was not recorded by observers on Earth. It lies about 10,000 light years away and appears optically as a faint nebula.

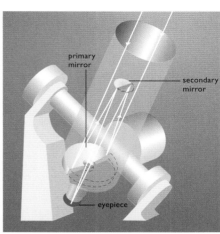

▲ *Cassegrain telescope* Light collected by the primary mirror is reflected from a hyperboloidal secondary mirror to the eyepiece through a hole in the center of the primary.

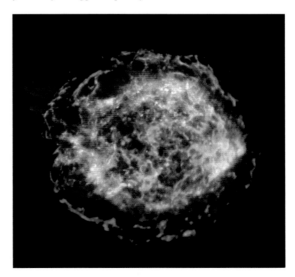

▲ *Cassiopeia A* This image of Cassiopeia A was obtained by the Chandra X-ray Observatory. The green ring marks the position of a shock wave generated by the supernova explosion. The blue areas are iron gas.

225

Castor The star Alpha Geminorum, magnitude 1.58, distance 52 light years. Castor is one of the finest visual binaries in the sky for small telescopes, consisting of main-sequence components of magnitudes 2.0 and 2.9, types A2 and A5, period 450 years. Each of the stars is itself a spectroscopic binary. A distant companion, Castor C, orbits the system. It is an eclipsing binary also known as YY Geminorum, made up of two faint red dwarfs, magnitude range 9.2 to 9.6.

cataclysmic variable A term used for a diverse group of stars that erupt in brightness, including **supernovae**, **novae**, **dwarf novae**, **symbiotic variables**, nova-like variables and some X-ray objects. With the exception of one type of supernova, these objects are very close binaries whose outbursts are caused by interaction between the two components. A typical system of this type has a low-mass secondary which fills its **Roche lobe** so that material is transferred over to the primary, which is usually a white dwarf. The transferred material has too much angular momentum to fall directly on to the primary, and so forms an **accretion disk**. A hot spot may arise on the outer edge of the accretion disk where the infalling material hits it. Outbursts occur when the flow rate varies, and take place at irregular, unpredictable intervals from weeks to many years – thousands of years, it is thought, for novae. In dwarf novae the thick disk brightens in outburst, but in novae (in particular) outbursts occur when the material that has accreted on to the white dwarf is ignited. SEE ALSO **eruptive variables**.

catadioptric A type of telescope in which a lens–mirror combination forms the image. The **Schmidt-Cassegrain telescope** is the commonest type of catadioptric design for amateur use; another is the **Maksutov telescope**. A catadioptric telescope is basically a reflector with a full-aperture lens at its top end. The lens corrects for **aberration** (2), and produces a divergent light-beam which is reflected on to the secondary mirror by a short-focus **spheroidal** or **paraboloidal** primary mirror. The design is compact and often used for portable instruments.

Cat's Eye A bright planetary nebula in the constellation Draco. Designated NGC 6543, it has a pronounced blue-green color. It was from observations of this object that William **Huggins** was able to demonstrate, in 1864, that planetary nebulae shine by emission rather than reflection. The Cat's Eye lies at a distance of 3000 light years.

CCD Abbreviation for *charge-coupled device*, a solid-state electronic imaging device. A CCD is an integrated circuit whose upper layer contains a square array of closely spaced electrodes. Photons of light from a celestial object falling on an electrode are converted into an electronic charge, and the charges from all the electrodes are "coupled" together and read off as a pattern – an instantaneous image of the object. A series of these images are fed to a computer and combined to create a picture, just as a photographic emulsion builds up a picture over the course of its exposure.

Images that formerly required hours of exposure on traditional photographic emulsion can be recorded in minutes by using a CCD. Fainter objects can be detected, the effects of **light pollution** can be compensated for, image processing by computer can be used to bring out detail, and images taken though colored filters can be

▲ **Centaurus A** *Crossed by a prominent dark lane and with jets of gas shooting out from a central black hole, this peculiar galaxy (also designated NGC 5128) is a powerful radio and X-ray source.*

combined to give true-color pictures. In professional astronomy CCD imaging has replaced traditional **astrophotography** techniques, and they are becoming increasingly cheaper and readily accessible to amateur astronomers.

celestial equator The projection of the Earth's equator on to the celestial sphere, dividing the sky into the northern and southern hemispheres.

celestial latitude, longitude SEE **coordinates**.

celestial mechanics The branch of astronomy dealing with the motions of heavenly bodies. It uses the laws of physics to explain and predict the orbits of planets, satellites and other celestial bodies.

celestial meridian The **great circle** passing through the observer's **zenith**, the **nadir**, and the celestial poles, cutting the horizon at the north and south points.

celestial poles SEE **celestial sphere**, **pole**.

celestial sphere The imaginary sphere, of infinite radius and with the Earth at its center, upon which all the celestial bodies appear to be projected as though they were at a uniform distance. The most obvious aspect of the celestial sphere is its apparent daily east-to-west rotation, caused by the axial spin of the Earth. The celestial sphere appears to rotate about the *celestial poles*, as though on an extension of the Earth's own axis, which points toward the celestial poles. Although the positions of the two celestial poles seem to be fixed, they are in fact slowly drifting because of the effect of **precession**.

The celestial sphere is divided into two halves by the **celestial equator**, which is equidistant from the poles. One can thus visualize a set of **great circles**, analogous to circles of latitude on Earth, called *declination circles*, each parallel to the equator. Declination circles allow the position of an object to be specified in terms of its angular distance north or south of the celestial equator.

Right ascension, the celestial equivalent of longitude, is also an extension of the terrestrial system. The zero of right ascension is related to the apparent motion of the Sun around the sky during the year, caused by the Earth's orbital motion around the Sun. The track traced by the Sun on the celestial sphere, called the **ecliptic**, indicates the plane of the Earth's orbit. The ecliptic is not the same as the celestial equator because the Earth's axis is tilted by about 23½°.

For half of each year the Sun is in the northern hemisphere of the sky, and for the other half it is in the southern hemisphere. On two dates each year the Sun crosses the celestial equator; these points are known as the **equinoxes**. It is the vernal equinox, when the Sun crosses into the northern hemisphere, that is chosen as the zero of right ascension. Right ascension is measured eastward from this point but, unlike terrestrial longitude, it is measured in units of time rather than degrees, minutes and seconds of arc (SEE **sidereal time**).

Centaur One of a class of small objects orbiting the Sun between Jupiter and Neptune. The first to be discovered was **Chiron**, and over 450 were known as of early 2015. Typically 100 km (60 miles)

▶ **celestial sphere** *A useful concept for describing positions of astronomical bodies is the celestial sphere. As shown, key points of reference – the celestial pole and celestial equator – are projections on to the sphere of their terrestrial equivalents. An object's position can be defined in terms of right ascension and declination (equivalent, respectively, to longitude and latitude).*

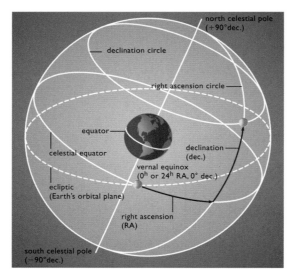

north celestial pole
(+90°dec.)

declination circle

right ascension circle

equator

celestial equator

declination
(dec.)

vernal equinox
(0ʰ or 24ʰ RA, 0° dec.)

ecliptic
(Earth's orbital plane)

right ascension
(RA)

south celestial pole
(−90°dec.)

in diameter, the Centaurs are probably icy **planetesimals** that have been perturbed inward from the **Kuiper Belt** and now occupy short-lived, unstable orbits.

Centaurus A brilliant southern constellation representing a centaur, with 13 stars above magnitude 3.5. The brightest are **Alpha Centauri** and **Beta Centauri** (also known as Hadar or Agena), which indicate the direction of the Southern Cross. Gamma is a close binary with an 85-year period; the components are equal, at magnitude 2.9, and their combined brightness is magnitude 2.2. R is a Mira-type variable, magnitude range 5.3 to 11.8, and the period is unusually long, 546 days on average. T is a semiregular variable (magnitude 5.5 to 9.0, period 90 days). The globular cluster **Omega Centauri** is the finest in the entire sky. There are also two bright open clusters, NGC 5460 and NGC 3766, and a fine planetary nebula, NGC 3918. The elliptical galaxy NGC 5128 is the radio source Centaurus A.

Centaurus A A strong radio source about 12 million light years away in the constellation Centaurus, identified with the galaxy NGC 5128. Optically it appears as an elliptical galaxy crossed by a prominent dust lane. It appears to be a massive elliptical galaxy in collision with a smaller spiral with a high content of dust.

central meridian (c.m.) The imaginary line joining the north and south poles of a planet or the Sun, bisecting its disk. It is used by observers as a reference line to estimate the longitude of features as the body rotates.

center of mass The point in a body, or a system of bodies, at which the total mass of the body or system may be regarded as concentrated. It is the point from which the gravitational attraction of the system as a whole appears to act. SEE ALSO **barycenter**.

Cepheid variable One of an important class of variable stars that pulsate in a regular manner, accompanied by changes in luminosity. They take their name from the first of the type to be discovered, Delta Cephei, whose variability was noted by John **Goodricke** in 1784. Cepheids periodically expand and contract, changing in size by as much as 30% in each cycle. The diagram shows the pulsations and light curve of Delta Cephei itself. A typical Cepheid variable has a surface temperature that varies between 6000 and 7500 K and a spectral type that ranges from G2 at minimum to F2 at maximum. The average luminosity is 10,000 times that of the Sun.

Cepheids became important in cosmology in 1912, when Henrietta **Leavitt** discovered a simple relationship between the period of light variation and the absolute magnitude of a Cepheid. This relationship, the **period–luminosity** law, enables the distances of stars to be ascertained, not only in our Galaxy but also in other galaxies.

There are two classes of Cepheid: type I, or *classical Cepheids*, are younger and more massive than type II, or **W Virginis stars**. Both types follow a period–luminosity relationship, but their curves are different. The classical Cepheids, such as Delta Cephei itself, are yellow supergiants of Population I, while the second type are older, Population II stars and are found in globular clusters and the center of the Galaxy (SEE **stellar populations**).

In using Cepheids to determine distances it is necessary to know which type is being observed. Classical Cepheids are one to two magnitudes brighter than type II Cepheids of the same period. The periods of Cepheids are in the range 1–50 days; classical Cepheids have a somewhat shorter period on average than those of type II.

Cepheus A constellation in the far north, adjoining Ursa Major, but not very distinctive. In mythology it represents the husband of Cassiopeia and the father of Andromeda. Alpha (Alderamin) is magnitude 2.4. Beta (Alfirk), magnitude 3.2, is variable over a small range, and is a prototype of the class known as either *Beta Cephei stars* or *Beta Canis Majoris stars*. Delta is the prototype **Cepheid variable**, with a range from magnitude 3.5 to 4.4 and a period of 5.37 days; it has a wide optical companion of magnitude 7.5. Mu, which William **Herschel** named the Garnet Star, is a semiregular red supergiant with a magnitude range of 3.4 to 5.1 and a period of about 2 years. T Cephei is a Mira variable (magnitude 5.2 to 11.3, period 388 days); VV is a giant eclipsing variable, magnitude 4.8 to 5.4, period 7430 days.

Ceres The first **asteroid** to be discovered, by Giuseppe **Piazzi** on 1 January 1801. Its diameter of 934 km (580 miles) makes it the largest asteroid. It orbits in the main asteroid belt, at an average distance from the Sun of 2.8 au (414 million km; 257 million miles). It is now also classified as a **dwarf planet**.

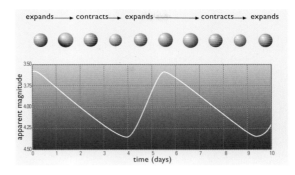

▲ **Cepheid variable** *Their extremely regular light variations make Cepheids valuable standard candles: the longer the period from one maximum to the next, the greater the star's intrinsic luminosity. As shown, the typical Cepheid light curve shows a rapid rise to peak brightness, followed by a slower decline to minimum.*

Cerro Tololo Inter-American Observatory An observatory on Cerro Tololo peak near La Serena, Chile, part of the US National Optical Astronomy Observatory. Its largest telescope is the Victor M. Blanco Telescope, a 4 m (158-inch) reflector, the twin of that at Kitt Peak. Other instruments include the 0.61 m (24-inch) Curtis Schmidt Telescope of the University of Michigan.

Cetus A large constellation, crossed by the celestial equator, but not very conspicuous; it represents the sea monster of the Andromeda legend. Its brightest star is Beta, magnitude 2.0. The most famous star in the constellation is the prototype long-period variable, **Mira**. Tau Ceti, magnitude 3.5, 11.9 light years away, is one of the closest Sun-like stars, being a main-sequence star of type G8. UV Ceti is a **flare star**; it is a pair of 12th-magnitude red dwarfs 8.7 light years away and exhibits large flares every few hours.

Chamaeleon A small constellation near the south pole, representing a chameleon. Its brightest stars are only of magnitude 4.1.

Chandrasekhar, Subrahmanyan (1910–95) Indian–American astrophysicist. He made significant contributions to the theories of stellar evolution, stellar atmospheres and relativity. His theory of white dwarf stars suggests there is a limit to their mass (SEE **Chandrasekhar limit**).

Chandrasekhar limit The maximum possible mass for a **white dwarf** star, first calculated in 1931 by Subrahmanyan **Chandrasekhar**. The value is about 1.4 solar masses, although more recent computations suggest that a higher value is possible for a rapidly rotating white dwarf. If a star has a mass which exceeds the Chandrasekhar limit it will collapse to become a **neutron star** or possibly a **black hole**.

Chandra X-ray Observatory (CXO) A NASA satellite for studying the sky at X-ray wavelengths, launched on 23 July 1999. It carries the highest-resolution **X-ray telescope** ever constructed. The main instruments are a CCD imaging spectrometer and a high-resolution camera. It is named in honor of Subrahmanyan **Chandrasekhar**.

Chang'e A series of Chinese lunar probes. Chang'e 1 and 2, launched in October 2007 and October 2010 respectively, went into orbit around the Moon. Chang'e 3, launched in December 2013, landed on the lunar plain called Sinus Iridum, delivering a small rover called Yutu ("Jade Rabbit") to the surface.

charge-coupled device SEE **CCD**.

Charon The largest moon of **Pluto**, discovered in 1978 by James Christy. Its diameter of 1186 km (737 miles) makes it the largest satellite in relation to its primary in the Solar System. Its mass is about 12% of Pluto's, and its density is about that of Pluto, 2.0 g/cm^3. It is unique among satellites in having a **synchronous orbit** with a period (6.4 days) matching the rotation period of its primary. Charon seems to have a grayish surface of water ice and some ammonia ice.

Chiron A small body in the outer Solar System, originally designated as an asteroid (no. 2060) following its discovery in 1977 by Charles Kowal. However, it has an eccentric, cometary orbit (period 50.7 years) which takes it from within Saturn's orbit out nearly as far

▶ **comet** *Comet Hale–Bopp (C/1995 O1) of 1997 demonstrated both types of cometary tail, one of gas (blue and straight) and one of dust (yellower and broader). Hale–Bopp was one of the most prominent comets of the 20th century.*

as Uranus. In 1989, seven years before reaching perihelion, it was found to have a comet-like **coma** and was given a periodic comet designation (95P/Chiron). It has a diameter of roughly 200 km (125 miles), much larger than any "ordinary" comet nucleus. It is now classified as a **Centaur**.

chondrite A type of **stony meteorite**, consisting of an agglomeration of *chondrules* – millimeter-sized globules believed to have remained unchanged since they condensed out of the nebula from which the Sun and the Solar System formed. They are therefore more ancient than **achondrites**. Chondrites consist mainly of iron, calcium, aluminum, magnesium and sodium silicates. They are classified into various types, including **carbonaceous chondrites**.

chromatic aberration An **aberration** (2) in lenses that results from the unequal refraction of the different colors that make up white light. Each color is brought to a focus at a slightly different distance from the lens, causing the image of an object to be fringed with prismatic colors. Chromatic aberration is reduced by the use of an **achromatic lens**.

chromosphere ("sphere of color") The layer of the Sun's atmosphere between the photosphere and the corona. The chromosphere is normally invisible because of the glare of the photosphere shining through it, but it is briefly visible near the beginning and end of a total solar eclipse as a spiky red rim around the Moon's disk. At other times it can be studied by spectroscopy or with a **spectroheliograph**. The chromosphere is about 10,000 km (6000 miles) thick, with a temperature of 10,000–25,000 K. **Filaments**, **flocculi**, **plages** and **prominences** originate in the chromosphere. SEE ALSO **supergranulation**.

Circinus A small constellation near Alpha and Beta Centauri, representing a pair of dividing compasses. Its only fairly bright star is Alpha, magnitude 3.2.

circumpolar stars Stars which from a given place of observation never dip below the horizon. Their polar distance (90° minus their declination) is less than the observer's latitude.

civil twilight SEE **twilight**.

closed universe In cosmology, a mathematical model that describes the behavior of a universe in which there is enough matter to halt the expansion triggered in the **big bang**. The universe collapses back on itself, ending in an event that is the opposite of the big bang, sometimes called the *big crunch*. In a closed universe, the **curvature of space** is positive, the **deceleration parameter** is greater than 0.5, and the **density parameter** is greater than 1 (i.e. the mean density of the universe is greater than the **critical density**). COMPARE **flat universe**, **open universe**.

cluster of galaxies A group of galaxies, which can range in size from a few dozen members, as in our own **Local Group**, to a rich cluster containing thousands of galaxies. Poor clusters are usually irregular in shape, but rich clusters tend to be spherical and condensed at the center, often with a giant elliptical galaxy or a cD galaxy dominating the central region. cD galaxies are thought to

have grown by *cannibalism*, their overwhelming gravity dragging in smaller galaxies. The nearest rich clusters of galaxies are the **Virgo Cluster**, 50 million light years away, and the Coma Cluster, 300 million light years away. Clusters are grouped into *superclusters* up to about 100 million light years across. The Local Supercluster, of which the Local Group is part, is centered on the Virgo Cluster.

CMB ABBREVIATION FOR **cosmic microwave background**.

CME ABBREVIATION FOR **coronal mass ejection**.

Coalsack A prominent dark nebula in the constellation Crux, the Southern Cross. It is seen in silhouette against the background stars of the Milky Way.

Coathanger A bright group of ten stars in the constellation Vulpecula. Six of the stars are in an east–west line extending for an angular distance of 1.5°, while the other four form a "hook" to the south. Cataloged as Collinder 399, the Coathanger is also known as *Brocchi's Cluster*; it is now known to be an **asterism** and not a genuine cluster, the component stars lying at various distances.

COBE ABBREVIATION FOR **Cosmic Background Explorer**.

coelostat Two mirrors arranged so that slowly rotating one of them counteracts the apparent rotation of the sky, enabling light from a celestial object to be directed along a fixed path. The first mirror is mounted equatorially and is clock-driven at half the rate of the Earth's rotation. The beam of light it receives is reflected on to a second (fixed) mirror, which in turn reflects the beam into the instrument being used for the observation. A coelostat is used with heavy or fixed equipment, for example a **spectroheliograph**. It is an improved form of **heliostat** and **siderostat**. SEE ALSO **coudé**.

collimation The process of aligning the optical components of a telescope so as to bring the rays of light from the object being observed to a focus at the correct position. A refracting telescope requires only occasional adjustment of the object glass in its cell. Collimation is more of a problem with reflectors because the primary and secondary mirrors both need to be adjusted, and the secondary can become misaligned quite easily. An incorrectly collimated telescope can suffer from **coma** (1).

color index The difference in brightness of a star at two different wavelengths, used as a measure of the star's color and hence its temperature. The brightness of the star is measured through colored filters, usually at blue (B) and yellow (V, for visual) wavelengths, and the difference between the two readings is the B−V color index. The B−V color index for cool red stars is positive (up to about +2.0), while for hot blue stars it is negative (about −0.5 maximum). A white star of spectral type A0 such as Vega has a B−V color index of exactly 0.0. A measurement at U (ultraviolet) wavelengths may also be taken, and a U−B color index derived. Other measurements in the red and infrared can also be used.

The color index of light from a star may change as the light passes through dust in interstellar space. Dust scatters blue light but allows red light to pass more freely, thereby reddening the star. This effect must be taken into account when determining the color index of a star. SEE ALSO **UBV system.**

Columba A southern constellation adjoining Canis Major and Carina, representing a dove. Alpha (Phakt) is magnitude 2.6, but the constellation contains little of interest.

colure A **great circle** on the celestial sphere that passes through both celestial poles. SEE ALSO **equinoctial colure**, **solstitial colure**.

coma (1) A defect in an optical system which results in the image of a point appearing as a blurred pear-shaped patch with a flared appearance resembling a comet. It is caused by incident light striking a lens obliquely. The defect increases with distance from the center of the field of view.

coma (2) The spherical cloud of gas and dust surrounding the nucleus of an active **comet**.

Coma Berenices A northern constellation adjoining Boötes, representing the hair of Queen Berenice of Egypt. Its brightest stars, of fourth and fifth magnitude, form a very wide star cluster. It is rich in faint galaxies; some are part of the **Virgo Cluster**, but others are part of the separate and more distant *Coma Cluster*, 300 million light years away.

Coma–Virgo Cluster SEE **Virgo Cluster**.

comes (plural **comites**) The fainter companion in a **binary star**.

comet A small, icy Solar System body in an independent, eccentric orbit around the Sun. The solid *nucleus* of a comet is small – that of **Halley's Comet**, for example, measures just 16 × 8 km (10 × 5 miles). The nucleus is made up of ices, largely water ice (plus some methane, carbon dioxide and carbon monoxide, with traces of organic substances), in which are embedded rock and dust particles, and is covered by a thin dark crust. As the comet approaches the Sun and gets warmer, sublimation begins, and jets of gas and dust escape through the crust to form the luminous, spherical *coma*. The coma can be huge, a million kilometers or so across, but extremely tenuous. Closer in, radiation pressure from the Sun and the **solar wind** may send dust and gas streaming away as a *tail*, tens of millions of kilometers long. The *gas tail* or *ion tail* is straight and long, up to 10 million km (6 million miles), while the *dust tail* is shorter, broader and curved.

Comets are thought to originate in the **Oort Cloud**, where there may be as many as 6 trillion (6 × 10^{12}) of them, and in the **Kuiper Belt**. They are believed to be **planetesimals** left over from the formation of the Solar System. A comet may be gravitationally perturbed (by a passing star, for example) toward the Sun, swinging past the Sun on a parabolic orbit to return to the outer Solar System, or entering an elliptical orbit as a *long-period comet*, with a period of thousands or even millions of years (SEE **parabola**). Further perturbation by one of the giant planets may drive it into a smaller orbit as a *short-period comet*, with a period of 200 years or less. Further such close encounters can subsequently modify the orbit: some comets may even be ejected from the Solar System on hyperbolic trajectories.

With their loss of material at every approach to the Sun, periodic comets have a limited lifetime. The debris a comet leaves in its wake spreads around its orbit to form a **meteor stream**. Some comets, once they have expended all their gas and dust, may be strong enough to hold together and continue to orbit the Sun – the **Apollo asteroids**, **Amor asteroids** and **Aten asteroids** may be extinct comets. Others, among them Comets **Biela**, **West** and **Shoemaker–Levy 9**, have been observed to break up. Some objects first classed as asteroids have later behaved as comets – **Chiron**, for example, developed a coma.

Over 3000 comets have had their orbits calculated in detail, of which some 300 are short-period comets (i.e. they have appeared more than once). Dozens of new comets are discovered every year. Once its orbit has been confirmed, a comet is given an official designation based on its year and half-month of discovery and is usually named after its discoverer, which in many cases can be a spacecraft or a survey telescope. A few, such as Halley's Comet and **Encke's Comet**, are named after the people who first calculated their orbit. In the case of co-discoverers, up to two names may be given, as in Comet **Hale–Bopp** (prior to 1995 three names were allowed). Single-apparition comets are prefixed by "C/" and periodic comets by "P/". Comets that have broken up or disappeared are prefixed "D/".

cometary globule A reflection nebula, which can superficially resemble a comet, associated with a young star such as a **T Tauri star**.

commensurable Describing two numbers with a common factor. If the orbital periods of two planets around the Sun, or of two satellites round a planet, can be expressed as a ratio of small whole numbers, then these periods are commensurable. Because of tidal interactions (SEE **tides**) between bodies of similar size, stable orbits tend to have (nearly) commensurable periods. Examples are the orbital periods of Saturn and Jupiter, which are in a ratio of nearly 5:2, and the periods of Jupiter's **Galilean satellites**. Orbital **resonances** may prevent bodies which are very different in size from having commensurable orbital periods; this explains the **Kirkwood gaps** in the main asteroid belt and the gaps in Saturn's ring system.

comparator A device with which two images of the same starfield obtained at different times can be compared. If any object has changed its position or varied in brightness during the interval between the times when the two images were taken, the comparator will show it up. In the *blink comparator* the two images are rapidly alternated in the field of view. The *stereo comparator* is a simple device for viewing the two images simultaneously with binocular vision; another arrangement uses two slide projectors and a rotating shutter. Hunters of novae and asteroids use comparators. Variable stars reveal themselves as a pulsation, because they give images of different sizes according to their brightness.

SOME PERIODIC COMETS AND ASSOCIATED METEOR SHOWERS						
Name	Discovery	Orbital	Perihelion period (years)	Eccentricity distance (million km)	Inclination (°)	Shower(s)
2P/Encke	1786, 1818	3.28	50	0.85	12	Taurids
21P/Giacobini–Zinner	1900, 1913	6.61	155	0.71	32	Draconids (Giacobinids)
8P/Tuttle	1790, 1858	13.5	149	0.82	55	Ursids
55P/Tempel–Tuttle	1865	32.9	147	0.90	163 R	Leonids
1P/Halley	240 BC	76.0	88	0.97	162 R	Eta Aquarids, Orionids
109P/Swift–Tuttle	1737, 1862	135*	143	0.96	113 R	Perseids
C/1861 G1 Thatcher	1861	415*	138*	0.98	80	Lyrids
* approximate values	R indicates retrograde orbit					

Compton Gamma Ray Observatory (GRO) A NASA satellite for studying the sky at gamma-ray wavelengths, launched on 5 April 1991. In its nine years of operation to June 2000, GRO increased the number of known gamma-ray sources tenfold, and recorded 2500 **gamma-ray bursts**.

Cone Nebula A V-shaped region of dark nebulosity seen in silhouette against brighter material in the constellation Monoceros. Visually elusive, but prominent in long-exposure images, this is part of a star-forming region, including the open cluster NGC 2264, at a distance of 2400 light years.

conjunction The alignment of the Earth with two other bodies in the Solar System so that from the Earth the other two appear in (nearly) the same position. The term is used mainly for alignments with the major planets (SEE the diagram). A **superior planet** is at conjunction when on the opposite side of the Sun from the Earth. An **inferior planet** is at *inferior conjunction* when it lies between the Earth and the Sun, and *superior conjunction* when it is on the far side of the Sun. Because planetary orbits are inclined at various angles to the plane of the ecliptic, alignments at conjunction are rarely exact, and conjunction is defined to occur when the two bodies have the same celestial longitude as seen from the Earth.

Two or more planets – including in this case the Moon – are also said to be in conjunction when they are close together in the sky.

constellation A region of the sky containing an arbitrary grouping of stars, forming an imaginary figure. The stars in a constellation lie at very different distances from us, so the groupings have no physical significance. The original number of 48 constellations given

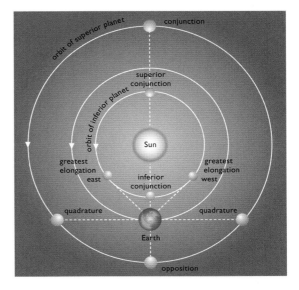

▲ *conjunction* Superior planets come to conjunction with the Sun when on the far side of it, as seen from Earth (and are therefore lost from view). The inferior planets, Mercury and Venus, can undergo conjunction at two stages in their orbit. At superior conjunction, they lie on the far side of the Sun from Earth, while at inferior conjunction they are between the Sun and the Earth. Under certain circumstances, Mercury and Venus can transit across the Sun's disk at inferior conjunction.

▶ *Copernicus, Nicholas*
Copernicus's model of the Solar System put the Sun at the center, with Earth and the other planets in orbit about it. Replacing the earlier geocentric system, the Copernican system removed Earth from any special position in the Solar System.

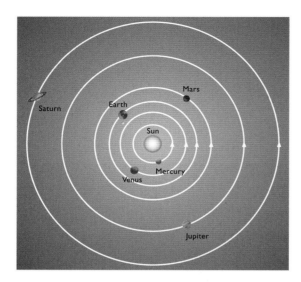

in Ptolemy's **Almagest** has grown to 88 (SEE page 26), which were assigned boundaries on the celestial sphere by the International Astronomical Union in 1930. Constellations are very unequal in size and prominence: they range from the vast Hydra down to the tiny but brilliant Crux. Many of the names are drawn from ancient mythology, while more recent mapping of the southern-hemisphere skies introduced some modern-sounding names such as the Octant and the Telescope. Prominent stars within a constellation are designated by Greek letters followed by the name of the constellation in the genitive (possessive) case, as with, for example, Alpha Lyrae; these are often abbreviated in the form α Lyr.

continuous spectrum The unbroken sequence of colors, merging one into the other, produced when white light is decomposed by refraction through a prism. An incandescent solid, liquid or dense gas emits a continuous spectrum. The spectrum of a star consists of a continuous spectrum crossed by **absorption lines**. SEE **emission spectrum**.

continuum SEE **spacetime**.

co-orbital Describing a satellite orbiting a planet at the same average distance as another satellite. Although the distances – and hence the periods – are the same, the orbital inclinations and eccentricities may differ slightly. An example is **Helene**, a small satellite of Saturn with a very similar orbit to the larger satellite Dione.

Coordinated Universal Time SEE **Universal Time**.

coordinates A pair of values which, in conjunction with a reference plane that serves as a zero, are used to define the position of a point or body on the **celestial sphere**. Several systems of coordinates are used in astronomy.

The most commonly used is the *equatorial system*, in which the reference plane is that of the celestial equator, and the coordinates are usually **right ascension** and **declination** (although **hour angle** and **polar distance** can be used alternatively).

In the *horizontal system* the reference plane is that of the observer's horizon, and the coordinates are **azimuth** and **altitude**.

In the *ecliptic system* the reference plane is that of the ecliptic, and the coordinates are celestial latitude and longitude. *Celestial latitude* is the angular distance between the object and the ecliptic, measured at right angles to the ecliptic. It is measured from 0° to 90°, positively toward the north ecliptic pole and negatively toward the south ecliptic pole. *Celestial longitude* is given by the angle between the object and the **vernal equinox**, measured along the ecliptic from 0° to 360° eastward from the vernal equinox. Celestial latitude and longitude are not measured with reference to the celestial equator; they are sometimes more correctly called *ecliptic latitude* and *ecliptic longitude*.

For specifying positions with respect to the Galaxy, the system of *galactic coordinates* is used, in which the plane of reference is the galactic plane (the plane of the Milky Way). Its coordinates are galactic latitude and longitude. *Galactic latitude* is measured from 0° to 90° from the galactic equator, positively toward the north galactic pole. *Galactic longitude* is measured from 0° to 360° eastward along the galactic equator. As defined by the IAU, the position of zero galactic longitude is RA 17h 46m, dec. −28° 56′ (epoch 2000.0).

Copernicus, Nicholas (1473–1543) Polish churchman and astronomer. As canon at Frauenburg Cathedral and a local administrator, he pursued astronomy as a side-interest. The accepted view in his day was the *geocentric* (Earth-centered) universe as expounded by **Ptolemy** nearly 1500 years before, with its complicated system of orbits. Through his study of planetary motions, Copernicus developed a *heliocentric* (Sun-centered) theory of the universe. In this *Copernican system*, as it is called, the planets' motions in the sky were explained by having them orbit the Sun. The motion of the sky was simply a result of the Earth turning on its axis, and the stars remained fixed as the Earth orbited the Sun because they were so very far away.

An account of his work, *De revolutionibus orbium coelestium*, was published in 1543. Because the new system was inferior at calculating planetary positions, most astronomers continued to believe in the Ptolemaic system. It was not until the discoveries by **Galileo** and Johann **Kepler** that the superiority of the Copernican system became apparent and its reality began to take hold.

Copernicus A large lunar crater, 93 km (58 miles) in diameter, in Oceanus Procellarum. It is one of the most conspicuous craters, and is surrounded by a system of rays stretching for well over 500 km (300 miles).

Cordelia The innermost small **satellite** of Uranus, discovered in 1986 during the Voyager 2 mission. Cordelia and Ophelia act as **shepherd moons** to the planet's Epsilon Ring.

Coriolis force A force that appears to act on a body (such as a mass of air) moving freely across the Earth so as to deflect it from a straight path. The deflection is to the right in the northern hemisphere, to the left in the southern. The effect is a consequence of us observing the motion from within a rotating frame of reference – the spinning Earth. In 1735 the English meteorologist George Hadley (1685–1768) recognized the effect of the Earth's rotation on the movement of air currents; a general deduction of the displacing force of the Earth's rotation was made by Gaspard de Coriolis (1792–1843). The Coriolis force accounts for the circulation of air around cyclones on the Earth, and is a driving force for the meteorology of the Earth and other planets with substantial atmospheres.

corona The outermost layer of the Sun's atmosphere, extending for many millions of kilometers into space. It springs into view as a white halo during a total solar eclipse; at other times it can be observed in visible light only by using a special instrument called a **coronagraph**. The corona emits strongly in the extreme ultraviolet and X-ray regions of the electromagnetic spectrum, and has been studied by spacecraft such as **SOHO** and the **Transition Region and Coronal Explorer**. Its overall shape changes during the 11-year solar cycle, from regular and symmetrical at solar minimum to uneven with long streamers at solar maximum. The corona has a temperature of 1–2 million degrees K. When lines of the Sun's magnetic field connect, large currents flow and heat the corona to this high temperature.

The corona may be divided into the inner *K corona*, which shines by light from the photosphere scattered by high-energy electrons, and the outer *F corona*, which shines by light scattered by lower-energy dust

▲ **corona** *Loops in the Sun's inner corona were imaged by the TRACE spacecraft in 1998. These loops are shaped by magnetic fields.*

particles. The K corona has a **continuous spectrum** (the "K" standing for the German word *Kontinuum*), on which are superimposed the **Fraunhofer lines** of the F corona. There are also emission lines in the spectrum contributed by the *E corona*, consisting of lower-energy metal ions such as calcium which are present out to about two solar radii. SEE ALSO **coronal hole, coronal mass ejection, flare, Sun**.

Corona Australis A small southern constellation representing a crown or wreath. It contains no star above fourth magnitude, but its arc of stars under the forefeet of Sagittarius makes it easy to identify.

Corona Borealis A northern constellation, representing the wedding crown of Ariadne, adjoining Boötes. Alpha (Gemma or Alphecca), magnitude 2.2, is the only bright star, but the curve of five stars makes the constellation easy to find. In the "bowl" of the crown is the famous variable star **R Coronae Borealis**, the prototype of a group of stars that fade suddenly and unpredictably. T Coronae Borealis is a recurrent nova, known as the Blaze Star, which erupted in 1866 and 1946.

coronagraph An instrument, used in conjunction with a telescope, which allows the Sun's corona and prominences to be viewed by producing an "artificial eclipse", masking the light from the Sun's disk with a circular obstruction. Before the invention of the coronagraph, such observations were possible only during the brief duration of a total eclipse.

coronal hole A relatively cool region of the solar **corona**. Coronal holes are associated with weak regions of the Sun's magnetic field, and are sources of high-speed streams in the **solar wind**.

coronal mass ejection (CME) The eruption into interplanetary space of millions of tonnes of plasma from the Sun's corona, triggered by disturbances in the Sun's magnetic field. CMEs are more frequent at solar maximum (several a day) than at solar minimum (about one a week). They are often associated with eruptive **prominences** in the chromosphere and **flares** in the inner corona above active sunspot groups. CMEs commonly travel at hundreds of kilometers per second, and produce shock waves in the slower **solar wind**. The arrival at Earth of energetic particles and strong magnetic fields associated with CMEs can cause *geomagnetic storms*, producing **aurorae** and interfering with radio communications.

Corvus A constellation adjoining Hydra and Crater, representing a crow. Its four brightest stars, of third magnitude, form a quadrilateral which is easy to find. Corvus contains the Antennae, a pair of tenth-magnitude interacting galaxies.

cosmic abundance The relative proportions of the various chemical elements (SEE **element, chemical**) in the Universe as a whole. It has been found from spectroscopic studies of the atmospheres of the Sun and stars that (with the exception of hydrogen and helium) the cosmic abundance is much the same as in the Sun. Stars are roughly two-thirds hydrogen and one-third helium, with other elements – chiefly carbon and oxygen – accounting for only 2%. The chemical makeup of a particular star depends on its age (SEE **stellar evolution**). Some rare types of star, for example **Wolf–Rayet stars**, show unusual abundances of certain elements; others, such as *technetium stars*, contain elements it is difficult to explain by currently accepted theories of how stars evolve. SEE ALSO **interstellar matter**.

Cosmic Background Explorer (COBE) A NASA satellite launched on 18 November 1989 to study the **cosmic microwave background** radiation. It determined that the microwave background has a **black body radiation** spectrum with a temperature of 2.73 K. In 1992 it detected slightly warmer and cooler spots in the background radiation, differing by about 30 millionths of a degree from the average; these *ripples* are attributed to density fluctuations in the early Universe that marked the first stage in the formation of galaxies.

cosmic microwave background (CMB) Weak electromagnetic radiation from the Universe, first detected in 1965 by Arno **Penzias** and Robert **Wilson** of the Bell Telephone Laboratories. The microwave background is **black-body radiation** at a temperature of 2.73 K and has an almost equal intensity in all directions in space. It is considered to be the remnant of the radiation from the hot, early Universe following the **big bang**. The background radiation increased its wavelength as the Universe expanded until today it peaks in the millimeter region of the spectrum. It is called the microwave background because it was first detected in the form of microwaves. Penzias and Wilson won the 1978 Nobel Prize for Physics for their discovery. Observations from the **Cosmic Background Explorer**, the **Wilkinson Microwave Anisotropy Probe** and **Planck**, as well as from ground-based and balloon-based instruments, have shown that the background varies on all angular scales by a few thousandths of 1%, and this requires the total average density of the Universe to be within 1% of the **critical density**.

cosmic rays High-energy atomic particles, moving at speeds approaching the speed of light, which enter the Earth's atmosphere from space. Cosmic rays (called "rays" because they were originally thought to be gamma rays) consist mainly of the nuclei of atoms, mostly hydrogen, stripped of their electrons. Heavier nuclei are present – practically every known element has been detected – but in exceptionally small numbers. While traveling through space they are called *primary cosmic rays*. When they enter the atmosphere they cause disintegration of the atoms they encounter and produce various atomic and subatomic particles, called *secondary cosmic rays*, which may be detected at ground level. The highest-energy cosmic rays come from X-ray sources such as the center of the Galaxy and the Virgo Cluster of galaxies. Cosmic rays of medium energy are now known to originate in supernovae; low-energy cosmic rays are ejected from the Sun by solar **flares**.

cosmogony The study of how the Universe and objects within it came into being. **Cosmology**, by contrast, aims to understand the structure, history, future and governing laws of the Universe as it now exists. Mythical cosmogony covers non-scientific descriptions of the world's creation, ranging from myths of antiquity to accounts from various belief systems. Scientific cosmogony covers theories of the origin of the Universe (SEE **big bang, cosmology**) and of the Solar System.

For the Solar System, the scenario now accepted by most astronomers is a development of the *nebular hypothesis* first proposed in the 18th century by Immanuel Kant and Pierre Simon de **Laplace**. As a **molecular cloud** collapsed under its own gravity, a central condensation grew into a **protostar** which later became the Sun. Around the protostar, material from the cloud spread into a broad disk (SEE **protoplanetary disk**). Small grains of material stuck to one another by **accretion**, gradually forming **planetesimals**. (**Comets** and objects in the **Kuiper Belt** are believed to be remnant populations of these planetesimals.) The planetesimals collided with one another and grew into larger bodies, **protoplanets**, which swept up most of the remaining bits and pieces, as evidenced by **impact features** visible on many planetary bodies today. Gravitational perturbations between the resulting planets then moved them into their present orbits.

The different chemical compositions of the planets thus arose as a result of substances of differing chemical composition condensing out at different distances from the Sun (SEE **volatile**), and by the proto-Jovian planets becoming large enough for their gravitational fields to scoop up the still gaseous hydrogen and helium. By this theory, planets are a normal consequence of star formation. Strong

◀ *cosmic microwave background* A detailed map of the cosmic microwave background obtained by the Planck spacecraft. The blue and orange areas are regions of slightly different temperature, corresponding to different densities in the early Universe from which grew the stars and galaxies of today.

▶ **Crab Nebula** *A supernova remnant, the Crab Nebula is an expanding gas cloud with filamentary structure and a pulsar at its heart.*

observational support for these ideas comes from the large number of extrasolar planets that have now been found and the detection of equatorial dust disks around many stars.

Earlier "catastrophe" theories had explained the formation of planets by a chance event, such as material being torn from the Sun by the close approach of a passing star. However, planetary systems would be extremely rare if they formed in this manner.

cosmological constant A constant arbitrarily added by Albert **Einstein** to the relativistic equations describing the behavior of the Universe so that they would have a static solution. When the Universe was shown to be expanding, Einstein regretted using the cosmological constant, but recent observations of the **accelerating Universe** suggest that it may be real. SEE ALSO **dark energy**, **inflation**.

cosmological principle The principle that the Universe is both *homogeneous* and *isotropic*: in other words, that it is the same (on a sufficiently large scale) in all directions. The principle was assumed to be true by **Einstein** and others as it helped to solve the equations of general relativity. The discoveries that the **cosmic microwave background** is indeed isotropic down to about the 0.001% level, and that the large-scale distribution of galaxies in the Universe appears to be homogeneous, show that the principle is real.

cosmological redshift SEE **redshift**.

cosmology The study of the structure, history, future and governing laws of the Universe on the largest scale. The starting point for modern cosmology was the discovery in the 1920s by Edwin **Hubble** of the large redshifts of galaxies, interpreted as evidence that the Universe is expanding. The rate of expansion is given by the **Hubble constant**. Other observational evidence shows that the Universe appears much the same in all directions (*isotropy*) and at all distances (*homogeneity*).

A *cosmological model* is a description of the Universe in terms of mathematical equations. These models aim to predict how the Universe behaves, and are tested by comparing them with observations. Cosmological models have included the **steady-state theory**, in which the Universe is not only the same in all places but also at all times. It therefore had no beginning, will have no end, and never changes at all when viewed on the large scale. This theory required matter to be created as the Universe expanded in order that the overall density of galaxies should not decrease. For this reason it is also referred to as the *continuous creation* model. On the other hand, according to the **big bang** theory, the Universe was created in a single instant about 13.8 billion years ago, and has been expanding ever since. In the future it may continue to expand or possibly collapse back on itself, depending on the total amount of matter and energy in the Universe – on whether it is above or below its **critical density**. An important consideration is the question of the **missing mass**: the amount of ordinary (or luminous) matter we see in the Universe is less than 5% of the amount

actually present. The remainder consists of so-called **dark matter** (about 27%) and **dark energy** (about 68%).

Cosmologists have made several important advances in discriminating between cosmological models. The discovery of the **cosmic microwave background** radiation provided strong evidence against the steady-state theory. This background radiation is very uniform in all directions, which has led cosmologists to postulate that there was a very rapid period of expansion, known as **inflation**, shortly after the big bang. The abundances of helium and other light elements also support the big bang model since these would have been formed naturally from protons and neutrons in the early stages of the explosion. The Hubble constant is currently thought to have a value of around 70 km/s/megaparsec. The **deceleration parameter** – which determines whether the Universe will expand for ever or collapse back on itself – has also been estimated, and seems to have a value close to 0.5, the value for a Universe which will just continue to expand for ever.

coudé An arrangement of auxiliary mirrors used with a telescope on an equatorial mount to direct the light path down the polar axis to a fixed focal point. The coudé system is used in conjunction with large, immovable instruments, such as spectrographs, which are too massive to be mounted on the telescope itself. (The word comes from the French *coudé*, meaning "bent like an elbow," and is not the name of a person, as is sometimes thought.) SEE ALSO **Nasmyth focus**.

Crab Nebula The nebula M1 (NGC 1952), about 6500 light years away in Taurus, the remnant of a supernova that was noted by Chinese astronomers in July 1054, when it shone as brightly as Venus and was visible even in daylight. The nebula, now about eighth magnitude, was discovered in 1731 by the English astronomer John Bevis and independently by Charles **Messier** in 1758. It gained its popular name of the Crab Nebula after its crab-like appearance in a sketch made by Lord **Rosse**.

The Crab Nebula is an intense source of radio emission, known as *Taurus A*, and is also a source of X-rays (Taurus X-1). Gas ejected in the supernova explosion forms filaments that appear red on color images, surrounding an ionized gas that emits a yellow glow. The glow is **synchrotron radiation**, caused by electrons spiralling in an intense magnetic field. This radiation extends from the radio domain into the visible, and the Crab Nebula is the only synchrotron nebula that can be seen in a small telescope. Inside the nebula lies the *Crab Pulsar*, the remaining core of the supernova of 1054. It spins 30 times a second, emitting electrons to replenish

▼ **coudé** *In this optical configuration, mirrors direct light from the telescope along the polar axis to a fixed observing position. This arrangement is advantageous for observations that require use of heavy or bulky detectors.*

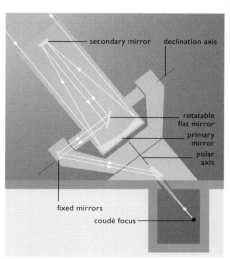

secondary mirror

declination axis

rotatable flat mirror

primary mirror

polar axis

fixed mirrors

coudé focus

the synchrotron radiation. The Crab Pulsar also flashes at visible wavelengths. SEE **pulsar**.

crater A circular formation, found on many bodies in the Solar System – planets, satellites and asteroids. Nearly all craters are **impact features**, although on some bodies there are volcanic craters. A typical impact crater has a raised rim, or wall, and its floor is below the level of the surrounding terrain. Some have a central peak. Impact craters range in size from the microscopic (as found on lunar rocks returned by the Apollo missions) to features a third of the diameter of the body (as on Saturn's moon **Mimas**). Heavily cratered bodies such as **Callisto** are geologically inactive and preserve a record of the bombardment that took place in the early Solar System, when there were still plenty of free **planetesimals**. Less cratered bodies are usually those on which geological activity has led to some degree of resurfacing, obliterating older impact sites.

Crater A small constellation adjoining Hydra and Corvus, representing a cup or chalice. Its brightest stars are of fourth magnitude and it contains no objects of note.

Crepe Ring A popular name for Ring C of **Saturn**. SEE ALSO **ring, planetary**.

Cressida An inner **satellite** of Uranus, discovered in 1986 during the Voyager 2 mission.

critical density The average density of matter in the Universe that would ensure that it would just barely continue expanding endlessly, without ever collapsing back on itself. For a value of the **Hubble constant** of 71 km/s/megaparsec, the critical density is about 10^{-26} kg/m^3, or 6 hydrogen atoms per cubic meter. The observed density of "normal" or luminous matter in the Universe is about 5% of the critical value. However, results from the **Planck** mission place the total density of the Universe to within 1% of the critical density. The deficiency is thought to be accounted for by **dark matter** (27%) and **dark energy** (68%). The ratio of the observed and critical densities is the *density parameter*, Ω. Hence, if the Universe has the critical density, then $\Omega = 1$. Ω is twice the value of q_0, the **deceleration parameter**. SEE ALSO **missing mass**.

Crux The smallest constellation in the sky, popularly known as the Southern Cross. It is surrounded on three sides by Centaurus. Its brightest star, Alpha (Acrux), is of magnitude 0.77 and lies about 321 light years away. It is actually a visual double, magnitudes 1.3 and 1.7, both type B. With Beta (Mimosa, or Becrux), magnitude 1.2, Gamma (Gacrux, magnitude 1.6) and Delta (magnitude 2.8), it makes a pattern that is more like a kite than a cross; the symmetry is disturbed further by Epsilon (magnitude 3.6), the only other star above magnitude 4. Interesting objects include the **Jewel Box** cluster and the **Coalsack**, a dark nebula.

C star SEE **carbon star**.

culmination The passage of a celestial body across the observer's meridian. At culmination, an object reaches the greatest altitude above, or least altitude below, the observer's horizon. For circumpolar stars both culminations are observable. *Lower culmination* occurs when a star transits between the pole and horizon; the star's **hour angle** is then exactly 12h. *Upper culmination* occurs when a star transits between the pole and zenith; the star's hour angle is then exactly 0h.

Curiosity A NASA Mars probe, also known as the Mars Science Laboratory rover, launched on 26 November 2011. It landed inside a crater called Gale, 154 km (96 miles) in diameter, on 6 August 2012. Its studies, which included drilling into the surface to take soil samples for analysis, produced evidence that the crater once contained a lake, probably billions of years ago. Photographs returned by Curiosity showed what appeared to be sedimentary rocks deposited by running water in the distant past.

curvature of space The distortion of **spacetime** in the neighborhood of matter, one of the consequences of **Einstein**'s general theory of relativity. This curvature makes rays of light and particles of matter follow curved paths called *geodesics* which, for the planets, for example, are their orbits around the Sun. If the Universe contains enough mass it will have enough curvature to be a **closed universe**; if not, it will be an **open universe** or a **flat universe** (SEE **cosmology**).

cusp One of the points or horns of the crescent Moon, or of an inferior planet seen at crescent phase.

cusp caps The bright areas at the **cusps** of the crescent Venus that are sometimes reported by observers.

CXO ABBREVIATION FOR **Chandra X-ray Observatory**.

Cygnus One of the most distinctive constellations, sometimes called the Northern Cross. Its brightest star is first-magnitude **Deneb**. Beta is Albireo, a beautiful colored double with a giant primary of type K3, magnitude 3.1, and a main-sequence companion of type B9, magnitude 5.1, separation 34 arc seconds. Chi has the largest visual range of any Mira variable, magnitude 3.3 to 14.2, period about 400 days. P Cygni is a B-type supergiant variable that has ranged between magnitudes 3 and 6 in the past, because of mass loss, and is now around fifth magnitude. **61 Cygni** is a binary pair of orange dwarf stars, magnitudes 5.2 and 6.0. The constellation also contains the radio galaxy **Cygnus A**, the probable black hole **Cygnus X-1**, the **North America Nebula** and the **Cygnus Loop** supernova remnant.

Cygnus A The strongest source of radio emission outside our Galaxy, situated in the constellation Cygnus. It is thought to be two galaxies in collision. Cygnus A is also a source of X-ray emission.

Cygnus Loop A **supernova remnant** consisting of a vast loop of gas ejected from a star that exploded about 5000 years ago. It lies 1400 light years away. The Loop is about 3° in diameter and is expanding at about 100 km/s. Different parts of it bear the designations NGC 6960, 6979, 6992 and 6995. The brightest part, NGC 6992, is known as the Veil Nebula.

Cygnus Rift A prominent dark nebula that splits the Milky Way in the constellation Cygnus, also called the *Northern Coalsack*; its extension through Aquila to Ophiuchus is known as the *Great Rift*. The rift is a string of **molecular clouds** lying in the **Orion Arm**, about 2200 light years away.

Cygnus X-1 A strong source of X-ray emission in the constellation Cygnus, believed to contain a **black hole**. At the position of the source lies a spectroscopic binary star, HDE 226868, with an orbital period of 5.6 days. The visible star is a blue supergiant. Its invisible companion is calculated to be around 10 solar masses or more, far too great for a white dwarf or neutron star, so a black hole seems the only possibility.

▼ **Crux** *The distinctive pattern of the Southern Cross, with the dark Coalsack Nebula to its lower left, is a prominent feature of the southern Milky Way.*

Danjon scale A rough measure of how bright the Moon appears in a **lunar eclipse**, devised by the French astronomer André Danjon (1890–1967). Eclipses are graded from very dark (0) to bright (4), the former occurring when Earth's atmosphere is more opaque, and therefore refracts less sunlight to the eclipsed Moon, as a result of clouds or volcanic dust.

dark adaptation (dark adaption) The process by which the human eye adjusts from vision under conditions of high illumination to night vision. Dark adaptation takes 20 minutes or more to develop fully, during which time the pupil widens and different receptors come into play. It is essential for observers leaving a brightly lit room to allow their eyes to adapt to the dark before they begin observing. Dark adaptation is instantly destroyed by any bright light, so observers use dim red lights when they need to consult star charts or record observations.

dark energy A hypothetical form of energy that may pervade the whole Universe and leads to a repulsive force between distant objects. It was postulated in the 1990s to explain the **accelerating Universe**, and its effects may be modeled mathematically using the **cosmological constant**.

The nature of dark energy is unknown, although vacuum energy – a consequence of quantum mechanics – has been suggested, even though this seems to be too large by a huge factor (10^{120}). An alternative to dark energy is *quintessence* (a fifth "essence," or force) that generates a negative pressure. It may be possible to distinguish between different forms of dark energy and quintessence when more accurate measurements of the changes in the rate of expansion of the Universe with time become available. If dark energy exists, then it may make up over two-thirds of the Universe. Observations by the **Planck** spacecraft show that the mean density of the Universe is within 1% of the **critical density**, yet ordinary matter and **dark matter** jointly account for only about 32% of that amount.

dark matter Unseen matter, inferred to exist in galactic haloes and in the space between galaxies, that is thought to make up about one-quarter of the mass of the Universe. Whatever form it takes, it must not be able to undergo **nucleosynthesis** reactions, otherwise elements heavier than helium would have been produced in the early stages of the **big bang**. Thus it could consist of objects such as **black holes**, **WIMPS** or other unknown atomic particles which are termed *cold dark matter*; or it could be in the form of a "sea" of fast-moving neutrinos, known as *hot dark matter*; or it could be a mixture of both. SEE ALSO **MACHO, missing mass.**

dark nebula A cloud of gas and dust that is not illuminated, and can thus be seen only in silhouette against stars or bright nebulae beyond. They range in size from minute **Bok globules** to the naked-eye clouds of the **Coalsack** nebula in Crux and the gigantic Rho Ophiuchi Dark Cloud, which covers about 1000 square degrees (2%) of the sky. Dust comprises only about 0.1% of the mass of a dark nebula, but is believed to play an important part in the formation of molecules in space, since the surface of the dust particles provides a site for atoms to adhere and combine into molecules. The interiors of these molecular clouds are very cold, typically only 10 K, which allows them to collapse under their own gravity into stars.

Dawes limit SEE **resolving power.**

Dawn A NASA space probe to the asteroids **Vesta** and **Ceres**. Dawn was launched on 27 September 2007 and went into orbit around Vesta on 16 July 2011. It studied the body's surface

▶ **Dawn** The asteroid Vesta, seen in a mosaic of images taken by the Dawn spacecraft. At the south pole is a mountain more than twice the height of Mount Everest on Earth. A chain of three craters at top left of this image forms a feature nicknamed "the snowman."

features and composition for over a year before leaving on 5 September 2012 and moving on to Ceres, which it went into orbit around in March 2015. Ceres and Vesta are two of the largest remaining building blocks of the planets and will shed light on the conditions and processes that operated in the first few million years of the Solar System's evolution.

day The time taken by the Earth to rotate once on its axis. There are various definitions. An *apparent solar day* is the interval between two successive transits of the Sun across the meridian. This interval is variable because the Sun's apparent motion over the course of the year is not uniform (SEE **apparent solar time**). For convenience, an imaginary *mean Sun* is defined as that which moves along the celestial equator at a constant rate, giving the *mean solar day*. A *sidereal day* is defined as the interval between two successive transits of the spring **equinox** across the meridian. This day is about four minutes shorter than the mean solar day owing to the Sun's apparent eastward movement of about 1° each day. A mean solar day is 24h 03m 56.555s of mean sidereal time. A mean sidereal day is 23h 56m 04.091s of mean solar time. SEE ALSO **timescale.**

deceleration parameter (symbol q_0) A figure describing the rate at which the expansion of the Universe is slowing down. If q_0 is greater than 0.5, the expansion will eventually stop and reverse. If q_0 is less than 0.5, the Universe will expand for ever. Theoretical considerations suggest that the true value of q_0 should be exactly 0.5, although in an **accelerating universe** it can be negative. SEE ALSO **critical density, missing mass.**

declination (dec., symbol δ) The angular distance of a celestial object north or south of the celestial equator. It is reckoned positively from 0° to 90° from the equator to the north celestial pole, and negatively from 0° to 90° from the equator to the south celestial pole. SEE ALSO **celestial sphere.**

decoupling era The time, about 300,000 years after the **big bang**, by which the Universe had cooled sufficiently for the first atoms to form; also known as the *recombination era*. Before decoupling, particles and photons underwent constant interactions (coupling); afterward, the Universe was effectively transparent and the **cosmic microwave background** originated from the radiation that was released then.

deep sky The Universe beyond the Solar System. Double stars, star clusters, nebulae and galaxies are known as *deep-sky objects*; the term is not usually applied to individual stars.

Deep Impact A NASA spacecraft launched on 12 January 2005 that had a successful rendezvous with Comet 9P/Tempel 1 in July 2005, releasing a 370 kg (816 lb) copper-tipped impactor that collided with the comet's nucleus, 7.6 km (4.7 miles) long by 4.9 km (3 miles) wide. The impactor created a crater in the nucleus and ejected a plume of hot gas and dust, the composition of which was analyzed by Deep Impact as it flew past at a safe distance. The ejecta contained more dust and less ice than expected and the comet nucleus was found to be extremely porous with up to 80% empty space. Deep Impact subsequently flew past Comet 103P/Hartley on 4 November 2010, photographing the comet's nucleus and discovering a swarm of icy particles around it.

Deep Space 1 The first space probe in NASA's New Millennium series, launched on 24 October 1998. The intention was for the probe to test advanced technologies, including an ion propulsion system. It achieved a rendezvous with asteroid 9969 Braille in July 1999 and with Comet 19P/Borrelly in September 2001.

deferent SEE **Ptolemy.**

degenerate matter A state of matter existing in stars in the final stage of their evolution (SEE **stellar evolution**), when they have ceased producing energy at their cores. Atomic nuclei and electrons are packed closely together at ultra-high densities, and the laws of classical physics no longer apply. Pressure ceases to be dependent on temperature and is a function only of density.

Deimos The smaller of Mars' two **satellites**, discovered in 1877 by Asaph **Hall**. It is a dark, irregular body, measuring about 15 × 12 × 10 km (9 × 7.5 × 6 miles), and may well be a captured asteroid.

Delphinus A small but distinctive constellation near Aquila, representing a dolphin. Its leading stars, of fourth magnitude, make up what looks like a widespread cluster.

Delta Aquarids A **meteor shower** which occurs between 15 July and 20 August. It has a double radiant in Aquarius, the more active radiant lying near the star Delta Aquarii, and peaks at about **ZHR** = 25, close to 28 July.

Delta Scuti star A short-period (0.01–0.2 day) pulsating variable of spectral type between A0 and F5. Delta Scuti stars lie at the lower end of the **instability strip**, either on the main sequence or nearby among the subgiants and giants, and are Population I stars (SEE **stellar populations**). They have very small amplitudes (0.003–0.9 magnitude), frequently with multiple periodicities. Their pulsations are driven by hydrogen-burning instabilities. They are related to the *AI Velorum stars*, which have greater amplitudes (0.3–1.2 magnitude) and higher luminosities, and to the *SX Phoenicis stars*, which are subdwarfs belonging to Population II and have multiple periods and amplitudes of about 0.7 magnitude.

Deneb The star Alpha Cygni, magnitude 1.25, a supergiant of type A2, luminosity half a million times the Sun's, distance about 1400 light years. It is the brightest star in the constellation Cygnus.

density parameter (Ω) The ratio of the actual total density of the Universe to its **critical density** – the density it would have if it contained just enough mass to halt expansion after an infinite time. The density parameter therefore determines whether the Universe will continue to expand for ever, or whether its expansion will halt and it will ultimately collapse back on itself. If Ω is greater than 1 the Universe will collapse; if it is less than 1 the Universe will continue to expand. Ω = 1 corresponds to the critical density. In cosmological models (SEE **cosmology**), it is the parameter that distinguishes between **open** and **closed universes**.

descending node The point at which an orbit crosses from north to south of a reference plane such as the celestial equator or ecliptic. SEE **node**.

Desdemona One of the small inner **satellites** of Uranus discovered in 1986 during the Voyager 2 mission.

de Sitter, Willem (1872–1934) Dutch cosmologist. His hypothetical *de Sitter Universe*, derived from **Einstein**'s general theory of relativity, contained no mass, but provided the first theoretical indication of an expanding Universe.

Despina One of the small inner **satellites** of Neptune discovered in 1989 during the Voyager 2 mission. It appears to be the inner **shepherd moon** to the planet's Le Verrier Ring.

diagonal (1) An optical device, used in conjunction with a telescope's eyepiece, for deflecting the light path through 90° in order to make it easier for the observer to view the image when the telescope is at an awkward angle. This can be a problem with amateur refracting telescopes. In this device, also called a *star diagonal*, either a mirror or a right-angled prism is set at 45° to the light path.

diagonal (2) Another name for the flat secondary mirror in a **Newtonian telescope**.

diamond-ring effect A short-lived phenomenon seen during a total solar eclipse, just before or after totality, when just the bright central part of the edge of the Sun's disk is visible, shining through a depression on the Moon's limb as an intense point of light, together with the faint inner corona, giving the appearance of a diamond ring.

dichotomy The phase of Mercury or Venus when it appears exactly half-illuminated by the Sun, and the **terminator** is a straight line. The term is also applied, less commonly, to the half Moon.

differential rotation The rotation of different parts of a non-solid body at different rates. For example, the deep atmospheres of the giant planets rotate slightly faster at the equator than at the poles. The Sun shows a similar differential rotation, but rather more pronounced.

differentiation The process by which a planetary body evolves a layered structure. In a body which has grown to a sufficient size by **accretion**, energy supplied by gravitational compression and radioactive decay, supplemented by the kinetic energy of impacting bodies, will cause the interior to melt. Under the action of gravity, denser materials will then sink toward the center, where they will eventually form the *core*, and less dense materials will rise toward the

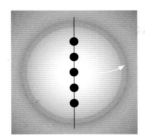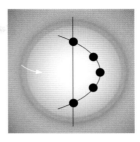

▲ **differential rotation** In a fluid body, such as a star or gas giant planet, the equatorial regions rotate more rapidly than the poles. As shown, a consequence of this is that a set of points lined up on the central meridian will become spread out in longitude over the course of a rotation. Points close to the equator will return to the central meridian earlier than those near the poles.

surface, where they will form the *mantle* and the *crust*. Differentiation has given the terrestrial planets, for example, cores of nickel–iron and outer layers of a predominantly rocky composition.

diffraction The slight sideways spreading of a beam of light as it passes by a sharp edge or through a narrow slit. Starlight passing through a telescope is diffracted by the edges of various components so that, instead of coming to a sharply focused point, it appears as a small disk of light surrounded by concentric *diffraction rings* of light, and radial *diffraction spikes*. The bright central disk is called the **Airy disk**.

diffraction grating A surface on which a very large number of equidistant parallel lines have been ruled very close together. Light striking a grating is dispersed by **diffraction** into a spectrum. There are metal gratings that function by reflection, and glass gratings that function by transmission. The lines are ruled at, typically, 100 to 1000 per millimeter. Diffraction gratings produce very high-quality spectra, and are used in astronomical **spectrographs**.

diffuse nebula A luminous cloud of gas in space. The term "diffuse" refers to the fact that through a telescope diffuse nebulae cannot be resolved into individual stars, unlike star clusters and galaxies. They come in two varieties: **emission nebulae**, also known as **H II regions**, which shine by fluorescence; and **reflection nebulae**, which shine by starlight reflected off dust particles.

Dione A medium-sized **satellite** of Saturn, diameter 1120 km (695 miles), discovered in 1684 by Giovanni Cassini. It has the second-highest density, at 1.44 g/cm³, of Saturn's main satellites. The surface is bright and icy, but one hemisphere is darker than the other. There are three main types of terrain: cratered terrain, cratered plains and smooth plains. Evidence of past geological activity includes a number of **tectonic** features such as Palatine Chasma – a trough over 600 km (nearly 400 miles) long and up to 8 km (5 miles) wide. Dione shares its orbit with a small **co-orbital** satellite, Helene.

direct (prograde) (1) Describing orbital or rotational motion in the same direction as the Earth's motion: counterclockwise as seen from above the Sun's north pole. The orbital motion of all planets and most satellites is direct. Orbital or rotational motion is direct if the orbital or axial inclination is less than 90°. COMPARE **retrograde** (1).

direct (prograde) (2) Describing the regular, west-to-east motion of Solar System bodies as seen from Earth relative to background stars. COMPARE **retrograde** (2).

disconnection event The shearing of a comet's ion tail in response to changes in the local solar wind. Disconnection is usually followed by the development of a new ion tail in a slightly different orientation.

Discovery A series of NASA space probes built on a "faster, better, cheaper" principle: low-cost missions that can be designed and launched quickly. The first was the **NEAR Shoemaker** probe; another was **Mars Pathfinder**.

disk galaxy A name given to any type of **galaxy** in the form of a thin disk of stars which are in roughly circular orbits around the center; the brightness of the galaxy tails off with distance from the center. Disk galaxies include spiral galaxies and lenticular galaxies.

▲ **Dione** This Voyager image clearly shows Dione's uneven surface. Central peaks can be seen in the larger craters.

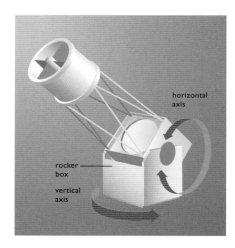

▲ **Dobsonian telescope**
This simple altazimuth mounting
has enabled many amateur
astronomers to construct large-
aperture instruments comparatively
cheaply. The Dobsonian offers
considerable advantages of
portability.

dispersion The separation of white light into its constituent colors by a lens, prism or diffraction grating. This happens because the different wavelengths of light are refracted more at the red end than at the violet end of the spectrum. Dispersion has to be countered in telescope lenses to avoid **chromatic aberration**, but it is exploited in astronomical instruments such as the **spectrograph** to study the spectra of stars.

distance modulus The difference between the apparent magnitude (*m*) and the absolute magnitude (*M*) of a star. It is a measure of the star's distance (*r*) in parsecs:

$$m - M = 5(\log r) - 5$$

diurnal motion The apparent daily motion of celestial bodies across the sky from east to west. It is a result of the Earth's axial rotation.

diurnal parallax SEE **parallax**.

Dobsonian telescope A Newtonian reflecting telescope on a simple form of **altazimuth** mount. The use of lightweight materials makes it possible to build a low-cost, large-aperture telescope that is portable, making the Dobsonian increasingly popular with amateur observers. A typical Dobsonian is mounted at its base, on a "rocker-box" assembly which rotates on a horizontal baseplate. It is named after its inventor, the American amateur John Dobson (1915–2014).

Dollond, John (1706–61) English optician. Formerly a silk-weaver, in 1752 he joined the optician's business founded by his son Peter Dollond (1730–1820). He found that the deviation of light rays could be achieved without **dispersion**, and combined crown and flint glass to make **achromatic lenses**. These he used in commercially successful refracting telescopes, which were further refined by Peter. John also invented the **heliometer**, a refractor modified so as to measure small angular distances. Peter Dollond was the first to make an achromatic triplet lens.

Dominion Astrophysical Observatory An observatory near Victoria, BC, Canada, opened in 1917. It has 1.83 m (72-inch) and 1.22 m (48-inch) reflecting telescopes. The Dominion Radio Astrophysical Observatory, at Penticton, BC, has a 26 m (85 ft) radio dish.

Doppler effect The apparent increase in frequency (and decrease in wavelength) of radiation from a source moving toward the observer, or the similar decrease in frequency (increase in wavelength) of radiation from a source moving away. In astronomy, its importance is that it explains the displacement of spectral lines. If a celestial object is moving toward the Earth, its light will be shifted to

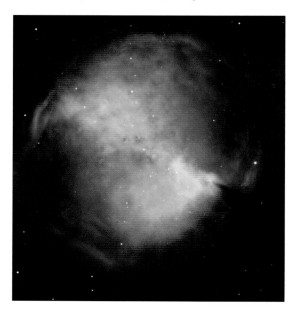

▶ **Dumbbell Nebula** The brightest of the planetary nebulae, the Dumbbell takes its name from the two prominent lobes of material to either side of the central, illuminating star.

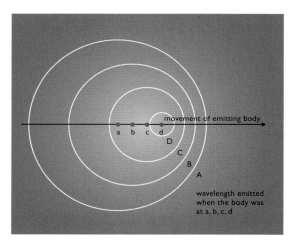

▲ **Doppler effect** Electromagnetic waves emitted from a moving point source appear compressed ahead of, and more spread out away from, the direction of motion. The result is that, for light, the wavelength is shortened (blueshifted) ahead of a source approaching the observer and lengthened (redshifted) behind a receding source such as a distant galaxy.

shorter (bluer) wavelengths. But if is moving away, then the shift will be toward longer (redder) wavelengths – a **redshift**.

This *Doppler shift*, as it is called, makes it possible to detect and measure relative motion in the line of sight, in other words **radial velocity**, and rotation of celestial objects. The redshifts of distant galaxies provided the first evidence that the Universe is expanding. Although galaxies and quasars do exhibit their own individual motions, the redshift is predominantly caused by the expansion of space itself. The effect is named after Austrian physicist Christian Doppler (1803–53), who published the first detailed discussion of it.

Dorado A southern constellation, representing a goldfish. It contains the main part of the Large **Magellanic Cloud**, including the **Tarantula Nebula** and the highly luminous eruptive variable star S Doradus. Alpha, its brightest star, is magnitude 3.3.

Double Cluster Two large star clusters in the constellation Perseus, visible to the naked eye. They bear the designations NGC 869 and NGC 884, and are also known as h and χ (chi) Persei. They lie about 7200 and 7500 light years away.

double star Two stars that appear close together in the sky. There are two types of double star. If the stars are genuinely close together in space and are connected by gravity they are known as a **binary star**. Two stars which are quite distant from each other, but appear close together as a result of chance alignment, are known as an **optical double**.

doublet A lens with two components, which are either cemented together or separated by an air-gap. This construction is used to reduce **chromatic aberration**.

Draco A long, winding northern constellation representing the dragon slain by Hercules, extending between Ursa Major and Ursa Minor, with the dragon's head near the star Vega. Gamma (Eltanin), magnitude 2.2, is its brightest star. Alpha (Thuban), magnitude 3.7, was the north pole star in ancient times. Draco is a constellation largely devoid of interest, but one of the stars in the head, Nu, is a double with a separation of 62 arc seconds; each component is of magnitude 4.9.

draconic month The time taken for one revolution of the Moon around the Earth relative to its ascending **node**. It is equal to 27.21222 days of **mean solar time**.

Draconids SEE **Giacobinids**.

Draper, Henry (1837–82) American pioneering spectroscopist. In 1872 he obtained the spectrum of Vega, the first time this had been done for a star other than the Sun. After his death, his widow made available the funds that made possible the **Henry Draper Catalog**. His father, the English scientist John William Draper (1811–82), was a pioneering astrophotographer who took the first images of the Moon and the Sun's infrared spectrum.

Dreyer, Johan Ludvig Emil (1852–1926) Danish astronomer, who from 1874 lived and worked in Ireland, initially at Lord **Rosse**'s observatory. He compiled the *New General Catalogue of Nebulae and Clusters of Stars* (SEE **NGC**) and its two supplementary *Index Catalogues*.

drive A means of moving a telescope to compensate exactly for the rotation of the Earth, so keeping the object being observed in the field of view. **Equatorial telescopes** are driven about their polar axis to track objects being observed. Computer control allows professional **altazimuth** mounted instruments to be driven simultaneously about their vertical and horizontal axes.

Dumbbell Nebula A large planetary nebula, also designated M27, about 1250 light years away in the constellation Vulpecula. It was named by Lord **Rosse** for its hourglass shape. The Dumbbell is visible in binoculars.

dwarf galaxy A galaxy that is much smaller than a normal galaxy and is of low luminosity. Such galaxies are usually elliptical or irregular. There are several dwarf ellipticals in the **Local Group** of galaxies.

dwarf nova A member of a class of irregular variable stars whose light curves resemble those of novae. Their luminosity stays the same for long periods, then rapidly increases, and finally returns slowly to normal. **U Geminorum stars** and **Z Camelopardalis stars** are the main types. Dwarf novae are close binary stars in which one component is a white dwarf and the other is a subgiant or dwarf.

dwarf planet A body in orbit around the Sun with sufficient mass for its own gravity to have pulled itself into a nearly round shape, but which is not massive enough to have cleared away other objects from around its orbit. This category of object was introduced by the International Astronomical Union in 2006 and includes **Pluto**, which was previously regarded as a major planet. Other bodies that fall within the definition of a dwarf planet are the asteroid **Ceres**, and the largest trans-Neptunian objects such as **Eris**, Haumea and Makemake.

dwarf star The commonest type of star in the Galaxy, constituting 90% of its stars and 60% of its mass. Dwarfs are also known as **main-sequence** stars, from their position on the **Hertzsprung–Russell diagram**. The term "dwarf" refers to luminosity rather than size, so dwarf stars should be thought of as normal rather than diminutive. The Sun is a typical dwarf, roughly midway in the range of properties of dwarf stars.

The hottest and most massive dwarfs (spectral types O and B) have relatively short lifetimes, a few hundred million years or less, so they are rare in the Galaxy. The lifetimes of the lowest-mass dwarfs (type K and M) are so long that none have evolved away from the main sequence since the Galaxy formed. Stars below 0.8 solar masses are known as **red dwarfs**, and those below 0.08 solar masses as **brown dwarfs**. SEE ALSO **stellar evolution**, **white dwarf**.

dynamical parallax The distance to a **binary star** determined from a relationship between the known masses of the components, the size of the orbit and the period of revolution.

▲ **dwarf galaxy** NGC 5477 is a dwarf irregular galaxy in Ursa Major, part of the same group that includes the spiral galaxy M101. The stars comprising NGC 5477 are so thinly spread that background galaxies can be seen through them.

Eagle Nebula A bright nebula surrounding a cluster of new-born stars, collectively designated M16 or NGC 6611, in the constellation Serpens. In the nebula (known also as IC 4703) are dark columns of gas and dust in which stars are being born. The nebula and stars are about 6500 light years away.

early-type star A hot star of spectral type O, B or A. The name was given when it was thought that the sequence of spectral types was an evolutionary one, hot stars being the youngest and cool stars the oldest. Although that is now known not to be so, the term has remained in use.

Earth The third major planet from the Sun, and the largest of the four inner, or terrestrial planets. Seventy per cent of the surface is covered by water, and it is this and the Earth's average surface temperature of 13°C (286 K) that make it suitable for life. The continental land masses make up the other 30%. Our planet has one natural satellite, the **Moon**. The main data for the Earth are given in the second table on the following page.

Monitoring the propagation of seismic waves from earthquakes has revealed the Earth's internal structure. Like all the terrestrial planets, the Earth has a dense *core* rich in iron and nickel, surrounded by a *mantle* consisting of silicate rocks. The thin, outermost layer of lighter rock is called the *crust*. Continental crust can be as much as 50 km (30 miles) deep, but oceanic crust has an average depth of only 10 km (6 miles). The mantle extends to a depth of 2890 km (1795 miles). The outer part of the core, down to 5150 km (3200 miles), is molten, but the inner core, 2460 km (1525 miles) in diameter, is solid because of the greater pressure there. The temperature and pressure at the Earth's center are estimated to be about 4000°C and 4 million bars (i.e. 4 million times the pressure at the surface). The solid inner core rotates two-thirds of a second faster than the layers outside it, and this, together with currents in the molten outer core, gives rise to the Earth's magnetic field.

The crust and the uppermost layer of the mantle together form the *lithosphere*, which consists of eight major and more than 20 minor tightly fitting slabs called *plates* that move slowly with respect to one another, and are supported on a semi-molten layer of mantle called the *asthenosphere*. Plates grow at *mid-oceanic ridges*, boundaries between adjoining plates where molten rock rises and solidifies into new oceanic crust, pushing the plates apart; an example is the Mid-Atlantic Ridge. Where two plates meet, one may be forced below the other, descending into the mantle and melting. This process of *subduction* produces earthquakes and volcanoes at the edge of the plate above. Where plates collide and no subduction occurs, they crumple up to form folds that develop into mountain ranges. All these interactions are known collectively as *plate tectonics*.

Because of tectonics, most of the Earth's surface is young compared with the planet's age of 4.6 billion years. The oldest rocks belong to the *Precambrian shields*, areas over 600 million years old, as in parts of Canada; some individual rocks have been dated to more than 4 billion years ago.

The Earth's atmosphere consists of 78% (by volume) nitrogen, 21% oxygen, 0.9% argon,

▶ **Eagle Nebula** One of the Eagle Nebula's dust pillars is shown in this Hubble Space Telescope image. These dark fingers of nebulosity are associated with star formation.

EARTH: LAYERS OF THE EARTH'S ATMOSPHERE				
Layer	Upper boundary	Altitude (km)	Temperature (°C)	Pressure (mbar)
Troposphere	Tropopause	20*	−80*	250
Stratosphere	Stratopause	50	0	0.9
Mesosphere	Mesopause	80	−90	0.007
Thermosphere	—	500	1500	10^{-9}

The altitude, temperature and pressure are values at the upper boundary.
* Above the equator. At the poles the altitude of the tropopause is 10 km (6 miles).

EARTH: DATA	
Globe	
Diameter (equatorial)	12,756 km
Diameter (polar)	12,714 km
Density	5.52 g/cm^3
Mass	5.974×10^{24} kg
Volume	1.083×10^{12} km^3
Sidereal period of axial rotation	23h 56m 04s
Escape velocity	11.2 km/s
Albedo	0.37
Inclination of equator to orbit	23° 26′
Surface temperature (average)	290 K
Orbit	
Semimajor axis	1 au = 149.6×10^6 km
Eccentricity	0.0167
Inclination to ecliptic	0° (by definition)
Sidereal period of revolution	365.256d
Mean orbital velocity	29.8 km/s
Satellites	1

and 0.03% carbon dioxide, other gases such as neon making up the remaining 0.07%. In addition there can be up to 3% water vapor, depending on geographical location and weather conditions. The atmosphere produces a **greenhouse effect**, raising the surface temperature by 35 K.

The atmosphere is divided into a number of layers according to the way in which its temperature varies with altitude (SEE the first table above). The *troposphere* contains most of the atmosphere, and is where weather systems operate. The *stratosphere* contains the *ozone layer*, which absorbs high-energy ultraviolet radiation from the Sun. **Meteors** occur just above the mesosphere, and **aurorae** in the thermosphere. The thermosphere is extremely rarefied, and its high temperature indicates the high kinetic energy of its molecules, rather than its heat content. Ionized atoms and molecules in the mesosphere and thermosphere constitute the **ionosphere**. Above the thermosphere is the exosphere, which contains the Earth's **magnetosphere** and **Van Allen Belts**, and merges into interplanetary space.

earthgrazer An alternative name for a **near-Earth asteroid**.

earthshine Sunlight reflected from the Earth on to the part of the Moon's near side that is in the Sun's shadow. It is easiest to see when the Moon is a thin crescent, when it appears as a pale grayish light over the remainder of the lunar disk.

eccentric (1) Describing an orbit with a high **eccentricity**.

eccentric (2) SEE **Ptolemy**.

eccentric anomaly SEE **anomaly**.

eccentricity (symbol e) One of the elements of an **orbit**. It indicates how much an elliptical orbit departs from a circle. The eccentricity is found by dividing the distance between the two foci of the **ellipse** by the length of its major axis. A circle has an eccentricity of 0, and a parabola has an eccentricity of 1.

echelle grating A **diffraction grating** designed to yield spectra of high **dispersion** and detail. Its lines are ruled farther apart than on an ordinary grating, and are shaped so as to produce high resolution over a narrow band of wavelengths at angles of illumination greater than 45°. Such a grating is used in an *echelle spectrograph*.

eclipse The partial or total obscuration of the light from a celestial body as it passes through the shadow cast by another body. A body may be eclipsed by the passage of another body between it

and the observer, as in a **solar eclipse**, or by the intervention of another body between it and the source of the light it reflects, as in a **lunar eclipse**. The eclipse of a star by the Moon or by a planet or other Solar System body is called an **occultation**. SEE ALSO **annular eclipse**, **transit**.

eclipse year The interval between successive passages of the Sun through a given node of the Moon's orbit, equal to 346.62003 days of mean solar time. It differs significantly from other types of year because the nodes of the Moon's orbit move westward by over 19° per year (SEE **regression of the nodes**). Nineteen eclipse years are equal to 6585.78 days, almost exactly the same length of time as a **saros**.

eclipsing binary A type of **binary star** in which the orbital plane of the two stars is viewed almost edge-on, leading to mutual eclipses and consequent variations in the two stars' combined light output. Two eclipses may be expected during an orbital cycle, one usually causing a bigger drop in light than the other since the two stars are rarely equal in size or brightness. The deeper minimum is called the primary minimum, and the shallower minimum is the secondary minimum. If the stars are very unequal in brightness the secondary minimum may hardly be noticeable.

Eclipsing binaries are classified into three types on the basis of their light curves: **Algol** type (EA), **Beta Lyrae stars** (EB) and **W Ursae Majoris stars** (EW). The EA types have a constant light level outside eclipses, but EB and EW types show continuous variability. This is because one or both stars in the binary are elliptical in shape, having expanded to fill their **Roche lobe**. (If only one star has filled its lobe, the system is called *semidetached*; a system in which neither lobe is filled is known as *detached*.) The EW stars show more or less equal depths of primary and secondary minima. In such systems, both stars are thought to be filling their Roche lobes and are thus touching each other. They are therefore known as *contact binaries*, and in most systems the two stellar cores are surrounded by a common convective envelope.

ecliptic The apparent yearly path of the Sun against the background stars. It intersects the celestial equator at the two equinoxes. The ecliptic is really the projection of the Earth's orbit around the Sun on to the celestial sphere. Because of the tilt of the Earth's axis the ecliptic is inclined to the celestial equator by an angle of about $23\frac{1}{2}$°; this tilt is known as the *obliquity of the ecliptic*. The *ecliptic poles* are the points 90° from the ecliptic, and lie in the constellations Draco and Dorado.

ecliptic coordinates SEE **coordinates**.

ecliptic limits The greatest angular distance the Sun or Moon can be from the Moon's nodes for which a **solar eclipse** or **lunar eclipse** can take place. The limits are 37° for the Sun at a solar eclipse, and 24° for the Moon at a lunar eclipse.

E corona SEE **corona**.

ecosphere The shell-shaped region of space surrounding the Sun or another star within which the temperature is such that any suitable planet would be capable of sustaining life. The dimensions of the ecosphere vary greatly with the type of star. The ecosphere of a small dwarf star would be narrow and lie close in, whereas that of a luminous giant would be broad and a considerable distance from the star.

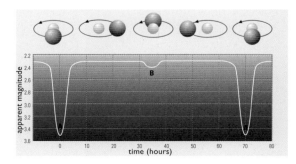

▲ *eclipsing binary* The light curve (bottom) of an Algol-type binary can be accounted for by partial eclipses of each star by the other (top). At the deep primary minimum, the brighter, smaller star is largely covered by the fainter. At the secondary minimum, the brighter star passes in front of the fainter.

Eddington, Arthur Stanley (1882–1944) English astronomer and physicist. He pioneered the use of atomic theory to study the internal constitution of stars, and explained the role of radiation pressure in preventing stars from collapsing under gravity. Among his many discoveries were the mass–luminosity relationship, and the fact that white dwarfs contain degenerate matter. He supported and popularized **Einstein**'s theory of relativity, and in 1919 he obtained experimental proof of the theory's prediction that gravity bends light by measuring star positions close to the Sun during a solar eclipse.

Eddington limit The maximum ratio of luminosity to mass that a star can have before radiation pressure overcomes the gravitational force holding it together, and the star consequently blows apart. It gives a theoretical limit to stellar mass of between about 50 and 100 solar masses.

Edgeworth–Kuiper Belt SEE **Kuiper Belt**.

E-ELT SEE **European Extremely Large Telescope**.

effective temperature (symbol T_{eff}) The temperature a star would have if its output were **black-body radiation** at the same energy and at the same wavelengths as the star. As stellar spectra approximate closely to black-body spectra, a star's effective temperature is a good approximation to its actual surface temperature.

Effelsberg Radio Telescope A radio telescope near Bonn, Germany, operated by the Max-Planck-Institut für Radio-astronomie, completed in 1970. With a diameter of 100 m (328 ft), it is the world's second largest fully steerable radio dish. It is used either alone or as part of a **very long baseline interferometry** network.

Einstein, Albert (1879–1955) German–Swiss–American theoretical physicist. His work had an enormous impact on 20th-century science. For astronomy, and particularly in cosmology, his theories of relativity had profound consequences. The special theory of relativity (1905) gave the relation $E = mc^2$ between mass and energy; the general theory of relativity (1916) extended the theory to encompass gravitation. The general theory has been borne out, for example, by its explanation of the advance of the perihelion of **Mercury**, the curvature of light by a gravitational field (SEE **Eddington, Arthur Stanley**), and the **gravitational redshift** of spectral lines. In the 1920s relativity made Einstein world famous, but he never again matched the scientific work of his early years. He spent much time unsuccessfully seeking a unified field theory which would link relativity with electromagnetic forces. SEE ALSO **grand unified theory**.

Einstein–de Sitter universe SEE **flat universe**.

Einstein ring A circular image of a distant source produced by a **gravitational lens**. Albert Einstein realized that if the lensing object has its mass concentrated in a single point and lies directly in our line of sight to a distant source, then in theory the source's image should be a perfect circle. In practice this is rarely seen, but incomplete circles – arcs – are often observed.

ejecta SEE **rays**.

Elara A small **satellite** in Jupiter's intermediate group, discovered by Charles Perrine in 1905.

electromagnetic radiation (em radiation) The emission and propagation of energy in the form of periodic waves that can travel through a vacuum. It originates when charged atomic particles are accelerated, and consists of oscillating electric and magnetic fields. There is a continuum of electromagnetic radiation – from long-wavelength radio waves of low frequency and low energy, through visible light waves, to short-wavelength X-rays and gamma rays of high frequency and high energy – known as the *electromagnetic spectrum*. The speed at which em radiation travels in a vacuum (c) is given by its wavelength (λ) times its frequency (ν):

$$c = \lambda\nu$$

This speed is constant, and is commonly known as the speed of light (SEE **light, speed of**). In the *quantum theory*, an em wave (particularly if its wavelength is that of visible light or shorter) may also be regarded as a stream of particles called *photons*, each having an energy (E) related to the frequency (or wavelength) of the radiation:

$$E = h\nu = hc/\lambda$$

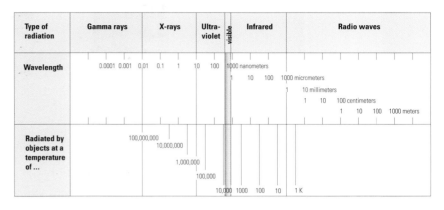

Type of radiation	Gamma rays	X-rays	Ultra-violet	visible	Infrared	Radio waves
Wavelength	0.0001 0.001 0,01 0.1 1 10 100		1000 nanometers 1 10 100		1000 micrometers 10 millimeters 1 10 100 centimeters 1 10 100 1000 meters	
Radiated by objects at a temperature of ...	100,000,000 10,000,000 1,000,000 100,000		10,000 1000 100 10 1 K			

▲ *electromagnetic radiation*
Spanning from short, gamma-ray wavelengths to long-wavelength radio emissions, the electromagnetic spectrum covers a wide range. Selection of appropriate wavelengths allows different processes and classes of objects to be observed. The visible region comprises only a tiny part of the electromagnetic spectrum.

With X-rays and gamma rays, this energy, expressed in **electronvolts**, is a more important property than their wavelength. Photons also possess momentum, which means they can exert a pressure, called **radiation pressure**.

Almost all of our information about the Universe has been learned from studies of the electromagnetic radiation that reaches us from celestial objects. Much of the radiation that arrives at the Earth is absorbed by its atmosphere, but there are some bands of the electromagnetic spectrum that can reach the surface of the Earth and be studied by Earthbound instruments. The two main ones are known as the *optical window* and the *radio window*. Other wavelength bands of cosmic radiation are detected and analyzed by instruments carried above the atmosphere by satellites or rockets.

electronvolt (symbol eV) A unit of energy used mainly in atomic physics for the energies of atomic particles. One electronvolt is defined as the kinetic energy acquired by a particle carrying a charge equal to that on one electron when it passes through a potential difference of 1 volt. 1 eV is equivalent to 1.6021×10^{-19} joules, and, from the equation $E = h\nu$ (SEE **electromagnetic radiation**), corresponds to a frequency of 2.42×10^{14} hertz, or a wavelength of 1240 nm. An electron with a kinetic energy of 1 eV has a velocity of about 580 km/s.

element, chemical One of the basic materials of which everything in the Universe is composed. Elements are chemically homogeneous substances – that is, they cannot be decomposed into any simpler substances by chemical means. The atoms of a particular element all have the same number of protons in their nucleus, but the number of neutrons can vary, giving rise to different *isotopes* of the same element. The number of known elements is over 110. Of these, 98 occur in nature, and the remainder have been prepared artificially, in laboratories. SEE ALSO **cosmic abundance**, **nucleosynthesis**.

element, orbital SEE **orbit**.

ellipse A closed curve, for which the sum of the distances of any point on it from two points within it – the *foci* (singular *focus*) of the ellipse – is a constant (SEE the diagram). In mathematics the ellipse is a type of conic section, so called because it is one of the intersections of a plane with a cone. The ellipse is important in astronomy because the closed **orbits** of planets, satellites, companion stars, comets, and so on are all close approximations to ellipses (**perturbations** make orbits depart from true ellipses) – SEE **Kepler's laws**. Important parameters of the ellipse, and therefore of orbits, are the **semimajor axis** (a) and the **eccentricity** (e). The shape of the ellipse is governed by its eccentricity, which is given by $e = c/a$, where c is the distance from the center of the ellipse to one of its foci. For an ellipse, e is always less than 1.

elliptical galaxy A **galaxy** that is ellipsoidal in shape. Elliptical galaxies are composed of old stars and contain little gas or dust. They are often the central dominant galaxies in very rich clusters, and many are powerful radio sources. Ellipticals are classified according to their degree of ellipticity (flattening), and given a designation from E0 to E7. They range widely in size,

▼ *ellipse The principal features of an ellipse are its two foci, its semimajor axis (a), its short (minor) axis, its eccentricity (e), and the distance of the foci from the center of the ellipse (c). In the case of an ellipse used to define a planetary orbit, the Sun occupies one of the foci while the other is empty. The planets' orbits are reasonably close to circular, but for most comets the orbital ellipse is eccentric, with a very long major axis.*

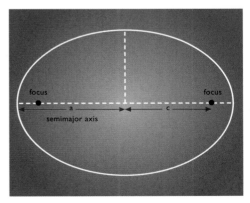

focus   focus
a   c
semimajor axis

▲ **Enceladus** *While heavily cratered in some regions, the surface of Enceladus has also been smoothed over by episodes of cryovolcanism. Various types of terrain can be seen in this image from Cassini.*

from 100 million stars for *dwarf ellipticals* to 10 million million stars for *supergiant ellipticals*.

ellipticity Another word for **oblateness**.

elongation The angular distance between the Sun and a planet or other Solar System body orbiting the Sun (or the angular distance between the Sun and the Moon). More accurately, it is the difference in the two bodies' celestial longitudes, measured in degrees. A planet with an elongation of 0° is at **conjunction**; at 90° or 270° it is at **quadrature**; and at 180° it is at **opposition**. The inferior planets, Mercury and Venus, when at their maximum angular distance from the Sun, are said to be at *greatest elongation east* when following the Sun (setting at their latest time after it) and at *greatest elongation west* when preceding the Sun (rising at their earliest time before it). SEE the diagram at **conjunction** on page 229.

emersion The reappearance of a celestial body from the shadow of another after it has been eclipsed or occulted.

emission nebula A cloud of gas in space that emits its own light. Ultraviolet radiation from nearby stars heats the gas of the nebula, causing its atoms to become ionized (lose their electrons). These electrons gain energy from the ultraviolet radiation. When they eventually recombine with the atoms, they re-emit the energy they gained, some of it at visible wavelengths. The free electrons in the gas also lose energy in the form of radio waves, so emission nebulae can also be detected by radio telescopes.

Interstellar gas is mostly hydrogen, which is easily ionized by ultraviolet light. The symbol for ionized hydrogen is H II, and such emission nebulae are therefore also called **H II regions**. Hydrogen gives off its strongest light in the red, so emission nebulae appear red in images. However, to the eye they often look green because the eye responds more readily to a prominent pair of spectral lines of this color emitted by oxygen.

Hot stars that emit ultraviolet light are usually young, having recently formed from the gas and dust cloud around them. Thus the association between stars and gas clouds is not by chance. A famous example is the **Orion Nebula**, with the **Trapezium** stars inside it. Emission nebulae are very tenuous. Typically, every gram of material is spread through a volume of a million cubic kilometers.

emission spectrum The **spectrum**, consisting of a series of bright lines, produced by a highly energetic source. Energy in the form of heat or electromagnetic radiation can be absorbed by an electron orbiting the atomic nucleus, causing it to jump to a higher orbit. When it returns to its old orbit, the electron emits radiation at a characteristic wavelength. The series of lines made up of emissions from different types of atom in the source is the emission spectrum. The hottest stars have many strong emission lines in their spectra. SEE ALSO **forbidden lines**.

Enceladus A **satellite** of Saturn, diameter 498 km (309 miles), discovered in 1789 by William **Herschel**. It is the most reflective body of any significant size in the Solar System, with an albedo approaching 1. Its surface consists of older, cratered terrain and more recently formed plains traversed by straight and curved grooves, and there are indications that tectonic and volcanic processes may still be operating. Tidal interaction with the larger satellite **Dione** probably explains why such a small body has been so geologically active. Material from Enceladus may be the source of Saturn's tenuous E Ring, in which the satellite is embedded; the satellite does possess a tenuous atmosphere.

Encke, Johann Franz (1791–1865) German mathematician and astronomer. In 1818 he computed the orbit of **Encke's Comet**, and predicted its return, and in 1837 he was the first to record the division in **Saturn**'s rings now named after him. Encke also calculated the **solar parallax** from measurements of the 1769 transit of Venus.

Encke's Comet (Comet 2P/Encke) A periodic **comet** discovered by Pierre Méchain in 1786 and independently by Jean Louis Pons in 1818. In 1819 **Johann Encke** computed its orbit and period, and proved that it was identical with a comet that had previously been observed on three returns. It was the second periodic comet to be discovered, and has the shortest-known period, at 3.3 years. It is the parent of the **Taurid** meteor shower.

ephemeris (plural **ephemerides**) A table giving the predicted positions of a celestial object, such as a planet, satellite or comet, at given intervals.

epicycle SEE **Ptolemy**.

Epimetheus A small inner **satellite** of Saturn discovered during the Voyager missions. It is **co-orbital** with **Janus**.

epoch (1) A point in time used as a fixed reference for comparison of astronomical data in, for example, star catalogs. **Precession** and the **proper motions** of stars cause their positions on the celestial sphere to change gradually, and observations made over any considerable period of time have to be reduced to a common epoch for them to be comparable. SEE ALSO **standard epoch**.

epoch (2) One of the elements of an **orbit**, defined as the time of perihelion passage.

Epsilon Aurigae An **eclipsing binary** star with a very long period, 27 years, magnitude 2.9 at maximum and a minimum of 3.9 during eclipse. The primary component is a supergiant of type A8. The eclipses last 700 days, the total phase taking 400 days; the last one was in 2009–11; the next will start in 2036. No light is detected from the secondary component, which is thought to be surrounded by an obscuring disk of gas and dust which causes the eclipses; the secondary itself may actually be a close binary consisting of two white dwarfs. Epsilon Aurigae is estimated to lie around 2000 light years away, making it one of the most distant of the brighter naked-eye stars.

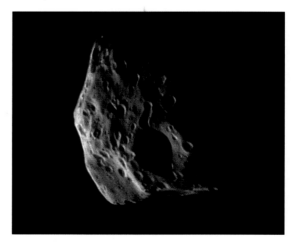

▲ **Epimetheus** *The Cassini spacecraft obtained this image on 30 March 2005. The large crater just below center is called Hilairea, and has a diameter of 33 km (21 miles). The moon itself is 116 km (72 miles) across.*

equant SEE **Ptolemy**.

equation of the center An irregularity in the motion of a body in an elliptical orbit. It is the difference between the true anomaly and the mean anomaly (SEE **anomaly**).

equation of time The difference between mean solar time, as shown on a clock, and apparent solar time, as shown on a sundial, caused by the combined effect of the eccentricity of the Earth's orbit and the **obliquity of the ecliptic**. The difference is zero four times a year: on 15 April, 14 June, 1 September and 25 December. The maximum difference is 16 minutes and occurs in early November.

equatorial coordinates SEE **coordinates**.

equatorial telescope A telescope mounted such that it has one axis, the *polar axis*, parallel to the Earth's axis and therefore pointing to the celestial pole, and the other axis, the *declination axis*, at right angles to the polar axis and therefore parallel to the plane of the Earth's equator. Once an object is in the field of view the telescope needs only to be moved about the polar axis to follow the object. If the telescope is driven at the sidereal rate to counter the apparent rotation of the celestial sphere, the object will remain in the field of view.

There are various forms of equatorial mounting. The *German mounting*, so named because the first of the type was made by Joseph von **Fraunhofer**, is still used for small instruments. The telescope is attached at its center of mass to one end of the declination axis, which pivots about the top end of the polar axis; balance is achieved by attaching counterweights to the other end of the declination axis. Large, professional reflecting telescopes before the age of computer-controlled **altazimuth** mountings required massive equatorial mounts in which the telescope could be supported on both sides of the declination axis and sometimes at the top and bottom of the polar axis, yet still have enough freedom of movement to access most of the sky. Such designs included the *fork mounting* and the *horseshoe mounting*.

A departure from traditional equatorial mounts is the equatorial platform or *Poncet mounting*. This is a flat surface on whose underside are casters that sit on inclined planes. It allows amateur instruments like the traditionally altazimuth **Dobsonian telescope** to be given an equatorial capability.

equinoctial colure The **great circle** that passes through the north and south celestial poles and the equinoxes (which are also, in this context, called the equinoctial points).

equinox Either of the two points at which the Sun crosses the celestial equator, or the times at which these events occur. The *vernal equinox* (or *spring equinox*) is when the Sun crosses the equator from south to north on or near 21 March each year. The point at which this happens is called the *First Point of Aries*. The other equinox is the *autumnal equinox*, which is on or near 23 September, when the Sun crosses the equator from north to south. The point at which this happens is called the *First Point of Libra*. At the equinoxes, the Sun rises due east and sets due west.

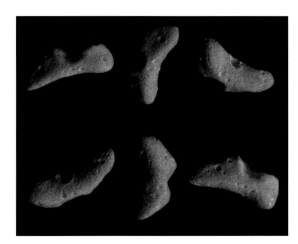

▲ **Eros** The rotation of Eros can be seen in this series of images obtained from the NEAR Shoemaker spacecraft as it made its close approach to the asteroid in February 2000.

Equuleus A very small and obscure constellation, representing a foal, next to Pegasus. Its brightest star, Alpha (Kitalpha), is only of magnitude 3.9. It has no features of particular interest.

Eratosthenes (*c.* 275–195 BC) Greek geographer, mathematician and astronomer. He is best known for making the first scientific calculation of the Earth's circumference, for which he measured the altitude of the Sun at Alexandria on a day when he knew the Sun to be overhead at a place a considerable distance away. How accurately he knew this distance, and hence how accurate his result was, is uncertain, but the method was sound. Eratosthenes also compiled a star catalog and measured the **obliquity of the ecliptic**.

Erfle eyepiece A telescope eyepiece with low power and wide field of view – over 60°. It consists of three elements, at least one of which is a doublet. (After German optician Heinrich Erfle, 1884–1923.)

Eridanus One of the largest constellations in the sky, representing a river; it extends from near Rigel in Orion to the south polar area. **Achernar** is its brightest star. Theta (Acamar) is a fine double, with components of magnitude 3.4 and 4.3. Epsilon, magnitude 3.7, is a nearby K2 main-sequence star, 10.5 light years away.

Eris A large trans-Neptunian object, discovered in 2005. It has a diameter of about 2400 km (1500 miles), slightly larger than Pluto. Eris orbits the Sun every 560 years in an eccentric orbit ($e = 0.44$) ranging in distance between 38 and 98 astronomical units. It has a satellite called Dysnomia about one-tenth its diameter. Eris is classified as a **dwarf planet**.

Eros Asteroid no. 433, discovered by Gustav Witt in 1898. It measures 33×13 km (21×8 miles) and rotates in 5.3 hours. Its orbit crosses that of Mars, and it is the second largest of the **Amor asteroids**. The **NEAR Shoemaker** probe entered orbit around Eros in February 2000 and studied the asteroid for 12 months.

eruptive variable Any of a broad group of intrinsic variable stars that show abrupt changes in magnitude with no obvious period. The class includes many different types, such as **flare stars**, **R Coronae Borealis stars**, **shell stars**, **T Tauri stars**, certain young stars just settling on to the **main sequence**, as well as various other **irregular variables**, some associated with nebulosity.

ESA ABBREVIATION FOR **European Space Agency**.

escape velocity The minimum velocity that a body, such as a rocket or a space probe, must attain to overcome the gravitational attraction of a larger body and leave on a trajectory that does not bring it back again. It depends on the mass (m) and radius (r) of the larger body, and is given by $\sqrt{(2GM/r)}$, where G is the gravitational constant (see **gravity**). It is escape velocity rather than mass that determines the gases that a body is able to retain as an atmosphere.

Eskimo Nebula A bright, compact planetary nebula, designated NGC 2392, in the constellation Gemini. Lying at a distance of 3000 light years, it was discovered by William **Herschel** in 1787. The Eskimo Nebula takes its name from its telescopic appearance of a "face" surrounded by a hazy periphery resembling a fur hood.

ESO ABBREVIATION FOR **European Southern Observatory**.

Eta Aquarids A **meteor shower** active from 24 April to 20 May, with a broad peak centered on May 4–5 reaching a **ZHR** of about 50. Its radiant is in Aquarius, near the star Eta Aquarii. Best seen from southerly latitudes, the shower is, like the Orionids, associated with **Halley's Comet**.

Eta Carinae A peculiar variable star lying within the nebula NGC 3372 – the Eta Carinae Nebula – in the constellation Carina, about 8000 light years away. It was originally cataloged at third magnitude, but from 1833 began to vary irregularly, becoming at its brightest second only to Sirius. It then faded to just below naked-eye visibility. Material thrown off in the outburst has formed the *Homunculus Nebula*, so called because to early observers its shape resembled a human figure, which obscures its light. Eta Carinae is

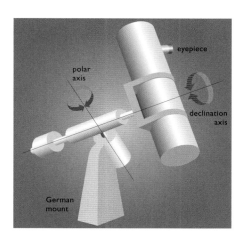

▲ *equatorial telescope*
The equatorial mounting allows a telescope to be driven so that it can follow the apparent motion of the stars due to Earth's rotation. The equatorial is based around two main axes. The polar axis, around which the driving motor turns the telescope, is aligned parallel to Earth's axis of rotation. North and south movement is made along the declination axis.

ESCAPE VELOCITIES OF BODIES IN THE SOLAR SYSTEM	
Body	Escape velocity (km/s)
Sun	617.0
Mercury	4.4
Venus	10.4
Earth	11.2
Moon	2.4
Mars	5.0
Jupiter	59.5
Saturn	35.5
Uranus	23.5
Neptune	21.3
Pluto	1.1

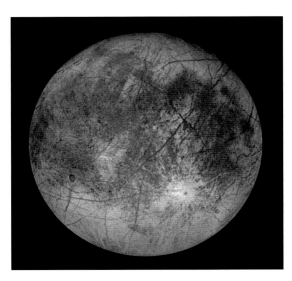

► **Europa** *Darker patches on Europa's icy surface are thought to be caused by minerals oozing up from a subsurface ocean. The bright impact crater at lower right is about 25 km (15 miles) wide. Long dark fractures can also be seen.*

the brightest infrared source in the sky, showing that it is still shining as brightly as in the 19th century. Studies of the light reflected from the dust cloud show that the central object varies in its emission lines with a period of 5.52 years. This suggests that the central object is in fact a binary system consisting of two stars, each of about 80 solar masses, orbiting each other at a distance of about 20 astronomical units.

Eudoxus of Cnidus (*c*. 408–355 BC) Greek mathematician and astronomer. His theory of 27 "homocentric" (centered on the Earth) spheres was the first ever attempt to account for the movements of the Sun, Moon and planets on what we would now call a properly scientific basis. This picture of "heavenly spheres" would survive for 2000 years. According to the Roman historian Pliny, Eudoxus fixed the length of the year at 365.25 days.

Europa The smallest of Jupiter's **Galilean satellites**, with a diameter of 3122 km (1940 miles), and the second closest to the planet. Its density of 2.97 g/cm^3 indicates that it is a predominantly rocky body. Europa's smooth crust of water ice is crisscrossed by a network of light and dark linear markings. There are very few craters, so old ones must have been removed by some form of geological activity.

A kind of ice **tectonics**, driven by internal heating resulting from tidal forces, could be operating on Europa. Images from the **Galileo** probe support this idea: they suggest that the dark lines may be deposits from geysers that erupted along long fissures, and that a layer of liquid water or "slush" about 100 km (60 miles) deep lies 15–25 km (10–15 miles) beneath the surface. Some scientists have suggested that a subsurface liquid environment on Europa might have favored the development of primitive life, though this remains a matter of debate. Europa orbits at the outer edge of **Io**'s plasma torus, in which energetic particles break down water molecules to release oxygen, producing a thin atmosphere. SEE ALSO **satellite**.

European Extremely Large Telescope (E-ELT) An optical/infrared telescope with a mirror 39 m (128 ft) in diameter, being built by the **European Southern Observatory** at Cerro Armazones, a peak in the Chilean Atacama Desert, about 20 km (12 miles) from ESO's **Very Large Telescope** on Cerro Paranal. Its mirror will consist of 798 hexagonal segments, each 1.4 m (55 inches) wide. The secondary mirror will be 4.2 m (165 inches) across. When complete in the 2020s it will be the largest optical telescope in the world.

European Southern Observatory (ESO) A consortium of 15 member states, including the UK, that operates telescopes at four sites in Chile: La Silla, home of a 3.6 m (142-inch) reflector opened in 1976 and also of the **New Technology Telescope**; Paranal, site of the **Very Large Telescope**; Chajnantor, where lies the **Atacama Large Millimeter Array**; and Cerro Armazones, where the **European Extremely Large Telescope** is being built. ESO's headquarters are at Garching near Munich, Germany.

European Space Agency (ESA) An organization of European nations to promote space research and technology for peaceful purposes, founded in 1975. ESA has 20 member states: Austria, Belgium, the Czech Republic, Denmark, Finland, France, Germany, Greece, Ireland, Italy, Luxembourg, the Netherlands, Norway,

Poland, Portugal, Romania, Spain, Sweden, Switzerland and the UK. Its headquarters are in Paris. ESA has designed and built the Ariane series of launch rockets, and operates a launch site at Kourou in French Guiana, on the Atlantic coast of South America.

EUVE ABBREVIATION FOR **Extreme Ultraviolet Explorer**.

evection An irregularity in the Moon's motion caused by perturbations by the Sun and planets. It amounts to a maximum of 1° 16′ in a period of 32 days.

Evening Star Name sometimes given to the planet Venus when it is visible in the west after sunset.

event horizon The boundary surface of a **black hole** from within which nothing can escape. Observers outside the event horizon can therefore obtain no information about the black hole's interior. The radius of the event horizon is known as the **Schwarzschild radius**.

exit pupil For a telescope, the image of the primary mirror or lens as seen through the eyepiece. The point at which this image is formed is the optimum position for the observer's eye, so for comfort it should not be too close to the eyepiece itself (SEE **eye relief**).

exobiology Another name for **astrobiology**.

ExoMars A two-part program to explore Mars being undertaken jointly by the **European Space Agency** and the Russian Federal Space Agency, Roscosmos. The first ExoMars mission, due for launch in January 2016, will consist of a Trace Gas Orbiter which will sample gases in the Martian atmosphere such as methane and a lander called Schiaparelli which will test landing techniques. A second mission, to be launched in May 2018, will include a rover to study the surface of Mars and look for possible signs of life.

exoplanet Another name for **extrasolar planet**.

Exosat A European Space Agency satellite for X-ray astronomy, launched on 26 May 1983. It operated for three years, studying objects such as supernova remnants, X-ray binaries and galactic nuclei.

expanding Universe SEE **cosmology**, **redshift**.

extinction SEE **atmospheric extinction**, **interstellar matter**.

extrasolar planet (exoplanet) A planet orbiting a star other than the Sun. In 1992 the first confirmed extrasolar planets were detected, orbiting the pulsar PSR 1257+12. In 1995 the first planet orbiting another main-sequence star, 51 Pegasi, was identified. By early 2015 over 1200 planetary systems had been discovered around a wide variety of stars. Many of these *substellar* objects are in very close orbits and are many times the mass of Jupiter. Their masses range from similar to that of Mercury to some 60 times greater than Jupiter, and their orbital periods from less than a day to hundreds of years. The lower theoretical mass limit for a **brown dwarf** is about 0.013 solar masses (approximately 13 Jupiter masses), which by one convention is taken as the upper mass limit for a giant planet.

Extreme Ultraviolet Explorer (EUVE) A NASA satellite launched on 7 June 1992 to survey the sky and study individual sources at very short ultraviolet wavelengths, a region of the spectrum not covered by previous satellites. It operated until December 2000.

extrinsic variable SEE **variable star**.

eyepiece (ocular) The system of lenses in an optical instrument nearest the observer's eye. Its function in a telescope used for visual observation is to magnify the image formed at the focus by the primary lens or mirror. Telescope eyepieces usually have two lenses, the *field lens*, farthest from the eye, and the *eye lens*. The magnification yielded by an eyepiece is given by the ratio of the **focal length** of the telescope's primary lens or mirror to the focal length of the eyepiece. There are numerous designs of eyepiece, varying in focal length, field of view and magnification: SEE **Erfle eyepiece**, **Huygenian eyepiece**, **Kellner eyepiece**, **Nagler eyepiece**, orthoscopic eyepiece, **Plössl eyepiece**, **Ramsden eyepiece**. SEE ALSO **Barlow lens**.

eye relief In an optical instrument such as a telescope or binoculars, the distance between the eye lens of the **eyepiece** and the **exit pupil**. When the observer's eye is placed at the exit pupil, the whole of the eyepiece's field of view is visible. The eye relief must be sufficiently large for comfortable viewing.

Fabry–Pérot interferometer A type of **optical interferometer** for making accurate measurements of close spectral lines (SEE **spectrum**). The two parts of the incoming beam are recombined after passing through an arrangement of parallel plates called a *Fabry–Pérot etalon*. The beams undergo multiple reflections between the plates, which are highly reflective and accurately parallel. The instrument was developed by the French physicists Charles Fabry (1867–1945) and Alfred Pérot (1863–1925).

faculae (singular **facula**) Bright areas in the Sun's upper photosphere that herald the appearance of **sunspots**. They appear some days before the spots, in the same place, but can remain for months after the sunspots have gone. Faculae are at a higher temperature than their surroundings, and so appear brighter. The word is Latin for "little torches."

falling star A popular (US) name for a **meteor**.

False Cross A pattern of stars made up of Iota and Epsilon Carinae, and Kappa and Delta Velorum. It can be mistaken for the Southern Cross, but it is larger, less bright and more symmetrical.

Far Ultraviolet Spectroscopic Explorer (FUSE) A NASA satellite launched on 24 June 1999 to study the sky at wavelengths of 90–120 nm, in the ultraviolet. One objective was to determine the amounts of deuterium in various cosmic sources in order to refine our knowledge of conditions in the very early Universe. It ceased operations on 18 October 2007.

F corona SEE **corona**.

Fermi Gamma-ray Space Telescope A gamma-ray satellite launched on 11 June 2008 to study highly energetic phenomena such as active galactic nuclei, supernova remnants, merging neutron stars and solar flares. Fermi carries a Large-Area Telescope that scans the entire sky every three hours and a Gamma-Ray Burst Monitor to analyze transient sources. The Fermi Gamma-ray Space Telescope is a collaborative venture between the US, Japan and several European nations and is named after the Italian–American physicist Enrico Fermi.

field of view The angular diameter of the area of sky visible through an optical instrument. For a telescope or binoculars it depends on the eyepiece that is used: a higher-power eyepiece will increase the magnification and thus decrease the field of view. Wide-angle eyepieces such as the **Nagler eyepiece** have fields of 80° or more.

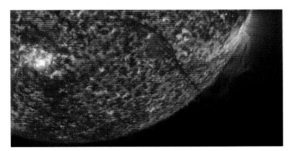

▲ *filament* A dark filament snakes over the Sun, becoming visible as a bright prominence when it reaches the Sun's edge at right of this image, which was taken in ultraviolet light by the Solar Dynamics Observatory.

filament A solar **prominence** seen in silhouette against the Sun's bright disk as a dark, thread-like marking in images taken in the light of hydrogen alpha or the K line of calcium.

filar micrometer A device, used in conjunction with the eyepiece of a telescope, that incorporates cross-wires for measuring the angular size of an object or the separation between two objects. Superimposed on the field of view the observer sees two parallel wires and a third wire perpendicular to them. One of the parallel wires is movable, and the observer adjusts its position until the intersections of the two parallel wires with the third wire mark off the distance being measured. The angular distance may then be read from the micrometer scale.

filter A device, used in conjunction with an instrument for receiving electromagnetic radiation, that allows some wavelengths to pass but blocks others. Filters for use with optical instruments such as telescopes are at their simplest a thin sheet of material placed over the full aperture of the instrument. Simple color filters made of gelatin bring out planetary features such as Jupiter's Great Red Spot. More sophisticated filters that transmit only a narrow band of wavelengths (said to have a narrow *passband*) are used for observing the Sun, or objects such as nebulae. Multilayer or interference filters are used to combat **light pollution**. In **CCD** imaging, color pictures can be created by combining separate images obtained through blue, green and red filters. SEE ALSO **Sun**.

finder A small, low-power telescope mounted on a larger one and having a much larger field of view than the main telescope, enabling the observer to locate celestial objects more easily.

fireball An exceptionally bright **meteor**. Fireballs have been defined only loosely as meteors brighter than the planets; with the modern estimate of the maximum brightness of Venus, this can be taken to mean that all meteors brighter than magnitude −5 should be classified as fireballs. Some meteor showers, such as the Geminids, produce more fireballs than others.

first contact In a **solar eclipse**, the moment when the leading edge of the Moon's disk first touches the Sun's disk. In a **lunar eclipse**, it is the moment when the Moon begins to enter the umbra of the Earth's shadow. The term is also used for the corresponding stage in eclipses involving other bodies.

first light The occasion on which a newly constructed telescope is first used to observe a celestial object.

First Point of Aries An alternative name for the vernal **equinox**, so called because it originally lay in the constellation Aries. Because of **precession**, it has now moved into neighboring Pisces.

first quarter One of the Moon's **phases**, when it is at quadrature on the way to being full and is half illuminated as seen from the Earth.

Flamsteed, John (1646–1719) English astronomer. He was appointed the first **Astronomer Royal** by King Charles II, with the task of obtaining accurate measurements of the Moon and stars for use in navigation. From Greenwich Observatory he used equipment fitted with sighting telescopes to measure with unprecedented precision the positions of thousands of stars. Flamsteed was a perfectionist and slow to release his results. Much to his anger, the Royal Society published his uncorrected observations. The catalog of his corrected observations was not published until 1725, after his death; the numbers assigned to stars in this catalog are still used (SEE **Flamsteed numbers**).

Flamsteed numbers A sequence of numbers allocated to the stars in each constellation in order of right ascension, for purposes of identification. The numbers were allocated by later astronomers to stars in John **Flamsteed**'s star catalog.

flare A sudden and violent explosion in the Sun's inner atmosphere, usually above an active group of **sunspots**. In the *flash stage* a flare builds to a maximum in a few minutes, after which it gradually fades within an hour or so. Flares emit radiation across the whole electromagnetic spectrum, from gamma rays to kilometer-wavelength radio waves. Particles are emitted, mostly electrons (some at half the speed

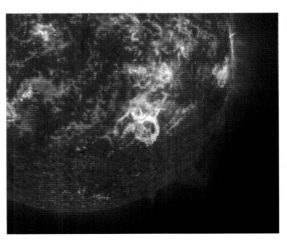

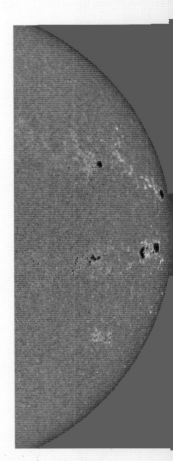

▲ *faculae* Bright regions marking the location of clouds of hot hydrogen above active regions on the Sun, faculae are best seen, as here, close to the limb. Faculae are commonest around sunspot maximum.

◄ *flare* Violent releases of magnetic energy in the inner solar atmosphere above active regions are seen in spectroheliograms as bright patches or ribbons, described as flares. Flares are most common at times of high sunspot activity.

of light) and protons, and smaller numbers of neutrons and atomic nuclei. A flare can cause material to be ejected in bulk, most spectacularly in the form of **coronal mass ejections**, at speeds that can exceed the Sun's escape velocity; **prominences** disturbed by the shock waves from flares may become detached and their material ejected into space. When energetic particles from flares reach the Earth they cause radio interference, magnetic storms and more intense **aurorae**. Although not well understood, the origin of flares is believed to be connected with local discontinuities in the Sun's magnetic field. It has even been suggested that flares are a result of coronal mass ejections rather than vice versa. SEE ALSO **flare star**.

flare star A variable star, usually a red dwarf (spectrum K–M), whose luminosity from time to time increases unpredictably by tenths of a magnitude to as much as seven magnitudes in a very short time – only a few seconds or tens of seconds – and then decreases to its normal value, in a few minutes to tens of minutes. Flare stars are also known as *UV Ceti stars*, after the best-known example, UV Ceti, which is a pair of 13th-magnitude red dwarfs and lies just 8.7 light years away. Outbursts of flare stars are sometimes accompanied by an increase in radio emission. Large flares may give out ten to a hundred times as much light as the rest of the star. A form of flare star, the *BY Draconis stars*, show additional variations of a few tenths of a magnitude attributed to large "starspots" passing across their disk as they rotate. They are therefore regarded as belonging to both **eruptive variables** and **rotating variables**. Flaring appears to be a consequence of instabilities in these young stars, allied with complex magnetic fields.

flash spectrum The **emission spectrum** of the solar chromosphere observed the few seconds just before and after totality during an eclipse of the Sun.

flatness problem A problem with the standard **big bang** model in that the **curvature of space** is observed to be improbably close to zero. Unless the Universe were flat to within 1 part in 10^{60} during the early stages of the big bang, space should now be strongly curved. **Inflation** solves the problem because the enormous expansion associated with inflation has flattened out any curvature on the "small" scale of the visible Universe.

flat universe In cosmology, a mathematical model that describes the behavior of a universe in which there is just enough matter to halt the expansion triggered in the **big bang**. It is often called the *Einstein–de Sitter model*. The universe will continue to expand, but at an ever-decreasing rate, theoretically coming to a halt an infinite time into the future. In a flat universe, the **curvature of space** is zero, the **deceleration parameter** equals 0.5 and the **density parameter** equals 1 (i.e. the mean density of the universe equals the **critical density**). This model closely matches our observations of the real Universe. COMPARE **open universe**, **closed universe**.

flocculi (singular **flocculus**) Small features that give a mottled or granulated appearance to the Sun's chromosphere.

focal length The distance between the center of a lens or curved mirror and its focus.

▼ **Fraunhofer lines** *Dark lines interrupt the continuum of the solar spectrum. These result from absorption of light at specific wavelengths by atomic species in the Sun's atmosphere.*

focal ratio The ratio of the **focal length** of a lens or curved mirror to its diameter. A focal ratio of, say, 8 is written as *f*/8.

focus (1) For a lens or curved mirror, the point at which the image is formed of a distant source lying on the axis of the lens or mirror.

focus (2) SEE **ellipse**.

following (abbreviation *f*) Describing the trailing edge, feature or member of an astronomical object or group of objects. For example, the *following limb* of the Moon is the edge facing away from its direction of motion; and the *following spot* (or *f*-spot) of a group of sunspots is the last of the group to be brought into view by the Sun's rotation.

Fomalhaut The star Alpha Piscis Austrini, magnitude 1.16, distance 25 light years, luminosity 17 times that of the Sun. It is a main-sequence star of type A3 surrounded by a **protoplanetary disk**.

forbidden lines Emission lines (SEE **emission spectrum**) in the spectra of some celestial objects which, when they were first observed, could not be identified with lines in spectra produced under laboratory conditions on Earth. It is now known that they are produced by atoms in what is called a metastable state. On Earth this state is very short-lived, but in highly rarefied nebulae, for example, it can last long enough for "forbidden" lines to be emitted.

fork mounting A type of mounting used for a large **equatorial telescope**.

Fornax A southern constellation originated by **Lacaille**, representing a chemical furnace. The only star above fourth magnitude is Alpha, magnitude 3.9, a double. Fornax includes a cluster of galaxies.

fourth contact (last contact) In a **solar eclipse**, the moment when the trailing edge of the Moon's disk last touches the Sun's disk. In a **lunar eclipse** it is the moment when the Moon's trailing edge leaves the umbra of the Earth's shadow. The term is also used for the corresponding stage in eclipses involving other bodies.

Fraunhofer, Joseph von (1787–1826) German physicist and optician, the founder of astronomical spectroscopy. He studied the diffraction of light through narrow slits and developed the earliest form of diffraction grating. In 1814 he observed and began to map the dark lines in the Sun's spectrum that are now called **Fraunhofer lines**, and was the first to appreciate their significance. He later found similar but differently distributed lines in the spectra of other stars. Fraunhofer solved many of the scientific and technical problems of astronomical telescope-making. He manufactured achromatic lenses, was the first to build an **equatorial telescope** that realized the full potential of the design, and built the heliometer with which Friedrich **Bessel** measured the first stellar parallax.

Fraunhofer lines The dark **absorption lines** in the solar spectrum, caused by absorption at specific wavelengths in the upper, cooler layers of the Sun's atmosphere. The most prominent ones at visible wavelengths are caused by the presence of neutral hydrogen (the **hydrogen alpha** or Hα line), singly ionized calcium (the calcium H and K lines), sodium and magnesium. First observed in 1802 by William Hyde Wollaston, they were first carefully studied from 1814 by Joseph von **Fraunhofer**.

Fred Lawrence Whipple Observatory An observatory located on Mount Hopkins, near Amado, Arizona, operated by the Smithsonian Astrophysical Observatory (SAO). It was opened in 1968, and later renamed to honor Fred **Whipple**, a former SAO Director. The main instruments are a 1.5 m (60-inch) and a 1.2 m (48-inch) reflecting telescope. The observatory runs the **MMT Observatory** jointly with the University of Arizona.

F star A star of spectral type F, with moderately strong hydrogen lines and strong ionized calcium and other metallic lines in its spectrum. Such stars have a surface temperature between 6100 and 7400 K on the main sequence; giants and supergiants of the same spectral type are about 300 K cooler. Main-sequence F stars range in mass from 1.2 to 1.7 solar masses, but most F-type giants and supergiants have evolved from considerably higher-mass stars. Bright examples are Canopus and Procyon.

full Moon The Moon's **phase** when it is at opposition, and its illuminated hemisphere is fully visible from the Earth.

fundamental stars Reference stars whose positions and **proper motions** have been determined with the greatest possible accuracy. Their coordinates are published in a *fundamental catalog* and serve as the points to which positional measurements of other bodies can be related. The leading fundamental catalogs are the FK series (short for *Fundamentalkatalog*), produced in Germany. The latest version, the FK6, was published in two parts in 1999 and 2000; it contains a combination of ground-based data and **Hipparcos** satellite data for a total of 4150 stars.

FUSE ABBREVIATION FOR **Far Ultraviolet Spectroscopic Explorer**.

Gaia A spacecraft launched by the **European Space Agency** on 19 December 2013 to measure the parallaxes, proper motions, colors and brightnesses of a billion stars, roughly 1% the population of our entire Galaxy. Such data will give astronomers an accurate three-dimensional map of our galactic surroundings. In addition, Gaia will discover new small bodies in our Solar System, extrasolar planets, distant supernovae and quasars. Gaia observes from the L_2 **Lagrangian point** of the Earth's orbit. It is a successor to the **Hipparcos** mission.

galactic coordinates SEE **coordinates**.

galactic halo The spheroidal distribution of old stars and globular clusters that surrounds a galaxy. The halo around our Galaxy is about 50,000 light years in diameter, and there appears also to be a much more extensive halo of **dark matter** extending out to a radius of at least 150,000 light years.

Galatea One of the small inner **satellites** of Neptune discovered in 1989 during the Voyager 2 mission.

galaxy A huge assembly of stars, dust and gas, an example of which is our own Galaxy. There are three main types, as originally classified by Edwin **Hubble** in 1925. *Elliptical galaxies* are round or elliptical systems, showing a gradual decrease in brightness from the center outward. They are given a designation from E0 to E7 in increasing degree of ellipticity. *Spiral galaxies* are flattened disk-shaped systems in which young stars, dust and gas are concentrated in spiral arms coiling out from a central bulge, the *nucleus*; they are designated S, with lower-case letters added to show how tightly the arms are wound, from tight (a) to very loose (d). *Barred spiral galaxies* are distinguished by a bright central bar from which the spiral arms emerge; they are designated SB with letters from a to d appended, as for ordinary spirals. These three main classes are represented in the **tuning fork diagram**. In addition there are *lenticular galaxies*, systems intermediate between ellipticals and spirals, having a disk and nucleus, similar to spiral galaxies, but with no apparent spiral arms; they are classified S0. **Irregular galaxies** (designated Irr or Ir) are systems with no symmetry.

Current theories suggest that all galaxies were formed from immense clouds of gas at roughly the same time, soon after the **big bang**. In elliptical galaxies star formation took place rapidly over perhaps several hundred million years, using up all the interstellar gas. Spirals, however, are the result of a two-stage formation process. Gas was left over from the initial star formation which produced the bulge at their centers. The remaining gas rapidly settled into a disk, in which density waves formed spiral arms.

Galaxies can exist singly or in clusters that contain anywhere from just a few to thousands of members (SEE **cluster of galaxies**). Galaxies in clusters have a good chance of one near-collision in their lifetimes. Interactions between gas-rich galaxies such as spirals can produce bursts of star formation, and such interactions are a possible trigger mechanism for generating **active galactic nuclei**, as in **Seyfert galaxies** and **quasars**. These objects are generally believed to be galaxies with nuclei in extreme states of activity. The most massive elliptical galaxies are now believed to have been formed by repeated collisions and mergers of smaller, individual galaxies. SEE ALSO **disk galaxy**, **low-surface-brightness galaxy**.

Galaxy, the The star system that contains the Solar System; the capital "G" distinguishes it from other **galaxies**. The Galaxy is spiral in shape and about 100,000 light years in diameter. Our Sun and

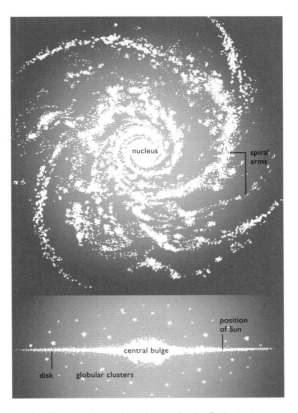

▲ *galaxy* The upper diagram shows the general outline for the structure of a spiral galaxy, of which our Galaxy is an example. Seen face-on (top), the galaxy has a fairly compact nucleus surrounded by an extensive disk containing star-rich spiral arms. Seen edge-on (below), the nucleus is revealed as a bulge, with the spiral arms confined to the flat plane of the disk. The galaxy as a whole is surrounded by a halo of globular clusters. As indicated, the Sun is located midway out in the disk of our Galaxy.

Solar System are located at the edge of one of the spiral arms, 26,500 light years from the center. The spiral arms form a disk-shaped system, with a bulging core (or nucleus) in the direction of Sagittarius. Old stars, of Population II, are found in the Galaxy's core; younger, hotter Population I stars, together with interstellar dust and gas, make up the spiral arms (SEE **stellar populations**). The stars of the spiral arms form the Milky Way, running around the sky. It is uncertain how many spiral arms the Galaxy has, but there is good evidence, from infrared studies for example, that there is a modest bar at the center, and that it should be classified as a barred spiral galaxy. Surrounding it is the spheroidal **galactic halo**, which is about 100,000 light years in diameter. In it are found very old Population II stars, including those in **globular clusters**.

Between us and the Galaxy's central bulge passes a prominent spiral arm, seen as the bright areas of the Milky Way in Centaurus, Crux and Carina. This arm curls away from us in Carina where we see it virtually end-on, which accounts for the great accumulation of stars and nebulae in that constellation. Immediately outside this arm is a band of gas and dust, a region where stars have yet to form, which is visible as the dark rift in the Milky Way running from Cygnus to Crux. Beyond Carina is a fainter arm which curves round outside the Sun. It crosses the northern sky, gaining in brightness as it moves inward toward Cygnus.

The center of the Galaxy is marked by the radio source **Sagittarius A**, which has been adopted as the zero point of galactic longitude. The *galactic equator* is the plane of the Galaxy, running along the Milky Way. The whole Galaxy is rotating, but the rotational rate varies with the distance from the center. Our Sun circles the center at about 250 km/s, taking 200 million years per orbit. The Galaxy's age is thought to be about 10,000 million years, during which time the Sun would have completed approximately 50 revolutions.

The Galaxy is a member of a cluster known as the **Local Group**, which includes the Andromeda Galaxy and the Magellanic Clouds.

Galaxy Evolution Explorer (GALEX) An orbiting space telescope launched on 28 April 2003 to observe galaxies in ultraviolet light. GALEX's observations have helped astronomers understand

THE GALAXY: DATA	
Diameter of disk	100,000 l.y.
Thickness of nucleus	6000 l.y.
Average density	7×10^{-24} g/cm^3
Total mass	2×10^{45} g = 10^{12} solar masses
North pole of galactic plane (epoch 2000.0)	RA 12h 51.4m, dec. +27° 7′
Point of zero longitude (epoch 2000.0)	RA 17h 45.6m, dec. −28° 55′
Galactic longitude of north celestial pole	123°
Distance of Sun from center	26,500 l.y.
Distance of Sun above galactic plane	14 l.y.
Rotational velocity of Sun	250 km/s
Period of Sun's revolution around center	200 million years
Absolute magnitude	−20.5

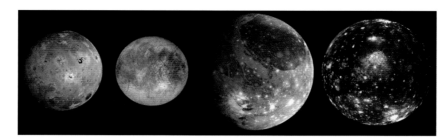

◄ *Galilean satellites* *The four large satellites of Jupiter are arranged in order of distance from the planet: (left to right) Io, Europa, Ganymede and Callisto. The images were obtained by the Galileo orbiter.*

▶ *Galileo* *Repeated passes close to the four large satellites of Jupiter by the Galileo orbiter since 1996 have revealed these in more detail than ever before. This image shows the volcano Zal Patera on Io.*

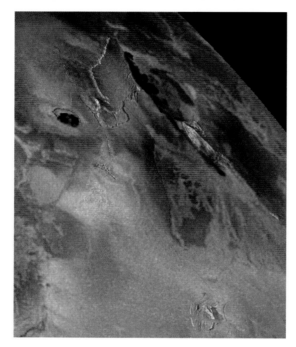

how galaxies and the stars within them formed and evolved. GALEX also performed an ultraviolet all-sky survey, identifying objects for follow-up study by other spacecraft. GALEX was turned off on 28 June 2013.

Galilean satellites The four chief satellites of Jupiter – **Io**, **Europa**, **Ganymede** and **Callisto** – named collectively after **Galileo Galilei**, who observed them in 1610. All are in low-inclination, near-circular, **synchronous orbits**, and their orbital periods are in simple ratios to one another as a result of **resonance**. Apart from the Earth's Moon, they are the brightest satellites in the Solar System, and all would be naked-eye objects were it not for the glare from Jupiter itself. They pass through cycles of **occultations** by Jupiter and eclipses in its shadow, and transits across its disk. SEE ALSO **Jupiter**, **satellite**.

Galileo Galilei (1564–1642) Italian astronomer, physicist and mathematician. He was one of the first to use the telescope for astronomical observations, and he improved its design. He discovered mountains and craters on the Moon, and found the four satellites of Jupiter now known as the **Galilean satellites**. He observed the phases of Venus and studied sunspots, from whose motion he deduced that the Sun rotates. He found that some celestial objects, such as the Milky Way and the Pleiades, are resolvable into many more stars than can be distinguished by the naked eye.

Galileo concluded that **Aristotle**'s picture of the world, then still widely believed, was wrong, and instead championed **Copernicus**'s heliocentric theory. This brought him into conflict with the Catholic Church, and led to his trial and house arrest for the last eight years of his life. His many advances in physics included the law of the pendulum and how to create a vacuum. His chief books were *Sidereus nuncius* (on discoveries with the telescope), *Dialogue on the Two Chief World Systems* (heliocentric Universe), and *Discourses and Mathematical Demonstrations on Two New Sciences* (mathematical physics).

Galileo A space probe to Jupiter, launched on 18 October 1989. *En route*, Galileo flew past and imaged the asteroids **Gaspra** and **Ida** in October 1991 and August 1993 respectively. Galileo entered

orbit around Jupiter in December 1995, after dropping off a sub-probe which entered the planet's atmosphere. This probe detected a hydrogen/helium ratio similar to the Sun's, and an intense radiation belt 50,000 km (30,000 miles) above the upper cloud deck. From 1995 to 2002 Galileo made over two dozen flybys of the Galilean satellites, returning highly detailed views. Its mission ended in 2003.

Galileo National Telescope (TNG) A 3.5 m (138-inch) reflecting telescope at **Roque de los Muchachos Observatory**, opened in 1998 as a national facility for Italian astronomers. (The initials are those of the Italian version of the name.)

Galle, Johann Gottfried (1812–1910) German astronomer. In 1846 he and Heinrich D'Arrest were the first to locate the planet Neptune, close to positions calculated independently by John Couch **Adams** and Urbain **Le Verrier**. Galle discovered Saturn's faint C Ring, and determined the solar parallax (used to fix the scale of the Solar System) by observing the asteroid Flora.

gamma-ray astronomy The study of very high energy **electromagnetic radiation** from space with wavelengths of about 0.01 nm and shorter. Gamma rays are absorbed by the Earth's atmosphere, so all studies must be made by high-altitude balloons or spacecraft. Early surveys found a general concentration of emission in the galactic plane, with a peak toward the galactic center, plus several strong individual sources such as the Crab Nebula, the Vela Pulsar and the quasar 3C 273. In 1991 NASA launched the **Compton Gamma Ray Observatory**, the **Swift Gamma Ray Burst Explorer** in 2004 and the **Fermi Gamma-ray Space Telescope** in 2008, while ESA launched **Integral** (the International Gamma Ray Astrophysics Laboratory) in 2002.

Their very short wavelengths mean that gamma rays have very high energies, and they are emitted from regions of very high temperature associated with some of the most violent events in the Universe. Most of the gamma radiation from the Galaxy is produced by cosmic rays interacting with gas in the interstellar medium. Certain gamma-ray lines at specific energies have been discovered, one attributed to the mutual annihilation of electrons and positrons in the galactic center, and another to the isotope aluminum-26, which is produced by supernovae. Sporadic, low-energy (soft) gamma-ray flares, whose sources are known as *soft gamma repeaters* (SGRs), are now thought

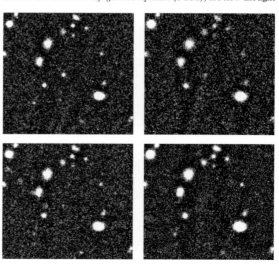

▲ *gamma-ray burst* *GRB 00131 was detected by several spacecraft on 31 January 2000. This series of images obtained with the Very Large Telescope (VLT) shows the fading visible-wavelength afterglow, just left of center.*

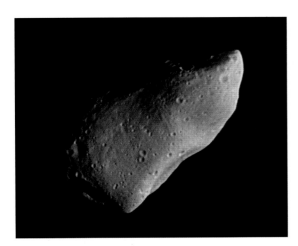

▲ *Gaspra* En route to Jupiter, the Galileo spacecraft made a flyby of asteroid (951) Gaspra in 1991, revealing it to be an irregular-shaped, heavily cratered body. Gaspra lies in the main belt.

to be produced by magnetic-field events on **magnetars** – neutron stars with exceptionally strong magnetic fields.

Outside the Galaxy, the main sources of gamma rays are those that produce **gamma-ray bursts**. There is also a gamma-ray background, a diffuse radiation from all over the sky, like the cosmic microwave background; its origin is unknown.

gamma-ray burst (GRB) An intense flash of gamma rays, which may last from about a second to a few minutes. GRBs were first detected in 1967 by a satellite monitoring nuclear explosions on Earth. Because they occur randomly across the sky, at the rate of a few per day, they could not originate in our Galaxy (they would then be concentrated near the Milky Way). They had to be either relatively near or very distant. In 1997 it was found that GRBs are followed by an X-ray "afterglow," which enabled their positions to be pinpointed with greater accuracy. They turned out to occur in young galaxies with high redshifts and hence at great distances, making the bursts extremely energetic events.

It has now been confirmed that the longest GRBs, lasting up to a few minutes, result from the collapse and explosion of extremely massive stars to form a black hole, while those lasting less than a second or two are thought to be caused by the merger of two neutron stars or a neutron star and a black hole. The **Swift Gamma Ray Burst Explorer** was launched in 2004 to react rapidly, relaying an accurate position to a network of ground-based telescopes within seconds of detecting a burst.

Gamow, George (1904–68) Russian–American nuclear physicist and cosmologist. His most famous work, co-authored with Ralph Alpher and Hans **Bethe**, pictured a hot and dense early Universe in which the nuclei of elements were built up from protons, and was a cornerstone of **big bang** theory. Gamow's greatest prediction, borne out by the detection of the **cosmic microwave background** radiation, was that the Universe should by now have cooled to a few degrees above absolute zero.

Ganymede The largest of Jupiter's **Galilean satellites**, with a diameter of 5262 km (3270 miles), and the largest satellite in the Solar System – bigger, though less massive, than the planet Mercury. With a density of 1.94 g/cm³, Ganymede contains a high proportion of water ice. Its surface is a mixture of dark, heavily cratered terrain and a brighter, more lightly cratered terrain covered with meandering, intersecting grooves that suggest recent geological activity. The most prominent feature is the dark Galileo Regio, some 4000 km (2500 miles) across. The **Galileo** probe detected a weak magnetic field, probably originating in a liquid layer deep below the surface. Ganymede's magnetosphere interacts with that of Jupiter. The satellite has an extremely thin oxygen atmosphere. SEE ALSO **satellite**.

gas giant SEE **giant planet**.

Gaspra Asteroid no. 951, discovered by Grigorii Neujmin in 1916. In 1991 it became the first asteroid to be imaged from close quarters when the probe **Galileo** passed it *en route* to Jupiter. It is irregular, with a maximum dimension of about 20 km (12 miles), and is pitted with small craters.

Gauss, Carl Friedrich (1777–1855) German mathematician. His many achievements include valuable contributions to astronomy, geodesy and physics. In 1801 he worked out how to calculate the orbit of the newly found asteroid **Ceres** from observations spanning just 9° of its orbit. This led him to the study of **celestial mechanics**, and on to a comprehensive study of the determination of cometary and planetary orbits, taking into account the effects of perturbations, and then to the theory of errors of observation, for which he developed the mathematical method of least squares. From 1807 Gauss was director of the observatory at Göttingen.

gegenschein A faint luminous patch on the ecliptic, visible usually only from the tropics on dark clear nights, directly opposite the Sun. It is caused by the scattering of sunlight back toward the Earth by particles in the zodiacal dust cloud, and is part of the **zodiacal light**. The word is German for "counterglow," by which name it was formerly also known.

Geminga One of the most powerful gamma-ray sources in the sky, in the constellation Gemini (hence the name). Geminga's energy is emitted as pulses of gamma rays (and X-rays) with a period of 0.237 seconds, so it is most likely a spinning neutron star: a short-wavelength version of a **pulsar**.

Gemini A large, prominent constellation of the **zodiac**. Its brightest stars are **Castor** and **Pollux**, named after the twins of Greek mythology. Eta is a semiregular variable, magnitude 3.2 to 3.9, with a period averaging 233 days. Zeta is a Cepheid variable, magnitude range 3.6 to 4.2, period 10.15 days. The open cluster M35 contains about 200 stars.

Geminids A **meteor shower** which occurs between 7 and 15 December, peaking at **ZHR** in excess of 100 close to 13–14 December. Its radiant is in the constellation Gemini, near the star Castor. The shower, one of the most prolific, is associated not with a comet but with **Phaethon**, an asteroid.

Gemini Telescopes A pair of 8.1 m (320-inch) reflectors jointly funded by the USA, UK, Canada, Australia, Chile, Brazil and Argentina. Gemini North is sited at Mauna Kea, Hawaii, and was dedicated in 1999. Gemini South is at Cerro Pachón, Chile, and opened in 2002. Together they provide complete sky coverage at optical and infrared wavelengths.

Genesis A NASA spacecraft launched on 8 August 2001 to collect samples of the **solar wind** and return them to Earth. In September 2004 the sample return capsule crash-landed in the Utah desert after its parachutes failed, but the solar wind samples were recovered for analysis of the chemical and isotopic composition of the Sun.

geocentric As viewed from or related to the Earth's center, as in *geocentric parallax* or *geocentric system*. COMPARE **heliocentric**.

geocentric parallax SEE **parallax**.

geodesic SEE **curvature of space**.

German mounting A type of mount used for an **equatorial telescope**.

Ghost of Jupiter A bright planetary nebula, designated NGC 3242, in the constellation Hydra at a distance of 2000 light years. The object was named by William **Herschel**, who discovered it in 1785, from its visual telescopic appearance as a very dim version of the giant planet. Many observers report a pronounced blue-green color.

◄ *Gemini Telescopes* Sunset over the dome of the Gemini North telescope on Mauna Kea, Hawaii. The sides of the dome can open to provide ventilation, as seen here.

247

▲ *Giacobini–Zinner,*
Comet 21P This comet was
imaged on 1 November 1998,
by the Kitt Peak 0.9 m (36-inch)
telescope. The teardrop-shaped
coma shows strong bluish carbon
monoxide emission.

Giacobinids A meteor shower associated with Comet **Giacobini–Zinner**. The shower is periodic, and produced significant displays on 9 October 1933, 9–10 October 1946 (a **meteor storm**) and 8 October 1985; a minor outburst occurred in 1998. Activity is seen only in those years when the parent comet is close to perihelion. The radiant lies close to the star Beta Draconis, and the shower is also known as the *Draconids*.

Giacobini–Zinner, Comet 21P/ A short-period **comet**, orbiting the Sun in 6.59 years. It was discovered in 1900 by Michel Giacobini (1873–1938) and recovered in 1913 by Ernst Zinner (1886–1970). In 1985 it became the first comet to be studied at close hand when the International Cometary Explorer space probe passed through the comet's ion tail. Debris from the comet produces the periodic **Giacobinid** meteor shower.

Giant Magellan Telescope (GMT) An optical/infrared reflecting telescope with a primary made of seven mirrors with an equivalent aperture of 24.5 m (80.4 ft). The design is based on a Gregorian telescope and uses **active optics** for the primary mirrors, with **adaptive optics** for the secondary mirror. It is being built at **Las Campanas Observatory** in Chile by a consortium of US, Australian and Korean institutions.

giant molecular cloud (GMC) SEE **infrared astronomy**.

giant planet Any of the planets Jupiter, Saturn, Uranus and Neptune, also known as *gas giants*. They are so called because their masses are much greater than the Earth's. The terms are also used for large **extrasolar planets**.

giant star A star that is placed well above the **main sequence** in the **Hertzsprung–Russell diagram**. Giant stars are larger than main-sequence (**dwarf**) stars of the same temperature, but giants of intermediate temperatures are smaller than the hottest dwarfs. Thus the term "giant" really refers to the relative luminosity of a star rather than its dimensions. Giant stars evolve from dwarf stars that have run out of hydrogen in their central regions, and so represent later stages in the lives of stars. Because of their high luminosity, giants are quite common among naked-eye stars, examples being Arcturus, Aldebaran and Capella, but they are relatively rare overall. SEE ALSO **red giant**, **blue giant**, **supergiant**.

gibbous A **phase** of the Moon or a planet when more than half but less than all of its illuminated hemisphere is visible from the Earth.

Giotto A European Space Agency mission to investigate **Halley's Comet**. The probe was launched on 2 July 1985 and flew to within 600 km (375 miles) of the comet's nucleus on 14 March 1986. Giotto imaged the nucleus and analyzed the gas and dust given off. After the encounter, the probe went on to rendezvous with Comet **Grigg–Skjellerup**, 10 July 1992.

glitch SEE **pulsar**.

globular cluster A near-spherical cluster of very old Population II stars (SEE **stellar populations**) in the halo of our Galaxy. Globular clusters contain a large number of stars (10^5 to 10^7), packed closely into a roughly spherical space typically 50 to 150 light years across. The stars are strongly concentrated toward the center.

▲ *Giotto* The ESA Giotto probe
recorded this close-up view of
the nucleus of Halley's Comet on
13 March 1986. Shortly after this
image was obtained, collision with
a dust particle from the comet
disabled the camera.

Most globulars are at least 10,000 million years old, as deduced from the highly evolved state of their stars. This extreme age, combined with their distribution in the galactic halo, indicates that globulars formed while our Galaxy was condensing from a huge cloud of gas. Globulars can also be seen around some other galaxies. They are more plentiful around elliptical galaxies than spirals – our Galaxy has about 150 known globular clusters, whereas the largest elliptical galaxies can have thousands.

Although they occupy the spheroidal galactic halo, most globulars lie no farther from the center of the Galaxy than does the Sun. As a result they congregate in the general area of sky containing the center of our Galaxy, in particular in the constellations Sagittarius, Ophiuchus and Scorpius. Globular clusters orbit the galactic center, making periodic passages through the plane of our Galaxy. They suffer a slow loss of stars (known as *cluster evaporation*), because gravitational interactions accelerate some individual stars, mostly of low mass, into hyperbolic escape orbits. SEE ALSO **star cluster**.

globule SEE **Bok globule, cometary globule**.

GMC ABBREVIATION FOR giant molecular cloud (SEE **molecular cloud**).

GMT ABBREVIATION FOR **Giant Magellan Telescope** or **Greenwich Mean Time**.

Gold, Thomas (1920–2004) Austrian–American astronomer. With Hermann **Bondi** and Fred **Hoyle** he developed the **steady-state theory**, in which the continuous creation of matter drives the expansion of the Universe. In the 1960s he explained that the signals from pulsars were beams of synchrotron radiation from rapidly rotating neutron stars.

Goodricke, John (1764–86) English astronomer, born in Holland. With the encouragement of his neighbor and collaborator Edward Pigott, the deaf-mute Goodricke began at the age of 17 to study the brightest variable stars. In 1782 he announced the cause of Algol's variability (that it was what is now called an eclipsing binary), and the length of its period. Two years later he discovered the variability and periods of Beta Lyrae and Delta Cephei, the prototype Cepheid variable. He died aged just 21 from pneumonia contracted while observing Delta Cephei.

Gould's Belt A region of bright stars and gas inclined at about 20° to the galactic plane. It contains the greatest concentration of naked-eye stars of spectral types O and B, and probably represents part of the local spiral arm of which the Sun is a member. Named after American astronomer Benjamin Gould (1824–96), who studied the region.

GRAIL The Gravity Recovery and Interior Laboratory, a pair of NASA spacecraft launched together on 10 September 2011. Called Ebb and Flow, they orbited over the Moon's poles, measuring variations in the Moon's gravitational field from which its internal structure can be deduced. At the end of their mission, the twin probes were crashed into the Moon's surface on 17 December 2012.

▲ *globular cluster* Containing at least 300,000 ancient stars, the globular
cluster G1 orbits the Andromeda Galaxy. This Hubble Space Telescope image
shows G1 to have a rich, compact core and less densely packed periphery.

grand unified theory (GUT) A theory that attempts to "unify" three of the four forces of nature, and thus to demonstrate that they are different manifestations of a single force. In descending order of strength, the four forces of nature are: the *strong nuclear force*, which binds atomic nuclei; the *electromagnetic force*, which holds atoms together; the *weak nuclear force*, which controls the radioactive decay of atomic nuclei; and **gravitation**. The weak and strong nuclear forces operate over tiny distances within atomic nuclei, while the other two forces are infinite in range.

In the 1970s and 1980s, the unification of the electromagnetic and weak nuclear forces was first predicted and then demonstrated experimentally. GUTs are attempts to unify this *electroweak force* with the strong nuclear force. Many physicists believe that a more elaborate theory, a so-called *theory of everything* (TOE), will eventually show how all four forces are unified.

Gran Telescopio Canarias (GTC) A Spanish 10.4 m (34 ft) reflecting telescope at the **Roque de los Muchachos Observatory**, opened in 2009. With its segmented mirror, the instrument closely resembles the 10 m telescopes of the W. M. Keck Observatory.

granulation The mottled appearance of the Sun's **photosphere** caused by gases rising from the Sun's interior. When seeing conditions are very good the granulations can be resolved into small, nearly circular patches, called *granules*, surrounded by darker areas. Individual granules are convection cells, typically 1000 km (600 miles) in diameter, and last only a few minutes.

graticule A system of reference marks or a measuring scale consisting of parallel vertical wires, or cross-wires, or a reticule (a grid of squares) placed in the focal plane of a telescope and covering the entire field of view. SEE ALSO **filar micrometer**.

grating SEE **diffraction grating**.

gravitation (gravity) The universal force of attraction between all particles of matter. *Newton's law of gravitation* states that the force of attraction (F) between two masses (m_1 and m_2) is inversely proportional to the square of their separation (r):

$$F = Gm_1m_2/r^2 ,$$

where G is the *gravitational constant*. This law is the basis of celestial mechanics.

Although at close range gravity is far, far weaker than the nuclear or electromagnetic forces, it has a greater range and is the most important force on the large, cosmological scale. The whole concept of gravitation was reinterpreted by Albert **Einstein** in his general theory of relativity in terms of the curvature of **spacetime**.

gravitational lens A massive body, such as a star, black hole or galaxy, that bends light or other radiation from a more distant object in the same line of sight. The effect, which is predicted by **Einstein**'s theory of relativity, was first detected during the total solar eclipse of 1919 by Arthur **Eddington**, who found that light from distant stars was bent as it passed the Sun. In 1979, the first example of gravita-

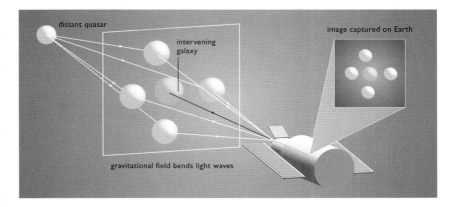

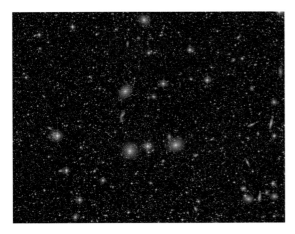

▲ **Great Attractor** *The central region of ACO 3627, a huge galaxy cluster thought to form part of the Great Attractor, is seen here in an image obtained by the European Southern Observatory. The Great Attractor's gravitational pull is dragging all the galaxies in our local cluster toward it.*

◄▲ *gravitational lens (top) Light from a distant quasar may be bent in the gravitational field of a foreground galaxy to produce a distorted or multiple image. In this example, the quasar's light is split into four separate images, each arriving at the telescope along a slightly different path. (left) The massive galaxy cluster Abell 2218 in Draco acts as a gravitational lens, producing distorted, arc-shaped images of more distant objects in the same line of sight.*

tional lensing of light from a quasar was discovered. In such cases, the quasar's light may be split into two or four separate images by an intervening massive galaxy or cluster of galaxies. Other distant objects may have their light lensed into arcs, or even a complete **Einstein ring**. Gravitational lensing of quasars can be used to measure their distances. SEE ALSO **microlensing**.

gravitational redshift SEE **redshift**.

gravitational waves Waves emitted as the result of a violent disturbance in a gravitational field, such as a massive body being accelerated or distorted with a velocity that is an appreciable fraction of the velocity of light. Gravity waves are predicted by **Einstein**'s general theory of relativity, but have proved elusive. Just as an oscillating charged particle emits electromagnetic waves, so a violently disturbed mass should emit gravity waves. Likely sources of gravity waves include close binary systems where at least one component is a neutron star or black hole, or the merger of two such dense bodies. Because gravity is by far the weakest of the forces of nature, gravity waves are extremely feeble. Detectors based on **interferometry** have been developed and are being commissioned, and should be sensitive enough to pick up gravitational waves.

Gravity Probe B A NASA spacecraft, launched on 20 April 2004, to test **Einstein**'s general theory of relativity by measuring the precession of gyroscopes in Earth polar orbit. The results, published in 2011, were consistent with Einstein's theory.

Great Attractor A hypothetical distant concentration of matter with a mass about 100,000 times that of the Milky Way, 150 million light years away in the direction of the constellations Hydra and Centaurus. It was suggested to account for the observed motions of galaxies, including those of the **Local Group**. Although there is a cluster of galaxies in the place in question, it now seems more likely that the gravitational pulls of several different concentrations of **dark matter** are responsible, and there is no single Great Attractor.

great circle A circle on a sphere whose plane passes through the center of the sphere. On the **celestial sphere** the celestial equator is a great circle, as are meridians (circles passing through both celestial poles).

greatest elongation SEE **elongation**.

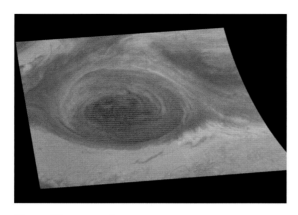

Great Observatories Umbrella term for four major NASA spaceborne telescopes: the **Hubble Space Telescope**, the **Compton Gamma Ray Observatory**, the **Chandra X-ray Observatory** and the **Spitzer Space Telescope**.

Great Red Spot (GRS) A large oval area in the southern hemisphere of Jupiter. Since it was first observed, by Heinrich Schwabe in 1831 (and possibly by Robert Hooke in 1664, though this may have been a different but similar spot), it has varied in size and hue – sometimes deep red, at other times pale gray. Even when the GRS is faint, its position is marked by a *hollow* on the southern edge of Jupiter's dark South Equatorial Belt. The GRS's red color may be produced by phosphorus.

Green Bank The location in West Virginia, USA, of the **National Radio Astronomy Observatory** and its Green Bank Telescope.

green flash An atmospheric phenomenon, observable under ideal conditions during the last seconds before sunset or the first seconds after sunrise, in which the uppermost segment of the Sun separates and flashes with a green color. At these moments most of the sunlight has been absorbed in its long passage through the atmosphere, and only two narrow bands of wavelengths in the red and green are left. The atmospheric refraction of light is greater for short wavelengths (green rather than red), so the green flash appears.

greenhouse effect The warming of a planet's surface by heat trapped by its atmosphere. Incoming sunlight is absorbed at the surface and re-radiated at longer wavelengths, in the infrared – that is, as heat. If the atmosphere contains *greenhouse gases*, such as carbon dioxide, which are not transparent to the infrared, the heat is reflected back to the ground. The effect is the same as is produced by the glass enclosing a greenhouse. The result is to raise the planet's average surface temperature by an amount that depends on the atmospheric composition and density. On **Venus** the dense carbon dioxide atmosphere has raised the surface temperature to around 750 K. On Earth a smaller greenhouse effect has led to a surface temperature that suits many life forms, but some people believe that industrial emissions of carbon dioxide will increase the greenhouse effect, producing *global warming*, and should be halted.

Greenwich Mean Time (GMT) Mean solar time for the meridian of Greenwich. In 1928 the International Astronomical Union recommended that GMT should be known as **Universal Time** for scientific purposes. Today, that usage has extended to legal and other purposes, although the term GMT is still used in many contexts. Greenwich was chosen as the world's prime meridian for timekeeping and navigation at the International Meridian Conference in 1884.

Gregorian calendar SEE **calendar**.

Gregorian telescope A reflecting telescope described by Scottish mathematician James Gregory (1638–75) in 1663, five years before Isaac **Newton** constructed his first reflector. Light received by the parabolic primary mirror is reflected to a concave spheroidal secondary mirror, and then reflected back through a hole in the center of the primary to an eyepiece. It differs from the **Cassegrain telescope** in having a concave instead of a convex mirror for the secondary. The configuration, long disused, has been revived in designs for giant instruments such as the **Giant Magellan Telescope**.

Grigg–Skjellerup, Comet 26P/ A periodic comet with one of the shortest-known periods, 5.3 years. It was discovered in 1902 by New Zealander John Grigg (1838–1920) and recovered in 1922 by Australian John "Frank" Skjellerup (1875–1952). In 1992 it became the third comet to be encountered by a spacecraft when the **Giotto** probe passed within 200 km (125 miles) of it.

GRO ABBREVIATION FOR **Compton Gamma Ray Observatory**.

GRS ABBREVIATION FOR **Great Red Spot**.

Grubb, Howard (1844–1931) Irish telescope-maker. In 1868 he took over the telescope-manufacturing business established by his father, Thomas Grubb (1800–1878), and went on to make seven of the so-called astrograph refractors used for the *Carte du Ciel* photographic sky survey, and many other medium-sized and large telescopes for leading observatories.

Grus A southern constellation, representing a crane. Its brightest star is Alpha (Alnair), magnitude 1.7. Mu and Delta are both optical pairs, wide enough to be separated with the naked eye.

G star A star of spectral type G, characterized by weak hydrogen lines, but strongly ionized calcium lines, and (at lower temperatures) strong neutral calcium and sodium lines in its spectrum. A G star has a surface temperature of between about 5000 and 6000 K if it is on the **main sequence**, and between 4600 and 5600 K if it is among the giants. The coolest G-type supergiants are about 4200 K. G-type dwarfs range in mass from 0.9 to 1.1 solar masses. The giants and supergiants have generally evolved from more massive stars further up the main sequence. Almost all G stars are dwarfs with compositions similar to the Sun. A few are subdwarfs with lower metal abundances. A small fraction of the G-type giants have peculiar chemical compositions that give them the name *barium stars*. Examples of G stars are the Sun, Alpha Centauri and Capella.

Gum Nebula An immense emission nebula spanning 30° to 40° of sky in the constellations Puppis and Vela, containing the Vela Pulsar. The nearest part is 450 light years away, and its farthest part is 1500 light years. It is named after the Australian astronomer Colin Gum (1924–60).

GUT ABBREVIATION FOR **grand unified theory**.

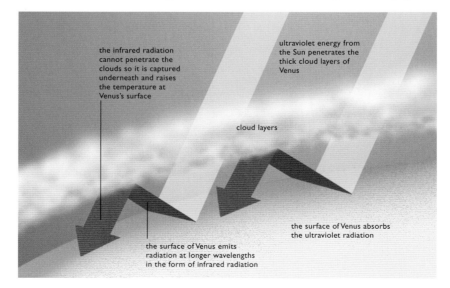

the infrared radiation cannot penetrate the clouds so it is captured underneath and raises the temperature at Venus's surface

ultraviolet energy from the Sun penetrates the thick cloud layers of Venus

cloud layers

the surface of Venus absorbs the ultraviolet radiation

the surface of Venus emits radiation at longer wavelengths in the form of infrared radiation

H I region A cloud of neutral (that is not ionized) atomic hydrogen gas in interstellar space. H I (pronounced "h one") regions are typically of 10 to 100 solar masses and measure between 10 and 20 light years across, with a temperature of around 100 K. They are about ten times as dense as interstellar matter, which is too low for molecular hydrogen to form in significant quantities.

H I regions do not emit visible radiation, but may be detected from their radio emission at a wavelength of 21 cm (SEE **twenty-one centimeter line**). Radio telescopes can thus detect H I regions even in other galaxies. Different kinds of galaxy contain different proportions of H I. Our Galaxy and other spiral or irregular galaxies contain an appreciable fraction of their total mass in H I. Elliptical galaxies contain no (or extremely little) detectable H I.

In our Galaxy, H I regions are heavily concentrated in a flattened disk in which a spiral pattern occurs. Spiral arms are made conspicuous, even in very distant galaxies, by hot, massive blue stars lying along the arms. These represent recent star-forming activity from the H I regions in the arms. Not all H I regions are capable of forming stars, only the densest ones that also contain molecular cores (SEE ALSO **giant molecular cloud**). Vast H I clouds, apparently devoid of luminous matter, have also been detected in the space between galaxies (SEE **Magellanic Clouds**).

H II region A cloud of ionized hydrogen in interstellar space. H II regions (pronounced "h two") are often embedded within **H I regions**, and have densities 10 to 100,000 times greater. Because of the high temperatures (8000–10,000 K) and densities of H II regions, the pressure inside them is far higher than in the surrounding H I region or interstellar gas, so all H II regions are expanding, at around 10 km/s. Visible radiation from the H II region results from the *recombination* of protons and electrons to form hydrogen atoms; hot dust emits at infrared wavelengths; and *free–free radiation* at radio wavelengths is emitted as electrons curve around protons. On images, H II regions often appear red because of the **hydrogen-alpha line** at 656 nm, but they can look greenish when seen visually because of a prominent pair of **forbidden lines** of oxygen at 496 and 501 nm.

Regions of ionized hydrogen can be produced by ultraviolet emission from old hot stars, from nova or supernova explosions, or from high-velocity **jets** or stellar winds impacting on interstellar gas. Conventionally, however, these are not classified as H II regions but as, respectively, **planetary nebulae**, **supernova remnants** and **Herbig–Haro objects**.

H II regions are stellar nurseries, the stars within them having recently formed through gravitational collapse within a **giant molecular cloud**. The life of H II regions is only a few million years, and their presence along with the hot stars marks out the arms in spiral galaxies. Complex H II regions are found in **starburst galaxies**. SEE ALSO **emission nebula**.

HA ABBREVIATION FOR **hour angle**.

Hα SEE **hydrogen-alpha line**.

Hadar Another name for **Beta Centauri**.

Hale, George Ellery (1868–1938) American astronomer. He planned and secured finance for the 1 m (40-inch) refracting telescope at **Yerkes Observatory**, completed in 1897; the 2.5 m (100-inch) Hooker reflector at **Mount Wilson Observatory**, completed in 1917; and the 5 m (200-inch) reflector at the **Palomar Observatory**, which entered regular service in 1949. Each was in turn the largest telescope of its day. The 5 m (200-inch) was named the **Hale Telescope** in his honor. Hale was an accomplished solar observer, discovering among other things the magnetic fields of sunspots via the Zeeman effect in their spectra. He also invented the **spectroheliograph**.

Hale–Bopp, Comet C/1995 O1 A long-period comet discovered in 1995 by American amateur astronomers Alan Hale (1958–) and Thomas Joel Bopp (1949–). Its nucleus was exceptionally large – about 50 km (30 miles) across – so that even though it approached the Earth no closer than 1.3 au, it was still a brilliant object (magnitude −0.7 at maximum) in April 1997. Hale–Bopp emitted dust at a prodigious rate. As the nucleus of the comet rotated, with a period of 11.5 hours, the dust created a series of concentric shells, which could be seen in the comet's inner coma. The gas and dust tails extended to 20° and 25° respectively, which corresponds to actual distances of more than 1 au. It will not return for another 2500 years.

Hale Telescope The 5 m (200-inch) reflecting telescope at the **Palomar Observatory**.

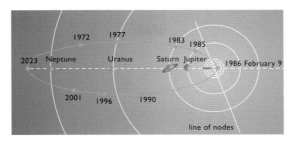

▲ **Halley's Comet** *The elongated elliptical orbit of Halley's Comet takes it from aphelion beyond the orbit of Neptune to perihelion between the orbits of Venus and Mercury. The comet's orbit is retrograde, and it was last at perihelion on 9 February 1986.*

Hall, Asaph (1829–1907) American astronomer. His planetary studies covered the orbital elements of satellites, the rotation of Saturn and the mass of Mars. On 11 and 17 August 1877 he discovered the two tiny satellites of Mars, which he named Deimos and Phobos. The largest craters on Phobos have been named Hall and Stickney (the maiden name of Hall's wife).

Halley, Edmond (1656–1742) English scientist. His most famous achievement in astronomy was to realize that comets could be periodic, following his observation of **Halley's Comet** in 1682. He cataloged the stars of the southern hemisphere, discovered the proper motions of stars, suggested that the nebulae were clouds of gas and speculated on the infinity of the Universe. He made the first complete observation of a transit of Mercury and showed how the results could be used to measure the Sun's distance. As the second **Astronomer Royal**, he embarked on a long series of lunar studies. It was Halley who persuaded Isaac **Newton** to write the *Principia*, and he also financed its publication. He founded modern geophysics, charting variations in the Earth's magnetic field, and establishing the magnetic origin of the aurora borealis. In meteorology he showed that atmospheric pressure decreases with altitude, and studied monsoons and trade winds.

Halley's Comet (Comet 1P/Halley) The brightest and best-known periodic **comet**. It takes 76 years to complete an orbit that takes it from within Venus's orbit to outside Neptune's. The orbit's inclination is 162°, making it **retrograde** (1). It was observed by Edmond **Halley** in 1682; later he deduced that it was the same comet as had been seen in 1531 and 1607, and predicted it would return in 1758. There are historical records of every return since 240 BC. Several space probes were sent to Halley on its 1986 return. The closest approach was by the probe **Giotto**, whose cameras showed the nucleus to be an irregular object measuring 15 × 8 km (9 × 5 miles). The main constituent was found to be water ice, but the surface was covered with a dark deposit, giving an albedo of only about 0.03. Some craters were imaged. Its next perihelion will be in 2061.

halo (1) A bright whitish ring, complete or partial, seen around a celestial body, usually the Sun or the Moon. It is caused by the refraction of light by ice crystals suspended in the atmosphere.

halo (2) The **galactic halo**, a spheroidal distribution of stars and globular clusters around our Galaxy.

◄ **H II region** *Pink patches in the spiral arms of the galaxy M83 in Hydra are H II regions, clouds of gas where stars are forming. They appear pink because of emission from hydrogen, their principal constituent.*

halo population stars SEE **stellar populations**.

Harvard classification The system of spectral classification of stars developed in the late 19th century by E. C. **Pickering** at Harvard College Observatory. Stars were originally placed in a sequence from A to P, indicating decreasing strength of hydrogen absorption lines in their spectra. The rearrangement and pruning of these classes gave the sequence O, B, A, F, G, K, M, based on decreasing surface temperature, used in the *Henry Draper Catalog*. It was later refined as the **Morgan–Keenan classification**.

harvest Moon The full Moon nearest the autumnal **equinox** (on 23 September). The Moon's path is then almost parallel to the horizon, and so it rises only about 15 minutes later each evening instead of the usual half-hour or more, providing a succession of moonlit evenings formerly of great help to harvesters. In the southern hemisphere the harvest Moon is the full Moon nearest the vernal equinox (on 21 March). SEE ALSO **hunter's Moon**.

Hawking, Stephen William (1942–) English theoretical physicist. He has used the general theory of relativity and quantum mechanics to study the **big bang** and **black holes**. He found that small black holes should lose energy – by emitting *Hawking radiation* – and eventually "evaporate."

Hayashi track The evolutionary path on the **Hertzsprung–Russell diagram** of a star beginning to form (SEE **protostar**) after it has collapsed from a gas cloud and before it reaches the main sequence. The Hayashi track lies above and to the right of the main sequence, and is nearly vertical. The star descends on its Hayashi track, decreasing in luminosity but with little change in effective temperature. (After Japanese astrophysicist Chushiro Hayashi, 1920–2010.)

heavy elements SEE **metals**.

Hektor Asteroid no. 624, discovered in 1907 by August Kopff. It measures roughly 300 × 150 km (200 × 100 miles) and is the largest and brightest of the **Trojan asteroids** sharing Jupiter's orbit. Its elongated profile may indicate that it is a binary asteroid.

Helene A small **satellite** of Saturn, discovered in 1980 by Pierre Lacques and Jean Lecacheux. It is an irregular body, and is co-orbital with Dione, orbiting near the L_4 **Lagrangian point** of Dione's orbit around Saturn.

heliacal rising In modern usage, the rising of a celestial body simultaneously with the rising of the Sun. Formerly it meant the moment when a bright star, such as Sirius, could just be seen rising before the Sun. Observation of this event was once of great importance for agricultural purposes, the heliacal rising of Sirius heralding to the ancient Egyptians the annual flooding of the Nile.

▼ **Herbig–Haro object** *High-speed jets from the young object HH 34, near right center, impact on the surrounding interstellar material to produce V-shaped shockwaves. HH 34 lies in Orion, in a region of ongoing star formation.*

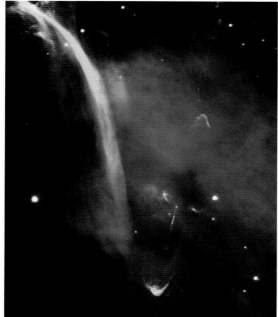

heliocentric As viewed from or in relation to the Sun's center, as in *heliocentric coordinates* and the *heliocentric system*. COMPARE **geocentric**.

heliocentric parallax SEE **annual parallax**.

heliometer A refractor with the objective lens split across its diameter to give two images, formerly widely used for measuring very small angular distances.

heliosphere The region of space in which the Sun's magnetic field and the solar wind dominate the interstellar medium. It is similar in form to a planet's **magnetosphere**, with an outer *bow shock* and a boundary called the *heliopause*, where the energy of the solar wind particles has fallen to the level of galactic cosmic ray particles. In 2012 the space probe Voyager 1 exited the heliosphere at a distance of over 125 au from the Sun.

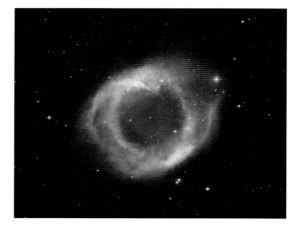

▲ **Helix Nebula** *One of the closest planetary nebulae, the Helix Nebula (NGC 7293) is visually faint because its light is spread over a fairly large area of sky.*

heliostat A mirror on an equatorial mount driven so that it follows the Sun's apparent motion across the sky and directs its light into a fixed instrument such as a solar telescope. A more sophisticated version is the **coelostat**.

helium flash The sudden commencement of helium fusion in the core of a red giant after the star's supply of hydrogen fuel has been exhausted. The helium flash is restricted to stars of less than about 2 solar masses; in more massive stars, helium fusion commences more gradually.

Helix Nebula A planetary nebula in Aquarius, NGC 7293. It has one of the largest angular diameters of any planetary nebula, a quarter of a degree, and at 300 light years away it is one of the nearest to the Sun.

Henry Draper Catalog A catalog of stellar spectra, compiled by Annie **Cannon** at Harvard College Observatory. It was named after Henry Draper (1837–82), an American pioneer of astrophotography, whose widow supported the work financially. The catalog, completed in 1924, introduced the Harvard system of classifying stellar spectral types: O, B, A, F, G, K, M, in order of decreasing surface temperature. Stars are still widely known by their HD numbers from this catalog. SEE ALSO **spectral classification**.

Herbig–Haro object (HH object) One of a class of small, faint nebulae independently discovered by George Herbig (1920–2013) and Guillermo Haro (1913–88). They are irregular in outline, contain bright knots, and are found in regions rich in interstellar material. HH objects are thought to be produced by very young stars hidden behind clouds of gas and dust. Powerful stellar winds from the young stars excite the gas, producing emission lines, while the light from the star is reflected by dust in other parts of the cloud.

Hercules A constellation named after the great hero of Greek mythology, but not very conspicuous. Its brightest star is Beta (Kornephoros), magnitude 2.8. Alpha (Rasalgethi) is a semiregular red giant variable, magnitude range 2.7 to 4.0, with a period that varies from 50 to 130 days; there are also longer-term variations with a period of about 6 years. It lies 382 light years away. It is also a double star, with a magnitude 5.4 companion, period about 3600 years. Zeta is a fine close binary, with a period of 34 years, magnitudes 2.9 and 5.5. The most interesting deep-sky objects are the two globular clusters M13 and M92.

Hermes An **asteroid**, about 1 km (0.6 mile) in diameter, discovered by Karl Reinmuth in 1937. On 28 October of that year it came within 800,000 km (500,000 miles) of the Earth, up to then the closest known approach of any asteroid. Hermes was then "lost" until October 2003, when Brian Skiff rediscovered it with the Lowell Observatory Near-Earth Object Search telescope. Hermes has now been assigned a number (69230). Radar observations suggest it may be a binary asteroid with components 300–450 m in diameter.

Herschel, Caroline Lucretia (1750–1848) German-born English astronomer, sister of William **Herschel**. In 1772 she joined her brother at Bath in England as his housekeeper and astronomical assistant. In the 1780s she became an observer in her own right, discovering many nebulae (as they were then called, including the galaxy NGC 253 in Sculptor) and eight comets. When her brother died in 1822, Caroline returned to her native Hanover, where she received many honors. She was the first woman to achieve real distinction in astronomy.

Herschel, John Frederick William (1792–1871) English scientist and astronomer, son of William **Herschel**. He extended his father's work on double stars and nebulae, discovering over 500 more nebulae and clusters. In 1834 he took one of William's telescopes to the Cape of Good Hope and undertook a systematic survey of the southern sky, discovering over 1200 doubles and 1700 nebulae and clusters. He combined these and his father's observations into a *General Catalogue of Nebulae and Clusters*, which formed the basis for the later **NGC**. He made the first good direct measurement of solar radiation, and inferred the connection between solar and auroral activity. He was also a pioneer of photography and its application in astronomy.

Herschel, (Frederick) William (originally Friedrich Wilhelm Herschel) (1738–1822) German-born English astronomer. In 1757 he came to England as a musician and became a naturalized Englishman. He turned to astronomy in the 1770s, making telescopes and mirrors for his own use and later for sale. Herschel became famous in 1781 by discovering the planet Uranus, which he wanted to name after King George III; the next year he was appointed the king's private astronomer. He discovered two satellites of Uranus (1787) and two of Saturn (1789). He observed and cataloged many double stars, hoping to measure their distances by detecting the parallactic movement of the nearer against the farther if they were optical doubles. He observed over 2000 nebulae and clusters, and published catalogs of them which later would be incorporated in the **NGC**. Herschel realized that the Milky Way is the plane of a disk-shaped stellar universe, whose form he calculated by counting the numbers of stars visible in different directions, and also noted the motion of the Sun toward a point in the constellation Hercules.

Herschel Space Observatory A European Space Agency satellite launched on 14 May 2009. It carried a cooled 3.5 m (138-inch) Cassegrain telescope for photometry and spectroscopy at infrared and submillimeter wavelengths (55 to 672 μm) and operated from orbit around the Earth's L_2 **Lagrangian point**. It ceased observations on 29 April 2013.

Hertzsprung, Ejnar (1873–1967) Danish astronomer. He showed that the colors of the stars were related to their luminosities, plotting one against the other on what is now called the **Hertzsprung–Russell diagram** and discovering the existence of giant and dwarf stars. All studies of stellar evolution stem from this diagram, but Hertzsprung's work remained largely overlooked until Henry Norris **Russell** independently developed the diagram in a slightly different form.

Hertzsprung gap The region on the **Hertzsprung–Russell diagram** between the main sequence and the giant branch where there are few stars. The gap exists because stars evolve rapidly through this region.

Hertzsprung–Russell diagram (HR diagram) A plot of the **absolute magnitude** of stars against their **spectral type**; this is equivalent to plotting their **luminosity** against their surface temperature or their **color index**. Brightness increases from bottom to top, and temperature increases from right to left. The diagram was devised by Henry Norris **Russell** in 1910, independently of Ejnar **Hertzsprung**, who had had the same idea some years earlier.

On the HR diagram most stars, including our Sun, lie on a diagonal band known as the **main sequence**. A region populated by **giant stars**, mainly K and M spectral types, lies above the main sequence. There are further groupings above the main sequence called **supergiants** and **subgiants**. Below the main sequence are **subdwarfs** and **white dwarfs**.

The HR diagram is an important aid for interpreting astrophysical data. For example, the total range of stellar brightness is 27 magnitudes, corresponding to a factor of 10^{11} in luminosity, and stars' surface temperatures range from 2200 to 50,000 K. Luminosity and surface temperature are both related to the size of a star. Most main-sequence stars are about the size of the Sun. White dwarfs are about the size of the Earth, while red giants and supergiants can be as big as the Earth's orbit, and supergiants even exceed that of Saturn. About 90% of the stars in our neighborhood are main-sequence stars, about 10% are white dwarfs, and only about 1% are red giants or supergiants.

The existence of the three main groupings on the HR diagram indicates that a typical star passes through three very different stages in its life, and the progress of an evolving star can be followed on an HR diagram, from **protostar** to main sequence, then to giant star, and ending as a white dwarf (SEE **stellar evolution**). The diagram can also provide information about the ages of stars, and such dating is especially useful for establishing the ages of star clusters. Because the main sequence is a progression in stellar mass, the stars at the upper (most massive) end of the main sequence are the first to leave. As the cluster ages, its main sequence "burns down" like a candle as stars of progressively decreasing mass and luminosity become red giants. The stage which this process has reached reveals the cluster's age. SEE ALSO **Hayashi track**, **Hertzsprung gap**, **instability strip**.

HET ABBREVIATION FOR **Hobby–Eberly Telescope**.

Hevelius, Johann (1611–87) German astronomer. He was the last great astronomer to make observations with naked-eye sighting instruments for positional measurements, from which he produced a catalog of 1500 stars. He also made a map of the Moon, and discovered the Moon's **libration** in longitude.

Hewish, Antony (1924–) English radio astronomer. He and Martin **Ryle** developed the technique of **aperture synthesis** in 1960. In 1967 his student Jocelyn **Bell Burnell** obtained the first signal from a pulsar, now known as CP 1919. For subsequent work on pulsars, Hewish shared (with Ryle) the 1974 Nobel Prize for Physics.

Hidalgo Asteroid no. 944, discovered by Walter **Baade** in 1920. Its diameter is about 50 km (30 miles), and its eccentric comet-like orbit takes it from just outside Mars' orbit out as far as Saturn. Although it has an asteroidal designation, it could well be a large extinct cometary nucleus.

high-velocity star A star in the galactic halo that does not share the rotation of the majority of stars in the Galaxy, including the Sun, around the galactic center. Consequently, such stars seem to have a large proper motion and a high velocity relative to the Sun. The orbits of high-velocity stars are usually eccentric, with a large inclination to the galactic plane. Most of them are Population II stars (SEE **stellar populations**).

Himalia The largest of Jupiter's intermediate group of **satellites**. It was discovered by Charles Perrine in 1904.

Hipparchus of Nicea (c. 190–125 BC) Greek astronomer, geographer and mathematician. He made many accurate astronomical observations and compiled a star catalog (now lost but incorporated in Ptolemy's *Almagest*) giving coordinates and magnitudes. Hipparchus discovered the **precession** of the equinoxes. He found irregularities in the motions of the Sun and Moon, and from observations of eclipses made commendable estimates of their distances and sizes.

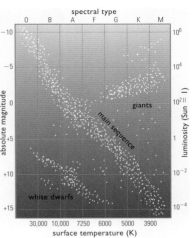

▲ *Hertzsprung–Russell diagram* A plot of stars' luminosity against temperature. Most stars lie on the main sequence running from top left to bottom right. This diagram shows the principal regions of the Hertzsprung–Russell diagram, with giant stars above the main sequence to the top right, and white dwarfs to the lower left.

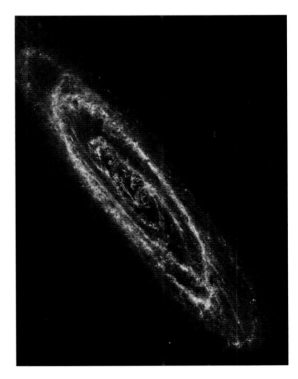

▶ *Herschel Space Observatory* The Andromeda Galaxy, M31, seen at far-infrared wavelengths by the European Space Agency's Herschel Space Observatory. Dust between the stars in the spiral arms of M31 glows brightly in the infrared, revealing details of the galaxy's structure.

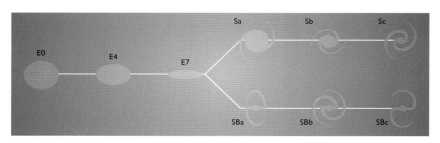

Hipparcos A satellite launched on 8 August 1989 by the European Space Agency, designed for **astrometry**. The name is a contraction of High Precision Parallax Collecting Satellite, and is also an allusion to **Hipparchus of Nicea**, who recorded accurate star positions more than two millennia previously. Hipparcos operated until August 1993, measuring parallaxes and proper motions of 120,000 stars to a precision of 1 to 2 milliarcseconds, and of a million further stars (though not to such a high accuracy). These two sets of observations were published in 1997 as the *Hipparcos Catalogue* and the *Tycho Catalogue*. A reanalysis of the observations, providing improved accuracy, was published in 2007.

Hirayama family A group of **asteroids** having closely similar orbital elements (SEE **orbit**), in particular the **semimajor axis**. About a hundred such families are known, named usually after the largest asteroid in them. The *Flora*, *Eos* and *Themis* families are three of the main ones. The fact that in many families the members have similar spectra suggests that each family originates from one larger body which has been broken up by impact. These groupings were discovered by Japanese astronomer Kiyotsugu Hirayama (1874–1943) in 1928.

Hobby–Eberly Telescope (HET) A reflecting telescope opened in 1997 at McDonald Observatory, Texas, with an effective aperture of 9.2 m (360 inches). The main mirror consists of 91 hexagonal segments 1 m (40 inches) across, the images from which are directed to a common focus by a complex corrector system. The mounting is considerably simplified by having the telescope permanently aimed at an altitude of 55°, and moving only in azimuth. Even so, 70% of the sky is visible to it over the course of the year. It is named after William P. Hobby and Robert E. Eberly, American supporters of public education.

homogeneity SEE **cosmology**.

Homunculus Nebula SEE **Eta Carinae**.

horizon The **great circle** on the **celestial sphere** 90° from the observer's overhead point, the zenith.

horizontal coordinates SEE **coordinates**.

horizon problem A problem with the standard **big bang** model in that the **cosmic microwave background** radiation is too uniform. Areas of the sky now separated by as little as 2° would not have been causally connected to each other at the time of the origin of the radiation, so there would be no reason for them to have the very similar temperatures they are indeed observed to have. **Inflation** solves the problem because the enormous initial rapid expansion means that all parts of the visible Universe were causally connected when the radiation originated.

Horologium An obscure constellation of the southern sky, representing a pendulum clock, introduced by **Lacaille**. Alpha, which is of magnitude 3.9, is the constellation's only star above magnitude 4.

Horsehead Nebula A dark nebula in the constellation Orion with a very distinctive shape, like the head and mane of a horse. It is silhouetted against the bright emission nebula IC 434 south of Zeta Orionis. IC 434 is colliding with an extensive dusty cloud on the east side of Orion, and the Horsehead is the most distinctive intrusion of that dark cloud. The Horsehead Nebula is 1600 light years away and measures some 3 light years across.

horseshoe mounting A type of mounting used for a large **equatorial telescope**.

Horseshoe Nebula SEE **Omega Nebula**.

hot dark matter SEE **dark matter**.

hour angle (HA) The angle that the **hour circle** of an object makes with the **celestial meridian**. It is measured westward from the meridian, along the celestial equator to the point where the equator and hour angle first intersect.

hour circle A **great circle** on the **celestial sphere** passing through both celestial poles. **Declination** is measured along hour circles. On each half of an hour circle running from pole to pole, all points have the same **right ascension**.

Hoyle, Fred (1915–2001) English astrophysicist and cosmologist. With Hermann **Bondi** and Thomas **Gold** he developed the **steady-state theory**, in which the continuous creation of matter drives the expansion of the Universe. Although it was subsequently displaced by the **big bang** theory (which owes its name to a disparaging remark by Hoyle), it sparked research by Hoyle and others into **nucleosynthesis** in stars. Hoyle attracted controversy with unorthodox ideas, such as his suggestion that life was brought to Earth by comets.

HR diagram ABBREVIATION FOR **Hertzsprung–Russell diagram**.

HST ABBREVIATION FOR **Hubble Space Telescope**.

Hubble, Edwin Powell (1889–1953) American astronomer. At **Mount Wilson Observatory** near Pasadena, California, he used the 2.5 m (100-inch) Hooker Telescope to examine nebulae. He found that although some nebulae were clouds of gas lying within our Galaxy, many others – the so-called spiral nebulae – were resolvable as independent star systems. His **Hubble classification** of galaxies is still in use. In 1924, he measured the distance to the Andromeda Galaxy by identifying **Cepheid variable** stars in it. He attributed the **redshift** in the spectra of galaxies to indicate that they were receding, which was interpreted by others as evidence that the Universe was expanding. Hubble established the velocity–distance relationship of recession (SEE **Hubble constant**, **Hubble law**).

Hubble classification A classification of **galaxies** according to their shape or structure, introduced by Edwin Hubble in 1925. The three major categories are elliptical (E), spiral (S) and barred spiral (SB); each category has several subdivisions depending on the observable shape or structure of the galaxy. Irregular galaxies (Ir) were not included in Hubble's original classification. SEE ALSO **tuning fork diagram**.

Hubble constant (symbol H_0) The rate at which the velocity of recession of galaxies increases with distance from us, according to the **Hubble law**. The zero subscript specifies that it is the expansion rate at the present time that is meant, for, according to the theory of the **big bang**, the rate changes with time (SEE **deceleration parameter**). Depending on what assumptions are made, observations using a variety of methods indicate that H_0 is close to 70 km/s/megaparsec.

The inverse of the Hubble constant is called the *Hubble time*, which gives a maximum age for the Universe on the assumption that there has been no slowing of the expansion. The Hubble time for $H_0 = 70$ km/s/megaparsec is 14 billion years, assuming that all the galaxies started to recede from a common point in the past, at the time of the big bang (SEE **age of the Universe**).

Hubble diagram A graph in which the apparent magnitude of galaxies is plotted against the redshift of their spectral lines. It is a straight line, demonstrating the linear relation between redshift and distance, as embodied in the **Hubble law**.

Hubble law The law proposed by Edwin **Hubble** in 1929 claiming a linear relation between the distance (r) of galaxies from us and their velocity of recession (v), deduced from the redshift in their spectra. The figure linking velocity with distance is the **Hubble constant**. Hence $v = H_0 r$.

Hubble Space Telescope (HST) An astronomical telescope operating at visible, near-ultraviolet and near-infrared wavelengths, of Cassegrain design with a main mirror 2.4 m (94 inches) in diameter. It was launched by the Space Shuttle *Discovery* in April 1990 into an orbit about 600 km (375 miles) high. It was found to suffer from **spherical aberration**, attributed to a manufacturing error in its main mirror. Astronauts fitted corrective optics to the telescope during a servicing mission in December 1993.

The HST was designed to be repaired and serviced in orbit by Space Shuttle crews, and to allow its main and auxiliary instruments to be replaced with new ones. Servicing missions took place in 1997, 1999, 2002 and 2009. Its final instrumentation consists of the Advanced Camera for Surveys (ACS), a battery of three cameras capable of imaging from the mid-ultraviolet to the near-infrared regions of the spectrum; the Wide Field Camera 3 (WFC3); the Space Telescope Imaging Spectrograph (STIS); the Cosmic Origins Spectrograph (COS); and the Near-Infrared Camera and Multi-Object Spectrometer (NICMOS). Its successor, the **James Webb Space Telescope**, is not planned for launch until 2018.

Hubble's Variable Nebula An emission nebula in Monoceros, associated with the variable star R Monocerotis. Variations in the light output of this young star, and motions of dark obscuring material in the star's vicinity, cause the nebula (designated NGC 2261) to vary in both brightness and outline on timescales of weeks. Discovered by William **Herschel** in 1783, NGC 2261 lies 2600 light years away, and was the subject of detailed study by Edwin **Hubble** in the 1910s.

Huggins, William (1824–1910) English amateur astronomer and spectroscopist. He built an observatory and constructed the first stellar spectroscope, which he used in a series of investigations aided by neighbor and chemist William Miller (1817–70). In 1863 he showed that the stars contain the same chemical **elements** that are present on Earth, and the next year he confirmed the gaseous nature of a diffuse nebula. He investigated the motion of Sirius, and made the first discovery of a **moving cluster** (the Ursa Major cluster). In 1876 he took one of the first photographs of a star's (Vega's) spectrum. Huggins used spectroscopy to investigate many other objects, including comets, meteors and the 1892 nova in Auriga.

Hulst, Hendrik Christoffel van de (1918–2000) Dutch astronomer. He carried out research on interstellar matter and the solar corona. In 1944 he predicted that hydrogen should emit radio waves with a wavelength of 21 cm (SEE **twenty-one centimeter line**), and in 1951 he and Jan **Oort** used Doppler shifts at this wavelength to map the Galaxy.

hunter's Moon The first full Moon after the **harvest Moon**, providing a succession of moonlit evenings in early October. In the southern hemisphere the hunter's Moon is in April or May.

Huygenian eyepiece A basic telescope lens commonly found on small refractors. It consists of two simple elements and is relatively free of **chromatic aberration**. (After Christiaan **Huygens**.)

Huygens, Christiaan (1629–95) Dutch mathematician, physicist, and astronomer. In 1655 he discovered Saturn's largest satellite, Titan, and the next year explained that the planet's telescopic appearance was due to a broad ring surrounding it. He helped to develop the telescope and introduced the convergent eyepiece. Huygens' many contributions to physics include the idea that light is a wave motion, and the theory of the pendulum; he built the first pendulum clock.

Huygens SEE **Cassini–Huygens**.

Hyades A large open cluster several degrees across, 151 light years distant in the constellation Taurus. The brightest members are visible to the naked eye; in all it contains about 200 stars, most within a radius of 20 light years. Its estimated age is 650 million years. It is a **moving cluster**, receding from us at 43 km/s. The star Aldebaran lies in the same field as the Hyades but is less than half as distant.

Hyakutake, Comet C/1996 B2 An exceptional comet, discovered by Japanese amateur astronomer Yuuji Hyakutake (1950–2002) in January 1996, which passed within 0.1 au of the Earth. At its best, around 25 March, it was of zero magnitude. The gas tail, which some observers reported to be over 90° long, underwent several **disconnection events**, and activity on the Sunward side of the nucleus was apparent in the form of jets and fans.

Hydra The largest constellation in the sky, representing the water snake killed by Hercules; it extends from Canis Minor to the south of Virgo. Its brightest star is Alpha (Alphard), magnitude 2.0. The Mira variable R Hydrae can reach magnitude 3.5 at its brightest, and fades to 10.9; its period is 389 days. There is one bright open cluster, M48.

hydrogen-alpha line (Hα line) The absorption or emission line of hydrogen in the red portion of the spectrum at a wavelength of 656.3 nm. The first line in the **Balmer series**, it is the third most conspicuous absorption line in the Sun's spectrum, and is responsible for the pink color of the solar chromosphere when viewed during a total eclipse. Light of this wavelength is often used for monochromatic study of the Sun with the spectrograph, and Hα filters are used by amateur observers for solar imaging.

hydrogen spectrum The **spectrum** of atomic hydrogen. The hydrogen atom has the simplest structure of any atom (one proton as nucleus with a single orbiting electron), and the hydrogen spectrum is the simplest to interpret. As over 90% of the atoms in the Universe are hydrogen atoms, hydrogen lines feature prominently in the spectra of many celestial objects. The spectrum contains five principal series of lines associated with different levels of energy in the hydrogen atom, among them the **Balmer series** and the **Lyman series**. Each line corresponds to a quantum-mechanical process called a transition in which the electron "jumps" between orbits (SEE **emission spectrum**).

Hydrus A far southern constellation representing a small water snake. There are few interesting telescopic objects. Beta Hydri, magnitude 2.8, is the brightest star in the constellation and the nearest fairly bright star to the south celestial pole.

Hygeia Asteroid no. 10, discovered by Annibale de Gasparis in 1849. It orbits in the main belt, and is the fourth largest asteroid, with a diameter of 408 km (254 miles).

hyperbola An open curve. In mathematics the hyperbola is a type of conic section, so called because it is one of the intersections of a plane with a cone, and is formed when a plane cuts a cone at an angle to the cone's axis that is less than the cone's slope. In astronomy a hyperbolic **orbit** is the path followed by a celestial body which passes another body but is not captured into an orbit around it. For a hyperbolic orbit the **eccentricity** is greater than 1.

hyperboloidal Having the form of the surface obtained by rotating a **hyperbola** about its axis. The hyperboloid is the shape used for the convex secondary mirror in a **Cassegrain telescope**.

Hyperion An outer **satellite** of Saturn, discovered by William Cranch Bond and his son George Philips Bond, and independently by William **Lassell** in 1848. Measuring 370 × 225 km (230 × 140 miles), it is irregular in shape and displays chaotic rotation – tumbling about no fixed axis as it follows an eccentric orbit which is in a 4:3 resonance with that of **Titan**. Although Hyperion is a low-density, icy body, its surface is dark. There are several large craters and a scarp, named Bond–Lassell, which extends for 300 km (200 miles). It seems most likely that Hyperion is a fragment of a larger body.

hypernova An exceptionally violent **supernova** event that appears to be an order of magnitude greater than a normal supernova explosion. Such events have been observed in starburst regions of irregular galaxies.

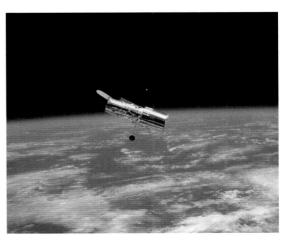

▼ **Hubble Space Telescope**
Launched in 1990, the Hubble Space Telescope has provided astronomers with many results of fundamental importance. Here it is seen soon after release following the second servicing mission, carried out by Space Shuttle astronauts in 1997.

▶ **Iapetus** *Cassini flew past Iapetus on 31 December 2004. The resulting images show the heavily cratered region named Cassini Regio; the topographic ridge extending along the moon's equator is conspicuous.*

Iapetus A large outer **satellite** of Saturn discovered by Giovanni Domenico Cassini in 1671. Its diameter is 1440 km (895 miles), and its density of 1.21 g/cm^3 indicates that it is predominantly icy. Like most of Saturn's satellites, Iapetus has **synchronous rotation**. Its trailing hemisphere is bright, but its leading hemisphere (the one facing the direction of orbital travel) is extremely dark, being completely coated by a deep-red material. It seems more likely that the dark material was deposited from outside rather than expelled from within. It may be dust ejected from **Phoebe**, the next satellite out; or material from the satellite itself which has fallen back to the surface following ejection by major impacts. The **Cassini** spacecraft revealed at least three large impact structures on Iapetus, along with a narrow, steep mountain range reaching a height of 20 km (12 miles) and extending for 1300 km (800 miles) along the satellite's equator in the dark hemisphere. As a result of the difference in **albedo** between the hemispheres, Iapetus appears markedly brighter when west of Saturn than when east from our terrestrial viewpoint.

IAU ABBREVIATION FOR **International Astronomical Union**.

IC ABBREVIATION FOR the *Index Catalogue*, consisting of two supplements to the *New General Catalogue* (SEE **NGC**) produced in 1895 and 1908 by J. L. E. **Dreyer**. Objects in the *Index Catalogues* are still referred to by their IC numbers.

Icarus Asteroid no. 1566, discovered by Walter **Baade** in 1949. It is an **Apollo asteroid** with a diameter of about 1.5 km (1 mile), and has a perihelion distance of 30 million km (19 million miles), within the orbit of Mercury.

Ida Asteroid no. 243, discovered by Johann Palisa in 1884. In 1993 it became the second asteroid to be examined from close quarters, by the probe **Galileo** *en route* to Jupiter. It is an elongated, cratered body measuring 56 × 24 × 21 km (35 × 15 × 13 miles), and has a 1.3 km (0.8-mile) satellite named Dactyl.

Ikeya–Seki, Comet C/1965 S1 The brightest comet of the 20th century, independently discovered in September 1965 by the Japanese amateur astronomers Kaoru Ikeya (1943–) and Tsutomu Seki (1930–). It is the type of comet known as a **sungrazer**. Perihelion was on 21 October 1965 at a mere 470,000 km (290,000 miles) from the Sun's surface, when it reached an estimated magnitude of −10. Around the time of perihelion the comet's nucleus broke into three pieces, creating a bright dust tail. The orbital period of the main fragment is 880 years.

immersion The entry of a celestial body into the shadow of another at the beginning of an eclipse or occultation.

impact feature Any surface feature of a planet, satellite or asteroid, or other body with a solid surface, that results from the past impact of another body. The most obvious are impact **craters**. *Basins* are large shallow depressions marking the impact of a large body, which have often been filled in by upwelling fluid from below; examples are the lunar maria (SEE **mare**). *Multi-ringed basins*, such as **Valhalla** on Callisto, are surrounded by concentric ringed structures representing "ripples" from the impact. Other impact features include **rays** radiating from craters, and *palimpsests* – the "ghosts" of craters flattened out or filled in since the impact that caused them. The term was also applied to the visible atmospheric disturbances that were caused by the collision of Comet **Shoemaker–Levy 9** with Jupiter.

inclination (1) (symbol *i*) The angle between the orbital plane of a body orbiting the Sun and the plane of the ecliptic.

inclination (2) (symbol *i*) The angle between the orbital plane of a satellite and the equatorial plane of the planet it orbits.

inclination (3) (symbol *i*) The angle between the orbital plane of a double star and the plane at right angles to the line of sight. The angle is zero if the orbit is seen in plan, and 90° if seen in profile.

inclination (4) (symbol *i*) The angle between the equatorial plane of a planet or satellite and its orbital plane, equivalent to the angle between the axis of rotation and the perpendicular to the orbital plane.

Indus A small southern constellation representing an American Indian. The only star above magnitude 3.5 is Alpha (3.1). Epsilon is the least luminous star visible to the naked eye: magnitude 4.7, distance 11.8 light years, type K4, absolute magnitude 6.9.

inequality The departure from uniform orbital motion of a body, due to **perturbations** by other bodies and the eccentricity of its orbit.

inferior conjunction SEE **conjunction**.

inferior planet Either of the planets Mercury and Venus, both of whose orbits lie inside the Earth's.

inflation A proposed early stage of the Universe, immediately after the **big bang**, in which for a brief fraction of a second the Universe expanded faster than the speed of light. Relativity is not contravened

▶ **Ida** *Asteroid (243) Ida was imaged by the Galileo spacecraft in August 1993. Its tiny satellite Dactyl is visible at right.*

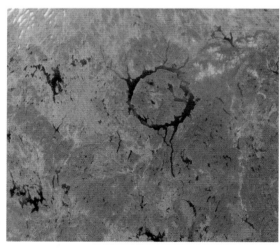

▲ **impact feature** *The circular Manicouagan impact structure in Quebec, Canada, is revealed in this satellite image. The 70 km (44-mile) diameter feature was formed by an asteroid impact about 212 million years ago.*

because it does not involve the movement of objects through space faster than light, but rather the expansion of space itself.

Inflation was proposed in 1981 by Alan Guth to explain the **flatness, horizon** and **monopole problems** that arise with the standard big bang model. Inflation requires the Universe to expand exponentially by at least a factor of 10^{50} and perhaps by as much as $10^{1,000,000,000,000}$ over the interval from about 10^{-34} to 10^{-32} seconds after the start of the big bang. After inflation the Universe could be as large as a football or many light years in diameter, in either case continuing to expand "normally." The driving mechanism for inflation is uncertain, but one suggestion is that immense quantities of energy were released when the strong nuclear force came into existence as a separate entity (SEE **grand unified theory**), and this not only drove the expansion but also created all the matter and radiation in the Universe.

Although inflation may seem an unlikely process, observational support for it came with the observation by the **Wilkinson microwave anisotropy probe** of polarization in the **cosmic microwave background** radiation. Inflation predicts that polarization with exactly the observed properties should be produced during the last scatterings of photons at the end of the **decoupling era**. SEE ALSO **cosmology**.

Infrared Astronomical Satellite (IRAS) A satellite, a collaboration between the USA, Netherlands and UK, that surveyed the sky at infrared wavelengths. It was launched on 26 January 1983 and operated for nine months, cataloging nearly 250,000 sources. Its discoveries included several comets and asteroids, possible planetary systems forming around nearby stars, and **starburst galaxies**.

infrared astronomy The study of **electromagnetic radiation** falling between the red end of the visible spectrum and radio wavelengths, between about 700 nm (or 0.7 μm) and 1 mm. The Earth's atmosphere is opaque to much of this wavelength range, although there are a number of atmospheric *windows* through which infrared (IR) observations can be made from high mountains. To avoid the Earth's atmosphere, observations are made from high-flying aircraft (SEE **Stratospheric Observatory for Infrared Astronomy**), from uncrewed balloons and from satellites such as the **Infrared Space Observatory**, the **Herschel Space Observatory** and the **Spitzer Space Telescope**.

Astronomical sources of IR radiation include the planets and other Solar System bodies, **extrasolar planets**, the stars and dusty regions of the Milky Way, and other galaxies (including **starburst galaxies**). Most of the interstellar dust and gas in our Galaxy is contained in

▲ **infrared astronomy** *Produced using data from IRAS, this image shows the constellation Orion. The bright patch at lower center is the Orion Nebula. Gas and dust glow brightly in the infrared region.*

giant molecular clouds (GMCs, SEE **molecular cloud**), such as the one in Orion, part of which is visible as the Orion Nebula. Maps of GMCs at IR wavelengths show the warm or hot regions where star formation is in progress or is about to begin. On a much smaller scale, highly compact and opaque **Bok globules** with just a few tens of solar masses are also potential star-forming regions. Infrared observations show the temperature of the globules to be very low, 10–20 K, indicating that they are at an early stage of gravitational collapse. Faint, wispy structures in interstellar space termed *infrared cirrus*, discovered by IRAS, are probably caused by infrared emissions from dust grains. These grains are composed of carbon in the form of graphite at a temperature of 30–40 K, ejected by the stellar winds of cool stars or supernova explosions.

Infrared wavelengths can penetrate the otherwise opaque clouds of dust in the plane of our Galaxy, so we can see sources toward the galactic center that are completely obscured at optical wavelengths. The observations show a high density of late-type stars and a large ring of molecular gas and dust, as well as many regions of active star formation. Beyond the Milky Way, IR observations are a key to understanding the energy output of active galaxies and quasars and the physical processes operating within them.

In the Solar System, IR observations allow us to work out the basic chemistry of the planets, their satellites and the asteroids. Molecules in planetary atmospheres exhibit detailed IR absorption spectra, which permit the composition and temperature of the atmospheres to be established. Spectral reflectance observations of asteroids, planetary satellites and small planets without substantial atmospheres have been used to discover the mineralogical compositions of their surfaces. IR techniques have established the presence of water and carbon dioxide ices on Mars, water ice on many of the satellites of Jupiter, Saturn and Uranus, and methane ice on Triton and on Pluto. Measurement of the black-body emission of small Solar System bodies has given us information on their sizes and surface albedos where no other technique would succeed.

Infrared Space Observatory (ISO) A European Space Agency satellite with a 0.6 m (24-inch) telescope for infrared astronomy, launched on 17 November 1995. It operated until 1998.

inner planet Any of the planets Mercury, Venus, Earth and Mars, which orbit inside the main asteroid belt.

instability strip The narrow region of the **Hertzsprung–Russell diagram** where pulsating stars are located. It extends from the brightest Cepheid variables down to pulsating white dwarf stars, and also includes RR Lyrae stars, Delta Scuti stars and dwarf Cepheids.

Integral A joint European, US and Russian satellite for gamma-ray astronomy launched on 17 October 2002; the name is short for International Gamma-Ray Astrophysics Laboratory. It carries a gamma-ray imager and spectrometer for identifying and analyzing gamma-ray sources such as active galaxies, supernovae and gamma-ray bursters.

interferometry The study of a celestial object by analyzing its interference pattern. When electromagnetic radiation emitted by a source is split into two and recombined, a pattern of interference fringes is produced that can be used to reveal more detailed information about the source than is possible from a single beam.

An *interferometer* is an instrument consisting of two or more detectors (telescopes or antennas) which collect signals from a single source. The signals are then combined to form the interference pattern. The longer the wavelength of the radiation being detected, the farther apart the detectors need to be.

Various types of **optical interferometer** are used to measure the diameters of stars and resolve fine detail in their spectra. Interferometry is particularly important in radio astronomy, where the long wavelengths of the radiation mean that a single detector gives only a coarse resolution (SEE **radio interferometer**).

intergalactic matter Matter in the space between galaxies. There are no significant quantities of intergalactic dust, but there are clear signs of intergalactic gas. X-ray emission from clusters of galaxies comes from a hot, tenuous gas with a temperature of 10 to 100 million K between the galaxies in the cluster. Radio-emitting clouds of electrons ejected by radio galaxies are drawn out into a tadpole shape as if they were plowing through a resisting medium. How much intergalactic gas is primordial hydrogen and helium from the big bang that is falling into the clusters, and how much is matter

▲ **Infrared Space Observatory** *(ISO) The Eagle Nebula is shown here in a composite ISO image. Shorter infrared wavelengths are rendered in blue, longer wavelengths in red.*

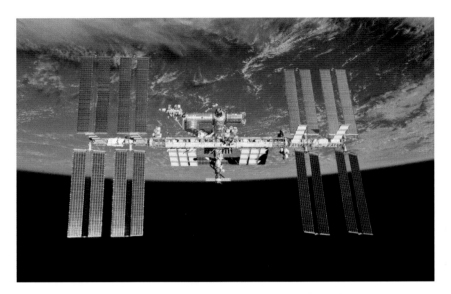

◀ *International Space Station*
Photographed by a crew member on the Space Shuttle Atlantis in May 2010, the International Space Station is seen with the Earth in the background. Huge flat solar panels on extended trusses provide electrical power to the station. Following retirement of the Space Shuttle fleet in 2011, astronauts have been ferried to and from the station in Russian Soyuz craft.

expelled from the galaxies, is uncertain. Tidal interactions (SEE **tides**) between galaxies are common, and radio observations reveal streams of neutral hydrogen around interacting galaxies, as, for example, with M81 and M82. A similar gas stream, the Magellanic Stream, links the **Magellanic Clouds** with our Galaxy. It is possible that **dark matter** exists in the space between galaxies and galaxy clusters.

International Astronomical Union (IAU) The controlling body of world astronomy, founded in 1919, with its headquarters in Paris. It is the authority for naming celestial bodies and their surface features.

International Atomic Time (TAI) The international reference scale of atomic time, formed by comparing data from around 200 atomic clocks around the world. (The abbreviation TAI is derived from the name in French.) The fundamental unit of time is the SI second, which is defined in terms of the resonance of the caesium atom. SEE ALSO **Universal Time**.

International Space Station (ISS) A space station built as a collaborative venture between NASA, the Russian Space Agency, ESA, Japan and Canada. The first section was launched in 1998 and additional small modules continue to be added from time to time. It has a total mass of nearly 420 tonnes, a length of 109 m (357 ft) from one end of its solar arrays to the other and a habitable volume larger than a six-bedroom house. The ISS orbits just over 400 km (250 miles) above the Earth and can be easily seen as a brilliant star when passing overhead. It is expected to remain in operation until at least 2020.

International Ultraviolet Explorer (IUE) A joint NASA, ESA and UK satellite, carrying a 0.45 m (18-inch) telescope. It was launched on 26 January 1978 and operated until 1996, taking ultraviolet spectra of stars and galaxies.

interplanetary matter The material in the space between the planets. It is made up of atomic particles (mainly protons and electrons) escaping from the Sun via the **solar wind**, and dust particles (mainly from **comets**, but some possibly of cosmic origin) and neutral gas in the plane of the ecliptic. The particles in the inner Solar System constitute the *zodiacal dust cloud*, scattered sunlight from which is responsible for the **zodiacal light**. An extremely tenuous ring of dust, over 50 million km (30 million miles) wide, orbits the Sun just outside the Earth's orbit. There is also the dust in **meteor streams**.

interstellar matter The matter contained in the regions between objects in the Galaxy. It includes clouds of *interstellar gas*, mainly hydrogen with some helium, and a small percentage (about 1% by mass) of interstellar dust grains. The hydrogen is mainly in a relatively cool neutral form at a temperature of 10–100 K (SEE **H I region**). Some very much hotter and more diffuse regions of ionized gas (**H II**) also exist, together with dense clouds of molecular hydrogen and other molecules. The dust grains are elongated in shape, about 100 nm long, and aligned by the galactic magnetic field. Most grains are thought to consist of silicates or forms of carbon, and some have icy coatings.

The clouds of gas and dust are largely confined to the plane of the Galaxy and tend to be concentrated in the spiral arms. Our Galaxy contains about 10^{10} solar masses of material distributed between the stars, making up about 10% of its visible mass. The density of these clouds is less than that of a laboratory vacuum on Earth, but they nevertheless have an obscuring effect on the light of stars behind them. The clouds scatter and absorb starlight passing through them, a phenomenon known as *interstellar extinction*. On average, starlight is dimmed by one magnitude for every kiloparsec it travels in the plane of the Galaxy. Red light is dimmed less than blue light, which causes a reddening of starlight.

The spectra of remote stars show **absorption lines** superimposed on the stellar spectra, produced by interstellar dust and gas. Sodium, calcium, potassium, iron, titanium and other elements have been detected from these absorption lines. These elements are ejected from stars. Heavier elements, including lead, gallium and krypton, are manufactured in the extreme conditions of a **supernova** explosion. SEE ALSO **nebula**.

interstellar molecules Molecules that exist in interstellar **molecular clouds**. Most have been identified by their emission lines at millimeter and centimeter wavelengths, although the first, including the cyanogen radical (CN) and the methylidyne radical and ion (CH, CH^+), were detected in the 1940s at optical frequencies. The hydroxyl radical (OH) was the first (in 1963) to be identified by radio astronomy. The list of interstellar molecules has since been extended to more than 200, and includes hydrogen cyanide (HCN), ammonia (NH_3), methane (CH_4) and methyl alcohol (CH_3OH), as well as more complex organic molecules such as the amino acid glycine.

intrinsic variable SEE **variable star**.

inverse-square law A reduction in the intensity of a physical quantity in proportion to the square of the distance from its source. For example, a planet three times the Earth's distance from the Sun

▼ *International Space Station* Backdropped by the Earth, Russian cosmonaut Sergey Ryazanskiy waves to the camera during extravehicular activity (EVA) outside the International Space Station in November 2013.

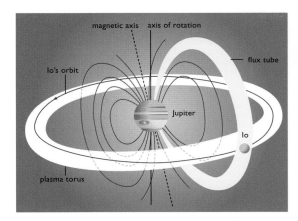

▲ **Io** *Material ejected from Io's volcanoes forms a torus around Jupiter. A magnetic flux tube links Io and Jupiter: particles ejected during the satellite's volcanic eruptions can lead to enhancements of Jupiter's aurorae.*

▲ **Io** *A composite of images from the Galileo orbiter shows the volcanically active surface of Jupiter's moon Io. Intense volcanism driven by tidal stresses leads to rapid resurfacing of Io.*

would experience one-ninth the gravitational attraction of the Sun and receive one-ninth as much sunlight.

Io The innermost of Jupiter's four **Galilean satellites**, and the second smallest, with a diameter of 3643 km (2264 miles). It has the highest density, at 3.53 g/cm^3, indicating a lack of water ice in its makeup. Io's tidal interaction with Jupiter, **Europa** and **Ganymede** raises a bulge up to 100 m (330 ft) high in its surface, creating sufficient flexing and heating to make it by far the most volcanically active world in the Solar System. The surface, multicolored by the presence of different forms of sulfur, is being continually renewed. The active regions contain volcanic vents and calderas, lava flows and volcanic plumes which send material to heights of up to 300 km (200 miles). Plumes prominent when the Voyager probes imaged Io included those named Pele, Loki and Prometheus. In the active regions there may be lakes of molten sulfur or even molten rock; elsewhere there are plains formed by volcanic deposits, and mountains. A thin atmosphere of sulfur dioxide surrounds Io, and a donut-shaped *plasma torus* of ionized (electrically charged) sulfur and oxygen surrounds its whole orbit and interacts with Jupiter's **magnetosphere** to create a *flux tube* carrying an electric current of about 5 million amperes (SEE the diagram above). In 1996 the **Galileo** probe detected an iron core and magnetic field, and found the surface features to have changed significantly since the satellite was first imaged during the 1979 Voyager missions. SEE ALSO **satellite**.

ionosphere A region in the atmosphere of a planetary body in which there are free electrons and ions produced by ultraviolet radiation and X-rays from the Sun. The degree of ionization is greatly affected by solar activity.

The Earth's ionosphere extends from about 60 km to 300 km above the surface, and reflects radio waves of long wavelengths, although shorter wavelengths can pass through undisturbed. The reflecting power of these layers makes long-range broadcasting and tele-communications possible up to frequencies of about 30 MHz. The radio waves and microwaves from around 30 MHz to about 100 GHz that pass through the ionosphere are used in radio astronomy and satellite communications; they make up the *radio window*. The free electrons formed as a result of ionization do at times have serious adverse effects on the reception of radio signals.

IRAS ABBREVIATION FOR **Infrared Astronomical Satellite**.

Iris Asteroid no. 7, discovered by John Russell Hind in 1847. Although it ranks only about 25th in size, with a diameter of around 200 km (125 miles), it is the fourth brightest.

iron meteorite A meteorite consisting mainly of iron (90–95%) and some nickel, with traces of other metals. Iron meteorites are classified into a number of groups according to the proportions of these other metals that they contain, and each group is thought to correspond to a different parent asteroid.

irradiation A physiological phenomenon in which a bright object viewed against a dark background appears larger than it really is. The effect can introduce errors into measurements made by an observer using a telescope.

irregular galaxy A galaxy that shows little or no symmetry. Some irregular galaxies are smaller than spirals and ellipticals, contain much gas and are undergoing star formation. Some are classified as irregular for the only reason that they do not fit into other categories of galaxy.

irregular variable A pulsating variable star whose variation in brightness does not follow any regular or predictable pattern. Irregular variables are divided into two broad groups: **eruptive variables** such as **shell stars**, **T Tauri stars** and young stars associated with nebulosity; and **pulsating stars** that are red giants or supergiants of late spectral class, and exhibit slow variations.

Isaac Newton Group of Telescopes (ING) Group of instruments consisting of the 4.2 m (165-inch) **William Herschel Telescope**, the 2.54 m (100-inch) Isaac Newton Telescope and the 1.0 m (39-inch) Jacobus Kapteyn Telescope, at the **Roque de los Muchachos** observatory on the island of La Palma in the Canaries. The ING is funded by the UK, the Netherlands and Spain.

ISO ABBREVIATION FOR **Infrared Space Observatory**.

isotropy SEE **cosmology**.

ISS ABBREVIATION FOR **International Space Station**.

IUE ABBREVIATION FOR **International Ultraviolet Explorer**.

▼ **iron meteorite** *An iron meteorite found on the surface of Mars by the NASA rover Opportunity. This basketball-sized object has the typical appearance of an iron meteorite, with pits on its surface and a composition of iron and nickel.*

James Webb Space Telescope (JWST) An infrared telescope to succeed the Hubble Space Telescope equipped with three main instruments, to study the earliest moments after the **big bang**. It is scheduled to be launched in 2018 into a special orbit 1.5 million km (900,000 miles) from the Earth. The telescope's 6.5 m (20 ft) diameter mirror will be shaded by a sunshield the size of a tennis court.

Jansky, Karl Guthe (1905–50) American radio engineer. In 1931, while looking for sources of "noise" that interfered with radio communications, he discovered one located in the Milky Way, in Sagittarius. Jansky himself did not follow this up, but it led directly to the development of radio astronomy. The unit of flux density used in radio astronomy is named the *jansky* in his honor.

Janssen, Pierre César Jules (1824–1907) French astronomer. He specialized in solar spectroscopy and developed special instruments, including a spectrohelioscope for observing prominences without waiting for an eclipse. He discovered the helium line in the solar spectrum, and produced a photographic atlas of the Sun.

Janus A small inner **satellite** of Saturn, **co-orbital** with Epimetheus. The two were discovered in 1978 when John Fountain and Stephen Larson re-examined photographs taken some years before. They are both irregular in shape, and are probably fragments of the same larger body. Either satellite may be the object reported in 1966 by Audouin Dollfus and given the name Janus, but not subsequently confirmed.

Jeans, James Hopwood (1877–1946) English mathematician and physicist. In astronomy he worked on stellar dynamics, the formation and evolution of stars, binary systems, and spiral galaxies. In 1917 he proposed that the Solar System was formed from matter pulled out of the Sun by the gravitational attraction of a passing star, a theory later developed by Harold Jeffreys (1891–1989), but since abandoned. Jeans wrote many popular books on astronomy.

jet A stream of very high-energy particles emitted by objects such as **active galactic nuclei** and protostars. They are detectable by their radio emissions and may also be visible at optical wavelengths. In galaxies there are often two jets, emerging in opposite directions from the nucleus perpendicular to the galactic plane. SEE ALSO **bipolar flow**.

Jewel Box The open cluster NGC 4755 in the constellation Crux, around the star Kappa Crucis. The name derives from the varied colors of the stars in the cluster. It lies 7600 light years away.

Jodrell Bank The site at Macclesfield, Cheshire, UK, of the University of Manchester's radio astronomy observatory, which was founded after World War II by Bernard **Lovell**. Its main instrument is the 76 m (250 ft) fully steerable dish opened in 1957 and named the Lovell Telescope in 1987; a major upgrade was completed in 2003. A second dish, elliptical in shape, 38 × 25 m (125 × 83 ft), was opened in 1964. These telescopes can be used individually or as part of the **MERLIN** array.

▲ *Jupiter* This view of the planet Jupiter was assembled from four images obtained by the Cassini spacecraft on 7 December 2000. The long-lived Great Red Spot can be seen at lower right. The shadow of Europa, the smallest of the Galilean satellites, is the dark spot at lower left.

Jones, Harold Spencer (1890–1960) English astronomer. He was concerned with refining values for the distances and periods of Solar System bodies, and in 1941 obtained what was then the most accurate value of the Sun's distance. As the tenth Astronomer Royal, he planned the Royal Greenwich Observatory's move to Herstmonceux in Sussex.

Jovian The adjective for Jupiter. *Jovian planet* is another term for **giant planet**.

Julian calendar SEE **calendar**.

Julian Day A system of numbering days continuously, without division into years and months. It is used in astronomy when calculating events that recur over long periods, such as the solar cycle of sunspot variations. The Julian Day Number is the number of days that have elapsed since noon GMT on 1 January 4713 BC, which is defined as day zero. The idea was conceived in 1582 by the French chronologist Joseph Justus Scaliger. The name "Julian" was given by Scaliger in memory of his father, Julius Caesar Scaliger (1484–1558), and is not to be confused with the Julian calendar devised by the Roman Emperor Julius Caesar.

The Julian Date is given by the Julian Day Number for the preceding noon at Greenwich followed by the fraction of the mean solar day that has elapsed. Thus, the Julian Date of an observation made at 6 p.m. on 1 January 1995 is 2,449,719.25, the Julian Day Number being 2,449,719.

In historical usage the year 1 BC is followed by the year AD 1, whereas in astronomical usage AD 1 is preceded by year 0, which in turn is preceded by year −1 (corresponding to the historical year 2 BC). Thus to convert years BC to astronomical years, subtract 1 and prefix a minus sign.

Juliet One of the small inner **satellites** of Uranus discovered in 1986 during the Voyager 2 mission.

Juno Asteroid no. 3, discovered by Karl Harding in 1804. It is the eighth largest, with a diameter of 268 km (167 miles).

Juno mission A NASA space probe to Jupiter, launched 5 August 2011. It will go into polar orbit around Jupiter in July 2016 to study the planet's atmosphere and its gravitational and magnetic fields, as well as photographing its clouds. Juno will orbit Jupiter for a

▼ *Jodrell Bank* The Lovell Radio Telescope at Jodrell Bank was built in 1957 and originally known as the Mark I Telescope. It is still one of the world's largest fully steerable radio telescopes.

JUPITER: DATA	
Globe	
Diameter (equatorial)	142,800 km
Diameter (polar)	133,500 km
Density	1.33 g/cm^3
Mass (Earth = 1)	317.8
Volume (Earth = 1)	1320
Sidereal period of axial rotation (equatorial)	9h 50m 30s
Escape velocity	59.5 km/s
Albedo	0.73
Inclination of equator to orbit	3° 07′
Temperature at cloud-tops	125 K
Surface gravity (Earth = 1)	2.69
Orbit	
Semimajor axis	5.203 au = 778.3 × 10^6 km
Eccentricity	0.048
Inclination to ecliptic	1° 18′
Sidereal period of revolution	11.86y
Mean orbital velocity	13.06 km/s
Satellites	60+

period of a year, eventually burning up in the planet's atmosphere in October 2017.

Jupiter The fifth major planet from the Sun, and the largest of the giant planets. It is one of the brightest objects in the sky, with a magnitude at opposition of between −2.3 and −2.9. Its maximum apparent diameter is 50 arc seconds. Following centuries of telescopic observation, our knowledge of the planet and its satellites owes most to visits by space probes, including Pioneers 10 and 11, Voyagers 1 and 2, Cassini and most notably **Galileo**. SEE above for the main data for Jupiter.

Telescopically, Jupiter shows an oblate disk, somewhat flattened by its rapid rotation, crossed by a pattern of dark *belts* and light *zones*. Most conspicuous are the North Equatorial Belt and South Equatorial Belt, between whose faster rotation (9 hours 51 minutes, known as *System I*) is observed than in the rest of the atmosphere (*System II*, 9 hours 55 minutes): in other words, Jupiter's atmosphere shows **differential rotation**. Other atmospheric features include short-lived dark and light spots and streaks. More persistent features have included white ovals in the South Temperate Belt, and also in this region the long-lived **Great Red Spot**. The ovals are cyclonic and anticyclonic storms.

Turbulence can cause major disturbances, particularly in the South Tropical Belt and Zone, and in the South Equatorial Belt. Jupiter's dynamic atmosphere is influenced by the planet's internal heat (Jupiter emits about 1.7 times as much heat as it receives from the Sun), and shows wind speeds of up to 400 km/h (250 miles/h). The bright zones are high-altitude clouds supported by upwelling gases, while the dark belts are regions of descent.

Jupiter's predominant atmospheric constituents are hydrogen (almost 90%) and helium. Chemical species detected include ammonia, methane, water and hydrogen cyanide. Chemical reactions may be driven by ultraviolet radiation from the Sun, lightning and auroral discharges. At 1000 km (600 miles) below the cloud-tops the atmosphere gives way to an ocean of liquid molecular hydrogen. At a depth of 20,000 to 25,000 km (12,500 to 15,000 miles) the immense pressure causes the hydrogen to behave as a metal. At the center of the planet there is probably an iron–silicate core perhaps 15 to 20 times the Earth's mass.

The deep metallic hydrogen "mantle" gives Jupiter a powerful magnetic field (nearly 20,000 times as strong as the Earth's) and a very extensive **magnetosphere**. High-energy plasma is funneled into extremely intense radiation belts. Most of Jupiter's trapped plasma comes from volcanoes on **Io** and occupies a *plasma torus* and a *flux tube* linking satellite and planet. The magnetosphere is a source of the planet's powerful radio emissions.

Jupiter has over 60 known **satellites**. The four major ones – Io, Europa, Ganymede and Callisto – are often referred to collectively as the **Galilean satellites**. Most of the others are minor objects, perhaps captured asteroids or comets, orbiting at distances of up to 28 million km (18 million miles) from the planet. Jupiter became the focus of public attention in 1994 when the fragmented Comet **Shoemaker–Levy 9** plunged into its atmosphere, producing a succession of impact scars. Jupiter also possesses a modest ring system (SEE **ring, planetary**), the brightest component of which is about 50,000 km (30,000 miles) above the cloud-tops.

JWST ABBREVIATION FOR **James Webb Space Telescope**.

Kapteyn's Star A red subdwarf, 12.8 light years away in the southern constellation Pictor, which has the second largest proper motion known: 8.7 arc seconds per year. It is of magnitude 8.9, type M1, and is over 250 times less luminous than the Sun. It was discovered by the Dutch astronomer Jacobus Kapteyn (1851–1922).

K corona SEE **corona**.

Keck Telescopes A pair of identical telescopes at the W. M. Keck Observatory on Mauna Kea, Hawaii. Both have mirrors 10 m (396 inches) in diameter, consisting of 36 hexagonal segments. Keck I was completed in 1992, and its twin, Keck II, in 1996. They are run by the University of California and Caltech.

Kellner eyepiece An eyepiece with large **eye relief**, and so is widely used for binoculars as well as telescopes. It is essentially a **Ramsden eyepiece** with one element replaced by an achromatic doublet (SEE **achromatic lens**). (Invented in 1849 by German optician Carl Kellner, 1826–55.)

Kepler, Johann (1571–1630) German mathematician and astronomer. He supported the Sun-centered Solar System put forward by **Copernicus**. In 1600 he went to Prague, becoming assistant to Tycho **Brahe**, on whose death in 1601 he succeeded as Imperial Mathematician to the Holy Roman Emperor, Rudolf II, with the task of completing tables of planetary motion begun by Tycho. From Tycho's accurate observations he concluded in *Astronomia nova* (1609) that Mars moved in an elliptical orbit, and went on to establish the first of his laws of planetary motion (SEE **Kepler's laws**). Erroneous reasoning led Kepler nevertheless to the second of these laws, and his desire to match celestial and musical harmony led to the third law, published in *Harmonices mundi* (1619). The *Rudolphine Tables*, based on Tycho's observations and Kepler's laws, appeared in 1627 and remained the most accurate until the 18th century. Other works of his include *De stella nova*, on the supernova of 1604 (SEE **Kepler's Star**), and *Dioptrice* (1611), on optics and the theory of the telescope.

Kepler mission A NASA orbiting observatory launched on 6 March 2009 to detect extrasolar planets. Kepler observed about 150,000 stars in a single patch of sky in Cygnus, looking for slight but regular variations in their brightness that would be caused by the passage of planets in front of them. As of early 2015, Kepler had detected over 1000 planets, some of them the size of the Earth or smaller.

Kepler's laws The three fundamental laws governing the motions of the planets around the Sun (and other celestial bodies in closed orbits), first worked out by Johann **Kepler** between 1609 and 1619, and based on observations made by Tycho Brahe (SEE the diagram).

Law 1 deals with the shape of a planetary orbit: the orbit of each planet is an **ellipse** with the Sun at one of the foci.

Law 2 explains the varying speed of planetary motion such that the planet moves faster the nearer it is to the Sun: the line joining the planet to the Sun (the *radius vector*) sweeps out equal areas in equal times. This law is known as the *law of areas*.

▼ *Kepler's laws* Planetary motions around the Sun are described by Kepler's three laws. The laws apply to all bodies in closed orbits around the Sun, and to satellites orbiting planets.

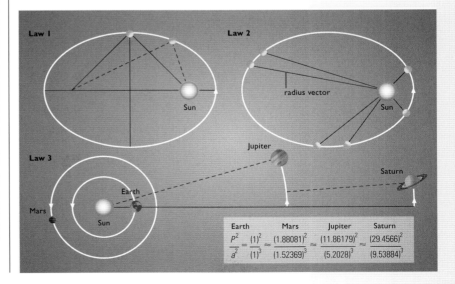

$$\frac{P^2}{a^2} = \frac{(1)^2}{(1)^3} \approx \frac{(1.88081)^2}{(1.52369)^3} \approx \frac{(11.86179)^2}{(5.2028)^3} \approx \frac{(29.4566)^2}{(9.53884)^3}$$

▲ **Kitt Peak National Observatory** *This aerial view of Kitt Peak shows the many telescopes sited there. From bottom left to top right are: the 3.5 m (138-inch) WIYN Telescope; the 0.9 m (35-inch) WIYN Telescope; and the 2.1 m (83-inch) KPNO Telescope. The triangular building at top right is the McMath–Pierce Solar Facility.*

Law 3 relates the size of the orbit to the period of revolution: the square of the period (P) is directly proportional to the cube of the mean distance of the planet from the Sun (a). In practice this means that P^2/a^3 is approximately constant for all bodies in orbit around the Sun.

Kepler's Star A **supernova** that appeared in October 1604 in the constellation Ophiuchus, reaching magnitude -3 and remaining visible to the naked eye until March 1606. It was studied by Johann **Kepler**. The smooth light curve is like that of a Type I supernova. The remnant is a radio source, and some faint nebulosity is visible optically. It was the last supernova seen in our Galaxy.

Kirkwood gaps Regions in the main **asteroid** belt where, because of **resonances** with Jupiter, no stable orbit is possible, and from which asteroids are therefore absent (or nearly absent). The gaps are at distances at which the revolution period of an orbiting body would be **commensurable** with Jupiter's, and Jupiter would perturb the body. There are pronounced gaps at $\frac{1}{3}$, $\frac{2}{5}$ and $\frac{1}{2}$ of Jupiter's period, for example. The gaps in the ring system of **Saturn** are kept clear by a similar perturbing effect of its satellites, in particular Mimas, whose period is twice that of a body at the distance of the Cassini Division. The American astronomer Daniel Kirkwood (1814–95) explained how the gaps came about.

Kitt Peak National Observatory An observatory near Tucson, Arizona, part of the US National Optical Astronomy Observatory, containing the world's largest collection of telescopes. Its largest instrument is the 4 m (158-inch) Mayall Telescope, opened in 1973. It is also the site of the McMath–Pierce Solar Telescope, the world's largest solar telescope; the 3.5 m (138-inch) WIYN Telescope, opened in 1994; and the 2.3 m (90-inch) Steward Reflector of the University of Arizona.

Kohoutek, Comet C/1973 E1 A **comet** discovered in 1973 by the Czech astronomer Luboš Kohoutek (1935–). When first detected it was nearly as far from the Sun as Jupiter, and was predicted to become as bright as the Moon at perihelion in December 1973. Although it failed to live up to these expectations, it was studied extensively and advances in comet science were made as a result.

K star A star of spectral type K, characterized by numerous absorption lines, especially of iron, titanium and calcium, and the presence of molecular bands of CH and CN in its spectrum. Main-sequence K stars have surface temperatures ranging from 3550 to 4900 K. Giants are about 400 K cooler than this, and supergiants 300 K cooler still. At their lowest point on the main sequence, K stars are about half a solar mass, rising to 0.8 of a solar mass at their highest. Because K-type giants and supergiants may be either evolved old stars of about one solar mass or evolved stars of higher masses, a wide range is represented. However, most of the K-type giants are stars of a few solar masses. Only a small fraction of K stars are variable: the coolest Cepheids have K-type spectra, as do the hottest of the **Mira** stars. A few percent of K-type giants show

spectral peculiarities with an overabundance of carbon and certain **heavy elements**. Bright examples of K stars are Pollux, Arcturus, one component of Alpha Centauri and Epsilon Eridani.

Kuiper, Gerard Peter (1905–73) Dutch–American astronomer. After World War II he embarked on an observational program which revitalized lunar and planetary astronomy. He discovered the satellites Miranda (of Uranus) in 1948 and Nereid (of Neptune) in 1949, and found methane in the atmospheres of Uranus, Neptune and Titan, and carbon dioxide in the atmosphere of Mars. He suggested the existence of what is now called the **Kuiper Belt**. Kuiper was involved with many US missions to the Moon and planets, in particular with the Ranger and Mariner probes. He advocated carrying infrared telescopes on high-flying aircraft; the Kuiper Airborne Observatory was named in his honor.

Kuiper Belt (Edgeworth–Kuiper Belt) A flattish zone stretching from the orbit of Neptune at 30 au to about 1000 au from the Sun, populated by **planetesimals** currently classed as asteroids. Its existence was proposed by Kenneth Edgeworth (1880–1972) in 1949 and by Gerard Kuiper in 1951 as a repository for comets or planetesimals left over from the formation of the Solar System. The concept was resurrected in the 1980s as a possible source of short-period comets, to account for the generally low orbital inclinations of such comets, which the spherical **Oort Cloud** could not.

The first member of the Kuiper Belt was discovered in 1992, and by early 2015 some 1800 had been found. Most are between 100 and 500 km (60–300 miles) in size, although larger ones exist: for example, **Quaoar** is some 1200 km (750 miles) in diameter while **Eris** is larger than Pluto. There are almost certainly many smaller members. Members of the belt are often called *trans-Neptunian objects* (TNOs). Many, like Pluto, are in a 3 : 2 orbital resonance with Neptune, and are called *plutinos*.

Physically, Kuiper Belt objects are probably similar to comets. They show no comet-like activity because they are so far from the Sun. If they move inward into planet-crossing orbits they may become **Centaurs**, and then outgassing can start, as has been observed with **Chiron**.

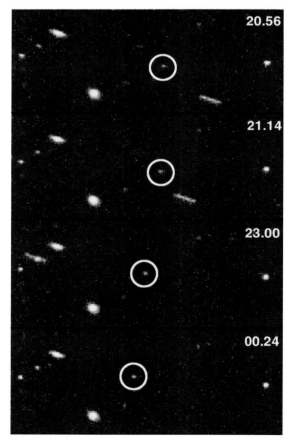

▲ **Kuiper Belt** *The first Kuiper Belt object to be discovered, 1992 QB1 (circled), revealed by its movement over 3½ hours. It orbits the Sun beyond the orbit of Neptune and is estimated to be about 200 km in diameter.*

Lacaille, Nicolas Louis de (1713–62) French astronomer. He spent the years 1751 to 1753 surveying the southern-hemisphere skies from the Cape of Good Hope, and introduced 14 new southern constellations. In 1761 he made an accurate measurement of the Moon's distance that took into account the Earth's oblateness; this made possible a more accurate method of determining terrestrial longitude.

Lacerta A small, obscure northern constellation between Andromeda and Cygnus, representing a lizard, introduced by **Hevelius**. Alpha, its brightest star, is magnitude 3.8.

Lagoon Nebula A bright nebula in Sagittarius, designated M8 or NGC 6523, surrounding the open star cluster NGC 6530. The nebula is cut by a dark cloud which gives it its name. The nebula and cluster lie at a distance of 5200 light years.

Lagrange, Joseph Louis de (1736–1813) French mathematician, born in Italy. He made many contributions to celestial mechanics, studying the Moon's **libration**, the motion of Jupiter's satellites, and perturbations and stability in the Solar System. From his investigation in 1772 of the **three-body problem** he found special solutions of the equations which suggested the existence of the equilibrium positions now known as **Lagrangian points**.

Lagrangian point One of the points at which a celestial body can remain in a position of equilibrium with respect to two much more massive bodies orbiting each other. At these points the forces acting on the smaller body cancel out.

There are five Lagrangian points, in the orbital plane of the two large bodies (SEE the diagram). Bodies at points L_1, L_2 and L_3, on the line joining M_1 and M_2, are less stable than bodies at points L_4 and L_5, which form equilateral triangles with M_1 and M_2. The **Trojan asteroids** lie in stable orbits near the L_4 and L_5 Lagrangian points of Jupiter's orbit around the Sun. (After Joseph Louis de **Lagrange**.)

Lalande, Joseph Jérôme (Le Français) de (1732–1807) French astronomer. His main achievement was a catalog of over 47,000 stars. He did much to popularize astronomy in France. Lalande observed Neptune in 1795, 51 years before it was located by Johann Galle, but he failed to realize that it was a planet.

Laplace, Pierre Simon de (1749–1827) French mathematician and astronomer. He used Isaac Newton's law of **gravitation** to interpret the perturbations of planets and satellites, the shape and rotation of Saturn's rings, and the stability of the Solar System. In 1796 he put forward his *nebular hypothesis* of the origin of the Solar System, which, although wrong in detail, is broadly in agreement with the presently accepted view. His five-volume *Traité de mécanique céleste* published over the period 1799–1825 summarized 18th-century advances in celestial mechanics.

Large Binocular Telescope A twin reflecting telescope, sited at **Mount Graham International Observatory**, completed in 2008.

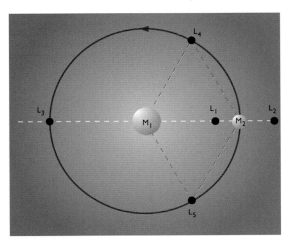

▲ **Lagrangian point** *Many spacecraft are stationed around the L_1 point between Earth and the Sun. The L_4 and L_5 (60° ahead of and behind the planet) Lagrangian points of Jupiter's orbit are occupied by Trojan asteroids.*

Each component has an 8.4 m (330-inch) primary mirror, giving a combined effective aperture of 11.8 m (465 inches). The centers of the mirrors are 14.4 m (567 inches) apart.

Large Magellanic Cloud (LMC) SEE **Magellanic Clouds**.

Large Synoptic Survey Telescope (LSST) An 8.4 m (330-inch) reflector with the unusually wide field of view of 3.5° being constructed on Cerro Pachón, Chile. It will survey the entire sky visible from its location at a range of wavelengths twice a week with a digital camera of 3.2 gigapixels. The LSST is expected to begin operation in the early 2020s.

Larissa One of the small inner **satellites** of Neptune, first detected in 1981 by Harold Reitsema and colleagues when it occulted a star, and confirmed by the 1989 visit of Voyager 2. Larissa appears to be the inner **shepherd moon** to the planet's Adams Ring. It is about 200 km (120 miles) in diameter.

Las Campanas Observatory An observatory near La Serena, Chile, run by the Carnegie Institution of Washington. Its main telescopes are the twin **Magellan Telescopes**, the **Giant Magellan Telescope** and the 2.5 m (100-inch) Du Pont Telescope, in operation since 1977.

Lassell, William (1799–1880) English brewer and amateur astronomer. From the observatory he built, he discovered several planetary satellites: Neptune's **Triton** – only 17 days after the discovery of Neptune itself, Saturn's **Hyperion**, and Uranus's **Ariel** and **Umbriel**. He also discovered 600 nebulae. Lassell was the first to use the equatorial system of mounting for reflecting telescopes.

last contact SEE **fourth contact**.

last quarter One of the Moon's **phases** when it has almost reached western quadrature and is half illuminated as seen from the Earth.

late-type star A cool star of spectral type K, M, C or S. The name was given when it was believed that the spectral types represented an evolutionary sequence, hot stars being the youngest and cool stars the oldest. Although that is now known not to be the case, the term has remained in use.

latitude A coordinate on the celestial sphere, or on the surface of a celestial body.

Celestial (or *ecliptic*) *latitude* is the angular distance of a celestial body from the ecliptic; it is measured at right angles to the ecliptic, from 0° at the ecliptic to 90° at the ecliptic poles, positive to the north and negative to the south.

Galactic latitude is the angular distance of a body from the galactic equator, measured from 0° to 90° at right angles to the galactic equator.

Heliocentric latitude is the celestial latitude of a body as seen by a hypothetical observer at the center of the Sun, rather than the center of the Earth.

Heliographic latitude is the latitude of a point on the Sun's surface, north or south of the Sun's equator.

leap second The adjustment of one second in the radio time signal at midnight on 30 June or 31 December. By international agreement the rate of radio time signals has been maintained since January 1972 in agreement with the International Atomic Time scale. However, the Earth's rotational period is slowing by about 0.003 of a second per day, so the radio time signals gradually diverge from Coordinated **Universal Time** (UTC), which is based on the Earth's daily rotation. To correct for this divergence, the radio time is retarded by the addition of one leap second as necessary.

Leavitt, Henrietta Swan (1868–1921) American astronomer. She specialized in variable stars, noticing that the brighter of the **Cepheid variables** tended to have the longer periods. From studies of Cepheids in the Small Magellanic Cloud she formulated the **period–luminosity law**, which was important in establishing the distance scale for other galaxies.

Leda A small **satellite** in Jupiter's intermediate group, discovered by Charles Kowal in 1974.

Lemaître, Georges Edouard (1894–1966) Belgian priest and astrophysicist. In 1927 he argued that an expanding Universe would have originated in the radioactive explosion of a "primeval atom" or "cosmic egg" into which all mass and energy was originally

▲ **Larissa** *Neptune's inner satellite Larissa was imaged during Voyager 2's flyby in August 1989.*

concentrated. This was the first statement of what would become known as the **big bang** theory.

lens An optical component of an instrument, made of a transparent material (such as glass) which, by the material's property of refraction, modifies an incoming beam of light to help form an image of the light source. Single lenses have either two curved surfaces, or one curved and one plane surface. Lenses are used in various combinations in astronomical instruments. SEE **aberration** (2), **eyepiece**, **objective**, **telescope**.

lenticular galaxy A type of **galaxy** intermediate in form between spiral and elliptical galaxies. Lenticular galaxies are designated S0, which indicates that they have the flattened form of spirals but lack spiral arms. They appear lens-shaped, hence their name.

Leo A large and easily identified constellation of the **zodiac**, in mythology representing the lion killed by Hercules. Its brightest star is first-magnitude **Regulus**, part of the curved figure of stars, known as the Sickle, that forms the lion's head and chest. Gamma (Algieba) is a fine, wide binary, magnitudes 2.2 and 3.5, one K1-type giant and one G7-type giant, period 620 years. The Mira variable R Leonis can reach magnitude 4.4 (minimum 11.3, period 310 days). There are five galaxies in Leo on Messier's list: M65, M66, M95, M96 and M105.

Leo Minor A small constellation between Leo and Ursa Major. It has no star above magnitude 3.8.

Leonids A **meteor shower** that occurs annually between 15 and 20 November. Its radiant is in the constellation Leo, in the "Sickle" asterism. The shower is periodic, with modest displays (**ZHR** = 15–20) in most years, but intervals of elevated activity, including

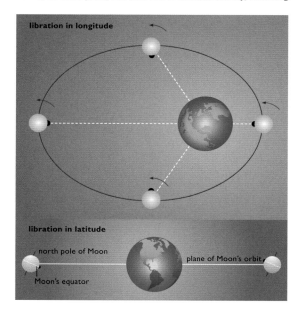

meteor storms, are seen for about a third of the 33-year orbital period of the parent comet, Comet 55P/**Tempel–Tuttle**, around the time of its perihelion. Notable storms occurred in 1799, 1833, 1866 and 1966. Most recently, intense Leonid activity was observed in 1999, 2001 and 2002; in addition to these storms, a spectacular display of Leonid fireballs occurred in 1998. SEE ALSO **meteor stream**.

Lepus A small constellation representing a hare hiding under the feet of Orion. Its brightest star is Alpha (Arneb), magnitude 2.6. It contains the very red Mira variable R Leporis (magnitude range 5.5 to 11.7, period 427 days) and the globular cluster M79.

Le Verrier, Urbain Jean Joseph (1811–77) French astronomer. In 1846, independently of John Couch **Adams**, he predicted that an unknown planet (Neptune) was responsible for discrepancies between the calculated and observed orbital motions of Uranus. In 1859 he suggested that a planet which he prematurely named Vulcan lay within the orbit of Mercury, but despite some claimed sightings its existence was never confirmed.

Libra An obscure constellation of the **zodiac**, representing the scales of justice, but formerly pictured as the claws of the scorpion, Scorpius, which it adjoins. The brightest star, Beta, is magnitude 2.6. Delta is an eclipsing binary of the Algol type, magnitude range 4.9 to 5.9, period 2.33 days.

libration A small oscillation of a celestial body about its mean position. The term is used most frequently in connection with the Moon. As a result of libration it is possible to see, at different times, 59% of the Moon's surface. However, because the areas that pass into and out of view are close to the limb and therefore extremely foreshortened, in practice libration has less of an effect on the features that can be clearly made out on the Moon's disk than this figure might suggest.

Physical libration results from slight irregularities in the Moon's motion produced by irregularities in its shape. Much more obvious is *geometrical libration*, which results from the Earth-based observer seeing the Moon from different directions at different times. There are three types of geometrical libration. *Libration in longitude* arises from a combination of the Moon's synchronous rotation and elliptical orbit (SEE the diagram). As a result, at times a little more of the lunar surface is visible at the eastern or western limb than when the Moon is at its mean position. *Libration in latitude* arises because the Moon's equator is tilted slightly from its orbital plane, so that the two poles tilt alternately toward and away from the Earth. A smaller effect is *diurnal libration*, by which the Earth's rotation allows us to see more of the Moon's surface at its western or eastern limb when it is rising or setting.

Lick Observatory The observatory of the University of California, sited on Mount Hamilton near San Jose, California. Its oldest telescope is a 0.91 m (36-inch) refractor, in operation since the observatory opened in 1888. The largest telescope is the 3 m (120-inch) C. Donald Shane Telescope, completed in 1959.

light, speed of (symbol *c*) The constant speed at which light and other **electromagnetic radiation** travels in a vacuum. Its value is 299,792,458.0 m/s. The first reasonably good estimate of *c* was made by Ole **Römer** in 1675. Albert **Einstein** showed with his special theory of relativity that the speed of light in a vacuum is always the same, whatever the relative speed of the observer and the source of the light.

light curve A graph of the change in magnitude with time of a variable star or other body showing brightness variations.

light pollution The detrimental effect of artificial lighting on astronomical observation and imaging. The problem has become much worse in recent years with the proliferation of such lighting and the development of new types of streetlamp. For amateurs, light pollution reduces the **limiting magnitude** (1) and makes deep-sky objects and comets, for example, difficult or impossible to see, but the development of special **filters** and of **CCDs** and related technology is providing solutions. For professional astronomers it can impede or distort scientific observation, particularly where the area surrounding a once-isolated observatory has become built up.

light-time The time required for light to travel from any celestial object to the Earth. For example, light takes 8.3 minutes to reach the Earth from the Sun.

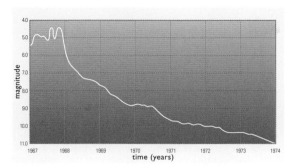

▲ **light curve** *An example of a light curve is this plot of the magnitude (brightness) of Nova Delphini 1967 against time. The nova was slow to arrive at peak brightness, showing several maxima over a 12-month period, before its light curve entered a long, gradual decline.*

light year (symbol l.y.) A unit of distance used in astronomy, equal to the distance traveled by light in a vacuum in one tropical year. 1 l.y. = 9.4607 × 10¹² km = 0.3066 parsecs = 63,240 au.

limb The edge of a celestial body which is visible as a disk, such as the Sun, the Moon or a planet. SEE ALSO **following, preceding**.

limb darkening The decrease in the brightness at visible wavelengths of a celestial body toward its limb, giving it a vignetted appearance. The term is used most often in connection with the Sun, where, toward the limb, the observer is looking at the deeper, brighter levels of the Sun through a greater thickness of its atmosphere. The giant planets also show limb darkening.

limiting magnitude (1) The faintest magnitude of objects visible or recordable by photography or electronic means on a given occasion. The limiting magnitude depends on sky conditions, instrument aperture and sensitivity of the eye or recording apparatus.

limiting magnitude (2) The faintest magnitude of objects recorded in a star atlas.

line of apsides SEE **apsides**.

line spectrum SEE **spectrum**.

Little Dumbbell A planetary nebula in the constellation Perseus, lying 3400 light years away. Discovered by Pierre Méchain in 1780, the object is designated M76, and takes its popular name from its twin-lobed appearance, resembling a smaller, fainter version of the **Dumbbell Nebula**.

LMC ABBREVIATION FOR **Large Magellanic Cloud**.

local arm SEE **Orion Arm**.

Local Group The small group of about 50 galaxies that includes our Galaxy, the two Magellanic Clouds, the Andromeda Galaxy and the Triangulum spiral; most are dwarf galaxies. They are distributed over a roughly ellipsoidal space about 5 million light years across. All the members are gravitationally bound so that, unlike more distant galaxies, they are not receding from us or from one another. The ten brightest (in absolute magnitude) are given in the table.

local time Time as measured at a given place on Earth, based on the mean solar day. Local time changes with longitude. A difference of 15° in longitude corresponds to a difference of one hour in local time. To avoid the inconvenience of a multiplicity of local times, the Earth has been divided into 24 **time zones**. In addition, *daylight saving time* (summer time) one hour in advance of standard time is kept in some countries for part of the year.

LOFAR The Low-Frequency Array, a radio interferometer consisting of many thousands of simple antennas arranged across Europe. It observes at the lowest radio frequencies detectable from the surface of the Earth. Two types of antennas are used: low band, operating at 10–90 MHz (30–3 m), and high band, 110–250 MHz (2.7–1.2 m). The core of the array is in the Netherlands, which has 38 antenna stations. There are in addition five stations in Germany and one each in the UK, France and Sweden, providing baselines up to 1500 km (930 miles).

longitude A coordinate on the celestial sphere or on the surface of a celestial body. *Celestial* (or *ecliptic*) *longitude* is measured eastward from the vernal equinox (the First Point of Aries) to the **great circle** passing through the pole of the ecliptic and the object to be measured. It is reckoned in degrees from 0 to 360°. *Galactic longitude* is the angle measured along the galactic equator from the direction of the galactic center to the point at which the galactic equator intersects the galactic circle of longitude of the object to be measured. *Heliocentric longitude* is the celestial longitude of a body as seen by a hypothetical observer from the center of the Sun. *Heliographic longitude* is the longitude of a point on the surface of the Sun. It is measured from the solar meridian that passed through the ascending node of the Sun's equator on the ecliptic on 1 January 1854, at Greenwich mean noon, from 0 to 360° in the direction of the Sun's rotation.

long-period comet SEE **comet**.

long-period variables Pulsating red giants or supergiants with periods ranging from about 80 to 1000 days, usually of spectral type M or C with characteristic emission lines. They are also known as **Mira** stars after the first of the type to be discovered. Each star has an average period which can vary by about 15%, so they are to some extent irregular. Their amplitudes range from 2.5 to 11 magnitudes, but the amplitude can also change appreciably from one cycle to the next. (Stars with lesser amplitudes are generally designated **semiregular variables**.) Long-period variables fall into three main divisions. Those that have a rise steeper than the fall tend to have wide minima and short, sharp maxima; as the asymmetry becomes greater, the period lengthens. Stars with symmetrical curves have the shortest periods. A third group show humps on their curves, or have double maxima and have long and short periods.

Lovell, (Alfred Charles) Bernard (1913–2012) English radio astronomer. From 1951 to 1981 he was director of the **Jodrell Bank** radio observatory, and oversaw the construction there of the world's first large steerable radio telescope. He used the telescope to pioneer the study of radio emission from space.

Lowell, Percival (1855–1916) American astronomer. In 1894 he built an observatory, now called the Lowell Observatory, at Flagstaff, Arizona. From there he observed Mars, producing intricate charts of the so-called canals, which he ascribed to the activities of intelligent beings. He calculated that the orbital irregularities of Uranus and Neptune were caused by an undiscovered Planet X, and his predictions of its position led to the discovery of the planet Pluto in 1930 by Clyde **Tombaugh**. It was at Lowell's urging that Vesto Slipher carried out the observations that led to the discovery of galactic **redshifts** in 1916.

lower culmination SEE **culmination**.

low-surface-brightness galaxy (LSB galaxy) A galaxy that is very faint at optical wavelengths either because it contains few stars or because its stars are spread out. LSB galaxies of many types are known, from dwarf irregulars to giant spirals. It is possible that, though difficult to detect, they are very common, and thus might account for some of the **missing mass** of the Universe.

▼ *Local Group Members of the Local Group are gravitationally bound to one another. As shown in this plot of the main members, which is centered on our Galaxy, the closest neighbors are the Magellanic Clouds.*

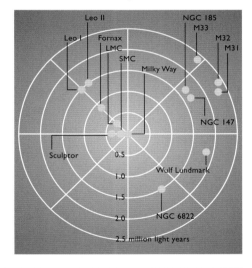

THE LOCAL GROUP: THE MAIN MEMBERS			
Galaxy	Apparent magnitude	Absolute magnitude	Distance (thousand l.y.)
Andromeda Galaxy (M31)	3.4	−21.2	2500
Milky Way	—	−20.9	—
Triangulum Galaxy (M33)	5.9	−18.9	2770
Large Magellanic Cloud	0.2	−18.5	163
Small Magellanic Cloud	2.0	−17.1	200
M32	8.1	−16.5	2510
IC 10	10.3	−16.4	2150
M110	8.4	−16.4	2710
Barnard's Galaxy (NGC 6822)	8.3	−16.0	1630
NGC 185	9.0	−15.6	2020

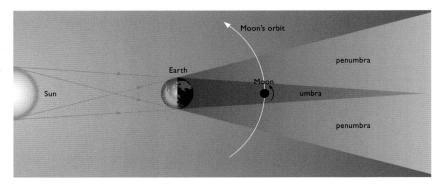

▲ **lunar eclipse** *Earth casts a shadow to the right in this diagram. The shadow has two components – a dark central umbra, and a lighter area of partial shadow, the penumbra. If the full Moon – opposite the Sun in Earth's sky – passes through the central cone of the umbra, a total lunar eclipse will result.*

LRO ABBREVIATION FOR **Lunar Reconnaissance Orbiter**.

LSST ABBREVIATION FOR **Large Synoptic Survey Telescope**.

L star A star of spectral type L, characterized by the presence of lithium lines in its spectrum. Because lithium is quickly destroyed in nuclear reactions, its presence indicates that no thermonuclear processes are taking place within these stars. The spectra also show absorption lines of metal hydrides, such as FeH and CrH, and alkali metals such as sodium and potassium. With surface temperatures typically of 1500–2000 K, L stars are cooler than M stars. The L type was introduced in 1998 for **brown dwarfs** and faint stars detectable at infrared wavelengths. Even cooler objects (below 1500 K) are classified separately, as **T stars**.

luminosity (symbol *L*) The absolute brightness of a star, given by the amount of energy radiated from its entire surface per second. Luminosity is expressed in watts (joules per second), or in terms of the Sun's luminosity. *Bolometric luminosity* is a measure of the star's total energy output, at all wavelengths. **Absolute magnitude** is an indication of luminosity at visual wavelengths.

luminosity class SEE **Morgan–Keenan classification**.

luminosity function A measure of the distribution of stars in space according to their **absolute magnitude**. It is usually expressed as the number of stars per cubic parsec for a given range of absolute magnitude. The value of the luminosity function in the solar neighborhood rises to a maximum at absolute magnitudes of 14–15, then falls away for fainter magnitudes, implying that faint M-type dwarfs of about one ten-thousandth of the Sun's luminosity are the most abundant type of star locally. Luminosity functions can be plotted for other categories, such as particular types of star, star clusters, galaxies or clusters of galaxies.

Luna A series of lunar probes launched by the former Soviet Union. The first three in the series were originally named Lunik. Luna 2 became the first man-made object to hit the Moon, on 13 September 1959. Luna 3, launched on 4 October 1959, obtained the first photographs of the far side of the Moon. Luna 9 achieved the first soft landing on the Moon on 3 February 1966. Luna 10, in March 1966, was the first craft to enter lunar orbit. In September 1970 Luna 16 made the first automatic return of a soil sample to Earth. Luna 17 carried the first remote-controlled lunar roving vehicle, named Lunokhod 1. The last in the series was Luna 24, which brought samples of lunar soil back to Earth in August 1976.

lunar eclipse An **eclipse** of the Moon by the Earth's shadow. Since the Moon's orbital plane is inclined to the plane of the ecliptic, a lunar eclipse can happen only when the Moon is at opposition (i.e. at full Moon) and at the same time is at or near one of its nodes (SEE **ecliptic limits**). The Sun, Earth and Moon are then very nearly in a straight line. Lunar eclipses are visible from anywhere on Earth where the Moon is above the local horizon. The Moon does not become completely invisible during an eclipse because it is partially lit by sunlight that has been refracted by the Earth's atmosphere, and it appears reddish because the blue component is removed from the refracted sunlight by **scattering**.

A *total lunar eclipse* occurs when the Moon is entirely within the umbra, the dark central part of the Earth's shadow (SEE the diagram). The eclipse is *partial* when the Moon is partly within the umbra, and *penumbral* when the Moon passes through the penumbra but completely misses the umbra. The overall duration of a total lunar eclipse, from **first contact** to **fourth contact**, can be 3½ hours; totality, the interval from **second contact** to **third contact**, can last for 1¾ hours.

Lunar Orbiter A series of US probes put into orbit around the Moon to photograph its surface in detail in preparation for the Apollo landings. There were five probes in the series, all successful, from Lunar Orbiter 1 (August 1966) to Lunar Orbiter 5 (August 1967).

Lunar Prospector A NASA probe to the Moon, launched on 7 January 1998. From polar orbit, its neutron spectrometer found strong evidence for the presence of water ice in the lunar regolith in both polar regions. The probe also found that the Moon has a small iron core.

Lunar Reconnaissance Orbiter (LRO) A NASA probe launched on 18 June 2009 that went into polar orbit around the Moon and took high-resolution images of the lunar surface. Among the details it saw were equipment left behind at the Apollo landing sites and the Russian Lunokhod rovers. Also launched on the same rocket was the Lunar Crater Observation and Sensing Satellite (LCROSS). This observed the ejecta produced when the rocket's upper stage crashed into the Moon inside the south polar crater Cabeus. The results confirmed the existence of frozen water in the lunar soil.

lunar transient phenomenon (LTP) A short-lived, localized change in the appearance of a lunar surface feature, such as a glow or an obscuration. LTPs were reported by lunar observers for many years but are now treated with scepticism.

lunation The time interval between two identical phases of the Moon, for example from one new Moon to the next new Moon. This corresponds to the Moon's **synodic period** or **synodic month** of 29.53059 days.

lunisolar precession The main component of **precession**, resulting from the gravitational effects of the Moon and Sun on the Earth. It amounts to about 50.4 arc seconds per year.

Lupus A southern-hemisphere constellation, adjoining Centaurus, representing a wolf. Its brightest star is Alpha, magnitude 2.3, but it has no objects of particular interest.

Lyman series A series of lines in the ultraviolet region of the **hydrogen spectrum**, analogous to the **Balmer series** in the visible part. The first Lyman line, Lα, is at 121.6 nm, and was first detected in the solar spectrum by photography from a rocket. The second, Lβ, is at 102.6 nm; Lγ is at 97.2 nm. In the spectra of **quasars**, Lyman lines are redshifted into the visible region. (After American physicist and spectotropist Theodore Lyman, 1874–1954.)

Lynx A dim northern constellation between Auriga and Ursa Major, introduced by **Hevelius**. Its leading star is Alpha, magnitude 3.1, but the only other star above fourth magnitude is 38 Lyncis, at magnitude 3.9.

Lyot, Bernard Ferdinand (1897–1952) French astrophysicist. He invented the **coronagraph** for observing the Sun's corona, and the **Lyot filter**. He also pioneered astronomical polarimetry, the measurement of the degree of polarization of light from celestial bodies, and invented a number of devices for the purpose.

Lyot filter A **filter** for observing the Sun at a particular wavelength, developed by Bernard **Lyot** in the 1930s. Sunlight is directed through a succession of alternating layers of quartz plates and polaroid sheets, and each separate passage reinforces some wavelengths and cancels others. The result is a number of very narrow (0.1 nm) and widely spaced wavebands which are easy to isolate.

Lyra A small but prominent constellation. In mythology, it represents the harp given to Orpheus by Apollo. Its brightest star is **Vega**. Lyra contains the famous "Double Double," the quadruple star Epsilon; the two main components, magnitudes 4.6 and 4.7, can be separated with the naked eye and each is again double. Beta (Sheliak) is an eclipsing variable, magnitude range 3.3 to 4.4, period 12.9 days; it is the prototype of the **Beta Lyrae stars**. Between Beta and Gamma (Sulafat), magnitude 3.2, lies the **Ring Nebula**, M57, a planetary nebula.

Lyrids A **meteor shower** active from 19 to 25 April, with its radiant on the border of the constellations Lyra and Hercules, southwest of the star Vega. Peak activity, around 22 April, is usually modest, with a **ZHR** of about ten. The Lyrids are associated with Comet C/1861 G1 Thatcher.

Lysithea A small **satellite** in Jupiter's intermediate group, discovered by Seth Nicholson in 1938.

McDonald Observatory The astronomical observatory of the University of Texas, on Mount Locke, Texas. It is home to the **Hobby–Eberly Telescope** and also houses a 2.7 m (107-inch) reflector, opened in 1968.

MACHO Abbreviation for "massive compact halo object," an as-yet undetermined type of invisible body assumed to pervade in the Galaxy's halo (outer regions) to account for the **dark matter** that it is supposed to contain. MACHOs could be brown dwarfs, extinct white dwarfs or Jupiter-sized planets. SEE ALSO **missing mass**.

Mach's principle The philosophical proposition that the properties of a body, particularly its inertia, depend on the existence of all the other objects in the Universe. Proposed by Ernst Mach (1938–1916), it is often illustrated by a "thought experiment" involving two cylindrical spacecraft partly filled with water. One of the spacecraft spins; the other does not. The water in the spinning spacecraft will accumulate against the spacecraft's walls, while in the non-spinning spacecraft the water will float around randomly. The only way the water in the spinning spacecraft can "know" that it is spinning is by reference to the rest of the Universe.

McNaught, Comet C/2006 P1 One of the most magnificent comets of recent years, discovered in August 2006 at Siding Spring Observatory by the British astronomer Robert McNaught (1956–). It reached perihelion on 12 January 2007 at a distance of 0.17 astronomical units, after which it became a naked-eye object in the southern hemisphere for several weeks, with a broad, streaked dust tail up to 35° long.

Maffei Galaxies Two nearby galaxies so close to the plane of the Milky Way that at optical wavelengths they are almost completely obscured by galactic dust. Maffei 1 is a lenticular or elliptical galaxy while Maffei 2 is a spiral. Both lie in a small group of galaxies about 10 million light years away. They were discovered as infrared sources in 1968 by the Italian astronomer Paolo Maffei (1926–2009).

Magellan A NASA space probe to Venus, launched on 4 May 1989. It went into orbit around Venus in August 1990 and made a detailed radar map of 98% of the planet, revealing impact craters, volcanic mountains, lava flows and other features. It operated until 1994.

Magellanic Clouds The two nearest galaxies to our own, visible to the naked eye as isolated patches of the Milky Way in the southern sky and named after the Portuguese explorer Ferdinand Magellan (1480?–1521), who described them. The *Large Magellanic Cloud* (LMC) lies in the constellations Dorado and Mensa and is about 6° across. The *Small Magellanic Cloud* (SMC) is in Tucana and is about 3° across. Their distances are about 163,000 and 200,000 light years. Although usually classified as irregular galaxies, both Clouds are somewhat bar-shaped with faint outer structure that can be interpreted as rudimentary spiral arms. The mass of the LMC is about 10^{10} solar masses, and that of the SMC about 2×10^9 solar masses.

At the northeast end of the LMC lies the **Tarantula Nebula**, one of the largest and brightest groupings of hot stars and bright nebulosity known in any galaxy. Near here erupted **Supernova 1987A**, the first naked-eye supernova since 1604. The irregular distribution of star clusters and diffuse nebulae in the LMC contrasts with the more uniform stellar distribution in the SMC. Young stars in the LMC lie in a thin disk, which is seen nearly face-on (about 27° to the plane of the sky). There is no evidence for a spherical halo of the kind that surrounds our Galaxy.

The SMC is greatly extended in our line of sight, with a depth of about 60,000 light years, over five times the dimension it presents to us. It appears to have been distorted by tidal forces, possibly as a result of a close passage of the SMC and LMC some 200 million years ago.

The *Magellanic Stream* is a vast cloud of neutral hydrogen (SEE **H I region**) stretching between the Clouds and extending toward the Galaxy. It was probably torn out from the Clouds by gravitational interaction with the Galaxy.

Magellan Telescopes A pair of reflecting telescopes, both of aperture 6.5 m (255 inches), at **Las Campanas Observatory** in Chile. The telescopes were completed in 2000 and 2002, and can be used independently or together.

magnetar A neutron star with an exceptionally strong magnetic field (10^{14}–10^{15} gauss). Magnetic-field events involving these exceptionally strong fields account for the previously unexplained gamma-ray outbursts observed in the *soft gamma repeaters* and the similar X-ray events observed in *anomalous X-ray pulsars*. SEE ALSO **gamma-ray astronomy**, **pulsar**.

magnetohydrodynamics (MHD) The study of the behavior of plasma (a mixture of ions and electrons) in a magnetic field. Magnetic-field lines are "trapped" by a plasma, so the motion of a plasma distorts a magnetic field and, conversely, a magnetic field exerts a force on a plasma, modifying its motion. Plasmas occur in many astronomical contexts, so MHD is significant in the study of solar phenomena such as sunspots, the interaction of the solar wind with planetary **magnetospheres**, neutron stars, star formation and supernovae.

magnetosphere The region of space surrounding a planet in which the planet's magnetic field predominates over the **solar wind** and controls the behavior of plasma (charged particles) trapped within it. The boundary of the magnetosphere is called the *magnetopause*, outside which is a turbulent magnetic region called the *magnetosheath*. Where the steady outward flow of the solar wind is first interrupted there is a **bow shock**. Downwind from the planet, the solar wind draws the magnetosphere out into a long, tapering *magnetotail*.

Mercury, the Earth and the giant planets possess magnetospheres. The Earth's contains the **Van Allen Belts** of charged particles. Jupiter's magnetosphere is many times larger than the Sun, and contains a vast disk of plasma and **Io**'s plasma torus.

magnification The factor by which an image produced by an optical system increases the angular size of an object. It is normally denoted by a multiplication sign, as in 50× or ×50 (SEE **binoculars**). The magnification given by a telescope is found by dividing its **focal length** by the focal length of the eyepiece being used with it. The higher the magnification, the dimmer the image and the more prone it is to the effects of bad **seeing** and defects of the instrument.

▲ *McNaught, Comet C/2006 P1 Comet McNaught spread an immense dust tail like a giant fan across the southern-hemisphere sky in January 2007. The bright object at lower right is the overexposed image of the young crescent Moon.*

magnitude A measure of the brightness of celestial objects. In the 2nd century BC the Greek astronomer **Hipparchus of Nicea** classified the stars into six magnitudes, first magnitude being the brightest and sixth magnitude the faintest visible to the naked eye. In the 19th century it was found that a star of first magnitude was about 100 times as bright as one of sixth magnitude. It was decided to adopt a scale of magnitudes in which this ratio is exactly 100:1. The ratio of brightness between one magnitude and the next is therefore the fifth root of 100, which is 2.512. Thus, differences of 2.5, 5 and 10 magnitudes correspond to brightness ratios of 10, 100, and 10,000 respectively. Very bright bodies, e.g. Sirius, Venus and the Sun, have negative magnitudes. The faintest objects detectable from ground-based telescopes are about magnitude 30.

These are *apparent magnitudes* – the brightness of those objects as seen from Earth. **Absolute magnitude** is the true brightness of an object, taking into account its distance from Earth. The difference between the apparent and absolute magnitudes is called the *distance modulus*. SEE ALSO **bolometric magnitude**, **photoelectric magnitude**, **photographic magnitude**, **photovisual magnitude**, **visual magnitude**.

main sequence The region on the **Hertzsprung–Russell diagram** where most stars lie, including the Sun. It runs diagonally from hot,

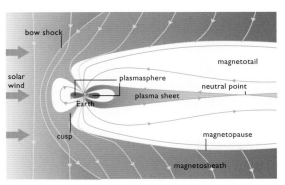

▲ *magnetosphere The Earth's moving iron core gives it an active magnetic field that extends well beyond the planet. The arrows indicate the direction of the magnetic field.*

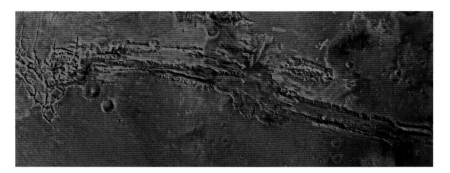

▲ **Mariner Valley** *This composite Viking image shows the huge, sprawling canyon system of the Mariner Valley (Valles Marineris) on Mars. Ancient dry river channels run northward from the chaotic terrain at the east (right).*

bright stars at the upper left down to cool, faint stars at the bottom right. The position of a star on the main sequence depends on its mass, the most massive stars being the brightest. The upper limit is about 60 solar masses, at stars of spectral types O and B. The lower limit is at 0.08 solar masses, with stars of spectral type M. Stars spend most of their lives on the main sequence, producing energy by the fusion of hydrogen to helium in their cores. The *zero-age main sequence* is where stars lie when they first start to burn hydrogen. As stars use up their internal hydrogen they move away from the main sequence. The more massive a star, the sooner it evolves off the main sequence. All stars on the main sequence are termed **dwarfs**, irrespective of whether they are larger or smaller than the Sun.

major planet Any of the eight largest planets of the Solar System – Mercury, Venus, Earth, Mars, Jupiter, Saturn, Uranus and Neptune. By contrast, the asteroids are known as *minor planets*. Objects of intermediate size between major planets and asteroids are known as *dwarf planets*.

Maksutov telescope A modification of the **Schmidt camera**, devised independently by Albert Bouwers (1940) and Dmitrii Maksutov (1944). The Schmidt's aspherical correcting plate is replaced by a concave lens which cancels out the **chromatic** and **spherical aberration** of the primary concave spherical mirror. A secondary mirror directs the light path through a hole in the primary to a Cassegrain focus, although various other configurations have been used. The Maksutov is a compact instrument once popular with amateur observers, but it has lost ground to the **Schmidt-Cassegrain telescope**. (After Dmitrii Dmitrievich Maksutov, 1896–1964.)

mantle SEE **differentiation**.

many-body problem (n-body problem) The problem in celestial mechanics of finding how a number (n) of bodies move under the influence of their mutual gravitational attraction only. It is an extension of the **three-body problem**. The many-body problem has no general solutions, but with modern computers good approximate solutions can be obtained. Analyzing how the planets of the Solar System perturb one another, and how space probes move between them, are examples of the many-body problem.

mare (plural **maria**) An extensive dark plain on the Moon, composed of basalt (solidified lava). Most of the maria were formed 3 to 4 billion years ago, after large meteorites struck the Moon and weakened the crust, and lava later erupted from below. They range from the vast Oceanus Procellarum, covering much of the northwestern quadrant of the Moon, down to areas less than 100 km (60 miles) across. Nearly all maria are on the Moon's near side, where the crust is thinner than on the far side. "Maria" is Latin for "seas," which is what early observers thought they were. Similar plains on Mars were also known as maria, but since detailed mapping by spacecraft these areas are now known as *planitiae*.

Mariner A series of US space probes to the planets. Mariner 2 flew past Venus in 1962, discovering that the surface temperature was hot. Mariner 4 flew past Mars in July 1965 and photographed craters on its surface. Mariner 5 passed Venus in October 1967, making measurements of the planet's atmosphere. Mariners 6 and 7 obtained further photographs of Mars in 1969. Mariner 9 became the first probe to enter planetary orbit when it went into orbit around Mars in November 1971. Mariner 10, the last of the series, was the first two-planet mission, passing Venus in February 1974 and then encountering Mercury three times, in March and September 1974 and March 1975.

Mariner Valley (Valles Marineris) A system of enormous, interconnected canyons on Mars, stretching for 4500 km (2800 miles) just south of the equator. Individual canyons extend for up to 200 km (125 miles) and are 7 km (4½ miles) deep. Mariner Valley is a fault structure whose formation is probably linked to that of the volcanic Tharsis region at its western end. The name commemorates the **Mariner** 9 probe, from which it was first imaged.

Markarian galaxy Any of the galaxies cataloged by the Armenian astronomer Benjamin Eghishe Markarian (1913–85), distinguished by their strong ultraviolet emission. They include **Seyfert galaxies** and **quasars**.

Mars The fourth major planet from the Sun, and the second smallest of the terrestrial planets. Mars is known popularly as the Red Planet, from its naked-eye appearance. Through a telescope it shows a small orange-red disk with lighter and darker markings, and white patches often visible at one pole or the other. Because of its markedly eccentric orbit, its distance from Earth at opposition varies from 56 to 101 million km (35 to 63 million miles), with corresponding variations in apparent size and brightness of 14 to 25 arc seconds and magnitude -1.0 to -2.8. Favorable *perihelic* oppositions occur at intervals of 15 to 17 years; that of 2003 was the closest for 60,000 years.

Mars shows a slightly **gibbous** phase at quadrature. Its axial tilt is similar to the Earth's, so it passes through a similar cycle of **seasons**, during which the polar caps vary in size, and the darker markings can vary in shape and extent as seasonal dust storms cover and uncover darker surface rocks. The planet was once imagined to be an abode of intelligent life, and some observers reported seeing linear surface markings which they believed were artificial **canals**. In 1996 it was announced that an **SNC meteorite** designated ALH 84001, recovered from Antarctica in 1984, appeared to contain the fossilized remains of primitive life forms, but most biologists remain sceptical. The main data for Mars are given in the table shown below.

Although similar geological processes have operated on Earth and Mars, the results on Mars have been dramatically different, because of its smaller size and gravity and the absence of plate **tectonics**. Mars' northern hemisphere is covered by largely smooth, lowland volcanic plains, while the more ancient terrain of the south comprises heavily cratered uplands crossed by many channels. Two large impact basins, Hellas (2200 km; 1400 miles) and Argyre (800 km; 500 miles) dominate the southern hemisphere. There is also a prominent bulge, 5000 km (3000 miles) across, in the volcanic Tharsis region, which straddles the equator. The biggest volcanic structure on Mars is **Olympus Mons**. It is possible that Mars has been volcanically active as recently as 100 million years ago.

Stretching east from Tharsis is the network of giant canyons known collectively as the **Mariner Valley**. While some Martian canyons and similar features are geological faults, others are clearly channels in which water once flowed. Some features have even been interpreted as dried-up seabeds. At some time in the past Mars must have had a very different climate, with a denser atmosphere. In 2001–2 the **Mars Odyssey** spacecraft found evidence for a permafrost layer, less than a meter below the surface, into which the water may now be frozen.

The **Mars Express** spacecraft confirmed that the variable polar caps are composed of solid carbon dioxide with underlying caps of water ice. The northern water cap is exposed in the summer when

MARS: DATA	
Globe	
Diameter (equatorial)	6792 km
Diameter (polar)	6752 km
Density	3.91 g/cm³
Mass (Earth = 1)	0.107
Volume (Earth = 1)	0.152
Sidereal period of axial rotation	24h 37m 23s
Escape velocity	5.0 km/s
Albedo	0.15
Inclination of equator to orbit	25° 11'
Surface temperature (average)	220 K
Surface gravity (Earth = 1)	0.38
Orbit	
Semimajor axis	1.524 au = 227.9 × 10⁶ km
Eccentricity	0.0934
Inclination to ecliptic	1° 51'
Sidereal period of revolution	686.980d
Mean orbital velocity	24.13 km/s
Number of satellites	2

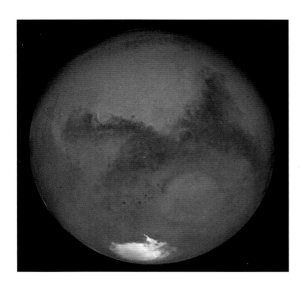

▲ **Mars** *The Hubble Space Telescope photographed the Red Planet during its exceptionally close approach to Earth in August 2003. At bottom is the south polar cap, with the impact basin Hellas just above it. The dark tongue-shaped area at upper right is Syrtis Major. The circular shapes of many impact craters can also be seen.*

the carbon dioxide warms and becomes part of the atmosphere, but the southern cap always retains a carbon dioxide covering.

The atmosphere consists of 95% carbon dioxide, $2\frac{1}{2}$% nitrogen and $1\frac{1}{2}$% argon, with smaller quantities of oxygen, carbon monoxide and water vapor. The atmospheric pressure at the surface is less than 1% of the Earth's. Winds of up to 300 km/h (200 miles/h) drive seasonal dust storms that may last for weeks. Water vapor forms into clouds and produces morning mists and ground frosts. The surface temperature on Mars varies between 130 and 290 K.

The crustal rocks of Mars are iron-rich, giving the planet its red color. The minerals present include olivine, and the presence of hematite, jarosite and goethite, confirmed by the twin **Mars Exploration Rovers**, provides strong evidence for liquid water having had a role in modifying Mars' surface chemistry in the past. The **Curiosity** rover found clear evidence of sedimentary rocks at its landing site, apparently laid down by running water. Much of the surface may be covered with a hard, ferrous, clay-like deposit. Basalts cover the volcanic plains. From the planet's density, its interior structure probably consists of a mantle overlying a core. However, the extremely weak magnetic field suggests that if there is an iron or nickel–iron core it is not hot and fluid.

Mars has two tiny **satellites** in very close orbits: Phobos and Deimos.

Mars Exploration Rover A NASA mission to land two roving vehicles on Mars. The first, *Spirit*, was launched on 10 June 2003 and landed in the crater Gusev, a possible lake-bed, on 4 January 2004. The second, *Opportunity*, was launched on 7 July 2003 and landed on 25 January 2004 in Meridiani Planum, an area rich in hematite – a mineral that forms under water. Both rovers were equipped with cameras and many mineralogical instruments. Spirit ceased working in 2010 but Opportunity continued operations.

Mars Express A European Space Agency probe, with international participation, launched on 2 June 2003. The main craft entered a polar orbit in January 2004, from where it carried out high-resolution imaging to examine Martian geology. The UK's Beagle 2 lander was a failure.

Mars Global Surveyor An American probe to Mars, launched on 7 November 1996. It arrived in September 1997, but problems with aerobraking manoeuvers delayed its achieving the desired polar orbit until 1999. The probe obtained over 300,000 high-resolution images of the surface, some of which showed conclusively that water had once flowed there. It stopped working in November 2006.

Mars Odyssey A NASA probe launched on 7 April 2001 which went into polar orbit around Mars on 24 October 2001, using aerobraking to refine its orbit. Mars Odyssey maps the chemical composition of the surface of the Red Planet. Its measurements of hydrogen distribution revealed that vast amounts of water

ice buried below the polar regions. The probe has also acted as a communications relay for the twin **Mars Exploration Rovers**.

Mars Pathfinder A NASA probe to Mars, part of the **Discovery** program, launched on 4 December 1996. It landed on 4 July 1997, using a combination of parachute and airbags. A miniature rover vehicle, Sojourner, analyzed rocks in the vicinity of the lander. Lander and rover operated for nearly three months. Images of the terrain at the landing site, in Ares Valles, showed that water had once flowed across the surface.

Mars Reconnaissance Orbiter A NASA craft, launched on 12 August 2005, which entered orbit around Mars in March 2006. It carries a high-resolution camera that is able to detect objects as small as 20 cm (8 inches) on the surface, searching for signs of water and locations for future landings. The craft is also equipped with a subsurface sounding radar to search for underground water.

Mars Science Laboratory SEE **Curiosity**.

mascon A high-density region below the Moon's surface; the word is derived from "mass concentration." Mascons cause small, localized increases in the Moon's gravitational field, which is how they were first detected by Lunar Orbiter 5 in 1967. The dozen or so mascons that are known have diameters of 50–200 km (30–125 miles) and lie about 50 km (30 miles) below the surface. They were probably formed when impacts from large meteorites blasted away the lighter crust, and the denser, underlying mantle material bulged upward.

maser A cloud of molecular **interstellar matter** in which **electromagnetic radiation** emitted by high-energy molecules at a particular wavelength in the microwave region stimulates other molecules to emit radiation at the same wavelength. The word is an acronym of "microwave amplification by stimulated emission of radiation." A maser therefore amplifies natural microwave emissions, just as a laser does for light.

Maskelyne, Nevil (1732–1811) English astronomer. He was appointed the fifth Astronomer Royal in 1765. Building on a method of his for determining longitude at sea by observing the Moon, in 1767 he published the first *Nautical Almanac*, an annual book for the same purpose. In 1774 he measured the mass of the Scottish mountain of Schiehallion by seeing how far it caused a plumb-line to deviate, from which he calculated the average density of the Earth to be 4.7 g/cm³ (the true value is 5.52 g/cm³).

massive compact halo object SEE **MACHO**.

mass–luminosity relation A relationship between luminosity and mass for stars on the **main sequence**, derived theoretically by Arthur **Eddington** in 1924. It is expressed approximately as

$$L \propto M^x ,$$

where L is the luminosity, M is the mass, and the exponent x varies between 2 and 5 according to the mass. The relationship is not obeyed by white dwarfs, or by giants or supergiants. The relationship enables the mass of a star to be determined if its absolute magnitude is known. The theoretical upper limit (SEE **Eddington limit**) of stellar mass is about 100 solar masses, but very few stars exceed 30 solar masses. The lower limit of a star's mass is 0.08 of a solar mass.

Mathilde Asteroid no. 253, discovered in 1885 by Johann Palisa. It was imaged in 1997 by the **NEAR Shoemaker** probe *en route* to Eros, and was found to have a diameter of 53 km (33 miles) and an unusually slow rotation period of 17.4 days. Mathilde is heavily cratered but with a very low density, about 1.3 g/cm³, which suggests

▲ **Mathilde** *Asteroid no. 253 was imaged by the NEAR Shoemaker probe. Its surface was seen to have several craters about 20 km (12 miles). It has a low albedo.*

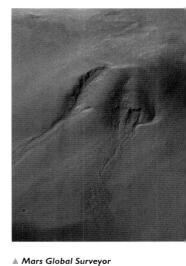

▲ **Mars Global Surveyor**
The existence of water on Mars, whether as subsurface liquid or as ice, is vital for the question of whether there was, or is, life on Mars. Several outflow channels in crater walls, including this example in Noachis Terra, were imaged by Mars Global Surveyor, and appear to resemble gullies on Earth.

▼ **Mars Exploration Rover**
Images taken by the panoramic camera on the rover Opportunity were combined to produce this approximately true-color picture of the crater "Fram" in Meridiani Planum. The crater is about 8 m (26 ft) in diameter. The images were obtained on 23 April 2004.

that it is a "rubble pile" – many fragments from a past impact held together by gravity.

matter era The period, from about 40,000 years after the **big bang** to the present day, in which the energy density of matter in the Universe exceeds that of radiation. SEE ALSO **radiation era**.

Mauna Kea Observatories A major observatory site in Hawaii, at an altitude of 4200 m (13,800 ft) – the world's highest and best observatory site for optical, infrared and millimeter-wave astronomy. Its main telescopes include the twin 10 m (396-inch) **Keck Telescopes**; the 3.6 m (142-inch) Canada–France–Hawaii Telescope; the 3.8 m (150-inch) UK Infrared Telescope; the 15 m (49 ft) James Clerk Maxwell Telescope, operated by the UK, Canada and the Netherlands and used for millimeter-wave astronomy; the Japanese 8.3 m (326-inch) **Subaru Telescope**; and one of the twin 8.1 m (319-inch) **Gemini Telescopes**.

Maunder minimum The period 1645–1715, when few sunspots were observed, as identified in the 1890s from historical records by British solar physicist E. Walter Maunder (1851–1928). This was not simply the result of a lack of data from those years: recent studies, especially of tree rings from the period, reveal a genuine lull in solar activity. Also, the Maunder minimum correlates well with the long succession of below-average annual temperatures known as the *Little Ice Age*. The Maunder minimum was caused by a number of much longer-term fluctuations acting together to damp down the 11-year **solar cycle**.

MAVEN The Mars Atmosphere and Volatile Evolution mission, a NASA probe launched on 18 November 2013 to study the upper atmosphere of Mars and its interaction with the solar wind. It went into orbit around Mars on 22 September 2014. MAVEN's goal is to understand how much of the planet's atmosphere has been lost into space over time and how this could have affected the planet's climate, including the existence of liquid water in the past.

mean anomaly SEE **anomaly**.

mean solar time The timescale based on the motion of the **mean Sun**, which travels at a uniform rate along the celestial equator.

mean Sun A fictitious Sun that moves along the celestial equator at a constant rate equal to the average rate of the true Sun along the ecliptic. It completes its annual course with respect to the vernal equinox in exactly the same time as the true Sun. The true Sun travels at a variable rate along the ecliptic, and cannot therefore give a uniform measure of time.

Mensa A far-southern constellation, introduced by **Lacaille**, representing the Table Mountain at the Cape of Good Hope, where he worked. Mensa has no star above the fifth magnitude, but part of the Large Magellanic Cloud extends into it.

Mercury The innermost planet of the Solar System, and the smallest of the terrestrial planets. Mercury is a challenge to observe

▲ **Mercury** *A false-color view of the Caloris Basin on Mercury, taken by the Messenger probe in 2008. Caloris, a huge impact feature about 1500 km across, appears orange because of lava flows that fill its floor. Brighter orange spots around its rim are thought to be volcanoes. The dark blue spots on the basin are older impact craters.*

because it is never more than 28° from the Sun, and its small angular diameter (8 arc seconds at greatest elongation) makes telescopic identification of surface detail (light and dark areas) very difficult. As an inferior planet, Mercury exhibits phases. It reaches magnitude −1.3 at brightest. A **transit** of Mercury across the Sun's disk occurs at intervals of either 7 or 13 years. The planet has no satellite. The main data for Mercury are given in the table.

Mercury rotates three times on its axis for every two orbits around the Sun. As a result, the planet has a "hot pole" in each hemisphere – areas that spend longer facing the Sun and get heated more. Also, the **synodic period** of 116 days is very close to twice the rotation period. Thus, at greatest elongation west, say, Mercury presents to Earth almost the same face as at the previous greatest elongation west. This led earlier observers into thinking that Mercury had **synchronous rotation**. The true rotation period was established in 1965 by **radar astronomy**.

Our first close-up views of Mercury's surface came from the Mariner 10 probe which made three close approaches to the planet in 1974 and 1975. Another probe, Messenger, went into orbit around Mercury in March 2011 to map its entire surface in detail. These probes showed Mercury to be a heavily cratered, lunar-like world. The largest impact feature, at 1500 km (950 miles) across, is the **Caloris Basin**, similar in size to the Moon's Imbrium Basin; the next largest is the 625 km (390-mile) diameter crater Beethoven. Like the Moon, Mercury has **ray** craters. Unlike the Moon, it has *lobate scarps*: cliffs up to 3 km (2 miles) high and hundreds of kilometers long, and sometimes cutting across craters. They may be compression faults which formed when the planet's crust cooled and contracted. The *intercrater plains* appear to be volcanic, like the lunar maria, but unlike the maria they contain many small craters. There are also valleys and ridges.

The surface temperature at noon on the equator can reach over 800 K at perihelion, while on the night side it can fall to 90 K. Radar mapping of Mercury's polar regions in 1991 and 1992 revealed what may be metallic deposits on the floors of craters permanently in shadow, where the temperature is estimated never to rise above 112 K. There is a very tenuous atmosphere, mainly of helium and sodium, deriving from the **solar wind**, but also from particles sputtered (knocked off) from the surface by solar radiation, and possibly from material slowly outgassing from the interior.

The presence of a significant magnetic field and the planet's high density for its size suggest that Mercury has a very large iron-rich core, proportionally far bigger than for any other Solar System body.

meridian SEE **celestial meridian**.

MERLIN (Multi-Element Radio-Linked Interferometer Network) An array of six radio telescopes linked to **Jodrell Bank**. Five of the telescopes are 25 m (82 ft) in diameter located at distances up to 127 km (79 miles) from Jodrell Bank; a sixth dish, of 32 m (105 ft) diameter at Cambridge, gives a maximum baseline of 217 km (135 miles).

Messenger A NASA Discovery series probe to Mercury, launched on 3 August 2004. After making three flybys of the planet in 2008 and 2009, it went into near-polar orbit around Mercury on 18 March 2011. It has mapped the planet's surface features, composition and magnetic field in detail. Its instruments found evidence for water-ice in permanently shadowed regions at the planet's poles. The mission

MERCURY: DATA	
Globe	
Diameter	4880 km
Density	5.43 g/cm^3
Mass (Earth = 1)	0.0553
Volume (Earth = 1)	0.0562
Sidereal period of axial rotation	58d 15h 30m
Escape velocity	4.4 km/s
Albedo	0.06
Inclination of equator to orbit	0°
Surface temperature	100–800 K
Surface gravity (Earth = 1)	0.38
Orbit	
Semimajor axis	0.387 au = 57.91 × 10^6 km
Eccentricity	0.206
Inclination to ecliptic	7° 00′
Sidereal period of revolution	87.969d
Mean orbital velocity	47.89 km/s
Satellites	0

SOME OF THE BEST-KNOWN MESSIER OBJECTS		
Messier number	Popular name	Description
1	Crab Nebula	Supernova remnant in Taurus
8	Lagoon Nebula	Emission nebula in Sagittarius
11	Wild Duck	Open cluster in Scutum
13	—	Globular cluster in Hercules
17	Omega, Swan or Horseshoe Nebula	Emission nebula in Sagittarius
20	Trifid Nebula	Emission nebula in Sagittarius
27	Dumbbell Nebula	Planetary nebula in Vulpecula
31	Andromeda Galaxy	Spiral galaxy in Andromeda
33	Pinwheel Galaxy	Spiral galaxy in Triangulum
42	Orion Nebula	Emission nebula in Orion
44	Praesepe or Beehive	Open cluster in Cancer
45	Pleiades	Open cluster in Taurus
51	Whirlpool Galaxy	Spiral galaxy in Canes Venatici
57	Ring Nebula	Planetary nebula in Lyra
64	Black Eye Galaxy	Spiral galaxy in Coma Berenices
97	Owl Nebula	Planetary nebula in Ursa Major
104	Sombrero Galaxy	Spiral galaxy in Virgo

ended on 30 April 2015 when the probe crashed into the surface of Mercury.

Messier, Charles Joseph (1730–1817) French astronomer. He kept a record of various fuzzy extended objects that were in fixed positions in the northern sky so that he should not confuse them with comets, of which he was a successful hunter, discovering more than 20. But it is for this record, the catalog of **Messier objects**, what we now know to be galaxies, clusters and nebulae, that he is remembered today.

Messier object Any of the nebulae, star clusters and galaxies, numbering over 100, which appear in the list begun by Charles **Messier**. Messier drew up the list so that he and other comet-hunters would not confuse these permanent, fuzzy-looking objects with comets. The first edition of the catalog was published in 1774, with supplements in 1780 and 1781. Not all the objects were discovered by Messier himself; several were found by Pierre Méchain. The numbered objects in the catalog are given the prefix M, and are still widely known by these *Messier numbers*. Some of the best-known Messier objects are listed in the table above.

metals (heavy elements) In astronomy, all elements heavier than helium. Population I stars (SEE **stellar populations**), which are young, have relatively high metal contents (up to 2%), while Population II stars, which formed first, have low metal contents (as little as 0.01%). The metal content of a star affects both its structure and its evolution.

meteor The brief streak of light in the night sky caused by a **meteoroid** entering the Earth's upper atmosphere at high speed from space. Meteors are popularly known as *shooting stars* or *falling stars*. They occur at altitudes of about 100 km (60 miles). The typical meteor lasts for a few tenths of a second to a second or two, depending on the meteoroid's speed of entry into the atmosphere, which can vary from about 11 to 70 km/s (7 to 45 miles/s).

Most meteor-producing meteoroids are low-density (about 0.3 g/cm^3) dust particles that originated in comets. They are rapidly eroded away by **ablation** and leave a trail of ionized and excited atmospheric atoms that is visible as the meteor. Most such meteors have a magnitude in the range 0–4. Larger, denser meteoroids that originated from asteroids produce brighter meteors, the brightest of which may rival the full Moon or, in rare cases, even the Sun (SEE **fireball**, **bolide**). Ionization trails are detectable by radar (SEE **radar astronomy**), enabling meteors to be studied in the daytime.

A few meteors per hour may be seen on any clear, moonless night at any time of year. But at certain times of the year there are **meteor showers**, when meteors are more numerous than usual. Showers occur when the Earth passes through a **meteor stream** – dust particles spread around the orbit of a comet. Most of the meteors appearing during the year are *sporadic meteors*, not associated with cometary orbits.

Meteor Crater (Arizona Meteor Crater, Barringer Crater) A circular crater, about 1200 m (4000 ft) in diameter and 180 m (600 ft) deep, near Flagstaff in the Arizona desert. It is the best known of the Earth's impact craters, and the first to be identified as such. The impacting body is estimated to have been a nickel–iron meteorite, about 40 m (130 ft) across and weighing a quarter of a million tonnes, which fell around 50,000 years ago. Quantities of meteoritic nickel–iron have been recovered from the plain around the crater, but most of the meteorite was probably vaporized on impact. The crater was discovered in 1891.

meteorite The part of a large **meteoroid** that survives passage through the Earth's atmosphere and reaches the ground. Hundreds of tonnes of meteoroidal matter is swept up by the Earth each day. Most of it burns up in the atmosphere to produce **meteors**, but about 10% reaches the surface as meteorites and **micro-meteorites**.

Meteorites whose fall is observed are known as *falls*. Their atmospheric passage produces a **fireball** or **bolide**, and analysis of the trajectories of some falls reveals them to have been in orbits like those of **near-Earth asteroids**. Over 99% of meteorite falls are unobserved. Meteorites discovered on the ground are known as *finds*. Many recent finds come from Antarctica, recovered from the ice surface, and also from desert and semidesert areas in, for example, Libya and Australia.

There are three main classes of meteorite: **stony meteorites** are composed mainly of silicates of iron, calcium, aluminum, magnesium and sodium; **iron meteorites** are composed of iron and nickel, and when polished and etched show characteristic **Widmanstätten patterns**; **stony-iron meteorites** have an intermediate composition. Stony meteorites, which make up 92% of all falls, are further subdivided into **chondrites** and **achondrites**. Iron meteorites account for just 7% of falls but over half of all finds, iron being much easier than stony material to pick out on the ground. Just 1% of falls are stony-irons. The exterior of all meteorites is a *fusion crust* of material, melted by friction during passage through the atmosphere and then resolidified.

The largest known single meteorite weighs about 60 tonnes and lies where it fell, at Hoba in Namibia. Stony bodies weighing several tonnes often fragment before impact; the fall at Allende, Mexico, in 1969 deposited 2 tonnes of material over 400 sq km (150 sq miles). The area over which fragments are scattered is called a *strewnfield*. Meteoroids weighing more than 100 tonnes that do not break up are not decelerated as much as lighter bodies, and produce impact craters such as **Meteor Crater**.

Most meteorites are fragments of asteroids, but a few are of lunar origin, and the **SNC meteorites** are believed to be of Martian origin. Analysis of meteorites can shed light on the geological histories of the parent bodies.

meteoroid A particle or body less than 10 m (30 ft) across in orbit around the Sun. Those following an Earth-crossing orbit have the potential to become a **meteor** or **meteorite**. Meteoroids are of cometary or asteroidal origin. Cometary dust particles spread around their parent comet's orbit can produce **meteor showers**. Asteroidal meteoroids are fragments produced by collisions in the main **asteroid** belt (or small asteroids in their own right), and may survive passage through the Earth's atmosphere to land as **meteorites**.

meteor shower The appearance of **meteors** from a common source at the same time each year. The particles that cause meteors are traveling on parallel paths, but the effect of perspective makes them appear to radiate from the same point in the sky, known as the **radiant**. Nearly all showers are named after the constellation in which their radiant lies. There are a dozen or so major showers and many minor ones. Meteors that belong to a shower are called *shower meteors* to distinguish them from the ever-present random background of **sporadic meteors**. Major showers have typical durations of one to two weeks, with activity showing a gradual rise to maximum followed by a rapid decline. The peak **zenithal hourly rate** for a major

MAJOR METEOR SHOWERS		
Shower	Dates	ZHR*
Quadrantids	Jan 1–6	100
Lyrids	Apr 19–25	10–15
Eta Aquarids	Apr 24–May 20	50
Delta Aquarids	Jul 15–Aug 20	25
Perseids	Jul 23–Aug 20	80
Orionids	Oct 15–Nov 2	30
Taurids	Oct 20–Nov 30	10
Leonids	Nov 15–20	—
Geminids	Dec 7–15	100

*Approximate zenithal hourly rate

▲ **Mimas** *This Cassini image of Saturn's tiny satellite Mimas was taken from 213,000 km (132,000 miles) on 16 January 2005. It shows a heavily cratered world, with the giant crater Herschel prominent.*

shower may be between 10 and 100; some minor showers produce only one or two meteors per hour, even at maximum.

A shower occurs when the Earth passes through a **meteor stream**. In an *annual shower* the meteor rates vary little from one year to the next because the meteoroids in the stream are spread evenly around the orbit. In other, usually younger streams, the meteoroids may be concentrated mainly in one part of the orbit, leading to a *periodic shower* from which strong activity, possibly including a **meteor storm**, is seen only in some years.

meteor storm By loose but common definition, a period of activity in a **meteor shower** during which the **zenithal hourly rate** exceeds 1000. Meteor storms have been seen from the **Leonids**, **Giacobinids** and **Andromedids**, and are usually of short duration (less than an hour).

meteor stream A large quantity of **meteoroids** sharing a common orbit around the Sun. With the exception of the **Geminids** (associated with asteroid Phaethon), meteor streams are associated with **comets** and consist of particles shed by comets at perihelion passage over the course of their active lifetimes. A **meteor shower** occurs when the Earth intersects a meteor stream. In a young meteor stream, cometary debris may orbit the Sun in narrow, concentrated strands or *filaments*, and an encounter between the Earth and such a filament can lead to a **meteor storm**. Over time, the debris becomes spread more evenly around the orbit, the orbits of individual meteoroids moving farther away from the primary orbit to produce a torus-shaped stream. The showers to which such streams give rise are low-activity showers lasting a number of weeks. Eventually, meteoroids in a stream become so spread out by gravitational perturbations by the planets and solar radiation effects that they merge into the "background" concentration of dust that pervades the inner Solar System (SEE **interplanetary matter**).

Metis One of the small inner **satellites** of Jupiter discovered in 1979 during the Voyager missions. It orbits within Jupiter's main ring.

Metonic cycle A 19-year cycle of **lunations**, after which the phases of the Moon begin to repeat themselves on the same days of the year. This is because 19 **tropical years** are equal to 235 **synodic months**, to within 2 hours. The discovery is often attributed to the 5th-century BC astronomer Meton of Athens. The cycle forms the basis of the Greek and Jewish calendars. SEE ALSO **saros**.

microlensing The action of a **gravitational lens** on a small scale. If **MACHOs** exist in the halo of our Galaxy they ought to make themselves apparent by microlensing the light of more distant stars. Such events would consist of temporary brightenings of the distant star lasting a few days. Stars of the Large Magellanic Cloud have been monitored since the mid-1990s, and a few hundred such brightenings have been observed.

▼ **Miranda** *The surface of Uranus's satellite Miranda is unlike anything else in the Solar System. Cliffs and craters can be seen.*

micrometeorite A micrometer-sized **meteorite**. Micrometeorites are continually settling on the Earth's surface. The *micrometeoroids* that give rise to them are too small to be burnt up in the atmosphere; they are quickly decelerated and drift gently downward. The quantity that falls greatly exceeds the total of all other meteorites. There are three main types: spheres of high density, irregular compacted particles and fluffy, non-compacted particles of lower density.

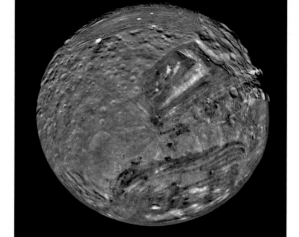

Microscopium A small southern constellation representing a microscope, introduced by **Lacaille**. Its two brightest stars are only of magnitude 4.7.

microwave background radiation SEE **cosmic microwave background**.

midnight Sun The name given to the Sun when it is visible at midnight from inside the Arctic or Antarctic Circle. At the North Pole the Sun is continuously visible for the six-month period for which it is

north of the celestial equator, and is continuously below the horizon for the six months it spends south of the celestial equator. A similar situation holds for the South Pole.

Milky Way The faint band of light visible on clear dark nights encircling the sky along the line of the galactic equator. The combined light of an enormous number of stars, in places it is obscured by clouds of interstellar gas and dust such as the **Coalsack**. It is in fact the disk of our **Galaxy** viewed from our vantage point. It is brightest in Sagittarius, toward the galactic center.

millimeter-wave astronomy Observations of **electromagnetic radiation** at wavelengths from about 1 to 10 mm, at the shortest end of the radio spectrum. Millimeter waves are given out by molecules in dense clouds where stars are forming. Waves with a wavelength shorter than 1 mm are known as *submillimeter waves* and bridge the gap between radio and infrared astronomy (SEE **submillimeter-wave astronomy**).

millisecond pulsar A **pulsar** that flashes every few thousandths of a second, the first of which was discovered in 1982. Unlike normal pulsars, millisecond pulsars are not slowing down, and they have weak magnetic fields. Over 250 are known, about half of them occurring in binary systems.

Mimas The smallest of Saturn's regular **satellites**, with a diameter of 398 km (247 miles), and the innermost, orbiting near the inner edge of the tenuous E Ring. It was discovered by William **Herschel** in 1789. Mimas is heavily cratered; the largest crater, named Herschel, has a diameter of 130 km (80 miles).

minor planet An alternative term for an **asteroid** which tends to be preferred by professional astronomers.

minute of arc SEE **arc minute, second**.

Mira The star Omicron Ceti, a red giant that is the prototype of the **long-period variables**, also known as *Mira stars*. It was the first star discovered to vary in a periodic manner. David Fabricius, a Dutch astronomer, first noticed Mira at third magnitude in August 1596. A few months later it vanished, but he saw it again in February 1609.

Mira's mean period is 332 days, but this is subject to irregularities. Its mean range is magnitude 3.5 to 9.1, but maxima as bright as 2.0 and minima as faint as 10.1 have been observed; its light curve is shown in the diagram below. Its spectrum varies from M5 to M9. Mira has a faint companion, VZ Ceti, a variable with a magnitude range of 9.5 to 12.0, possibly a white dwarf with an **accretion disk**. Mira lies 196 light years away.

Miranda The smallest and innermost of the five main **satellites** of Uranus, discovered in 1948 by Gerard **Kuiper**. Its diameter is 480 × 466 km (298 × 290 miles) and it has the lowest density of the five, at 1.35 g/cm³. Miranda displays one of the most complex and baffling surfaces of any planetary body. A rolling, cratered terrain containing both old and new craters is interrupted by well-defined regions called *coronae* containing parallel or concentric light and dark bands. One of these, Inverness Corona, contains a bright, tick-shaped feature known as the Chevron, and faults running from it lead to cliffs 20 km (12 miles) high. These features are probably the result of extensive ice **volcanism** and geological faulting, or **differentiation** affecting only part of the surface.

mirror An optical component of an instrument (which may be plane, **spheroidal**, **paraboloidal** or **hyperboloidal**) that redirects an incoming beam of light, and may also modify it, so as to aid the

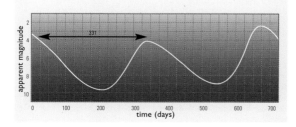

▲ **Mira** *The light curve of Mira (o Ceti) shows a slow fall in brightness followed by a long minimum and then a steep climb up to a short, sharp maximum.*

formation of an image of the light source. Mirrors are used in many astronomical instruments, in particular the **reflecting telescope**. They are made of glass or quartz, ground and polished to achieve the required accuracy of curvature, and coated with a thin layer of a highly reflective material. Aluminum is the usual coating material as it is cheap and easily deposited on the glass. A variety of sophisticated and exotic techniques, such as honeycomb structures and the use of ceramic materials, are now being used to fashion primary mirrors with diameters of 10 m (33 ft) or more for professional reflecting telescopes.

missing mass Invisible or low-luminosity matter that surrounds individual galaxies and pervades clusters of galaxies. Analysis of the rotation of our Galaxy strongly suggests that the galactic halo is embedded in an enormous *corona* that is roughly 600,000 light years in diameter and contains at least 10^{12} solar masses of low-luminosity matter. Other galaxies have similar rotation curves, so they too are presumed to be surrounded by large, massive coronae. In addition, clusters of galaxies must contain a substantial amount of non-luminous matter – otherwise there would not be enough gravity to hold the clusters together. The matter deduced to exist from the motions of galaxies is called dynamical matter as opposed to the luminous or ordinary matter that we can see directly. It is also known as **dark matter**. Results from the **Planck** satellite show that there is sufficient dynamical matter within the Universe to bring its mean density to about 27% of the **critical density**. Luminous matter adds about another 5% to this value.

The issue of the missing mass also arises on the cosmological scale. The measurements made by Planck place the total density of the Universe to within ±1% of the critical density, but luminous and dynamical matter account for only about 32% of this. The remainder may be in the form of **dark energy**.

Mizar A double star in Ursa Major, also known as Zeta Ursae Majoris, of magnitude 2.2. Its wide companion, Alcor (magnitude 4.0), can be distinguished with the naked eye. Mizar and Alcor are too far apart to be a genuine binary. However, with a small telescope a fourth-magnitude companion can be seen closer to Mizar, and this does form a very long-period binary with Mizar. In addition, Mizar, Alcor and Mizar's companion are all **spectroscopic binaries**.

MK system ABBREVIATION FOR **Morgan–Keenan classification**.

MMT Observatory An observatory on Mount Hopkins, near Amado, Arizona. It houses the 6.5 m (255-inch) MMT Telescope, which began operation in 2000. This instrument was originally opened in 1979 as the *Multiple-Mirror Telescope* (MMT), an innovative design in which six 1.8 m (72-inch) mirrors were mounted in a hexagonal array on an altazimuth mounting. The six images were collected at a common focus, giving the same light-gathering power as a 4.5 m (176-inch) telescope. The MMT closed for conversion to a single-mirror instrument in March 1998; the old initials have been retained.

mock Sun SEE **parhelion**.

molecular cloud An interstellar cloud in which the gas is in the form of molecules. Over a hundred different **interstellar molecules** have been identified in these clouds. Their temperature, about 20 K, is enough to make them visible to infrared telescopes (SEE **infrared astronomy**); optically they may show up as **dark nebulae**. They range in size from clouds of 1 solar mass to giant molecular clouds (GMCs) of 10^5 to 10^7 solar masses. GMCs are found in the spiral arms of galaxies, and are sites of star formation. The nearest one is part of the **Orion Nebula** complex.

Monoceros A constellation, representing a unicorn, which adjoins Orion. Its brightest star, Alpha, is only of magnitude 3.8, but it is crossed by the Milky Way and contains some famous clusters and nebulae, including the **Rosette Nebula** and the open cluster M50. Also in Monoceros is the massive binary known as **Plaskett's Star**.

monopole problem A problem with the standard **big bang** model, which predicts that enormous numbers of magnetic monopoles should have been formed – yet none are now observed. **Inflation** solves the problem because the enormous expansion involved has spread out the monopoles until they are now very widely separated.

month The period of revolution of the Moon around the Earth. Various months, defined according to the choice of reference point, are listed in the table. The *calendar months* are the 12 divisions of the

Gregorian **calendar**, and are approximately equal in length to one synodic month. SEE ALSO the entries for each type of month listed in the table.

moon (with lower-case "m") Another word for **satellite**.

Moon (with capital "M") The Earth's only natural **satellite**, orbiting at a mean distance of only 384,000 km (239,000 miles). Its apparent diameter is about half a degree. With an actual diameter of 3476 km (2160 miles) and a density of 3.34 g/cm^3, the Moon has 0.0123 of the Earth's mass and 0.0203 of its volume. The Earth and Moon revolve around their common center of gravity, the **barycenter**. Although its proximity makes it the brightest object in the sky apart from the Sun (the full Moon has a visual magnitude of −12.7), its surface rocks are dark, and the Moon's albedo is only 0.07.

As the Moon orbits the Earth, it is seen to go through a sequence of **phases** as the illuminated proportion of the visible hemisphere changes. One complete sequence, from one new Moon, say, to the next, is called a *lunation*. At the crescent phase, the dark side of the Moon is seen to be faintly illuminated by *earthshine* – sunlight reflected from the Earth. At new Moon or full Moon, **eclipses** can occur. An observer on Earth always sees the same side of the Moon (the *near side*) because it has **synchronous rotation**: its orbital period (the **sidereal month**) around the Earth is the same as its axial rotation period. **Libration** brings some of the *far side* into view at certain times.

The Moon has been studied by many space probes. The crewed **Apollo** missions and some of the later Luna probes returned samples of lunar material to the Earth for study. The surface features may be broadly divided into the darker maria (SEE **mare**), which are low-lying volcanic plains, and the brighter highland regions, which are found predominantly on the southern part of the near side and over the entire far side. There are **impact features** of all sizes. The largest are called *basins*, produced during the early history of the Moon when bombardment by impacting objects was at its heaviest. On the near side, where the crust is thinner, some basins were subsequently filled with upwelling lava to produce the maria, which in turn became cratered. In craters at the north and south poles, whose interiors are in permanent shadow, water ice exists just below the surface. The smaller basins are similar to the largest craters, which have flat floors and are surrounded by a ring of mountains. Other features are mountain peaks and ranges, valleys, elongated depressions known as **rilles**, **wrinkle ridges**, low hills called *domes* and bright **rays** of ejecta radiating from the sites of recent cratering impacts.

Early lunar cartographers assumed that the darker lunar features were expanses of water, and named them after fanciful oceans (oceanus), seas (mare), bays (sinus), lakes (lacus) and swamps (palus). Other features are named after famous people, principally astronomers and other scientists. The old names are still in use on lunar maps.

The most widely accepted theory for the Moon's origin, known as the *giant impactor* or *big splash* theory, is that a Mars-sized body collided with the newly formed Earth, and debris from the impact accreted (SEE **accretion**) to form the Moon. Subsequent large impacts produced the Moon's basins, and smaller impacts the craters. A very few lunar craters may have had a volcanic origin.

TYPES OF MONTH	
anomalistic month	27.55455 mean solar days
draconic month	27.21222 mean solar days
sidereal month	27.32166 mean solar days
synodic month	29.53059 mean solar days
tropical month	27.32158 mean solar days

◄ **Moon** When it is at its brightest, at full Moon, the differences between the bright lunar highlands and the dark, lava-filled basins become very apparent.

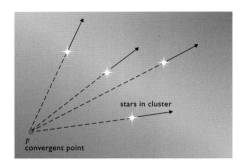

◄ **moving cluster** *The velocity of stars in an open cluster can be determined if they are close enough for their parallax to be measurable over a short period. Their parallax is combined with their line-of-sight rate of motion toward or away from Earth to obtain their speed of movement at right angles to the line of sight. Combined with the measurement of the angle of movement, this allows astronomers to judge their distances.*

Mild *moonquakes* occur at depths of roughly 700 km (450 miles); otherwise, the Moon is now geologically inactive.

Seismic measurements at Apollo landing sites yielded some information on the Moon's internal structure. The lunar crust is about 100 km (60 miles) thick in the highland regions, but only a few tens of kilometers under the mare basins. Under the largest basins the mantle has bulged upward to form **mascons**. A small iron core at the Moon's center accounts for about 4% of its mass, but there is no overall magnetic field.

Rocks brought back from the Moon have been found to be igneous. The maria have a basaltic composition, while the highlands consist largely of anorthosite. As a result of the constant bombardment, crustal rocks are in the form of **breccias**, perhaps to a depth of several kilometers, and the surface is covered with a **regolith**. The Moon has only the most tenuous of atmospheres. The atmosphere is probably made up of **solar wind** particles retained temporarily, and atoms sputtered (knocked off) from the surface by the solar wind. Consequently the surface temperature variation is extreme: from 100 to 400 K. Most recently, deposits of water ice at least 2 m (6 ft) thick have been detected in crater floors at the Moon's poles that remain permanently in shadow.

Morgan–Keenan classification The categorization of stellar spectra from their absorption features. The system was devised by William W. Morgan, Philip C. Keenan and Edith Kellman of Yerkes Observatory for their *An Atlas of Stellar Spectra*, published in 1943; it is also known as the *MK system*, *MKK system* or *Yerkes system*. It established formal rules for describing a star by spectral type (O, B, A, F, G, K, M, in sequence from hot to cool surface temperature) and luminosity class (I = supergiant, III = giant, V = dwarf; with interpolations Ia, Iab, Ib, II and IV). SEE ALSO **spectral classification**.

Morning Star Name sometimes given to the planet Venus when it is visible in the east before sunrise.

Mount Graham International Observatory An observatory on Mount Graham, near Safford, Arizona. It has three main instruments: the Vatican Advanced Technology Telescope (also known as the Alice P. Lennon Telescope), a 1.8 m (70-inch) reflector; the 10 m (400-inch) Submillimeter Telescope; and the **Large Binocular Telescope**.

mounting A means of supporting the weight of a telescope and aiming it at different points in the sky. Of necessity, all large professional instruments are on permanent mounts; most amateur telescopes and their mounts are portable. Permanent or portable, rigidity is a prime requirement.

To enable a telescope to be pointed at different parts of the sky, the mounting incorporates two axes, perpendicular to each other, about which the telescope can be rotated. In the **altazimuth** mount the axes are parallel and perpendicular to the horizon. The disadvantage of this mounting is that the telescope has to be moved about the two axes simultaneously in order to keep an object in view. This is overcome in **equatorial telescopes**, in which one axis (the polar axis) is parallel to the Earth's axis of rotation, and so points to the celestial pole. The telescope then only needs to be moved about the other axis, in the direction opposite to that of the Earth's rotation, to keep an object in the field of view.

Since the introduction of computer-controlled drive mechanisms, the altazimuth mount has become the choice for all large professional telescopes. A simple altazimuth platform is incorporated in the **Dobsonian telescope**, a large and portable amateur reflecting telescope.

Mount Palomar The location of the **Palomar Observatory**.

Mount Stromlo Observatory An observatory near Canberra, operated by the Australian National University. Founded in 1924 as a solar observatory, it changed its focus to stellar astronomy after World War II. Many of its facilities, including several historic telescopes, were destroyed in a bush fire in January 2003. It also runs telescopes at **Siding Spring Observatory**.

Mount Wilson Observatory An observatory in the San Gabriel Mountains near Pasadena, California. Its main instrument is the 2.5 m (100-inch) Hooker Telescope, opened in 1917, with which Edwin **Hubble** discovered the expansion of the Universe; it was renovated and reopened in 1993. The Mount Wilson Observatory was founded in 1904 by George Ellery **Hale**, originally for solar astronomy; a 1.5 m (60-inch) reflector was opened in 1908. There are also solar telescopes and interferometers, including the CHARA optical/infrared interferometer.

moving cluster A group of stars, often widely scattered, that share the same motion through the Galaxy. The stars in a moving cluster all follow nearly parallel paths. However, because of a perspective effect, they seem to be moving toward, or away from, a point known as the *convergent point*. This allows the distance of the cluster to be determined. First, the position of the convergent point must be established. Some clusters are very scattered, for example the Ursa Major moving cluster, which includes five stars of the Big Dipper or Plough (Beta, Gamma, Delta, Epsilon and Zeta). Several other bright stars show the same motion through the Galaxy as these five Ursa Major stars. They include Sirius, Delta Leonis, Beta Aurigae, Beta Eridani and Alpha Coronae Borealis. In total about 100 members of the group are known, scattered widely across the sky. Their separation in space is larger than in a normal cluster. Our Sun is moving through the outskirts of the group, which is why its members appear so widely spread across the sky.

M star A star of spectral type M, characterized by absorption bands of titanium oxide molecules in its spectrum. M-type dwarf stars, also known as red dwarfs, define the lower end of the **main sequence**. Proxima Centauri and Barnard's Star are examples. The surface temperatures of M-type dwarfs range from 3500 to 2500 K; their masses range from 0.08 to 0.5 solar masses, and their radii from 0.1 to 0.6 times that of the Sun. The most massive M-type dwarfs are nearly one-tenth as luminous as the Sun, while those at the bottom of the main sequence have less than one-thousandth of the Sun's luminosity. Most of their radiation is emitted in the infrared, so their visual luminosity is even fainter, from one twenty-fifth to one ten-thousandth that of the Sun. Many M-type dwarfs are **flare stars**; none are visible to the naked eye.

M-type giants and supergiants are all evolved stars with masses much greater than those of M-type dwarfs. Their surface temperatures are typically a few hundred degrees cooler than dwarfs of the same spectral type. The brightest M-type supergiants are up to a million times as luminous as the Sun. The largest of them are several thousand times the Sun's diameter – as large as the orbit of Saturn or Uranus. Even the coolest M-type giants are about 100 times the diameter of the Sun. Such distended stars are unstable and therefore variable; examples are **Betelgeuse** and **Mira**. The high luminosity and low surface gravity of M-type giants and supergiants give rise to a **stellar wind** of gas in which the mass loss can be as high as one solar mass every 100,000 years.

Mullard Radio Astronomy Observatory The radio astronomy observatory of the University of Cambridge, situated at Lords Bridge near Cambridge. It was there that the technique of **aperture synthesis** was developed. The Observatory's principal instrument is the Arcminute Microkelvin Imager (AMI), opened in 2008. This consists of two arrays of dishes, eight of 13 m (43 ft) aperture and ten of 3.7 m (12 ft) aperture, to study the **cosmic microwave background**. The Cambridge Optical Aperture Synthesis Telescope (COAST) was opened in 1995.

Multi-Element Radio-Linked Interferometer Network SEE **MERLIN**.

Multiple-Mirror Telescope SEE **MMT Observatory**.

multiple star A gravitationally connected group of more than two stars. **Castor**, for example, has six components. In multiple systems there is a recognizable hierarchy or order. In Castor, for instance, the two main pairs rotate about a common center of gravity, and a pair of cool red dwarfs rotates about the same center but at a larger distance. There is no clear distinction between a multiple star with many members and a small cluster. A good example is the **Trapezium** in the Orion Nebula, which consists of four young stars enmeshed in nebulosity with a further five possible members nearby.

Musca A small but distinctive southern constellation representing a fly, adjoining Crux. Its brightest star, Alpha, is magnitude 2.7. Musca contains the globular clusters NGC 4372 and NGC 4833.

nadir The point on the **celestial sphere** vertically below the observer. It is diametrically opposite the **zenith**.

Nagler eyepiece A telescope eyepiece with a very wide field of view, 80° or more, making it suitable for activities such as searching for comets. (After its designer, Albert Nagler, 1935– .)

Naiad One of the small inner **satellites** of Neptune discovered in 1989 during the Voyager 2 mission.

naked eye A term applied to observations made by the eye alone, without the aid of any optical instrument, or to celestial objects thus visible.

NASA ABBREVIATION FOR **National Aeronautics and Space Administration**.

Nasmyth focus A focus of a reflecting telescope on an **altazimuth** mount, situated to one side on the altitude axis. The light beam is directed there by an extra flat mirror angled at 45°. Heavy instruments positioned at a Nasmyth focus do not have to be moved bodily as the telescope tracks an object. It is named after Scottish engineer James Nasmyth (1808–90), who incorporated it in his 6-inch (150 mm) reflector. Its use has been revived with the present generation of large, altazimuthally mounted professional instruments. SEE ALSO **coudé**.

National Aeronautics and Space Administration (NASA) The US government agency for the exploration of space, formed in 1958. Its headquarters are in Washington, DC, and it operates various field stations. These include the Goddard Space Flight Center at Greenbelt, Maryland, which handles space science research and Earth satellite tracking; the Jet Propulsion Laboratory in Pasadena, California, which is mission control for NASA space probes and manages NASA's deep-space tracking stations around the world; the Ames Research Center, San Francisco, for aeronautics and planetary science; the John F. Kennedy Space Center at Cape Canaveral, Florida, the main US launch site; and the Lyndon B. Johnson Space Center in Houston, Texas, mission control for crewed space missions.

National Radio Astronomy Observatory (NRAO) The major US radio observatory, with headquarters at Green Bank, West Virginia. Its instruments include radio dishes of 42.7 m (140 ft), 26 m (85 ft) and 13.7 m (45 ft) aperture. The Green Bank Telescope was dedicated in 2000. With a dish measuring 100 × 110 m (330 × 360 ft), it is the world's largest fully steerable radio telescope. NRAO also operates the **Very Large Array** in New Mexico, the **Very Long Baseline Array**, and participates in the **Atacama Large Millimeter Array**.

nautical twilight SEE **twilight**.

n-body problem SEE **many-body problem**.

neap tide SEE **tides**.

NEPTUNE: DATA	
Globe	
Diameter (equatorial)	49,528 km
Diameter (polar)	48,686 km
Density	1.64 g/cm^3
Mass (Earth = 1)	17.14
Volume (Earth = 1)	57.67
Sidereal period of axial rotation	16h 07m
Escape velocity	23.5 km/s
Albedo	0.84
Inclination of equator to orbit	29° 34′
Temperature at cloud-tops	55 K
Surface gravity (Earth = 1)	0.98
Orbit	
Semimajor axis	30.06 au = 4497 × 10^6 km
Eccentricity	0.0097
Inclination to ecliptic	1° 46′
Sidereal period of revolution	164.79y
Mean orbital velocity	5.43 km/s
Satellites	14

near-Earth asteroid (NEA) A small **Amor asteroid**, **Apollo asteroid** or **Aten asteroid** whose orbit brings it relatively close to the Earth. The term *near-Earth object* is also used, to include close-approaching comets. *Potentially hazardous asteroids* (PHAs) are those whose orbits bring them particularly close to the Earth. Close approaches have been made by objects as small as 9 m (30 ft) across. Objects this size striking the Earth would burn up in the atmosphere; larger objects would penetrate further, causing a powerful low-level airburst as in the **Tunguska Event**, or producing a crater like **Meteor Crater**.

NEAR Shoemaker A NASA probe to the asteroid **Eros**, launched on 17 February 1996. The mission was originally called Near Earth Asteroid Rendezvous but was renamed to commemorate the American astrogeologist Eugene Shoemaker (1928–97). After a flyby of the asteroid **Mathilde**, NEAR Shoemaker became the first probe to orbit an asteroid when it entered orbit around Eros in February 2000. The spacecraft soft-landed on the surface of Eros on 12 February 2001, at the end of the mission.

nebula (plural **nebulae**) A region of interstellar gas and dust. The word nebula is Latin for "cloud." There are three main types of nebula.

Emission nebulae are bright diffuse nebulae which emit light and other radiation as a result of ionization (removal of electrons) and excitation of the gas atoms by ultraviolet radiation. The source of the ultraviolet is usually one or more hot stars. When gas ions recombine with free electrons, and forbidden transitions (SEE **forbidden lines**) occur in excited atoms, radiation is emitted, giving rise to emission lines in the spectrum. Examples of emission nebulae are **H II regions** such as the Orion Nebula, and **planetary nebulae** such as the Ring Nebula in Lyra. **Supernova remnants** are a form of emission nebula in which the gas is made to glow not by the ultraviolet radiation of the star within, but by frictional heating as it collides with surrounding interstellar gas; an example is the Cygnus Loop. The Crab Nebula is a type of supernova remnant that shines by **synchrotron radiation**.

In contrast, the brightness of **reflection nebulae** results from the scattering by dust particles of light from nearby stars.

Dark nebulae are not luminous: interstellar gas and dust absorb light from background stars, producing apparently dark patches in the sky. This third class includes a group known as **Bok globules**, which are dense absorption nebulae of nearly spherical shape.

Neptune The eighth major planet from the Sun, and the smallest and most distant of the four giant planets; it marks the inner edge of the **Kuiper Belt**. With a mean magnitude of 7.8, Neptune is invisible to the naked eye. Its average apparent size is 2.2 arc seconds. Through a telescope it appears as a small, featureless greenish-blue disk. In its size, mass, atmosphere and color, Neptune resembles Uranus. The main data for Neptune are given in the table.

Neptune was first identified by Johann **Galle**, assisted by Heinrich D'Arrest, on 23 September 1846, close to a position predicted by Urbain **Le Verrier**. The flyby of the Voyager 2 probe in 1989 provided most of our current knowledge of the planet. The upper atmosphere is about 85% molecular hydrogen and 15% helium. Its predominant blue color is due to a trace of methane, which strongly absorbs red light. Several different atmospheric features were visible at the time of the Voyager encounter. There were faint bands parallel to the equator, and spots, the most prominent being the Great Dark Spot, about 12,500 km (8000 miles) long and 7500 km (4500 miles) wide. It was about the same size relative to Neptune as the **Great Red Spot** (GRS) is to Jupiter, and like the GRS was a giant anticyclone. White, cirrus-type clouds of methane crystals cast shadows on the main cloud deck some 50 km (30 miles) below. Neptune's atmosphere shows **differential rotation**, some features taking 19 hours or more for one rotation. The atmosphere thus has **retrograde** (1) rotation with respect to the interior, which rotates in just over 16 hours. There are also the highest wind speeds recorded on any planet – over 2000 km/h (1250 miles/h) in places.

Neptune's dynamic and changing atmosphere, which since the Voyager encounter has been followed by the **Hubble Space Telescope**, probably has

▼ *Nasmyth focus* A number of large, professional telescopes have instruments at their Nasmyth foci. In the William Herschel Telescope there are two foci, one used for infrared and the other for optical observations. The third mirror (the Nasmyth flat) can be moved to direct light to either of them.

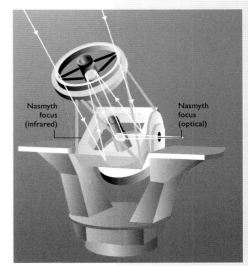

Nasmyth focus (infrared)

Nasmyth focus (optical)

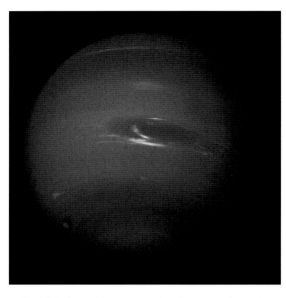

to do with its internal heat source – it radiates more than twice as much heat as it receives from the Sun. The amount of white cloud also varies with the **solar cycle**, which is hard to explain for such a distant planet.

Below the visible atmosphere the composition and structure are less certain. A chemical cycle, driven by ultraviolet radiation from the Sun, might operate in which methane is converted into other hydrocarbons and back again. At deeper levels there may be a mantle of water and ammonia about ten times the mass of the Earth, surrounding an iron–silicate core of about one Earth mass. Like Uranus, Neptune has a magnetic field in which the axis is markedly tilted (by 47°) to the axis of rotation and is displaced (by about 10,000 km/6000 miles) from the planet's center. In Neptune's case it could be that the mantle is at least partially fluid and the ammonia–water mixture is ionized, and therefore electrically conducting, and that this is the source of the magnetic field.

Of Neptune's 14 known **satellites**, **Triton** and **Nereid** were discovered before the Voyager 2 flyby, and another (**Larissa**) was suspected from an **occultation** measurement. Triton's orbital motion is **retrograde** (1), and Nereid's orbit is the most eccentric of any Solar System satellite. Seven inner satellites, including the second-largest, **Proteus**, lie in close, regular orbits.

Neptune's ring system was detected from Earth during an **occultation** in 1984, but the measurements suggested a number of incomplete "ring arcs." Voyager revealed four continuous rings: two narrow ones, and two faint broad ones. The outermost and brightest, the Adams Ring, contains "clumps" of material that appeared as arcs in the occultation data. SEE ALSO **ring, planetary**.

Nereid Neptune's third largest **satellite**, discovered by Gerard **Kuiper** in 1949. It has a highly eccentric orbit, taking it from 1.4 million km (850,000 miles) out to 9.7 million km (6 million miles). It is almost certainly a captured body. It is about 340 km (210 miles) in diameter.

neutrino astronomy The detection of neutrinos from astronomical sources. Neutrinos are atomic particles with no charge and very low (possibly zero) mass. They therefore interact hardly at all with matter. To maximize the chance of detecting one, neutrino "telescopes" consist of huge tanks filled with a suitable liquid. One type uses cleaning fluid, as an incoming neutrino has a chance of reacting with a chlorine atom to form a radioactive atom of argon, which can be detected. Another type contains water, as a neutrino may interact with an electron in a water molecule and emit a particular type of radiation. Neutrinos are emitted by nuclear reactions inside stars. Early attempts to detect neutrinos from the Sun found less than half the number that theorists expected. This *solar neutrino problem* was resolved when it became clear that neutrinos emitted by the Sun oscillated between different types *en route* to the Earth, and detectors were adapted to record the three types. Neutrinos are also produced by the complex reactions that take place in **supernovae**, and neutrinos from **Supernova 1987A** were detected. Neutrinos are also produced in the Earth's atmosphere by cosmic rays; these can be distinguished by their different energies.

neutron star An extremely small, dense star that consists mostly of neutrons. Neutron stars are observed as **pulsars** and are also thought to be one of the components in **X-ray binaries**. They are formed when a massive star explodes as a **supernova**, ejecting its outer layers and compressing the core so that its component protons and electrons merge into neutrons. Neutron stars have masses comparable to that of the Sun, but diameters of only about 20 km (12 miles) and average densities of about 10^{15} g/cm^3. A neutron star's gravity is so great that objects would weigh 10^{11} times more at its surface than on the surface of the Earth, and it also produces a significant **gravitational redshift** (z about 0.2). The maximum mass for a neutron star is about 3 solar masses, beyond which it would collapse further, into a **black hole**.

Being so small and having such strong gravity, neutron stars can spin rapidly without disintegrating – up to several hundred times per second (SEE **millisecond pulsar**). They retain the magnetic field of the original star but compressed into a much smaller volume, so the magnetic fields of neutron stars are about a million million times stronger than that of the Earth. A neutron star's structure is complicated because of the great variation in pressure from the surface to the center. The crust is made of iron about 10,000 times denser than terrestrial iron and a million times stronger than steel. Beneath this is a liquid of neutrons, and at the very center may be a core of exotic nuclear particles.

Newcomb, Simon (1835–1909) Canadian-born American mathematical astronomer. He calculated highly accurate values for astronomical constants, such as the solar **parallax**, and prepared extremely accurate tables of the motions of the Moon and the planets. It was Newcomb who calculated the discrepancy in the advance of **Mercury**'s perihelion that was to provide a proof of **Einstein**'s special theory of relativity.

New Horizons A NASA spacecraft launched on 19 January 2006 to fly past **Pluto** in July 2015 at a distance of about 10,000 km (6000 miles). It will then proceed into the **Kuiper Belt** to examine in close-up some of the most distant objects in the Solar System.

new Moon The **phase** of the Moon when at conjunction. The dark side of the Moon then faces the Earth, and the illuminated side is invisible to us.

New Technology Telescope A 3.6 m (142-inch) reflector at the **European Southern Observatory** in Chile, completed in 1989. It was the first large telescope to use a system of **active optics** to keep the main mirror in shape.

Newton, Isaac (1642–1727) English physicist and mathematician. His achievements were so important for astronomy that the phrase "Newtonian revolution" is often used. Most of his theories – on light, gravitation and calculus, for example – he developed in basic form while in his early twenties. In 1668 he built the first reflecting telescope in order to overcome the **chromatic aberration** inherent in the lenses used in refracting telescopes. In 1684, encouraged by Edmond **Halley**, he began work on his *Philosophiae naturalis principia mathematica*. Published in 1687, and usually referred to as

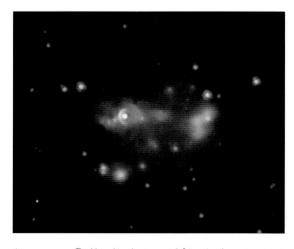

▲ **neutron star** *The blue-white dot at center left is a tiny, dense neutron star, seen in X-rays. Surrounding it, a wind of high-energy particles from the neutron star is shown in blue, while cooler dust ejected by the supernova is red.*

the *Principia*, this was one of the most epoch-making books in the history of science. In it he put forward his law of **gravitation** and laws of motion (SEE **Newton's laws of motion**), and showed that the orbit of a body moving under an **inverse-square law** of gravitation would be an ellipse. He went on to demonstrate how gravitation explained the second and third of **Kepler's laws** and also many outstanding problems in astronomy. In his *Opticks* (1704) he set out his theories of light, including how white light is made up of the colors of the spectrum.

Newtonian telescope The simplest form of **reflecting telescope**. The first, built by Isaac **Newton** in 1668, had a primary concave spherical mirror of speculum metal (an alloy of copper and tin) which reflected the light up the tube of the telescope. A flat speculum mirror near the top of the tube, angled at 45°, directed the light to a focus – the *Newtonian focus* – to one side of the tube, where the image of the object being observed was viewed through a small plano-convex eyepiece. This is still the basic design, although speculum has given way to glass, and for all but the smallest apertures the primary mirror is now **paraboloidal**. The Newtonian has long been a favorite with amateur astronomers.

Newton's law of gravitation SEE **gravitation**.

Newton's laws of motion The fundamental laws of dynamics, set out by Isaac Newton in his *Principia*:

1. *Law of inertia*: Every body continues in its state of rest, or of uniform motion in a straight line, until acted upon by some outside force.

2. *Law of acceleration*: The acceleration of a body is directly proportional to the force acting upon it, and is in the direction in which the force is acting.

3. *Law of action and reaction*: To any action there is an equal and opposite reaction.

NGC Abbreviation for the *New General Catalogue of Nebulae and Clusters of Stars*, published in 1888 by J. L. E. **Dreyer**, and incorporating previous catalogs by William and John **Herschel**. The NGC contains nearly 8000 objects – including galaxies, although their real nature was not then known – and another 5000 are listed in two subsequent *Index Catalogues* (IC). Objects are still referred to by their NGC or IC numbers.

noctilucent clouds (NLC) Clouds that form at very high altitude (average 82 km/51 miles) in the Earth's atmosphere, sometimes visible in the summer in twilight hours (noctilucent means "nightshining") from latitudes between 50° and 70°. NLCs have a pearly or blue color and resemble delicately interwoven cirrus cloud. They are made up of water vapor that has been carried aloft and then condensed and frozen on to particles, probably of meteoric origin; ions produced by solar radiation and even products of industrial pollution have also been suggested as the condensation nuclei. NLCs have become brighter and more common in recent years.

node The point at which one orbit cuts another. Specifically, a node is one of the two points in the orbit of a planet or comet at which it cuts the ecliptic, or at which the orbit of a satellite cuts that of its primary. The *ascending node* is the node at which a celestial body passes from south to north; the *descending node* is the node at which a celestial body passes from north to south. The *line of nodes* joins the ascending and descending nodes. The *regression of the nodes* is the backward motion of the Moon's nodes around its orbit. They move slowly westward so that the nodal line makes one complete revolution of the orbit in 18.6 years. The regression is caused by the Sun's gravitational attraction.

non-gravitational force Any force acting on a celestial body that is not gravitational, in particular one that produces similar effects to gravity. The term is most commonly used in the context of comets which, on their approach to perihelion, lose volatile material through one or several vents in their crust, producing a jet or jets in a particular direction. This can alter their course in much the same way as the gravitational **perturbations** to which they are prone. The **Poynting–Robertson effect** results from another type of non-gravitational force. SEE ALSO **magnetohydrodynamics**.

Norma A small southern constellation introduced by **Lacaille** between Ara and Lupus, representing a set square. It contains no star above magnitude 4.0.

North America Nebula A large emission nebula about 3500 light years distant in the constellation Cygnus. It is also known as NGC 7000. An associated region of nebulosity, IC 5070, is the *Pelican Nebula*.

Northern Cross A popular name sometimes given to the cross-shaped arrangement of the brightest stars in the constellation Cygnus.

North Star Popular name sometimes given to **Polaris**.

nova (plural **novae**) A star, usually quite faint, whose brightness suddenly increases by ten magnitudes or more before slowly returning to its pre-nova state. A typical nova rises to maximum in a few days, and thereafter declines in brightness by a factor of about ten in 40 days, although cases of both slower and faster declines are known. The diagram shows the light curve of the brightest nova (as opposed to **supernova**) of modern times, Nova Aquilae, in 1918. About 40 novae are estimated to explode in the Galaxy each year, although we see only a small percentage of them.

Novae occur in close binary systems where one component is a white dwarf and the other is a giant. Matter flows from the surface of the giant, where gravity is weak, to the surface of the white dwarf, where gravity is strong. When the infalling material reaches a sufficient density – after a long period of time – it suddenly ignites in a *thermonuclear runaway*, producing a violent explosion on the surface of the white dwarf. The energy of a typical explosion is about the same as the Sun emits in 10,000 years. Clouds of gas are ejected at speeds around 1500 km/s (900 miles/s) during the outburst. The total amount of material lost is estimated to be about one ten-thousandth the mass of the white dwarf. Some stars, known as **recurrent novae**, have been observed to undergo outbursts repeatedly. All novae probably recur over periods of thousands of years. SEE ALSO **dwarf nova**.

nucleosynthesis The production of chemical elements from other chemical elements via naturally occurring nuclear reactions. These reactions are of two types: *fusion*, in which heavier elements are built up from lighter ones, and *fission*, in which heavier elements are broken down into lighter ones. Fusion created most of the Universe's helium from hydrogen in the first few minutes following the **big bang**, and the **carbon–nitrogen cycle** and other fusion processes operate inside stars to create helium and heavier elements which are carried into interstellar space by stellar winds. A few light elements, such as lithium, are produced in the interstellar medium by cosmic rays. Both fusion and fission occur in the extreme conditions of novae and supernovae to produce many different elements which are then ejected into the surrounding space.

nucleus (1) The solid part of a **comet**.

nucleus (2) The central region of a **galaxy**.

NuSTAR (Nuclear Spectroscopic Telescope Array) A NASA X-ray observatory, launched on 13 June 2012. It detects higher energies than previous X-ray observatories, such as Chandra and XMM, coming from objects such as black holes, supernova remnants, clusters of galaxies and the Sun.

nutation The periodic oscillation in the precessional motion of the Earth's axis of rotation (SEE **precession**). It was discovered by James **Bradley** in 1747. It is in effect a slight "nodding" of the Earth's axis, caused by the combined effect of the attractions of the Sun and Moon upon the Earth. The effect is continuously varying as the relative positions of the three bodies are continuously changing. *Lunar nutation*, which causes the Earth's axis to trace out an ellipse, has a period of 18.6 years. *Solar nutation* has a period of 0.5 of a tropical year. The *fortnightly nutation* has a period of 15 days.

▲ **Newtonian telescope** In this form of reflecting telescope, light is reflected from a paraboloidal primary mirror to a flat secondary mirror placed diagonally within the tube, and directed to a side-mounted eyepiece.

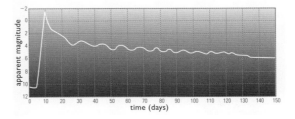

◀ **nova** The light curve of a nova typically shows a steep rise in brightness by a factor of many thousands followed by a slow, irregular decline. The rapid rise in brightness is caused by explosive hydrogen burning on the surface of the white dwarf.

OB association A region of space in which recent star formation has produced high-mass O and B stars. Such associations are less tightly bound than a normal star cluster and are slowly dispersing. Since O and B stars are very luminous, OB associations are recognizable at great distances, even through intervening dust. Tens or hundreds of these massive stars can occur in a single association. Fainter, less massive stars may also be present, but are harder to see at great distances. The Orion Nebula is part of an OB association which is relatively close to us, so young, low-mass stars are also detectable within it. OB associations are a feature of the gas-rich spiral and irregular galaxies, but not of elliptical galaxies, in which no young stars are found. COMPARE **T association**.

Oberon One of the five main **satellites** of Uranus, discovered in 1787 by William Herschel. It is 1520 km (945 miles) in diameter and consists largely of a mixture of rock and water ice, with a density of 1.58 g/cm³. Oberon has many craters, some with **rays** and some with dark floors, and mountains, one of which appears to be 20 km (12 miles) high. The high density of craters is evidence of a world unaltered by geological activity.

objective The main light-gathering element of an optical instrument. In a refracting telescope it is the lens or lens system (called the *object glass*, or OG, in older texts) that collects and focuses light from a celestial body to form an image of it.

objective prism A large, narrow-angle prism mounted across the aperture of a telescope so that the image of each star in the field of view is transformed into a low-resolution spectrum.

oblateness (ellipticity) A measure of how far a **spheroidal** body deviates from a true sphere, given by the difference between the equatorial and polar radii divided by the equatorial radius. The oblateness of a celestial body is an indication of the speed at which it is rotating.

obliquity of the ecliptic The angle between the plane of the ecliptic and the celestial equator. Alternatively, it may be defined as the angle between the axis of rotation of a planet and the pole of its orbit. The Earth's obliquity is currently about 23° 26′, and is decreasing by 0.47′ per year. The angle will start to increase again in about 1500 years; the value ranges between 21° 55′ and 28° 18′ in a period of 40,000 years. The obliquity is responsible for the seasons. It represents the greatest angular distance that the Sun can lie north and south of the equator.

observatory A structure built and equipped for making celestial observations and measurements, acting principally as a housing for a telescope. The best location for astronomical purposes is on a mountain or high plateau where there are many cloudless or near-cloudless nights. The most favorable locations include Hawaii, Chile, western parts of the USA and the Canary Islands.

Professional optical and infrared telescopes are installed in buildings with a rotating roof, usually dome-shaped, although square buildings are now becoming common, and in some cases the entire building rotates. Domes have a slit, closed by sliding shutters when not in use, through which observations are made. In addition there are control rooms, laboratories, workshops, living accommodation for personnel and other facilities. There are over 200 professional optical observatories throughout the world.

Sites for radio observatories are less critical than for optical observatories because radio waves are not affected so much by the atmosphere. Radio telescopes consisting of an array of antennas do, however, require a considerable area of land.

Observatories are also launched into Earth orbit, entirely above the absorbing and distorting effects of the Earth's atmosphere. In addition to optical and radio studies, these craft carry out measurements in those infrared, ultraviolet, X-ray and gamma-ray regions of the spectrum which do not penetrate the Earth's atmosphere and so cannot be studied from ground-based observatories. SEE ALSO the names of individual observatories.

occultation The temporary cutting off of the light from one celestial body as another, nearer one passes in front of it. For example, the Moon may pass in front of a star; a planet may do the same and may also occult its own satellites, as does Jupiter; at certain times Jupiter's **Galilean satellites** may undergo mutual occultations. The rings of **Uranus** were first detected when the planet occulted a star in 1977. Strictly speaking, a solar eclipse is an occultation of the Sun by the Moon. Accurate timings of lunar occultations are valuable for accurately determining the Moon's position and can also provide information about the occulted body.

▲ **Oberon** *This is the closest image obtained of Uranus's satellite Oberon during the Voyager 2 flyby of January 1986. Oberon's surface shows several impact craters surrounded by bright rays.*

▲ **Omega Nebula** *Also known as the Swan Nebula, M17 in Sagittarius is part of a huge complex of emission nebulosity. It contains about 800 solar masses of material.*

Grazing occultations, in which the star appears to skim the Moon's limb, are of particular interest because they reveal the topological profile of the lunar limb.

Octans A faint constellation, notable only because it contains the south celestial pole. The nearest naked-eye star to the pole is Sigma Octantis, at magnitude 5.4 the official south pole star. Octans was introduced by **Lacaille** and represents an octant, the forerunner of the sextant. The brightest star in the constellation is Nu, magnitude 3.7.

ocular An older term for **eyepiece**.

Olbers, Heinrich Wilhelm Matthäus (1758–1840) German doctor and amateur astronomer. He is best known for having posed the question now known as **Olbers' paradox**. He made the second and fourth asteroid discoveries, of Pallas (1802) and Vesta (1807). Olbers invented a method for calculating cometary orbits, and discovered several comets. He also proposed the theory that the tails of comets are highly rarefied matter expelled from the head of the comet by pressure of some kind from the Sun, thus anticipating the concept of **radiation pressure**.

Olbers' paradox The paradox that arises in attempting to answer the question, "Why is the sky dark at night?" If space is infinite and uniformly filled with galaxies, then an observer would end up looking at the surface of a star or galaxy at every point in the sky, so the entire night sky should be bright. In fact, space is not infinite. The Universe is believed to have been born in the **big bang** about 14,000 million years ago, and so light from the most distant objects has not yet reached us. In addition, the light from distant galaxies is weakened by the **redshift** caused by the expansion of the Universe. The paradox is named after Heinrich **Olbers**, who drew attention to it in 1823.

Olympus Mons An extinct volcano on Mars, and the largest known volcano in the Solar System, rising to 21 km (13 miles) above the planet's mean surface level. It is about 700 km (450 miles) across, including the surrounding area of eroded volcanic rock. The caldera (summit crater) is over 80 km (50 miles) across.

Omega Centauri The brightest globular cluster in the sky. It is visible to the naked eye as a fuzzy star of magnitude 3.7 in the constellation Centaurus. It has been designated NGC 5139 and is 17,000 light years away, making it one of the nearest globular clusters. It contains several hundred thousand stars in a volume 200 light years across.

Omega Nebula An emission nebula in the constellation Sagittarius, designated M17 or NGC 6618. Other names for it are the Horseshoe Nebula and the Swan Nebula. It lies 6800 light years away.

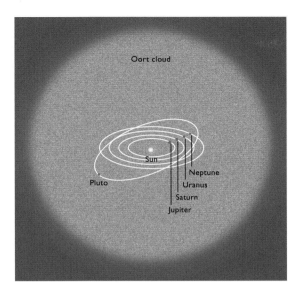

Oort, Jan Hendrik (1900–1992) Dutch astronomer. He carried out research on the structure and dynamics of stellar systems, especially the Galaxy, whose rotation he discovered in 1927. He was also a pioneer of radio astronomy, in particular with respect to the 21 cm radiation (SEE **twenty-one centimeter line**) of interstellar hydrogen, collaborating with Hendrik van de **Hulst** and Bart **Bok** to map the Galaxy at this wavelength. In 1950 he proposed the existence of what has come to be called the **Oort Cloud** of comets.

Oort Cloud (Oort–Öpik Cloud) The region of space surrounding the Solar System in which **comets** are thought to reside. It is spherical, extending from about 6000 au from the Sun to halfway to the nearest star, with the greatest concentration of comets in a torus-shaped region about 40,000 au from the Sun. The size and structure of the Oort Cloud have been worked out from statistical studies of comets' orbits, although there is no direct evidence for its existence. The idea was put forward by Ernst Öpik (1893–1985) in 1932, and first worked out in detail by Jan Oort in 1950. SEE ALSO **Kuiper Belt**.

open cluster A group, often irregularly shaped, of young stars in the spiral arms of our Galaxy, containing from a few tens of stars to a few thousand. Open clusters were formerly also known as *galactic clusters*. They contain young (Population I; SEE **stellar populations**) stars, are found in or near the plane of the Galaxy, and are typically several light years across. Examples are the **Hyades** and **Pleiades** in Taurus, and **Praesepe** in Cancer. Open clusters are susceptible to disruption by gravitational interactions between their individual stars, and also by tidal interaction with the Galaxy as a whole or encounters with interstellar clouds. Only the richest survive for more than 10^9 years, while the smallest and least tightly bound last no more than a few million years.

Related to open clusters are *associations*, loose groupings of stars of common origin. Often one or more open clusters are found in the central parts of associations. Because all stars in an open cluster are moving together, it is possible to find the distances of nearby clusters by the **moving cluster** method. SEE ALSO **star cluster**, **OB association**, **T association**.

open universe In cosmology, a mathematical model that describes the behavior of a universe in which there is not enough matter to halt the expansion triggered in the **big bang**. The universe thus continues to expand for ever. In an open universe the **curvature of space** is negative, the **deceleration parameter** is less than 0.5 and the **density parameter** is less than 1 (i.e. the mean density of the Universe is less than the **critical density**). COMPARE **closed universe**, **flat universe**.

Ophelia One of the small inner **satellites** of Uranus discovered in 1986 during the Voyager 2 mission. It and Cordelia act as **shepherd moons** to the planet's Epsilon Ring.

Ophiuchus A large constellation spanning the celestial equator, commemorating Aesculapius, a great healer. Its brightest star, Alpha (Rasalhague), is magnitude 2.1 and lies near Alpha Herculis. RS Ophiuchi, a **recurrent nova**, flared up to naked-eye brightness in 1898, 1933, 1958, 1967, 1985 and 2006. Rho Ophiuchi is a celebrated multiple star, and 70 Ophiuchi is a well-known binary, of magnitudes 4.2 and 6.0, period 88 years. The constellation contains several fairly bright globular clusters – M9, M10, M12, M14, M19, M62 and M107 – and **Barnard's Star**.

Opportunity SEE **Mars Exploration Rover**.

opposition The alignment of the Earth and Sun with another body in the Solar System so that the body lies exactly opposite the Sun in the sky, or the time when this alignment occurs. The term is used mostly for alignments with the major planets, although it is also applied to asteroids. Only a **superior planet** can come to opposition (SEE the diagram at **conjunction**). Opposition is the most favorable time to observe a body, for it is then closer to the Earth than at any other point in its orbit.

optical double Two stars that appear to be close together, but are not components of a binary system. They appear close together because they happen to be in almost the same line of sight, but are in fact separated by great distances.

optical interferometer An instrument for studying a celestial object by **interferometry** at optical wavelengths. The individual telescopes are linked either electronically or by laser beams, in conjunction with some form of **photometer**. In the *speckle interferometer*, an image enhancer or a CCD captures a succession of very short-exposure, high-magnification pictures. These individual pictures, which "freeze" the effects of atmospheric turbulence, are made up of a large number of "speckles." Computer image-processing combines these pictures and yields an image from which the effects of the atmosphere are removed electronically. Other types, such as the **Fabry–Pérot interferometer**, exploit interference to capture high-resolution spectra. SEE ALSO **radio interferometer**.

orbit The path of a celestial body in a gravitational field. The path is usually a closed one about the focus of the system to which it belongs, as with those of the planets about the Sun, or the components of a binary system about their common center of mass (SEE ALSO **ellipse**). Most celestial orbits are elliptical, although the eccentricity can vary greatly. It is rare for an orbit to be parabolic or hyperbolic. To define the size, shape and orientation of the orbit, seven quantities must be determined by observation. These are known as *orbital elements* (SEE the diagram). For a planetary orbit, only six elements are needed. They are:

a the *semimajor axis*, in au,
e the *eccentricity*,
i the *inclination* of the orbital plane to the plane of the ecliptic,
Ω the *longitude of the ascending node*,
ω the *argument of perihelion*,
T the *epoch* (the time of perihelion passage).

For the orbit of a binary star system a seventh element is needed, if the mass is not known – the *period* (P), or alternatively the *mean motion* (n).

In general, the same considerations apply when determining the orbit of a satellite as when determining the orbit of a planet, except that the inclination is usually referred to the equatorial plane of the primary planet, not to the plane of the ecliptic. SEE ALSO **parabola**, **hyperbola**.

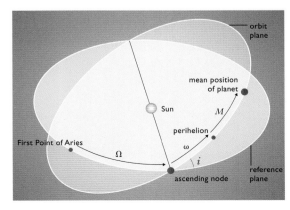

◄ **orbital elements** *The orbit of a planet around the Sun can be defined by a number of characteristics, including the inclination (i) relative to the reference plane of the ecliptic, the longitude (Ω) of the ascending node measured in degrees from the First Point of Aries, and the argument of perihelion (ω) measured in degrees from the ascending node.*

▲ **Orion Nebula** *A color composite of infrared images taken with the Very Large Telescope (VLT) at Cerro Paranal shows the central region of the Orion Nebula. The four bright, young Trapezium stars are prominent in the middle of this view.*

orbital element One of the quantities needed to define an **orbit**, or the position of a body in its orbit at a given time.

Orion Perhaps the most splendid of all the constellations, representing a hunter. Its brightest stars are **Betelgeuse** and **Rigel**. Gamma (Bellatrix) is a giant of type B2, magnitude 1.6, 243 light years away. Orion is so distinctive as to be unmistakable. A line of three stars – Zeta (Alnitak), magnitude 1.8; Epsilon (Alnilam), magnitude 1.7; and Delta (Mintaka), magnitude 2.2 – make up *Orion's Belt*. The celestial equator passes closely north of Mintaka. From the belt hangs *Orion's Sword*, marked by the **Orion Nebula**. Within the nebula lies Theta, a multiple star known as the **Trapezium**. The **Horsehead Nebula** is a dark nebula near Zeta Orionis.

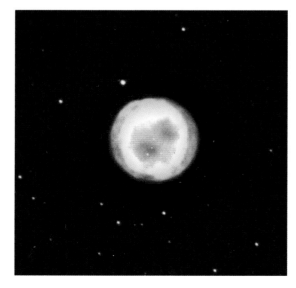

▲ **Owl Nebula** *The dark "eyes" of the Owl Nebula, M97, are well seen in this image taken with the Burrell Schmidt Telescope at Kitt Peak. The Owl is a planetary nebula, produced as an aged star puffed off its outer layers.*

Orion Arm A spiral arm of the **Galaxy**, known also as the *local arm*. The Sun is located on the inner side of a part of this arm known as the *Orion Spur*.

Orionids A **meteor shower** active between 15 October and 2 November, with a broad peak over 20–22 October when the **ZHR** can reach 30. The shower has a diffuse, probably multiple radiant between the stars Betelgeuse and Gamma Geminorum. Like the Eta Aquarids, the Orionids are associated with **Halley's Comet**.

Orion Nebula A gaseous emission nebula, visible to the naked eye as a diffuse glow in Orion's Sword. It is known as M42, and is separated from a smaller nebula, M43, by a dark nebula called the Fish's Mouth. It lies about 1350 light years away and is 25–30 light years across. It is an **H II region** lit up by a group of young stars called the Trapezium, probably less than 1 million years old. The Trapezium stars have destroyed most of the volatile dust in the nebula, etching out a roughly spherical hollow that is still growing. The visible Orion Nebula is therefore a cavity in a much larger giant **molecular cloud**. Within the densest portions of the dark cloud, star formation is still under way. The most recent stars to be formed lie concealed within it.

orrery A mechanical model of the Solar System. Orreries vary from simple ones with just the Earth, Moon and Sun to highly complicated representations of the whole Solar System in which planets and their satellites not only revolve in their orbits but also rotate on their axes. The name is from Charles Boyle, fourth Earl of Orrery (1676–1731), who commissioned one from John Rowley in 1712, although this was not the first to be made.

orthoscopic eyepiece A standard telescope lens with good definition and **eye relief**. It is a two-element lens, one element being a cemented triple achromatic (SEE **achromatic lens**).

oscillating Universe A variant of the **big bang** theory, in which the Universe passes through successive cycles of expansion and contraction (or collapse). At the end of the collapse phase, with the Universe packed into a small volume of great density, it is possible that a "bounce" would occur, leading to another phase of expansion. The Universe could thus oscillate between big bang and "big crunch" episodes and so be infinite in age. However, for this to happen the density of the Universe would have to be above a certain value (the **critical density**), which at present is not thought to be the case.

osculating orbit The path that an orbiting body, such as a planet or comet, would follow if it were not subject to any perturbations, but were governed solely by the **inverse-square law** of attraction by the Sun or other central body. The orbital elements, in this case, are called *osculating elements*, and are used in calculating perturbations.

O star A star of spectral type O, characterized by absorption lines of ionized helium in its spectrum. O stars are the hottest and brightest stars on the main sequence, with surface temperatures in the range 30,000 to 50,000 K and luminosities from 30,000 up to 2 million times that of the Sun. They are also the heaviest stars, ranging from about 15 to 130 solar masses or even more. O stars are rare – there are only about 300,000 in the Galaxy at present – because they remain on the main sequence for less than a million years. However, there has been time for 10,000 generations of O stars during the estimated life of the Galaxy. Because they return most of their mass to space via stellar winds and supernova explosions, they have been an important source of heavy elements (SEE **metals**). Together with the hotter B stars, they form **OB associations**. Most O and B stars form in binary systems, many with components of similar mass, an example being **Plaskett's Star**. When one star explodes as a supernova it may eject its companion from orbit at high velocity. As many as 20% of O stars may be such **runaway stars**. If the binary survives the supernova explosion intact, the result will be a neutron star or black hole orbiting through the stellar wind from its companion, producing an X-ray binary such as **Cygnus X-1**. Examples of O stars include Zeta Puppis, Delta Orionis and Zeta Orionis.

outer planets The planets Jupiter, Saturn, Uranus and Neptune, which orbit beyond the main asteroid belt.

Owl Nebula A planetary nebula, designated M97, in Ursa Major, of 11th magnitude and apparent diameter 3 arc minutes. It takes its name from two dark patches, which look like eyes, that give it the appearance of an owl's face. It lies 2600 light years away.

PA ABBREVIATION FOR **position angle**.

Paaliaq A small outer **satellite** of Saturn, discovered in 2000.

Pallas Asteroid no. 2, the second to be discovered, by Heinrich **Olbers** in 1802. It is the second largest, with a diameter of 580 × 470 km (360 × 290 miles).

Palomar Observatory An observatory located on Palomar Mountain, near San Diego, California, site of the 5 m (200-inch) Hale Telescope, a reflector opened in 1948, and the 1.2 m (48-inch) Oschin Telescope, a Schmidt camera. The Schmidt was used for the *Palomar Observatory Sky Survey*, an all-sky photographic survey. Palomar Observatory was founded by George Ellery **Hale** and is now run by the California Institute of Technology (Caltech).

Pan A small **satellite** of Saturn, identified in 1991 by Mark Showalter from images obtained during the Voyager missions. It orbits within the **Encke Division** in Saturn's rings.

Pandora A small inner **satellite** of Saturn discovered in 1980 during the Voyager missions. Pandora and Prometheus are the two **shepherd satellites** of Saturn's F Ring.

Pan-STARRS The Panoramic Survey Telescope and Rapid Response System, a group of four 1.8 m (70-inch) reflectors to be built on Mauna Kea, Hawaii, to survey the sky rapidly and repeatedly in search of near-Earth objects and other transient events. Each telescope will have a 3° field of view and a 1.4-gigapixel CCD camera, enabling the entire sky visible from Hawaii to be surveyed three times each month. A prototype telescope, PS1, began operation on Mount Haleakala, Hawaii, in 2009.

parabola An open curve. In mathematics the parabola is a type of conic section, so called because it is one of the intersections of a plane with a cone, and is formed when a plane cuts a cone parallel to its slope. The **eccentricity** of a parabola is 1. In astronomy, it is a type of open **orbit**, perhaps most familiar in the context of cometary orbits.

Half of all comets observed have an orbital eccentricity of approximately 1. Since these comets are observable over only a short arc of their orbits near perihelion, it is not possible to determine their orbital elements sufficiently accurately to distinguish between exactly parabolic orbits and extremely elongated elliptical orbits.

paraboloidal Having the form of the surface obtained by rotating a **parabola** about its axis. The paraboloid is the shape most often used for the primary mirror of reflecting telescopes, because a paraboloidal reflector, unlike a spherical one, brings an incoming beam of light to a single focus and is thus free from **spherical aberration**.

parallax (symbol π) The angular distance by which a celestial object appears to be displaced with respect to more distant objects, when viewed from opposite ends of a baseline, used as a measure of the object's distance. If the parallax is determined by direct means, using the principle of triangulation and a known baseline, it is known as **trigonometric parallax**. If the parallax is deduced by examining a star's spectrum, it is known as **spectroscopic parallax**.

In trigonometric parallax, the choice of baseline is made according to the remoteness of the object. For fairly close objects, the Earth's equatorial radius is used; this gives a *geocentric* (or *diurnal*) *parallax*. For more remote objects the radius of the Earth's orbit – the **astronomical unit** (au) – is used; this gives the *annual* (or *heliocentric*) *parallax*. The parallax of an object is therefore the angular size of the radius of the Earth or of its orbit as it would appear from that object.

The *solar parallax* is the Sun's geocentric parallax, defined as the angular size of the Earth's equatorial radius from a distance of 1 au. Its accurate determination has been historically very important in establishing the scale of the Solar System.

parhelion (mock Sun, sundog) A round patch of light at the same altitude as the Sun. Parhelia are images of the Sun produced when sunlight is refracted by ice crystals, most frequently in cirrus cloud. They often occur in pairs, one either side of the Sun, and may show elongated, spectrally colored tails.

Parkes Observatory A radio astronomy observatory in New South Wales, part of the **Australia Telescope National Facility**. Its main instrument is a 64 m (210 ft) dish aerial, completed in 1961, which is now a component of the Australia Telescope.

parsec (symbol pc) The distance at which a star would have a **parallax** of 1 arc second, equivalent to 3.2616 light years, 206,265 astronomical units or 30.857×10^{12} km.

Parsons, William SEE **Rosse, Third Earl of**.

partial eclipse SEE **lunar eclipse, solar eclipse**.

Pasiphae The largest (60 km/40 miles in diameter) of Jupiter's outer **satellites**, discovered in 1908 by P. J. Melotte. It is in a **retrograde** (1) orbit and may be a captured asteroid.

patrol A systematic survey aimed at recording certain transient astronomical phenomena. In a meteor patrol, for example, networks of cameras are used to photograph the sky to maximize the chance of recording meteors. In a supernova patrol, images of selected galaxies are obtained on as many nights as possible in the hope of discovering a supernova.

Pavo A southern constellation, representing a peacock; its brightest star, Alpha, magnitude 1.9, is called Peacock. Kappa is a **Cepheid variable**, magnitude range from 3.9 to 4.8 in a period of 9.1 days.

Pegasus A large constellation of the northern sky, representing the flying horse of Greek mythology. Its most prominent feature is the *Square of Pegasus*, made up of four stars, although one of these stars is actually part of Andromeda. Beta (Scheat) is a red giant irregular variable with a small magnitude range (2.3 to 2.7). Pegasus is not a rich constellation but it does contain a major globular cluster, M15, of magnitude 6.

Pelican Nebula SEE **North America Nebula**.

penumbra (1) The outer part of the shadow cast by a celestial body illuminated by an extended source such as the Sun. An observer in the penumbral region of a shadow sees only part of the source obscured, as, for example, in a partial **solar eclipse**. SEE ALSO **lunar eclipse**. COMPARE **umbra** (1).

penumbra (2) The lighter zone surrounding the dark central part of a sunspot. COMPARE **umbra** (2).

penumbral eclipse SEE **lunar eclipse**.

Penzias, Arno Allan (1933–) German–American physicist. In 1964 he was working with Robert **Wilson**, using a small radio telescope to measure noise that might interfere with satellite communications. They detected a strange signal that, after consultation with Robert Dicke and James Peebles, they realized was the **cosmic microwave background** – radiation left over from the **big bang**, as had been predicted by George **Gamow** and others in the 1940s. Penzias and Wilson shared the 1978 Nobel Prize for Physics for what was the most important discovery in modern cosmology.

peri- A prefix referring to the point in an object's orbit at which it comes closest to its primary, as in *perigee* and *perihelion*.

periastron The point of nearest approach of the components of a binary star in their orbit.

perigee The point of nearest approach to the Earth by the Moon or an artificial satellite.

perihelion The point in the orbit of a planet or comet at which it is nearest to the Sun.

periodic comet A common term for a short-period **comet** – one whose period is less than 200 years.

period–luminosity law A relationship between the period of light variation and the mean absolute magnitude (luminosity) of a **Cepheid variable** star: the absolute magnitude increases as the period increases. Once the period of a Cepheid is known, its absolute magnitude can be deduced from the period–luminosity law. Furthermore, the absolute magnitude taken in conjunction with the apparent magnitude then enables the distance of the Cepheid to be calculated. The law was discovered by Henrietta **Leavitt** in 1912.

▼ *penumbra (1)*
Cosmonauts aboard Russia's Mir space station took this photograph of the shadow cast by the Moon on to the Earth during the solar eclipse of 11 August 1999. The dark central umbra, under which the eclipse was total, is surrounded by a lighter outer penumbra.

▲ **Phobos** *Grooves and crater chains cross the surface of Phobos, the larger of the two moons of Mars, as seen in this close-up from ESA's Mars Express spacecraft. These linear features are now thought to be caused by debris thrown into space by impacts on the surface of Mars, which Phobos plowed into.*

Perseids A **meteor shower** active between 23 July and 20 August peaking around 12 August, at which time its radiant lies in the north of Perseus. The Perseids are one of the strongest, best known and most long-lived of all the meteor showers, showing a peak **ZHR** in excess of 80. The shower is associated with Comet 109P/**Swift–Tuttle**, and showed enhanced activity near maximum in the early 1990s around the time of the comet's perihelion passage. In the late Middle Ages they were known as the *tears of St Lawrence*, who was martyred in AD 258, on 10 August, close to the shower's maximum.

Perseus A prominent northern constellation, representing the Greek hero who rescued Andromeda. Alpha Persei (Algenib or Mirfak) is magnitude 1.8. Beta (**Algol**) is the prototype eclipsing binary and the brightest in the sky; the second brightest is Gamma, whose first eclipse was observed only in 1990 (components are magnitude 3.2 and 4.4, period 15 years). Rho is a red giant semiregular variable with a magnitude range from 3.3 to 4.0 and a very rough period of 50 days. Perseus is a rich constellation, crossed by the Milky Way; there are several open clusters, including the famous **Double Cluster** and M34.

personal equation A correction factor applied to visual observations made by a particular person. An individual observer may record timings of, for example, transits which are consistently early or late. Such systematic errors cannot be measured directly, but in any program of work carried out by several observers a personal equation can be found by performing a statistical analysis of all the results. It can then be applied to improve the accuracy of future observations.

perturbation An irregularity in the orbital motion of a celestial body, brought about by the gravitational attraction of other bodies. *Periodic perturbations* are small oscillations with periods similar to the orbital period of the perturbed body. *Secular perturbations* are slow, continuous changes in orbital motion, for example the slow rotation of the orbit of a planet (SEE **perihelion**), and in the orbit's eccentricity and inclination, that arise from commensurabilities (SEE **commensurable**). It is such perturbations that produce **precession**. Perturbations are responsible for dramatic changes in the orbits of **comets**, converting long-period orbits into short-period or hyperbolic orbits. SEE ALSO **non-gravitational force**.

Phaethon Asteroid no. 3200, discovered in 1983 by Simon Green and John Davies, using the **Infrared Astronomical Satellite**. Phaethon is a 5 km (2.5-mile) diameter **Apollo asteroid** with a perihelion within the orbit of Mercury. Its orbit coincides with the **Geminid** meteor shower.

phase The proportion of the illuminated hemisphere of a body in the Solar System (in particular the Moon or an inferior planet) as seen from the Earth. The phase of a body changes as it, the Sun and the Earth change their relative positions. All the phases of the Moon (SEE the diagram above right) – new, crescent, half, gibbous and full – are observable with the naked eye. The inferior planets, Mercury and Venus, show phases from a slender crescent to a fully illuminated disk when observed through a telescope. Of the superior planets, Mars can show quite a marked gibbous phase, and Jupiter a very slight gibbous phase, but with Saturn and the planets beyond no phase is discernible. Phase is sometimes expressed as the percentage of the visible disk that is illuminated.

Phobos The larger of the two **satellites** of Mars, discovered in 1877 by Asaph **Hall**. It is a dark, irregular body measuring 26 × 21 × 18 km (16 × 13 × 11 miles), and may well be a captured asteroid. It has two large craters, named Stickney and Hall.

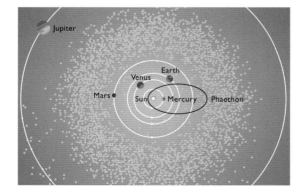

▶ **Phaethon** *Phaethon's orbit (red) carries it from the inner part of the main asteroid belt at aphelion, to a perihelion closer to the Sun than Mercury. As an Apollo asteroid, Phaethon has an orbit that cuts across that of Earth.*

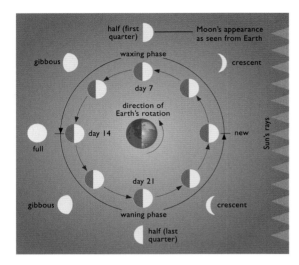

▲ **phase** *The changing appearance of the Moon's illuminated disk results from its orbital motion around the Earth. At new (right), the Moon lies in line with the Sun, and the hemisphere presented to Earth is completely dark. At full, the Moon lies opposite the Sun in Earth's sky and is completely illuminated. Between these two extremes, varying degrees of illumination are seen as shown.*

Phoebe The largest of Saturn's outer **satellites**, discovered in 1898 by William Pickering. Phoebe is quite small (diameter 230 × 210 km/145 × 130 miles), moves in a distant, inclined and **retrograde (1)** orbit, and has a non-**synchronous rotation** period of 9.4 hours, all of which suggests that it is a captured body, perhaps a comet nucleus or a **Centaur**. The **Cassini** spacecraft made a close flyby in 2004, imaging a heavily cratered body covered to a depth of several hundred meters with dark material.

Phoenix A southern constellation representing a phoenix. Alpha (Ankaa), magnitude 2.4, is the brightest star. Zeta is an eclipsing binary with a range from magnitude 3.9 to 4.4 in a period of 1.67 days.

photoelectric magnitude The **apparent magnitude** of an object measured by a **photoelectric photometer** attached to a telescope. The measurement of brightness is made through a number of colored filters in turn. SEE ALSO **UBV system**.

photoelectric photometer An instrument that measures the brightness of a celestial object by generating an electrical signal that is directly proportional to the intensity of the light falling on it. There are various types of detector. A *photomultiplier tube* is a device in which electrons that are given off when light falls on a special screen liberate more electrons in a multi-stage cascade, so producing a measurable electric current. A **CCD** may be used instead (strictly speaking, it is then not a photoelectric device, but it performs the same function). The brightness thus measured is an object's **photoelectric magnitude**; an accuracy of 0.01 magnitude is possible. Photoelectric photometers are therefore used for accurate determination of stellar magnitudes, especially in the study of variable stars, and also for measuring the brightness variations of asteroids and planetary satellites, and thus their rotation periods. The light falling on the detector can be confined to a particular part of the visible spectrum, or infrared or ultraviolet wavelengths can be selected.

photographic magnitude (symbol m_{pg}) The **apparent magnitude** of an object as measured on a traditional photographic emulsion which is more sensitive to blue light than the human eye is. Hence the photographic magnitude differs significantly from the visual magnitude.

photometer SEE **photoelectric photometer**.

photosphere ("sphere of light") The visible surface of the Sun. It is a layer of highly luminous gas about 500 km (300 miles) thick and with a temperature of about 6000 K, falling to 4000 K at its upper level, where it meets the **chromosphere**. Because this is the region of the Sun from which all of its output of visible light is emitted, the photosphere is the source of the Sun's visible spectrum. The lower, hotter gases produce the continuous emission spectrum, while the higher, cooler gases absorb certain wavelengths and so give rise to

▲ **Phoebe** *Two images taken by the Cassini spacecraft during its close flyby of Phoebe on 11 June 2004 were combined to produce the picture above. The bright craters are thought to be relatively young features.*

the dark **Fraunhofer lines**. This absorption contributes to **limb darkening**. Sunspots and other visible features of the Sun are situated in the photosphere.

photovisual magnitude (symbol m_{pv}) The **apparent magnitude** of an object measured photographically using emulsions and filters that mimic the spectral response of the human eye.

Piazzi, Giuseppe (1746–1826) Italian astronomer and monk. He discovered the first asteroid, Ceres, on 1 January 1801, during a lengthy program of taking accurate measurements of star positions (published as a catalog in two parts, in 1803 and 1814). Piazzi proposed the name "planetoid," but "asteroid" has prevailed.

Pickering, Edward Charles (1846–1919) American astrophysicist who was director of the Harvard College Observatory. His accurate stellar magnitudes, determined by photometry, were published in the *Revised Harvard Photometry*. He discovered spectroscopic binary systems, and invented a method of obtaining the spectra of several stars on one photographic plate. The spectroscopic work that Pickering instituted culminated in the *Harvard system* of **spectral classification** and the **Henry Draper Catalog** of stellar spectra. This work was largely carried out by a team of women, originally hired as assistants but who grew to be able astronomers in their own right. They included Annie Jump **Cannon**, Williamina Fleming and Antonia C. Maury. Edward's younger brother, William Henry Pickering (1858–1938), instituted the search for **Pluto**; he also discovered Saturn's satellite Phoebe, in 1898.

Pictor An unremarkable constellation in the southern sky representing a painter's easel and introduced by **Lacaille**. Its leading star is Alpha, magnitude 3.3, and the only other star above fourth magnitude is *Beta Pictoris*, magnitude 3.8, which has a disk of gas and dust around it from which a planetary system is in the process of formation.

Pinwheel Galaxy A large, face-on spiral galaxy in Ursa Major, also known as M101 and NGC 5457. It lies about 25 million light years away. The name is also sometimes used for the galaxies M33 in **Triangulum** and M99 in Coma Berenices, both also face-on spirals.

Pioneer A series of US space probes launched between 1958 and 1978. The first four were failed Moon probes. Pioneers 5 to 9 were put into orbit around the Sun to study the solar wind and solar flare activity. Pioneer 10, launched on 3 March 1972, was the first probe to travel through the asteroid belt; it passed Jupiter in December

1973, sending back photographs. Pioneer 11 followed it, flying past Jupiter in December 1974 and moving on to Saturn, which it reached in September 1979. Both probes are now on their way out of the Solar System; contact was maintained until 1997. The last two in the series were the **Pioneer Venus** probes.

Pioneer Venus Two NASA space probes to the planet Venus, launched on 20 May and 8 August 1978. Pioneer Venus 1 went into orbit around the planet in December 1978, making a radar map of the surface and photographing the planet's clouds. Pioneer Venus 2 dropped five small probes into the atmosphere of Venus in December 1978, which sent back data on the planet's clouds and atmospheric conditions.

Pisces A constellation of the **zodiac**, representing a pair of fishes. It is very obscure, consisting of a chain of rather faint stars south of Pegasus. Its brightest star, Eta, is magnitude 3.6. Alpha (Alrescha) is a close binary, magnitudes 4.2 and 5.2, with a period of over 900 years.

Piscis Austrinus The constellation of the southern fish. First-magnitude **Fomalhaut** is its only star brighter than fourth magnitude.

plage A bright cloud of hot gas in the Sun's **chromosphere**. Plages are intimately connected with **faculae**, which occur below them in the photosphere, and are associated with active regions, where the local magnetic field is enhanced.

Planck A European Space Agency spacecraft launched on 14 May 2009 which studied the **cosmic microwave background** from a position in orbit around the L_2 **Lagrangian point** of the Sun–Earth system. Planck provided a new and more accurate value of 13.8 billion years for the age of the Universe. The mission's results showed that normal matter in stars and galaxies comprises only 4.9% of the mass of the Universe. **Dark matter** contributes 26.8% of the mass, while the remaining 68.3% is **dark energy**. Planck was switched off on 23 October 2013.

Planck era The first 10^{-43} seconds after the start of the **big bang**, during which time quantum fluctuations probably dominated the Universe and almost nothing can be understood about it with present-day physics. It is also called the *Planck time*. The size and density of the Universe at that instant (10^{-35} m and 10^{97} kg/m^3, respectively) are called the *Planck scale* and *Planck density*.

planet A large non-stellar body in orbit around a star, shining only by reflecting the star's light. In our **Solar System** there are eight **major planets**, as opposed to the thousands of small bodies known as **asteroids** or minor planets. SEE ALSO **dwarf planet**, **extrasolar planet**, **Kuiper Belt**.

planetarium A structure in which a representation of the stars and planets as visible in the night sky is projected on to the inside of a dome for an audience seated below. They vary in size and sophistication from small portable projectors used in inflatable 3 m (10 ft) domes to highly complex installations in permanent theaters seating hundreds of people. In all planetaria the positions and

▼ **Pictor** *A false-color view of the disk of dust and gas around the star Beta Pictoris, seen at infrared wavelengths by the Very Large Telescope in Chile. In the outer part of the image, light from the star is reflecting off the dust disk, colored red and yellow. A bright dot in the inner part of the image is a planet several times larger than Jupiter about 8 astronomical units from the star.*

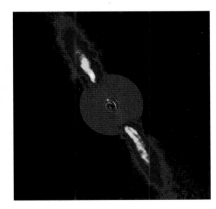

◄ **Pinwheel Galaxy** *The face-on spiral galaxy M101 in Ursa Major, popularly known as the Pinwheel Galaxy, seen in a composite image of exposures by the Hubble Space Telescope and ground-based observatories. The Pinwheel is about twice the diameter of our own Milky Way Galaxy.*

▲ **planetary nebula** *This Hubble Space Telescope image shows the Ant Nebula (Mz 3), a planetary nebula in the southern constellation Norma. The twin-lobed structure has been produced by ejection of gas from the aged star at the nebula's center. The Ant Nebula lies about 3000 light years away.*

motions of celestial bodies can be simulated. The first planetarium, built by the Carl Zeiss company, began operation in Munich in 1923. From then until the early 1980s, planetarium projectors consisted of many individual optical projectors mounted on a complicated rotating frame. In the new generation of projectors, a computer-generated image is either projected through a single, stationary fish-eye lens or through a series of projectors around the auditorium that each display a segment of the overall view.

planetary nebula An emission nebula formed when a red giant or supergiant sheds its outer layers, leaving a hot core. The core's stellar wind compresses the gas, and its ultraviolet radiation ionizes it and makes it shine. Planetary nebulae range in size from about that of our Solar System to a light year or so across. They were so named because the first ones to be discovered gave the impression of planetary disks when viewed in small telescopes; the Ring Nebula (M57) in Lyra is a well-known example. However, only 10% of the *c.* 3000 now known have a circular shape, and about 70% have two lobes.

The central stars of planetary nebulae are hot and blue, with temperatures of 30,000 K to over 100,000 K. They form a transitional stage at the end of a star's evolution between the giant stars and the white dwarfs. Ultraviolet light from the central star makes planetary nebulae fluoresce, as in other emission nebulae. In addition to hydrogen and helium, planetary nebulae contain small proportions of oxygen, nitrogen and neon, which accounts for their various colors seen on images.

Planetary nebulae are thought to result from stellar winds blown out from red giants and supergiants, leaving the red giant's core as the central star (SEE **RV Tauri star**). Over time, this star cools to become a white dwarf. Because red giants are so big, the gravity at their surfaces is very low and flares or mass ejections on the surface can easily throw off material. A star below about 8 solar masses can lose most of its mass into space before it turns into a white dwarf of about 1.4 solar masses. The visible nebula is simply the central region of a much larger object, the outer parts of which are not lit up by the central star. All planetary nebulae are expanding, at speeds of around 20 km/s (12 miles/s).

The curious shapes of planetary nebulae are probably linked to the magnetic field and rotation of the stars. The magnetic field lines might control the outflow, channeling it in particular directions, and the rotation of the star could add spiral twists. The double-lobed structure of most planetary nebulae must be connected with the star's direction of spin or magnetic axis (SEE **bipolar flow**). Some planetary nebulae, such as the Eskimo and Saturn nebulae, have a bright central nebula surrounded by fainter extensions, as though the central star has ejected a succession of shells. Because they are expanding as the central stars fade, planetary nebulae are relatively transitory objects with lifetimes of tens of thousands of years. Several new planetary nebulae form each year in our Galaxy.

planetary precession The smaller of the two components of **precession**. It results from the gravitational effects of other planets on the Earth and moves the equinoxes eastward by about 0.12 arc seconds per year.

planetesimal In theories of the origin of the Solar System (and other planetary systems), a body, between a few millimeters and a few kilometers in size, that condensed out of a cloud of gas and dust (the solar nebula), away from the center where the Sun was forming. Once they had reached a few kilometers in size, the gravitational attraction of planetesimals would allow them to combine with one another, by a process of **accretion**, to form **protoplanets**.

planisphere A simple device for showing the stars on view from a certain latitude at any time and on any date. It consists of two disks held together by a pivot at their centers. On the lower disk is printed a map of the night sky, with the celestial pole at the pivot, and into the upper disk is cut an oval window. The planisphere is set by rotating one disk with respect to the other so that the date, marked on the rim of one, is aligned with the time, marked on the rim of the other. The part of the map visible through the window then shows the stars on view in the sky. There is some distortion of constellation shapes, particularly near the rim.

Plaskett's Star A massive binary star in the constellation Monoceros. In 1922 the Canadian astronomer John Plaskett (1865–1941) showed it to be a spectroscopic binary with a period of 14 days. Each component is a blue supergiant of about 50 solar masses. It is the second most massive binary known, exceeded only by the recently discovered WR 20a in Carina, where the two stars are each of about 80 solar masses.

Platonic year The time taken for the celestial pole to describe a circle around the pole of the ecliptic, as a result of **precession**. Its value is about 25,800 years. (After the Greek philosopher Plato, *c.* 427–*c.* 347 BC.)

Pleiades A young open cluster in the constellation Taurus, popularly called the *Seven Sisters*. Although only six or seven stars are visible to the naked eye, there are in fact over a thousand embedded in a reflection nebula. The brightest member is Alcyone, a B-type giant over 300 times as luminous as the Sun. Another member of the cluster, Pleione, is a slightly variable **shell star** that throws off shells of gas as a result of its fast rotation. The other named Pleiades are Maia, Atlas, Electra, Merope and Taygeta. The cluster lies 410 light years away.

plerion SEE **supernova remnant**.

Plössl eyepiece A telescope eyepiece with good **eye relief** and a wider field of view (about 40°) than an orthoscopic lens. The commonest form has two identical achromatic doublets (SEE **achromatic lens**), but some quite different designs have been called "Plössls." (After Austrian optician Simon Plössl, 1794–1868.)

Plough Popular name for the shape formed by the seven main stars of the constellation **Ursa Major**. In the USA, the stars are known as the **Big Dipper**.

plutino The name given to members of the **Kuiper Belt** with similar orbital periods to that of Pluto, i.e. about 247 years. The name means "little Pluto."

Pluto A **dwarf planet** that crosses the orbit of Neptune. Discovered in 1930, it was considered one of the major planets until its status was redefined by the International Astronomical Union in 2006. It now bears the minor planet number 134340 and is regarded as a large member of the **Kuiper Belt**. It is in a much more inclined and eccentric orbit than are the major planets, and for a 20-year period around perihelion (as, for example, from 1979 to 1999) it comes within the orbit of Neptune. However, a **resonance** between the two planets' orbits (their periods are in the ratio 3:2) means they never get close. Pluto's small size and great distance give it a maximum apparent diameter of 0.1 arc seconds, and it is never brighter than magnitude 14.0. The main data for Pluto are given in the table.

The planet was found by Clyde **Tombaugh** on photographic plates taken in January and February 1930 during a search for a new planet originally mounted by Percival **Lowell**.

The first visit to Pluto by a space probe took place in July 2015, by **New Horizons**. Prior to this, most of our knowledge of it dated from the discovery in 1978 of its satellite, **Charon**, which allowed the planet's mass and unusual axial tilt to be determined. A series of

▲ **Pleiades** *The hot, young blue stars of the Pleiades illuminate a shroud of interstellar dust. Marking the shoulder of Taurus, the Bull, the Pleiades is a distinctive naked-eye object, prominent on winter evenings in the northern hemisphere.*

mutual **occultations** of planet and moon from 1985 until 1990 gave information on their relative brightnesses and sizes, and a very rough brightness map of Pluto's surface. The Hubble Space Telescope has discovered four smaller moons: Nix and Hydra in 2005, Kerberos in 2011 and Styx in 2012.

Charon has both a **synchronous orbit** and **synchronous rotation**, so it keeps the same face turned toward Pluto, and hangs motionless in Pluto's sky. This orbital coupling of Pluto and Charon could well have brought about sufficient tidal heating (SEE **tides**) to cause **differentiation** in both bodies, giving them rocky cores and icy mantles.

Pluto has a mottled surface with light and dark regions, and signs of recent geological activity. The surface is covered with icy deposits consisting of 98% nitrogen, with traces of methane, and also probably water, carbon dioxide and carbon monoxide. For a 60-year period in each orbit, around perihelion, the surface warms to above 50 K, sufficient to release a thin, temporary atmosphere of nitrogen. Interaction with the **solar wind** probably draws some of this atmosphere into a comet-like tail.

Pointers The stars Alpha and Beta Ursae Majoris (Dubhe and Merak), which point toward Polaris – the north pole star and the brightest star in Ursa Minor.

polar caps Deposits of ice in the polar regions of a planet or satellite. The Earth's permanent polar caps are of water ice, and show only a little seasonal variation. Mars has caps of water ice and carbon dioxide that show pronounced seasonal variation because of the planet's eccentric orbit. The axial inclination of Neptune's moon **Triton** with respect to the Sun produces a migration of its nitrogen-ice cap from one pole to the other over the course of the planet's 165-year orbital period.

polar distance The angular distance of a celestial body from the celestial pole. It is the complement of **declination** (i.e. polar distance plus declination equals 90°).

Polaris The star Alpha Ursae Minoris, lying within 1° of the north celestial pole. **Precession** is bringing Polaris closer to the pole, and it will be at its closest around AD 2100. Polaris is a supergiant of spectral type F, magnitude 2.0, distance 431 light years, luminosity nearly 3700 times that of the Sun. It is a **Cepheid variable** whose amplitude steadily decreased during the 20th century, stabilizing at 0.03 magnitude in the mid-1990s. Its period is 3.97 days. It is an optical double, with an eighth-magnitude companion 18 arc seconds away.

pole The points on a sphere that are 90° from the equatorial plane. The *celestial poles*, the points where the Earth's axis of rotation intersects the celestial sphere, are 90° north and south of the celestial equator. Similarly, the *ecliptic poles* are 90° from the plane of the ecliptic, and the *galactic poles* are 90° from the galactic plane.

pole star The nearest naked-eye star to either celestial pole. In the northern hemisphere it is **Polaris**, magnitude 2.0; in the south it is fifth-magnitude Sigma Octantis. Because of **precession** the positions of the celestial poles, and hence of the pole stars, are continually changing. In 3000 BC the north pole star was Thuban, in Draco; by AD 10,000 it will be Deneb, and by AD 14,000 it will be Vega.

Pollux The star Beta Geminorum, magnitude 1.16, distance 34 light years, luminosity 40 times that of the Sun. It is a giant of type K0.

Population I, **Population II** SEE **stellar populations**.

pore A very small and short-lived **sunspot**.

Portia One of the small inner **satellites** of Uranus discovered in 1986 during the Voyager 2 mission.

position angle (PA) The orientation in the sky of one celestial body with respect to another, measured from 0° to 360° from north via east. For a binary system it is the angle between the direction of the celestial north pole and the direction of the line joining the stars.

Poynting–Robertson effect A **non-gravitational force** produced by the action of solar radiation on small particles in the Solar System which causes them to spiral inward toward the Sun. The effect is most marked for particles between about a micrometer and a centimeter in size. A similar effect makes particles in planetary ring systems slowly spiral inward. The effect is named after John Poynting (1852–1914), who first described it, and Howard Robertson (1903–61), who proved its reality.

◀ **Pluto** *Sufficiently similar in size to be regarded as a "double planet," Pluto (left) and its satellite Charon (right) are locked in synchronous rotation about the barycenter between them. This picture is a composite of separate images taken by the New Horizons spacecraft as it approached the planet in July 2015.*

Praesepe A large open cluster in the constellation Cancer, known also as the *Beehive* or *Manger*, and designated M44. It was known to the Greeks as a hazy naked-eye patch; binoculars and telescopes show about 60 stars. The cluster is believed to be about 700 million years old, and lies 525 light years away.

preceding (abbreviation *p*) Describing the leading edge, feature or member of an astronomical object or group of objects. For example, the *preceding limb* of the Moon is the edge facing its direction of motion; and the *preceding spot* (or *p*-spot) of a group of sunspots is the first of the group to be brought into view by the Sun's rotation.

precession The circular motion of the celestial poles around the poles of the ecliptic, and the associated westward movement of the equinoxes with respect to the background stars. The main component of precession is caused by the gravitational pull of the Sun and Moon on the Earth's equatorial bulge. This is called *lunisolar precession*. A smaller, similar effect caused by the gravitational influence of the other planets is *planetary precession*, but it acts in the opposite direction. The total effect, *general precession*, amounts to about 50.3 arc seconds per year. The celestial pole therefore describes a circle on the sky of radius $23\frac{1}{2}°$ (the inclination of the Earth's axis), centered on the pole of the ecliptic, in about 25,800 years. Because of precession, the observed positions of all stars (including, of course, the pole star) are continuously changing. Precession also has the effect of moving the **equinoxes** around the ecliptic, and is therefore sometimes called the *precession of the equinoxes*. The First Point of Aries, for example, is now in Pisces. Star catalogs give positions for a given **epoch**, e.g. 2000.0, so these positions must be corrected for precession to the date of observation.

primary (1) The largest of a system of celestial bodies, around which the others orbit. The term is used for the more massive component of a binary or multiple star, and for a planet with respect to its satellites.

primary (2) **(primary mirror)** The largest mirror in a **reflecting telescope** which collects light from the object being observed.

prime focus The point at which the objective lens or primary mirror of a telescope brings light to a focus when there is no other optical component in the light path, such as a secondary mirror. In large telescopes, special instruments are sometimes mounted at the prime focus. In the **Schmidt camera** a photographic plate or detector is placed at the prime focus.

In *prime focus photography*, a technique which can be used with small telescopes, the lens of a camera is removed and the camera back attached to the telescope via an adapter, in place of the eyepiece; photography – or CCD imaging – does not, despite the name, actually take place at the prime focus.

PLUTO: DATA	
Globe	
Diameter	2370 km
Density	2.1 g/cm³
Mass (Earth = 1)	0.0021
Volume (Earth = 1)	0.0058
Sidereal period of axial rotation (retrograde)	6.387d
Escape velocity	1.1 km/s
Surface gravity (Earth = 1)	0.03
Albedo	0.9
Inclination of equator to orbit	122° 27'
Surface temperature (average)	45 K
Orbit	
Semimajor axis	39.54 au = 5914 × 10⁶ km
Eccentricity	0.249
Inclination to ecliptic	17° 09'
Sidereal period of revolution	248.54y
Mean orbital velocity	4.74 km/s
Satellites	5

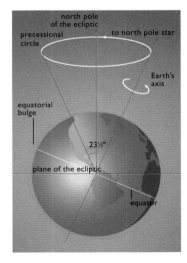

▲ **precession** *Over the course of a 25,800-year cycle, the direction in which Earth's axis of rotation points describes a 47° diameter circle relative to the star background. One result of this is that different stars will, over time, become the Pole Star.*

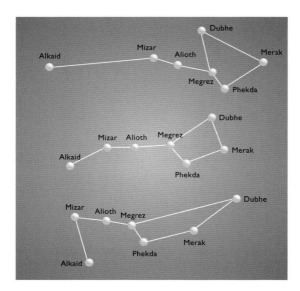

primordial black hole SEE **black hole**.

Procyon The star Alpha Canis Minoris, the eighth brightest star
in the sky at magnitude 0.40, 11.4 light years away and seven times
as luminous as the Sun. It is a main-sequence star of type F5 and
has a white dwarf companion of 11th magnitude which orbits it in
41 years.

prograde SEE **direct motion** (1).

Prometheus One of the small inner **satellites** of Saturn discovered in 1980 during the Voyager missions. Prometheus and Pandora
are the two **shepherd moons** of Saturn's F Ring.

prominence A cloud of matter extending outward from the
Sun's chromosphere, into the corona. Prominences are regions of
higher density and lower temperature than the surrounding corona.
Those near the Sun's limb are visible, during total eclipses or with
the aid of an instrument such as a **spectroheliograph** or a special
filter, as bright protrusions into the corona. Away from the limb,
they are silhouetted against the bright lower chromosphere as dark
filaments when viewed in the light of hydrogen-alpha or calcium.
Quiescent prominences, which extend for tens of thousands of kilometers, can last for weeks or months. *Active prominences* are short-
lived, high-speed, flame-like eruptions that can reach heights of
up to 700,000 km (450,000 miles) in just 1 hour. All prominences
have spectra showing lines of neutral hydrogen, helium and ionized
calcium. Their forms and behavior are very varied, and their direction of motion is controlled by the Sun's magnetic field.

proper motion (symbol μ) The apparent motion of a star on
the celestial sphere, as a result of its movement relative to the Sun.
About 300 stars are known to have proper motions greater than 1 arc
second per year, but most annual proper motions are smaller than
0.1 arc second. High proper motion is an indicator that a star lies
in the Sun's neighborhood. **Barnard's Star** has the largest known
proper motion, 10.3 arc seconds per year. Proper motion is determined by comparing the star's position on images obtained on
widely separated occasions, usually many years or decades apart, with
the aid of a **comparator** or similar device. Accurate proper motions
for over 100,000 stars were obtained by the position-measuring
satellite **Hipparcos**.

Prospero A small outer **satellite** of Uranus discovered in 1999.

Proteus Neptune's second largest **satellite**, discovered in 1989
during the Voyager 2 mission. It is a squarish, dark body, heavily
cratered, and its most prominent feature is the large crater Pharos,
250 km (150 miles) across.

proton–proton reaction The nuclear reaction thought to be
the main source of energy for main-sequence stars with masses
equal to or less than the Sun's. In more massive stars the main energy
source is the **carbon–nitrogen cycle**. The proton–proton reaction
is a three-stage process in which hydrogen is converted into helium
with the liberation of an enormous amount of energy. In the first

stage, two protons (hydrogen nuclei) combine to form a deuterium
nucleus, releasing a positron, a neutrino and radiation. In the
second stage, the deuterium nucleus combines with a proton to
form an isotope of helium (^{3}He), again releasing radiation. In the
third stage, two ^{3}He nuclei combine to form the normal helium
nucleus (^{4}He), releasing two protons and radiation. Overall, four
hydrogen nuclei have combined to form one helium nucleus and
give out radiation. The reaction requires a temperature of about
10 million K.

protoplanet In theories of the origin of the Solar System (and other
planetary systems), a body formed by the accretion of **planetesimals**.
Once the protoplanets attained the size of, say, the Moon, they could
grow no further by accretion. Instead, collisions between them caused
some to grow by accumulating fragments. The major planets of our
Solar System are the result of this second stage of growth, while the
collisional fragments that were not captured survive as bodies ranging in size from asteroids down to microscopic particles.

protoplanetary disk A disk of dust and gas around a young star
which is thought to represent the initial phase of planet formation.
Such disks are thought to form around **protostars** as they grow by
the gravitational collapse of interstellar matter in molecular clouds.
As the disks spread out and become gravitationally unstable, clumps
of matter within them form the progenitors of planets. More than
a hundred protoplanetary disks have been observed in the **Orion
Nebula**.

protostar An embryonic star formed when dense clumps of
gas and dust collapse within molecular clouds. Protostars are the
earliest, optically invisible stage of star formation. They are very red
and can be observed at infrared wavelengths. The Sun's protostellar
stage lasted about 0.1 to 1 million years. SEE ALSO **stellar evolution**.

Proxima Centauri The nearest star to the Sun, at 4.23 light
years slightly closer than **Alpha Centauri**. Proxima Centauri is an
M-type dwarf, of visual magnitude 11.0, and is also a **flare star**.
It is probably in a highly elongated orbit around the other two
members of the Alpha Centauri system with a period of a million
years or more.

Ptolemy (Claudius Ptolemaeus) (2nd century AD) Egyptian
astronomer and geographer. His chief astronomical work, known by
its title in Arabic translation, the *Almagest*, is largely a compendium
of contemporary knowledge, including a star catalog, drawing
heavily on the work of **Hipparchus of Nicaea**. His *Ptolemaic system*
is based on the geocentric universe of the ancient Greeks, with the
Earth fixed at the center, and the Moon, Mercury, Venus, the Sun,
Mars, Jupiter and Saturn revolving about it. Beyond these planets
lies the sphere of fixed stars. Each of the bodies moves around a small
circle called the *epicycle*, the center of which in turn revolves around
the Earth on a larger circle called the *deferent*. Ptolemy added to each
orbit two more points, the *eccentric* and the *equant*, equally spaced
either side of the Earth. He made the epicycle revolve around the
eccentric rather than the Earth, and the planet have uniform motion
with respect to the equant. The resulting model reproduced the apparent motions of the planets, including **retrograde** (2) loops, so well
that it remained unchallenged until the revival of the heliocentric
theory by Nicholas **Copernicus** in the 16th century.

Puck The largest of the small inner **satellites** of Uranus discovered
during the Voyager 2 mission at the end of 1985, shortly before
Voyager's encounter with the planet in January 1986. It is cratered
and approximately spherical.

pulsar An object emitting radio waves in pulses of great regularity.
They were first noticed in 1967 by Jocelyn **Bell**, working at the
Mullard Radio Astronomy Observatory, Cambridge. The first pulsar
to be discovered pulsed every 1.337 seconds.
  Pulsars are rapidly rotating **neutron stars**, produced either as the
result of a **supernova** explosion or by the collapse of a white dwarf
that has accreted material from a companion star in a binary system
and exceeded the **Chandrasekhar limit**. Acceleration of electrons
by the powerful rotating magnetic field generates a conical beam of
radio waves that sweeps across the Earth and is received in the form
of pulses, in the same way that a lighthouse is seen to flash. With
the Crab Pulsar, in the **Crab Nebula**, two pulses are observed
for each rotation, which suggests that the magnetic axis is pointed
almost directly at us so that we receive a beam from each pole as the
star spins. The Crab Pulsar was the first pulsar to be seen flashing
optically, in 1969. In some pulsars, the pulse amplitude and shape
can change with time, and the pulsation can fade out temporarily.

▲ **Proteus** Discovered by Voyager 2
during its 1989 flyby, Proteus is the
largest of the inner satellites of
Neptune. Proteus is notably dark,
with an albedo of only 0.07.

▲ **Puck** This distant view of
Uranus's inner satellite Puck was
taken from Voyager 2 during its
flyby of January 1986. Puck has
a diameter of about 154 km
(96 miles).

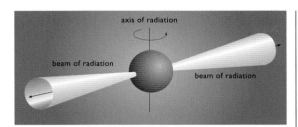

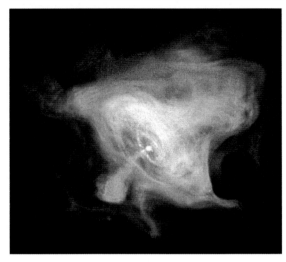

Quadrantids A **meteor shower** active between 1 and 6 January. The shower is one of the most prolific, reaching a narrow (less than 6 hours) peak **ZHR** in excess of 100 on 3 January. Its radiant lies in the constellation Boötes, near the borders with Hercules and Draco. The shower gets its name from an obsolete constellation called Quadrans Muralis which once occupied this region.

quadrature The position of the Moon or a planet when it makes an angle of 90° with the Sun (SEE the diagram at **conjunction**). The Moon is at quadrature at first and last quarter.

Quaoar A large **Kuiper Belt** object discovered by Chad Trujillo and Mike Brown on 4 June 2002. Quaoar has an estimated diameter of 1200 km (750 miles) and takes 286 years to orbit the Sun at a mean distance of 43 astronomical units. It has been assigned the minor planet number 50000. A moon was discovered in 2007.

quasar A compact and extremely luminous source of energy, radiated over a wide range of wavelengths from X-rays through optical to radio wavelengths. The name is derived from the term *quasi-stellar object*, which is what they were first called because of their starlike appearance on photographs. Quasars were discovered in 1963 when astronomers identified optical counterparts of certain strong radio sources. However, only a small proportion of quasars are actually radio sources. The most powerful of these are now known as *quasi-stellar sources* (QSSs), the original term "quasi-stellar object" (QSO) being reserved for radio-quiet objects.

Every quasar shows a very large **redshift** which, if it results from the expansion of the Universe, indicates that quasars are the oldest and most distant objects in the Universe. In some cases, galaxies that lie between us and a quasar cause **gravitational lens** effects. Quasars inhabit the centers of very remote galaxies. Because they may vary in brightness on timescales of a few days, they can be no larger than a few hundred astronomical units wide. A surrounding galaxy has been seen around the nearest quasars, almost lost in the glare from the brilliant center. Quasars are thus extreme examples of **active galactic nuclei**.

Quasar activity is thought to be caused by a **black hole** accreting material at the center of the host galaxy. The most luminous quasars would need to be powered by black holes of 100 million solar masses, swallowing stars at a rate of about one every year. Quasars are bright ultraviolet sources, and this hot radiation seems to come from a swirling **accretion disk**. Also associated with the inner regions of quasars are emission lines of hydrogen, helium, and iron that indicate motions as rapid as 5000 km/s, as would be expected near a massive black hole.

Quasar-like activity may be possible within every galaxy, but at any given time the black hole is accreting matter in only a small proportion of galaxies. Activity in the gas-free elliptical galaxies may produce a **radio galaxy** or **BL Lacertae object**. In galaxies such as our own, where the black hole is small, even at its most active it is swamped by the light of surrounding stars.

Because quasars are the most luminous objects in the Universe, they can be seen to greater distances, and hence further back in time, than anything else. At redshifts near $z = 4$ we see the Universe as it was about one-tenth of its present age. Therefore quasars give us a way of examining the Universe in its youth.

quintessence SEE **dark energy**.

▲ **pulsar** *The schematic illustration of a pulsar (top) shows the two beams of radiation directed from the collapsed star along its magnetic poles, which need not coincide with the axis of rotation. If the rotational axis is so aligned that the radiation beams sweep across our line of sight, the rapidly pulsing radio signal characteristic of a pulsar will be detected. The Chandra X-ray image (bottom) shows the heart of the Crab Nebula, the site of a pulsar remnant from the supernova explosion of 1054. A thin jet almost 20 light years long emerges from the pulsar's south pole toward lower left.*

Some highly energetic radio pulsars are now known to emit X-ray pulses. Of these, a few, the *anomalous X-ray pulsars* (AXPs) appear to be solitary neutron stars with exceptionally strong magnetic fields (known as **magnetars**). The magnetic field, rather than the star's rotation, provides the energy for the radiation.

Pulsars are gradually slowing down as their energy is dissipated by their radiation, but some, such as the Vela Pulsar, occasionally increase their spin rate abruptly. Such an event is called a *glitch* and is believed to be caused by a small, sudden adjustment in the internal structure of the star, producing a contraction of the stellar surface and increasing the rate of rotation. As of early 2015, nearly 2500 pulsars had been found, flashing at rates from 1.4 milliseconds (SEE **millisecond pulsar**) to more than 10 seconds. The commonest period is just under 1 second. Pulsars are concentrated toward the disk of our Galaxy, which is where supernovae most commonly occur. Most pulsars are thought to be over a million years old, so that the supernova remnants that once surrounded them would have long since dispersed and faded from view. SEE ALSO **binary pulsar**.

pulsating star A type of variable star whose fluctuating brightness arises from a regular cycle of expansions and contractions. **Cepheid variables**, **RR Lyrae stars** and **long-period variables** are important classes of pulsating stars. Numerous **semiregular variables** and giant and supergiant **irregular variables** are also of this type.

Puppis A large southern constellation, once part of the larger grouping of Argo Navis, the ship Argo, and representing the ship's stern. Its brightest star is Zeta (Naos), magnitude 2.2. L² is a red giant semiregular variable with a range from 2.6 to 6.2, and a period of 140 days or so. Puppis is a rich constellation, with several bright open clusters including M46, M47, M93, NGC 2451 and NGC 2477.

Pyxis A southern constellation adjoining Vela and Puppis, representing a ship's compass, introduced by **Lacaille**. There is no star above magnitude 3.7, but there is one object of particular interest, the **recurrent nova** T Pyxidis.

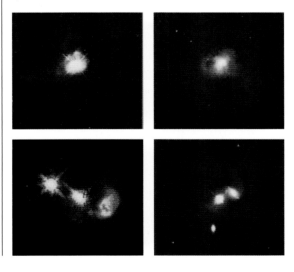

◄ **quasar** *This set of quasar images shows the bright nuclear activity that lies at the heart of these galaxies. Each of these quasars is probably fueled by material falling into a massive central black hole.*

R

RA ABBREVIATION FOR **right ascension**.

radar astronomy The use of radar techniques to study bodies in space, by transmitting continuous waves or pulses and observing the reflections. Because of the restrictions of signal strength, radar astronomy is limited to objects within the Solar System.

Radar astronomy has been used to study ionized **meteor** trails in the Earth's atmosphere, revealing the existence of daytime meteor showers. Radar reflections from Venus provided an accurate value of the solar **parallax** and hence of the scale of the Solar System. The giant dish at **Arecibo Observatory** was used to establish the true rotation rates of Venus and Mercury, mapped the surface of cloud-covered **Venus**, and revealed what could be water ice in craters near the poles of Mercury.

Radar altimeters have been carried aboard space probes to map Venus and, with the **Cassini** mission, the hidden surface of Titan. Other objects in the Solar System studied by radar include asteroids, Saturn's rings, planetary satellites and the nuclei of comets.

radial velocity The velocity of a celestial body along our line of sight, either toward us (positive) or away from us (negative). The radial velocity is determined from the **Doppler effect** on lines in the object's spectrum. SEE ALSO **redshift**.

radiant The point on the celestial sphere from which a **meteor shower** seems to radiate. Most meteor showers are named after the constellation in which their radiant is situated.

radiation belts (radiation zones) Donut-shaped regions of a planetary **magnetosphere** in which charged particles become trapped by the planet's magnetic field. The Earth's radiation belts are the **Van Allen Belts**. All the giant planets have radiation belts; Jupiter's are huge in extent and 10,000 times as intense as those of the Earth.

radiation era The period from about 10^{-32} seconds after the **big bang** to about 40,000 years later when the energy density of radiation in the Universe exceeded that of matter. It is subdivided into the *hadron era*, when most of the protons and neutrons now in the Universe were produced, and the *lepton era*, when most of the electrons were formed. SEE ALSO **matter era**.

radiation pressure The pressure exerted by **electromagnetic radiation**. A beam of electromagnetic radiation can also be considered as a beam of particles called photons. When a beam of photons strikes a surface, the photons transfer their momentum to the surface and so exert a pressure. It is radiation pressure that keeps a star's outer layers from collapsing inward under its own gravity. In interplanetary space, small particles are subjected to considerable radiation pressure from sunlight. It is what causes the dust tail of a comet to point away from the Sun.

radio astronomy The study of radio emissions – **electromagnetic radiation** with wavelengths from about 1 mm to many meters – reaching the Earth from space. Observations down to about 2 cm can be made at sea level, but at shorter wavelengths it is necessary to observe from high mountains to avoid atmospheric absorption, mainly by water vapor. At the long-wavelength end, from 10 to 20 m, radio waves are absorbed and scattered in the Earth's **ionosphere**. Between these extremes, interference from terrestrial communications is a problem. International agreements have set aside certain wavebands for the sole use of radio astronomy.

Radio noise from the Milky Way was discovered in 1931, by Karl **Jansky**. The first radio telescope purposely built to study such emission was constructed in the USA by Grote Reber, who made a radio map of the sky. In 1942, James Stanley Hey in England found that the Sun emitted radio waves. The subject grew rapidly after World War II once it was realized that significant information about the Universe could be obtained in this region of the spectrum.

Radio emission has been detected from an extremely wide range of sources: the Sun; some of the planets (particularly those, such as Jupiter, with powerful magnetic fields); radio flares from certain stars; general **synchrotron radiation** from high-speed electrons in the Galaxy's magnetic field; and interstellar molecular clouds. Many discrete radio sources have been discovered, including **pulsars**, **supernova remnants**, **radio galaxies** and **quasars**. In addition, the **cosmic microwave background** is strong evidence in favor of the **big bang** theory of the origin of the Universe.

Individual spectral lines can also be detected at radio wavelengths, similar to optical spectral lines. In 1951 the emission from hydrogen at 21 cm (SEE **twenty-one centimeter line**) was detected, and it has proved to be an especially useful means of studying the structure of the Galaxy. The first known **interstellar molecule**, hydroxyl (OH), was discovered in 1965 from radio observations. Spectral lines have now been observed from over 200 interstellar molecules. The study of these molecular lines has stimulated the extension of radio observations into the centimeter and millimeter range.

radio galaxy A galaxy that is an abnormally high emitter of radio waves, around a million times stronger than the weak emissions from galaxies such as our own. Radio galaxies are usually giant ellipticals, such as M87 in the constellation Virgo. The emission from a typical radio galaxy is concentrated in two *lobes* well outside the galaxy, often located symmetrically either side, which appear to have been ejected explosively from the galaxy. The biggest radio galaxies have lobes around 15 million light years across, comparable in size to a typical cluster of galaxies. The extended radio lobes are almost invariably connected to the main galaxy by jets which seem to originate from the galactic nucleus. Strong, point-like radio sources exist in the nuclei of many of these objects.

It is not understood why some elliptical galaxies emit such enormous amounts of radio energy. However, it does seem clear that the source of the activity lies in the galaxy's nucleus, and in this respect radio galaxies are similar to radio **quasars** and **BL Lacertae objects**. A radio galaxy's enormous energy output comes from **synchrotron radiation**. The central "engine" produces vast quantities of energetic electrons traveling near the speed of light. When these electrons encounter a magnetic field, they spiral around and, in so doing, lose energy. It is this energy that we detect as radio waves.

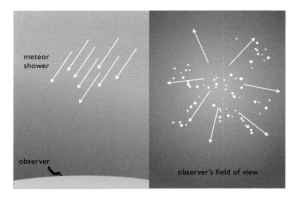

▲ **radiant** *Meteors from a common source, occurring during a shower such as the Perseids, enter the atmosphere along parallel trajectories, becoming luminous at altitudes of around 80–100 km (50–60 miles). If an observer on the ground plots the apparent positions of meteors on a chart of the background stars, they will appear to diverge from a single area of sky, the radiant. The radiant effect is a result of perspective.*

▼ **radar astronomy** *This Magellan orbiter radar image shows Maat Mons on Venus. Radar observations have allowed planetary astronomers to penetrate Venus's dense atmosphere and map the planet's surface topography.*

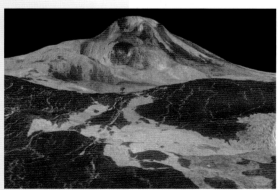

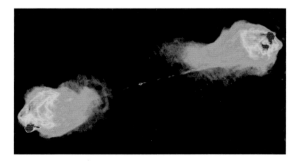

▲ **radio galaxy** *Jets emerging from the active nucleus of Cygnus A produce two lobes of strong radio emission to either side of the galaxy, as seen in this image from the Very Large Array. Many radio galaxies show similar patterns of emission as material they eject impacts on the surrounding intergalactic medium.*

▲ **rays** *The Moon's surface as seen by the Galileo spacecraft as it sped past Earth en route to Jupiter in 1992. Bright rays from the crater Copernicus spread across the dark surface of Oceanus Proceliarum at left of center.*

radio interferometer A type of radio telescope that operates on the same principle as an **optical interferometer**, achieving far higher resolution than is possible with a single dish. Single-dish radio telescopes typically have beam-widths of around 1 arc minute, which means in effect that they can only resolve angular sizes larger than this and measure positions to a few tens of arc seconds. A radio interferometer consists of two or more antennas which are spaced along a baseline and receive radio waves from the same source. The output signals from pairs of antennas are fed to a common radio receiver. The length of the baseline can be changed by varying the distance between the pairs of antennas. A small dish can be very effective as an interferometer in conjunction with a second dish of large collecting area. In **very long baseline interferometry** the radio telescopes can be on different continents, thereby providing a baseline the size of the Earth and achieving resolutions of about 1 milliarcsecond.

radio telescope An instrument used to collect and record radio waves from space. The basic design is a parabolic reflector, normally in the form of a large single dish similar to an optical reflector but much larger – up to 305 m (1000 ft) in diameter; some instruments employ a parabolic "trough" or a section of a much larger parabolic surface. The smoothness and accuracy of the reflecting surface determine the shortest wavelength that can be detected, because surface irregularities must be no more than a small fraction of this wavelength. However, for longer wavelengths, such as around 21 cm, it is adequate to use open wire mesh for the main reflector. Radio waves are reflected by the main reflector to a focus, sometimes via a secondary reflector, then converted into electrical signals that are amplified and recorded. A single-dish telescope does not produce a radio "photograph" of a source: it only collects radio energy in a narrow beam along its axis. To build up an image of a radio source, the dish must be scanned across it, both in altitude and azimuth. In some cases, the azimuth scan is carried out by utilizing the rotation of the Earth.

A radio telescope should have high *sensitivity* (the ability to detect faint signals) and high *resolution* (the ability to detect fine detail). The largest individual dishes, such as **Arecibo**, **Effelsberg**, **Jodrell Bank** and the **National Radio Astronomy Observatory**, are very sensitive. They collect radio energy as weak as one hundred-million-millionth of a watt (10^{-14} W). Resolution depends on wavelength and aperture. Meter wavelengths are about a million times longer than the wavelengths of light, and resolution is correspondingly poorer. At 1 m wavelength, large radio telescopes have a resolving power of around 1°, about 20 times inferior to that of the naked eye. At a few centimeters their resolution equals that of the eye – still much poorer than optical telescopes.

Better resolution can be obtained by linking two or more dishes to form a **radio interferometer**, or by combining many of them to form an **aperture synthesis** array, which gives resolutions of a few hundredths of an arc second, 10 to 100 times better than that of the largest optical telescopes on Earth. The sensitivity remains that of the individual dishes added together.

radius vector The imaginary straight line between an orbiting celestial body and its primary, such as the line connecting a planet and the Sun. SEE **Kepler's laws**.

Ramsden eyepiece A basic telescope eyepiece consisting of two simple elements. It suffers from **chromatic aberration**, so the **Kellner eyepiece** is usually preferred. (After English optician Jesse Ramsden, 1735–1800.)

Ranger The first series of US Moon probes. After several failures, Rangers 7, 8 and 9, in 1964 and 1965, transmitted thousands of photographs of the lunar surface before they impacted. The photographs showed details of boulders and craters as small as 1 m (3 ft), a much higher resolution than was possible with Earth-based telescopes.

rays Bright streaks radiating from a crater. They consist of *ejecta* – material ejected by the impact that produced the crater. They are bright because the shock wave accompanying the impact is sufficient to vaporize rock, which then resolidifies in a glassy form. Ray craters are numerous on the Moon. The most prominent is Tycho, whose rays are seen at full Moon to extend across much of the Moon's disk. Ray craters are also found on other bodies in the Solar System, such as the planet Mercury and Jupiter's moon **Ganymede**. Solar radiation gradually darkens rays, so the presence of a bright ray system around a crater indicates relatively recent formation.

R Coronae Borealis star A member of a class of variable stars that are subject to sudden, unpredictable fades. They normally spend most of their time at maximum, sometimes pulsating slowly by about half a magnitude over five to seven weeks. The fades can be anywhere between two and nine magnitudes and occur completely at random; the resultant deep minima can last for a few weeks or months, or for more than a year.

The type star, R Coronae Borealis, was first found to be variable in 1795. As shown in the diagram, it is normally at sixth magnitude, where it may remain for as long as ten years before suddenly fading. The minimum may be anywhere between seventh and 15th magnitude and can last for a few weeks or several years.

Only about three dozen stars of this sort are known. They are thought to be stars of 1 solar mass which have evolved past the red-giant stage and have ejected their hydrogen-rich outer layers to expose a core which is rich in carbon but depleted in hydrogen. Their declines are due to obscuration by clouds of carbon particles that have condensed in their outer atmosphere. Each cloud covers only a few percent of the star's surface, and we see a fade when a cloud lies in our line of sight.

recombination era SEE **decoupling era**.

recurrent nova A **nova** that is seen to erupt more than once. Recurrent novae are binary stars in which the primary is a late-type (G, K or M) giant filling its **Roche lobe** and transferring material to a white dwarf secondary. At intervals of years to tens of years this material, which has accumulated on the surface of the white dwarf, explodes in a thermonuclear reaction, causing the brightness of the star to increase by 7 to 10 magnitudes for about 100 days. The longer the time between outbursts, the brighter the outbursts are. Examples are T Coronae Borealis, which underwent major outbursts in 1866 and 1946, and T Pyxidis (1890, 1902, 1922, 1944 and 1966). Only six are known, but all novae probably recur, over periods of thousands of years.

reddening A phenomenon exhibited by starlight as it is selectively absorbed and re-radiated on passing through clouds of interstellar dust and the Earth's atmosphere. Light is re-radiated at longer wavelengths than the wavelengths of the incident light, and is thus redder. A correction factor for interstellar reddening must therefore be introduced when distances of remote stars are estimated from their spectra. This reddening is not to be confused with **redshift**.

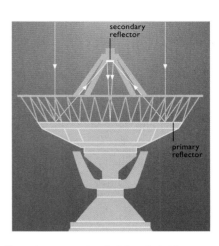

▲ **radio telescope** *Just like an optical reflector, a radio telescope depends on collection of electromagnetic radiation by a large-aperture parabolic reflector, often in the form of a dish. Radio waves are brought to a focus on the detector. Most radio telescopes can be steered in altitude and azimuth.*

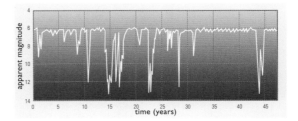

◄ **R Coronae Borealis star** *The light curve of a typical R Coronae Borealis star shows long intervals spent at maximum brightness, punctuated by irregular episodes of abrupt dimming caused by condensation of carbon in the stellar atmosphere.*

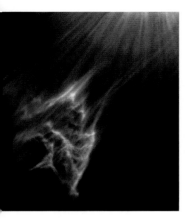

▲ reflection nebula The nebulosity surrounding the Pleiades star cluster is a well-known example of a reflection nebula. A small part of the nebula can be seen in this Hubble Space Telescope image. The star Merope is just out of frame at top right.

▼ refracting telescope The optical layout of a refractor is shown in this schematic cutaway. Light is collected by the main, objective lens at the front end of the telescope and brought to a focus. The observer views a magnified image of the focused light through a smaller lens, or set of lenses, making up the eyepiece. The telescope is shown on a German equatorial mount.

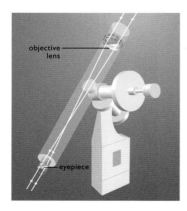

objective lens

eyepiece

red dwarf A star at the lower end of the main sequence, of spectral type K or M. Red dwarfs have masses of between 0.8 and 0.08 of a solar mass. They are of small diameter, relatively low surface temperature (between 2500 and 5000 K) and low absolute magnitude. **Barnard's Star** and **Proxima Centauri** are examples.

red giant A giant star of spectral type K or M, having a surface temperature of less than 4700 K, 10 to 100 times the Sun's diameter and 100 to 10,000 times as luminous. A red giant is a star in a late stage of evolution, having exhausted the hydrogen fuel in its core. Red giants lie at the upper right of the **Hertzsprung–Russell** diagram. Examples are Arcturus and Aldebaran.

Red Planet Popular name for the planet Mars, which has a distinctly red color.

redshift (symbol z) A lengthening of the wavelength of light or other electromagnetic radiation from a source, caused either by the source moving away (the **Doppler effect**) or by the expansion of the Universe. It is defined as the change in the wavelength of a particular spectral line, divided by the unshifted, or rest, wavelength of that line. For example, a redshift of $z = 0.1$ means that all the wavelengths of a source are lengthened by 10%; for $z = 0.2$ they are lengthened by 20%; and so on. The speed (v) at which a source is receding is found from its redshift by the formula $v = z \times c$, where c is the speed of light. Redshifts greater than 1 do not imply speeds greater than c; instead, for speeds greater than about $\frac{1}{3}c$, **Einstein**'s special theory of relativity has to be used. An object with a redshift of $z = 2$, for example, is not receding from us at twice the speed of light. Instead, a redshift of 2 corresponds to a speed of 80% of the velocity of light, and a redshift of 4 corresponds to 92% of the velocity of light. By early 2004, quasars had been detected out to a redshift of $z = 6.6$, and lensed galaxies (SEE **gravitational lens**) out to at least $z = 7.0$.

Redshifts caused by the expansion of the Universe, called *cosmological redshifts*, have nothing to do with the Doppler effect. They are caused by the expansion of space itself, which stretches the wavelengths of light traveling toward us. The longer the travel time of light, the more its wavelength is stretched, as embodied in the **Hubble law**. The most highly redshifted photons are those in the **cosmic microwave background** radiation. These photons have been traveling toward us since matter and radiation decoupled (SEE **decoupling era**) from each other some 300,000 years after the **big bang**. They exhibit a redshift of about $z = 1000$, meaning that their wavelengths have been stretched by over a thousandfold.

Gravitational redshift is a phenomenon predicted by Einstein's general theory of relativity. Light emitted by a star has to do work to overcome the star's gravitational field. As a result there is a slight loss of energy and a consequent increase in wavelength, so that all the spectral lines are shifted toward the red.

redshift survey A mapping of the three-dimensional distribution of galaxies in space, with the aim of determining the large-scale structure of the Universe. Current redshift surveys, such as the **Sloan Digital Sky Survey**, employ bundles of optical fibers, each precisely aimed at a different galaxy, to enable the spectra – and hence the redshifts – of several hundred galaxies to be obtained at one pointing of the telescope.

reflecting telescope (reflector) A telescope that forms an image by reflecting light from a primary concave mirror (usually parabolic) at the back of the tube, via a secondary mirror, to a focus whose position depends on the design of the instrument. The chief optical designs for smaller, amateur reflectors are the **Newtonian** and **Schmidt-Cassegrain**. Other configurations used for larger, professional instruments include the **Cassegrain**, **coudé** and **Schmidt camera**. Many large professional reflectors incorporate two or more different optical configurations, and with some it is possible to site instruments at the **prime focus**. Reflectors have the advantage of being free from **chromatic aberration**.

A reflector's mirrors are usually made of low-expansion glass or a ceramic material and coated with a thin layer of aluminum, which is highly reflective. Reflectors can be made larger than refractors because the primary mirror can be supported at the back, whereas a lens can be supported only around its rim, and is consequently liable to distort under its own weight. Also, new ways of building and controlling mirrors, as in the segmented mirrors of the **Keck Telescopes**, and the technique of **active optics**, have made possible further increases in the size and performance of big reflectors.

The Scottish mathematician James Gregory was the first to describe a reflecting telescope, in 1663; the first to build one was Isaac **Newton**, in 1668. In the late 18th century William **Herschel** demonstrated the instrument's potential by making large, accurately figured mirrors. SEE ALSO **Dobsonian telescope**, **Maksutov telescope**, **Ritchey–Chrétien telescope**.

reflection nebula A nebula in which light from a nearby star is reflected off the dust particles mixed in with the gas. Reflection nebulae generally appear blue in images; they are always bluer than the star that illuminates them because blue light is scattered more efficiently than red light. A good example is the Merope Nebula in the Pleiades.

refracting telescope (refractor) A telescope that forms an image by the refraction of light by a **lens** at the front of the tube. This lens, called the *objective*, is a compound lens, of perhaps more than one type of high-quality glass in order to minimize **chromatic aberration**, and it may be coated to reduce reflection and so increase the amount of light transmitted to the eyepiece. The tube is long so as to reduce **spherical aberration**. Refractors are still the first choice of some amateur observers for targets such as the planets, but the difficulties of making a large, optically perfect lens and supporting its weight mean that all large professional instruments are now **reflecting telescopes**.

The origins of the refracting telescope are uncertain. The Dutch spectacle-maker Hans Lippershey applied for a patent on one in 1608, although others were making similar instruments around the same time, and telescopes of some sort may have even been in use at the end of the 15th century. **Galileo** was among the first to use the refractor in astronomy. Improvements were made by a succession of opticians, including John **Dollond** and Joseph von **Fraunhofer**, who devised means of reducing aberrations, and Christiaan **Huygens** and Jesse Ramsden, who devised better eyepieces. The largest refractor, at the Yerkes Observatory, has an aperture of 1 m (40 inches) and was built in 1897.

refraction SEE **atmospheric refraction**, **refracting telescope**.

regolith The dust and debris covering the surface of the Moon and other bodies in the Solar System. It is produced by the disintegration of crustal rocks caused by the impact of meteorites.

regression of the nodes The slow westward motion of the **nodes** of the Moon's orbit, caused by the gravitational pull of the Sun. A full circuit takes 18.6 years.

Regulus The star Alpha Leonis, magnitude 1.36, distance 77 light years, luminosity 135 times that of the Sun. It is a main-sequence star of type B7.

relative sunspot number (symbol R) An index of sunspot activity which reduces the sunspot counts of different observers to a common, statistical basis. For a particular day, it is given by

$$R = k \, (f + 10g) \, ,$$

where g is the number of groups of sunspots – irrespective of the number of spots each contains – and f is the total number of spots in all the groups; k is a factor based on the estimated efficiency of observer and telescope. The work of recording sunspots in this way was begun by Swiss astronomer Rudolf Wolf at the Zurich Federal Observatory, and the terms *Wolf number* and *Zurich relative sunspot number* were both formerly used. It is now compiled at the Solar Influence Data Analysis Center in Brussels and is known as the International Sunspot Number.

relativity SEE **Einstein, Albert**.

residuals Differences between predicted and observed values of astronomical quantities. Residuals in, say, observations of the position of a planet, if collected over a sufficiently long period of time and then analyzed statistically, may reveal some long-term trend. Residuals may also result from inaccurate observations.

resolution (1) SEE **resolving power**.

resolution (2) The size of the smallest detail distinguishable by an imaging system at a particular distance, or of the smallest detail visible in an image (of, for example, an image obtained by a space probe, often expressed in kilometers per pixel).

resolving power (resolution) The ability of an optical system, such as a telescope or the human eye, to distinguish close but separate objects such as the components of a close binary star, or to image a single small object. It is measured in angular units. The resolving power of a telescope in arc seconds is found by dividing 116 by its aperture in millimeters (or 4.56 by the aperture in inches). This measure is called the *Dawes limit*, after the English amateur astronomer William Rutter Dawes (1799–1868). The resolving power of the human eye is about 1 arc minute. SEE ALSO **radio telescope**.

resonance An effect produced by the gravitational interaction between two bodies orbiting the same primary. Resonance between bodies of similar size can lock bodies into **commensurable** orbits, as has happened with Jupiter's **Galilean satellites**. Between bodies of highly dissimilar size, however, it can prevent small bodies from occupying commensurable orbits. This is what keeps clear the **Kirkwood gaps** in the asteroid belt and the various gaps in the rings of **Saturn**.

Reticulum A small but distinctive southern constellation, introduced by **Lacaille**, representing a reticle used for measuring star positions. Its brightest star, Alpha, magnitude 3.3, forms a compact pattern with Beta, Gamma, Delta and Epsilon.

retrograde (1) Describing orbital or rotational motion in the opposite sense to the motion of the Earth: clockwise as viewed from above the Sun's north pole. Some satellites, such as the outermost ones of Jupiter, have a retrograde orbital motion, and so do certain comets, notably Halley's Comet. Venus, Uranus and Pluto all have retrograde rotation. Orbital or rotational motion is retrograde if the orbital or axial inclination is more than 90°. COMPARE **direct** (1).

retrograde (2) Describing the temporary east-to-west motion traced out by a superior planet or asteroid on the celestial sphere as the Earth catches up and overtakes it. The curve traced out by this reversal is called a *retrograde loop*. The places at which the planet changes between easterly and westerly motion are termed *stationary points* (points 4 and 6 in the diagram). COMPARE **direct** (2).

revolution The movement of a planet or other celestial object around its orbit, as distinct from the rotation of the object on its axis.

PLANETARY RINGS		
Ring or gap	Distance from center (thousand km)	Width (km)
Jupiter		
Halo	100–122.8	22,800
Main	122.8–129.2	6400
Gossamer	129.2–214.2	85,000
Saturn		
D	67–74.5	7500
C (Crepe)	74.5–92	17,500
Maxwell Gap	88	270
B	92–117.5	25,500
Cassini Division	117.5–122.2	4700
A	122.2–136.8	14,600
Encke Gap	133.6	325
Keeler Gap	136.5	35
Roche Division	136.8–139.4	2600
F	140.2	30–500
G	165.8–173.8	8000
E	180–480	30,0000
Uranus		
6	41.8	1–3
5	42.2	2–3
4	42.6	2–3
Alpha	44.7	7–12
Beta	45.7	7–12
Eta	47.2	0–2
Gamma	47.6	1–4
Delta	48.3	3–9
Epsilon	51.1	20–100
Nu (R/2003 U 2)	67.3	3800
Mu (R/2003 U 1)	97.7	17,000
Neptune		
Galle	40.9–42.9	2000
Le Verrier	53.2	100
Lassell/Arago	53.2–57.2	4000
Adams	62.9	<50

▲ *Rhea* This Cassini image of Saturn's satellite Rhea was obtained from a distance of 343,000 km (213,000 miles) on 24 February 2006. It shows Rhea's leading hemisphere with the bright young ray crater called Inktomi.

Rhea Saturn's second largest **satellite**, diameter 1528 km (950 miles), discovered by Giovanni Cassini in 1672. It is a typical heavily cratered icy world, but with few large craters, the largest being the 250 km (150-mile) diameter Izanagi. Some resurfacing seems to have taken place in the past, but the surface is mostly ancient. The **Cassini** spacecraft also imaged a more recent, bright ray crater, now called Inktomi.

rich-field telescope (richest-field telescope) (RFT) A low-power telescope equipped with a wide-angle eyepiece (such as a **Nagler eyepiece**) designed to show a wide field of view at a relatively low magnification. Such an instrument is ideal for studying starfields and hunting for novae.

Rigel The star Beta Orionis, the brightest star in the constellation Orion, and marking Orion's left leg. It is of magnitude 0.18 (the seventh brightest star in the sky), and a supergiant of spectral type B8. Its distance is 773 light years, and its luminosity 40,000 times that of the Sun.

right ascension (RA, symbol α) A coordinate used to define the position of a celestial object. It is the angular distance measured eastward from the vernal equinox (also known as the First Point of Aries) to the point where the hour circle of an object meets the celestial equator. RA is usually expressed in hours, minutes and seconds, although it may be expressed as an angle, 1 hour of RA being equivalent to 15°. SEE ALSO **celestial sphere**.

Rigil Kentaurus Another name for the star **Alpha Centauri**.

rille A cleft on the surface of the Moon. Rilles that are curved or straight are geological faults. The winding *sinuous rilles* are most likely collapsed lava tubes. Lava was able to flow on the Moon for hundreds of kilometers. The top and sides of a flow can solidify, insulating the molten lava inside which continues to flow. When such a tube has been drained, the roof may collapse, leaving what we see today as a sinuous rille.

ring, planetary A narrow, circular filament or thin disk of large or small particles, or both, orbiting a planet. **Jupiter**, **Saturn**, **Uranus** and **Neptune** all have systems of rings. Saturn's rings, by far the most prominent and extensive, were discovered telescopically in the 17th century. Jupiter's were found in 1979 by Voyager 1; those of Uranus and Neptune were first observed during **occultations**, and were confirmed by Voyager 2 in the 1980s. Saturn's rings were found in 1849 to lie inside the **Roche limit**; any sizable, rigid satellite inside this limit would be broken up by gravitational forces, and any smaller bodies would be prevented from accreting into a larger one. In 1857 the Scottish physicist James Clerk Maxwell showed that on theoretical grounds the rings had to be made up of many separate particles.

The ring systems around the four giant planets show similarities and dissimilarities. For example, Saturn's main rings are made up principally of bright, icy particles, whereas the other three planets have dark rings. Some rings are very narrow and distinct, others are continuous sheets, but still a kilometer or less thick, and confined to

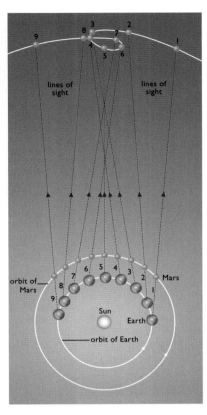

▲ *retrograde* (2) The cause of the apparent reversal of superior planets in their motion along the ecliptic close to the time of opposition is explained in the diagram. Here, Earth (inner orbit) is seen catching up Mars (outer orbit). Before opposition, Mars appears to move eastward (top). As Earth, moving more rapidly on its smaller orbit, catches up, Mars' eastward motion apparently slows, then reverses relative to the background stars as Earth overtakes. After opposition, the apparent eastward motion of Mars resumes.

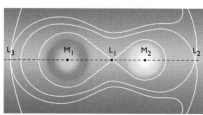

▲ **ring galaxy** *Cataloged as AM 0644-741, this ring galaxy lies 300 million light years away in the constellation Volans. Ring galaxies are thought to form when one galaxy passes through the disk of another. In this case the "intruder" galaxy is out of shot.*

the equatorial plane, while still others are large, tenuous, donut-shaped haloes. Small **shepherd moons** keep some rings in check (in Neptune's case, creating clumps of material – *ring arcs*), while larger satellites create concentric gaps. Particles range from micrometer-sized in Jupiter's main ring to as much as 10 m (30 ft) in Saturn's B Ring.

ring galaxy A galaxy having the form of an elliptical ring around a nucleus, like the rim of a wheel. Ring galaxies are thought to form from a collision in which a small galaxy passes through a larger one.

Ring Nebula The planetary nebula M57, 4100 light years away in the constellation Lyra, also known as NGC 6720. It appears as a hollow ellipse; in three dimensions it is shaped not like a spherical shell but like a toroid, or ring donut. The nebula, centered on a blue star of 0.2 solar masses, is expanding at 19 km/s (12 miles/s) and is estimated to be about 5500 years old.

Ritchey–Chrétien telescope A modified form of **Cassegrain telescope**, designed originally by George Ritchey (1864–1945) and Henri Chrétien (1879–1956). The primary and secondary mirrors are **hyperboloidal**, although there are variants with near-hyperboloidal, **paraboloidal** or **spheroidal** components. The system is free from **coma** (1) over a wide field, though there is some astigmatism and the field is curved.

robotic telescope A telescope designed to operate automatically under control from a remote site via an Internet connection. Examples are the 400 mm (18-inch) *Bradford Robotic Telescope* in West Yorkshire (the world's first); the two 2 m (6.6 ft) *Faulkes Telescopes*, one on Hawaii and the other at the Siding Spring Observatory in Australia; and the 2 m *Liverpool Telescope* at the Roque de los Muchachos Observatory in the Canary Islands.

Roche limit (symbol *R*) The minimum distance at which a planetary satellite can remain in orbit without being torn apart by gravitational forces; alternatively, the minimum distance at which smaller bodies in orbit round a planet can combine by **accretion** to form a larger one. A planet's Roche limit is about 2.5 times its radius (*r*). It depends, though not strongly, on the densities of the planet (d_p) and the satellite (d_s):

$$R = 2.44 \, r_3 \sqrt{(d_p/d_s)}$$

Actually the Roche limit applies only to "fluid" satellites. Rocky bodies with a high structural integrity, such as Neptune's innermost satellites, can exist inside the Roche limit. All planetary ring systems lie inside their planet's Roche limit. The concept was originated by the French astronomer and mathematician Edouard Roche (1820–83).

Roche lobe A surface defining the maximum size of a star in a binary system. If one star's surface overflows this size as it evolves, the companion's gravity will pull matter off it.

▼ **Roche lobe** *In a binary star system, the stars (here marked M_1 and M_2) have their own gravitational sphere of influence, defined by the Roche lobe. If one component fills its Roche lobe, material can escape through the inner Lagrangian point, L_1, on to its companion. If both stars fill their Roche lobes, material can escape entirely from the system through the outer Lagrangian points, L_2.*

Römer, Ole Christensen (1644–1710) Danish astronomer. He was the first to measure the speed of light. Working at Paris in 1675, he found that the times of occultations of the **Galilean satellites** predicted by Giovanni Domenico **Cassini** did not match the observed times: intervals between successive occultations

decreased as the orbital motions of the Earth and Jupiter brought them closer together, and increased as they moved farther apart. He deduced that the differences were caused by light taking time to travel the intervening distance. His timings gave the speed of light as 200,000 km (120,000 miles) per second, two-thirds of the true value. Römer also invented the first **transit instrument**.

Roque de los Muchachos Observatory A Spanish observatory on La Palma in the Canary Islands, the home of telescopes owned by several European nations, foremost among them the 10.4 m **Gran Telescopio Canarias**. Other major instruments include the 3.5 m Italian **Galileo National Telescope**, the UK's **William Herschel Telescope** and 2.5 m (100-inch) Isaac Newton Telescope, moved from England in 1984, and the 2.5 m Nordic Optical Telescope, opened in 1989.

Rosalind One of the small inner **satellites** of Uranus discovered in 1986 during the Voyager 2 mission.

Rosat A German satellite containing an X-ray telescope and a British wide-field camera that made the first survey of the sky at extreme ultraviolet wavelengths. It was launched on 1 June 1990, and operated until the end of 1998.

Rosetta A European Space Agency probe, launched on 2 March 2004 to investigate Comet 67P/Churyumov–Gerasimenko. It went into orbit around the comet's nucleus in September 2014, dropping a lander called Philae on to the surface in November 2014. On its way to the comet, Rosetta flew past the asteroids Steins and Lutetia, in September 2008 and July 2010 respectively, photographing them both.

Rosette Nebula A diffuse nebula 4900 light years away in the constellation Monoceros, surrounding the open cluster NGC 2244. The stars of the cluster were born within the past half million years from the Rosette Nebula and now illuminate it. Radiation pressure from the stars has created a cavity about 30 light years across in the central part of the nebula. Star formation continues in the nebula, which contains many **Bok globules**.

Rosse, Third Earl of (William Parsons) (1800–1867) Irish amateur astronomer. In the grounds of Birr Castle, near Parsonstown in Ireland, he built a 72-inch (1.83 m) reflector. Completed in 1845, it was the largest telescope in the world for the rest of its lifetime. Rosse studied in particular the hazy cloud-like objects then known collectively as nebulae. In 1845 he discovered the spiral structure of what is now called the **Whirlpool Galaxy** and other "spiral nebulae" (recognized as other galaxies in the 1920s) and was able to resolve certain nebulae into groups of stars. He also discovered the ring structure of planetary nebulae.

rotation The turning of a celestial body on its axis, as distinct from its orbital revolution.

rotating variable A star whose variations in magnitude arise either because of its non-spherical shape, as in *ellipsoidal variables*, where a binary system is not eclipsing but presents a changing apparent surface area throughout its orbit, or because a star has a non-uniform surface brightness. In the latter case, the variation may be caused by surface spots, chromospheric activity, or alterations in the surface brightness related to fluctuations in the magnetic field, arising either because of varying field strength or because the star's magnetic axis does not coincide with the rotational axis.

Royal Greenwich Observatory (RGO) The former UK national astronomical observatory, founded at Greenwich, London, in 1675. After World War II the observatory moved to Herstmonceux in Sussex, and in 1990 it moved again, to Cambridge. When it closed in 1998, some of its functions were transferred to the Royal Observatory, Edinburgh.

Royal Observatory, Edinburgh (ROE) An administrative center in Edinburgh which took on some of the functions of the **Royal Greenwich Observatory** when the RGO closed in 1998. Its two main departments are the Institute for Astronomy, run by the University of Edinburgh, and the UK Astronomy Technology Centre, which plans, designs and manufactures telescopes and instrumentation.

RR Lyrae star A pulsating variable star of a class formerly known as *short-period Cepheids* or *cluster variables* (because the first were found in globular clusters). RR Lyrae, the type star, was the first to be found outside a globular cluster. RR Lyrae stars are Population II giants

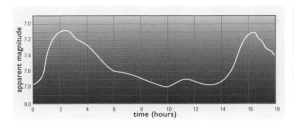

▲ **RR Lyrae star** *This is the light curve of a typical RR Lyrae variable. These stars show a rapid rise to peak light followed by a slower decline. They are commonly found in globular clusters.*

(SEE **stellar populations**) of spectral type between A and F. Their periods range from 0.2 to 1.2 days and their amplitudes from 0.2 to 2.0 magnitudes, although most have periods of between 9 and 17 hours. Their light curves differ from those of the so-called classical Cepheids; that of RR Lyrae itself is shown in the diagram. Most rise quickly to maximum in no more than a tenth of their total period. Their minima are comparatively long, so that for a few hours their light remains roughly constant. The absolute magnitude of all RR Lyrae stars is about +0.5, so they can be used as "standard candles" for distance finding. However, they are too faint to be seen beyond the nearest galaxies. SEE ALSO **Cepheid variable**.

runaway star A star of spectral type O or B that has an unusually high space velocity. Such stars are former members of massive binaries, one member of which has exploded as a supernova, flinging off the other like a slingshot. Three runaway stars with a common origin in Orion – 53 Arietis, AE Aurigae and Mu Columbae – may have belonged to a quadruple star system in which the fourth star exploded as a supernova 3 million years ago. As many as 20% of O stars may be runaways.

Russell, Henry Norris (1877–1957) American astronomer. In 1910 he produced a diagram plotting the absolute magnitude of stars against their spectral type which showed the division between main-sequence stars and giant stars. Unknown to him, Ejnar **Hertzsprung** had done the same some years previously. The diagram, now called the **Hertzsprung–Russell diagram**, is the starting-point for all modern theories of stellar evolution, although Russell first thought, erroneously, that stars evolved along the main sequence. In 1928 he determined the composition of the Sun's atmosphere from a study of its spectrum.

RV Tauri star A member of a class of pulsating variable stars with alternating primary and secondary minima. The depth of the minima also vary, such that the primary and secondary minima occasionally exchange places. RV Tauri stars are luminous supergiants of spectral type between F and M, with periods ranging from 30 to 150 days and amplitudes of up to 4 magnitudes. The diagram shows the light curve of AC Herculis, which has a constant mean magnitude, but with some other stars of this class, such as RV Tauri itself and DF Cygni, the mean varies by up to 2 magnitudes with a secondary periodicity of 600 to 1500 days. Apart from their pulsations with multiple periods, RV Tauri stars are undergoing pulsations of more than one period, and appear to be giving off material in a strong **stellar wind** in what may be the early stages of formation of a **planetary nebula**.

Ryle, Martin (1918–84) English radio astronomer. In 1946 he discovered **Cygnus A**, the first radio galaxy to be found. In 1955 he built the first **radio interferometer**, and in 1960 developed the technique of **aperture synthesis**. For his pioneering work in radio astronomy he received a share of the 1974 Nobel Prize for Physics.

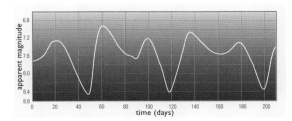

▲ **RV Tauri star** *The light curve of a typical RV Tauri star shows alternating deep and shallow minima. The deep minima result from different modes of pulsation in the star's outer layers coming into phase with each other.*

Sagan, Carl Edward (1934–96) American astronomer. He studied many aspects of the Solar System, including the physics and chemistry of planetary atmospheres and surfaces, particularly of Venus and Mars. He investigated the origins of terrestrial life (discovering in 1963 that ATP, an important biochemical, is formed when chemicals thought to be present on the early Earth are mixed together) and the possibility of life existing elsewhere. Sagan did much to popularize astronomy and science in general.

Sagitta A small constellation representing an arrow, adjoining Aquila. Its brightest star is Gamma, magnitude 3.5. The arrow shape is completed by the fourth-magnitude stars Alpha, Beta and Delta.

Sagittarius The southernmost of the zodiacal constellations, representing a centaur with a bow, and notable for containing the dense star clouds in the direction of the center of our Galaxy. The outline of its main stars is sometimes likened to a teapot. Sagittarius's brightest star is Epsilon (Kaus Australis), magnitude 1.8, followed by Sigma (Nunki), magnitude 2.0. There are plenty of variable stars, including the Cepheids X (magnitude range 4.3–5.1) and W Sagittarii (magnitude range 4.2–4.9). The Milky Way is at its richest in Sagittarius, and there are no fewer than 15 Messier objects – globular clusters, open clusters and gaseous nebulae, including the **Omega Nebula**, the **Lagoon Nebula** and the **Trifid Nebula**.

Sagittarius A A strong radio source at the center of our Galaxy. The exact center of the Galaxy is marked by a small part of this source, referred to as *Sagittarius A**, less than 4 au across. It is most likely a black hole surrounded by an **accretion disk**. Observations of stars in orbit around this black hole lead to an estimated mass of 4 million solar masses.

Saha, Meghnad (1894–1956) Indian astrophysicist and nuclear physicist. His theoretical work on spectra, the solar corona, radiation, and ionization resulted in the 1920s in an equation, now known as the *Saha equation* which relates the degree of ionization of atoms (i.e. how many electrons they have lost) in a star's atmosphere to the temperature. It provides an important insight into the differences in the spectra of stars of different spectral types, and allows their temperatures to be deduced from the lines present in their spectra.

Salpeter process SEE **triple-alpha process**.

Sandage, Allan Rex (1926–2010) American astronomer. Initially an assistant to Edwin **Hubble**, he continued to work toward establishing the value of the **Hubble constant** and thus fixing the age of the Universe. In 1960 Sandage made the first identification at optical wavelengths of what would prove to be a **quasar**, and five years later he found the first radio-quiet quasar.

saros A cycle of lunar and solar eclipses lasting just over 18 years. A saros consists of the same sequence of eclipses repeated at nearly the same time intervals. This is because 223 **lunations** (6585.32 days) have almost the same duration as 19 **eclipse years** (6585.78 days), after which time the Sun, the Moon and the Moon's **nodes** return to very nearly the same positions relative to one another. This period, known to several ancient peoples, enabled astronomers to predict eclipses, since, if an eclipse of the Sun or the Moon is observed, then after a complete saros there will be a similar eclipse. The difference of 0.46 of a day between the number of lunations and eclipse years in one saros means that each eclipse in the sequence will recur about 165° further west. In predicting a date, allowance must be made for leap days. SEE ALSO **Metonic cycle**.

satellite (moon) A celestial body in orbit around a planet. Of the major planets in the Solar System, Mercury and Venus have no known satellite. The Earth has one satellite (the Moon), Mars has two, Jupiter and Saturn over 60 each, Uranus 27 and Neptune 14 (as of mid-2015). The dwarf planet Pluto has five moons. There are almost certainly further small satellites of the giant planets awaiting discovery. Some asteroids, such as **Ida**, are known to have a satellite.

The satellites of the Solar System vary enormously in their size, orbit, surface features and supposed origin. The largest moon of all, Jupiter's Ganymede, is larger than Mercury, as is Saturn's largest moon, Titan, while the smallest are tiny bodies less than a kilometer across. Their orbital periods vary just as much, from Phobos, which circles Mars' equator more than three times a day, to the remotest satellite of Neptune, which takes over 25 Earth years to complete one orbit. In their surface features they are remarkably diverse, ranging from dead, densely cratered worlds to ones that show a fascinating variety of geological processes. Indeed, Jupiter's Io is the most geologically active body known. As to their origin, some, such

PRINCIPAL SATELLITES OF THE MAJOR PLANETS			
Satellite	Diameter (km)	Distance from center of planet (thousand km)	Orbital period (days)	Mean opposition magnitude
Earth				
Moon	3,475	384	27.32	212.7
Mars				
Phobos	22	9.4	0.32	11.4
Deimos	12	23.5	1.26	12.5
Jupiter				
Metis	43	128	0.29	17.5
Adrastea	16	129	0.30	18.7
Amalthea	167	181	0.50	14.1
Thebe	99	222	0.67	16.0
Io	3,643	422	1.77	5.0
Europa	3,122	671	3.55	5.3
Ganymede	5,262	1,070	7.15	4.6
Callisto	4,821	1,883	16.69	5.7
Leda	20	11,165	240.9	19.5
Himalia	170	11,461	250.6	14.6
Lysithea	36	11,717	259.2	18.3
Elara	86	11,741	259.6	16.3
Ananke	28	21,276	629.8 R	18.8
Carme	46	23,404	734.2 R	17.6
Pasiphae	60	23,624	743.6 R	17.0
Sinope	38	23,939	758.9 R	18.1
Saturn				
Pan	25	134	0.58	19.4
Atlas	32	138	0.60	19.0
Prometheus	100	139	0.61	15.8
Pandora	84	140	0.63	16.4
Epimetheus	116	151	0.69	15.6
Janus	178	152	0.69	14.4
Mimas	397	186	0.94	12.8
Enceladus	499	238	1.37	11.8
Tethys	1,060	295	1.89	10.2
Telesto	24	295	1.89	18.5
Calypso	19	295	1.89	18.7
Dione	1,118	377	2.74	10.4
Helene	32	377	2.74	18.4
Rhea	1,528	527	4.52	9.6
Titan	5,150	1,222	15.95	8.4
Hyperion	283	1,464	21.28	14.4
Iapetus	1,436	3,561	79.33	11.0
Phoebe	220	12,944	548.2 R	16.4
Paaliaq	19	15,199	686.9	21.3
Albiorix	26	16,404	783.4	20.5
Siarnaq	32	18,160	893.1	20.1
Ymir	16	23,096	1,312 R	21.7
Uranus				
Cordelia	40	50	0.34	23.6
Ophelia	23	54	0.38	23.3
Bianca	51	59	0.43	22.5
Cressida	80	62	0.46	21.6
Desdemona	64	63	0.47	22.0
Juliet	94	64	0.49	21.1
Portia	135	66	0.51	20.4
Rosalind	72	70	0.56	21.8
Belinda	81	75	0.62	21.5
Perdita	40	76	0.64	23.6
Puck	162	86	0.76	19.8
Mab	24	97.7	0.92	24.6
Miranda	472	130	1.41	15.8
Ariel	1,158	191	2.52	13.7
Umbriel	1,169	266	4.14	14.5
Titania	1,578	436	8.71	13.5
Oberon	1,523	584	13.46	13.7
Francisco	22	4,283	267.1 R	25.0
Caliban	89	7,231	579.5 R	22.4
Stephano	20	8,004	677.4 R	24.1
Sycorax	190	12,179	1,288 R	20.8
Prospero	30	16,243	1,977 R	20.8
Setebos	30	17,501	2,234 R	23.3
Neptune				
Naiad	58	48	0.29	24.6
Thalassa	80	50	0.31	23.9
Despina	148	53	0.33	22.5
Galatea	158	62	0.43	22.4
Larissa	192	74	0.55	22.0
Proteus	416	118	1.12	20.3
Triton	2,707	355	5.88 R	13.5
Nereid	340	5,514	360.1	19.7
Halimede	54	16,600	1,874 R	24.2
Sao	44	22,619	2,919.2	25.5
Laomedeia	42	23,613	3,175.6	25.5
Neso	43	48,600	9,412 R	24.7

R indicates retrograde orbit.

There are many minor satellites of the giant planets; only those with diameters greater than 10 km (Jupiter), 15 km (Saturn), 20 km (Uranus) and 40 km (Neptune) are included here.

as Jupiter's **Galilean satellites**, may have condensed out of the same part of the solar nebula as their parent planet, while several small ones are believed to be captured asteroids or cometary nuclei. The **Moon** is currently thought to have formed from debris flung out when a Mars-sized body collided with the Earth.

Data for the largest planetary satellites are given in the table at left. SEE ALSO **artificial satellite**.

Saturn The sixth major planet from the Sun, and the second largest of the four giant planets. Saturn was the remotest planet known in the pre-telescopic era. Viewed through a telescope it appears as a flattened golden yellow disk encircled by white rings. Because of Saturn's axial inclination, the angle at which the ring system is presented to us depends on the planet's position in its orbit, and this has a considerable effect on its brightness. At perihelic opposition with the rings fully open, Saturn is magnitude −0.3; at aphelic opposition with the rings edge-on, it is magnitude 0.8. The disk's maximum apparent size is 21 arc seconds (and the rings span up to 48 arc seconds). The main data for Saturn are given in the table on the facing page.

Saturn's magnificent system of rings was first identified as such by Christiaan **Huygens** in 1656. As seen from the Earth, the ring system consists of a few principal components, separated by a number of gaps. The three main rings, starting from the outside, are Ring A, which is grayish-white, Ring B, which is bright white, and Ring C, which is fainter and blue-gray. Ring D, the innermost, was discovered in 1969; Rings E, F and G, which lie outside Ring A, were discovered by space probes. The most prominent gap is the Cassini Division, between Rings A and B; the Encke Gap in Ring A is also quite prominent. These gaps are caused by the perturbing effect of Saturn's inner satellites. The rings are made up of components ranging from subcentimeter-size particles to objects many meters across, all in individual orbits around Saturn. The main rings are only a kilometer or so thick. Ring A's diameter is about 275,000 km (170,000 miles).

Following Pioneer 11 in 1979, the two Voyager probes flew past Saturn, Voyager 1 in 1980 and Voyager 2 in 1981. They revealed the ring system to be made up of thousands of separate ringlets. The gaps were not empty, but simply regions with a lower ring density. Some rings were eccentric, not circular. Ring F has a braided structure, being made up of intertwined ringlets and controlled by **shepherd moons**. *Spokes*, dark radial features in Ring B, apparently mark the trajectories of ring particles moving outward along magnetic field lines. The **Cassini** probe, which reached Saturn in 2004, confirmed these findings, and suggested that only small changes had occurred in the rings since the Voyager probes' exploration.

Like Jupiter's disk, Saturn's shows a pattern of darker *belts* and lighter *zones* parallel to the equator. These atmospheric features appear rather muted in comparison with Jupiter's because of the presence of a high-altitude haze layer. Storms and associated lightning have been detected, but there is no equivalent of Jupiter's Great Red Spot, although short-lived white spots occasionally erupt in the northern hemisphere. The most recent was in 2010. The largest of these grow until they spread around the entire planet before dispersing. They are thought to be due to seasonal heating of the atmosphere. There is **differential rotation** on Saturn: features at the equator rotate in about $10\frac{1}{4}$ hours, and those near the poles in nearly $10\frac{3}{4}$ hours (the rotation period of the interior). Winds at the equator can reach speeds of 1800 km/h (1100 miles/h).

Saturn has an internal heat source, which probably drives its weather. Its density, 0.7 g/cm^3, is much lower than for any other planet (and its oblateness is higher than for any other). Saturn is thought to be composed predominantly of hydrogen, and to have an iron–silicate core about five times the Earth's mass, surrounded by an ice mantle. Near the mantle the hydrogen should be in metallic form (so compressed it behaves as a conducting metal), above this existing as liquid molecular hydrogen. The upper atmosphere contains 97% hydrogen and 3% helium, with traces of other gases including methane, ethane, ammonia and phosphine.

The magnetic field of Saturn is 500 to 1000 times the strength of the Earth's, and its magnetic axis is nearly coincident with the axis of rotation. The magnetic field generates a **magnetosphere** intermediate between those of the Earth and Jupiter in its shape and extent and in the intensity of its radiation belts of trapped particles. Aurorae have been detected around Saturn's polar regions.

As of mid-2015, Saturn had over 60 known **satellites**.

Saturn Nebula The planetary nebula NGC 7009 in Aquarius, about 2000 light years away. Handle-like protuberances give it the appearance of Saturn. The nebula is small, of eighth magnitude, and its appearance probably indicates an internal structure consisting of a series of shells ejected by the central star.

SATURN: DATA	
Globe	
Diameter (equatorial)	120,000 km
Diameter (polar)	107,100 km
Density	0.69 g/cm^3
Mass (Earth = 1)	95.2
Volume (Earth = 1)	764
Sidereal period of axial rotation (equatorial)	10h 14m
Escape velocity	35.5 km/s
Albedo	0.70
Inclination of equator to orbit	26° 43′
Temperature at cloud-tops	95 K
Surface gravity (Earth = 1)	1.19
Orbit	
Semimajor axis	9.539 au = 1427 × 10^6 km
Eccentricity	0.056
Inclination to ecliptic	2° 29′
Sidereal period of revolution	29.46y
Mean orbital velocity	9.65 km/s
Satellites	60+

scattering The deflection of light or other types of electromagnetic wave by particles. For particles very much larger than the wavelength of the light, the scattering is a mixture of reflection and **diffraction**, and the amount of scattering depends very little on wavelength. For particles very much smaller than the wavelength, the amount of scattering (*d*) is inversely proportional to the fourth power of the wavelength (λ): $d \propto 1/\lambda^4$. Thus blue light is scattered by small particles ten times as much as red light. Scattering of sunlight by atoms and molecules in the atmosphere is what makes the sky blue. It is the main cause of **atmospheric extinction**, making the Sun appear red at sunrise and sunset by preferentially scattering blue light out of the line of sight.

Schiaparelli, Giovanni Virginio (1835–1910) Italian astronomer. He was an assiduous observer of the inner planets, Mars in particular. He prepared a map of the surface of Mars, introducing a nomenclature for the various features, that remained standard until the planet was mapped by space probes. His use of the term *canali* (SEE **canals**) was the cause of much subsequent controversy. Schiaparelli discovered the connection between **meteor streams** and comets, and also studied double stars.

Schmidt, Maarten (1929–) Dutch–American astronomer. He is known for his work on **quasars**, in particular for his discovery in 1963 of the large **redshift** of the lines in the spectrum of the quasar designated 3C 273. This discovery, and his subsequent finding that the number of quasars increases with distance, provided important support for the big bang theory. Schmidt also ascertained the spiral structure of the Galaxy and the distribution of mass within it.

Schmidt camera (Schmidt telescope) A wide-field **catadioptric** telescope used mainly for photography, developed in 1930 by the Estonian instrument-maker and astronomer Bernhard Schmidt (1879–1935). Schmidt telescopes use a spherical primary mirror with a correcting plate, placed at the mirror's center of curvature, figured so as to minimize optical aberrations in the image. The image is formed on a curved focal surface in front of the mirror (necessitating the use of a special plate-holder to curve the photographic plate or film) and is sharp over a diameter of 10° or more.

The Schmidt camera and its numerous variations enabled whole-sky photographic surveys to be carried out, and significant advances to be made in photographic astronomy. Modifications of the Schmidt camera design that have made popular amateur telescopes have been the **Maksutov telescope** and, more recently, the **Schmidt-Cassegrain telescope**.

Schmidt-Cassegrain telescope (SCT) A short-focus telescope combining features of the **Schmidt camera** and the **Cassegrain telescope**. The SCT has the Schmidt's spherical primary mirror and specially figured correcting plate, but light is reflected back down the tube by a convex secondary mirror mounted behind the plate and then through a hole in the primary to a Cassegrain focus. This makes for a highly compact and portable telescope which, since the 1970s, has become very popular with amateur astronomers.

Schmidt telescope SEE **Schmidt camera**.

Schröter effect A difference between the theoretical and observed dates of dichotomy for Venus. At western (morning) elongation dichotomy comes a day or two late, while at eastern elongation it is a day or two early. The effect, the cause of which is unknown, was first observed in the 1790s by the German astronomer Johann Schröter (1745–1816).

Schwarzschild, Karl (1873–1916) German astronomer. He worked on methods of measuring stellar magnitudes from photographs, and studied stellar atmospheres and motions. In the last year of his life he solved equations in **Einstein**'s general theory of relativity that dealt with the gravitational field of a point mass, from which he developed the concepts of what are now called the **Schwarzschild radius** and the **black hole**. His son, Martin Schwarzschild (1912–97), became a naturalized American and an astrophysicist, working on stellar evolution.

Schwarzschild radius The critical radius at which a very massive body under the influence of its own gravitation becomes a **black hole**. It is the radius of the event horizon of a black hole, from which nothing can escape, not even light. The radius is given by the expression

$$R = 2GM/c^2 ,$$

where *G* is the gravitational constant, *M* is the mass of the body and *c* is the speed of light. The Schwarzschild radius for the Sun is about 3 km, and for the Earth it is about 1 cm. (After Karl **Schwarzschild**.)

scintillation The twinkling of the stars – rapid changes in brightness and color, particularly noticeable in stars low in the sky; through a telescope, the positions of stars are seen to undergo short, rapid changes. The effect is caused by the non-uniform density of the Earth's atmosphere, producing uneven refraction of starlight. In general, planets do not exhibit this phenomenon because the scintillations from different points on their surface are not in phase, and the fluctuations are lost in the general illumination. The inconvenience of the effects of scintillation is reduced by siting observatories at high altitudes and by the use of **adaptive optics**. It is almost entirely avoided by using telescopes carried by balloons, rockets or artificial satellites.

Radio sources also scintillate: the non-uniformity of the refractive indices of the interstellar medium, the interplanetary medium and the ionosphere causes the strength of radio waves to fluctuate.

SEE ALSO **seeing**.

Scorpius A constellation of the **zodiac**, in mythology representing the scorpion that killed Orion. Its long line of bright stars has the appearance of a scorpion and its sting, with its heart marked by first-magnitude **Antares**, the constellation's brightest star. The "sting" is Lambda (Shaula, magnitude 1.6). Both Mu and Zeta are optical doubles. In addition to being immersed in a very rich part of the Milky Way, Scorpius contains some magnificent clusters, notably the globular cluster M4 near Antares, and M6 and M7, open clusters both visible with the naked eye.

Scorpius X-1 The first cosmic X-ray source, discovered in 1962, and the brightest known apart from occasional transients (SEE **X-ray burster**). It is a low-mass X-ray binary with an orbital period of 19.2 hours; one component is a neutron star and the nature of the other is unknown. X-rays are thought to originate from the neutron star and from a thin **accretion disk** around it.

SCT ABBREVIATION FOR **Schmidt-Cassegrain telescope**.

Sculptor A barren southern constellation representing a sculptor's studio, introduced by **Lacaille**. It contains no star above magnitude 4.3, but is rich in faint galaxies, including a dwarf member of our Local Group.

Scutum A small constellation adjoining Aquila, representing a shield. Alpha, magnitude 3.9, is the only star above magnitude 4, but Scutum contains the splendid open cluster M11, nicknamed the **Wild Duck Cluster**, and is crossed by the Milky Way. Delta Scuti is the prototype of a class of variable stars that pulsate in periods of 0.01 to 0.2 days with very small amplitudes, usually only several hundredths of a magnitude.

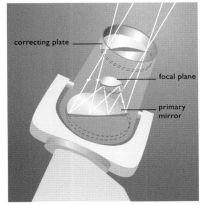

▼ *Schmidt camera* This type of reflecting telescope gives a wide field of view. They are used only for astrophotography and are useful in survey work.

correcting plate

focal plane

primary mirror

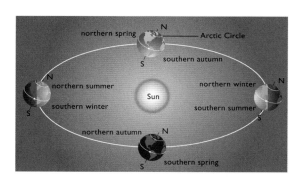

SDO ABBREVIATION FOR **Solar Dynamics Observatory**.

season One of a number of periods making up a cycle of changes in the surface conditions of a planet or satellite (average temperature and "weather") as its axial tilt gives different parts of its surface longer or shorter periods of sunlight over the course of one revolution around the Sun. The Earth's seasons are defined by the different positions of the Sun with respect to the ecliptic. In the northern hemisphere, spring is reckoned from the vernal equinox (21 March) to the summer solstice (21 June), summer from the summer solstice to the autumnal equinox (23 September), autumn from the autumnal equinox to the winter solstice (22 December) and winter from the winter solstice to the vernal equinox. Spring in the northern hemisphere corresponds to autumn in the southern hemisphere, summer to winter, autumn to spring, and winter to summer.

Of the other planets, Mars has a similar axial inclination to the Earth's (25° 11′ compared to 23° 27′), and shows seasonal variations such as changes in its **polar caps**.

Secchi, (Pietro) Angelo (1818–78) Italian astronomer and priest. He was a pioneer of spectroscopy, which he used to study the Sun and the stars. In the 1860s he carried out the first spectroscopic survey of stars, cataloging the spectra of over 4000 and dividing them into four classes according to their color and spectral characteristics (SEE **spectral classification**).

secondary (1) Any of the smaller members of a system of celestial bodies that orbit around the largest, the **primary** (1).

secondary (2) **(secondary mirror)** The small mirror in a **reflecting telescope** that diverts the converging beam of light from the primary mirror toward the eyepiece.

second contact In a total **solar eclipse**, the moment when the leading edge of the Moon's disk covers the last visible part of the Sun's disk, and totality begins. In a **lunar eclipse** it is the moment when the trailing edge of the Moon enters the umbra of the Earth's shadow. The term is also used for the corresponding stage in eclipses involving other bodies.

second of arc SEE **arc minute, second**.

secular acceleration A gradual long-term change in the orbital motion of one body around another. For example, tidal friction (SEE **tides**) between the Earth and the Moon is slowing the Earth's rotation and transferring angular momentum from the Earth to the Moon.

secular parallax The apparent displacement of a celestial body in the sky over time as a result of the Sun's motion in space.

Sedna A distant Solar System object discovered by Mike Brown, Chad Trujillo and David Rabinowitz in November 2003. Sedna's orbit is highly elliptical (eccentricity 0.85) with a period of around 12,500 years. Its perihelion is 76 au, well over twice the distance of Neptune, while its aphelion takes it out to around 1000 au, the farthest edge of the **Kuiper Belt**. How Sedna arrived in such an unusual orbit is unknown, but it could have been scattered inward from the **Oort Cloud**. Sedna's diameter is estimated to be about 1000 km (600 miles). It has been assigned the minor planet number 90377.

seeing The quality of the image produced by a telescope, as affected by the steadiness of the air. Seeing depends on the calmness of the atmosphere immediately overhead, and also on air movements at ground level, and even in the telescope tube itself. The evaluation of seeing is largely subjective, though values assigned according to

the **Antoniadi scale** can help achieve a degree of uniformity in the reporting of conditions under which observations are made.

seleno- Prefix referring to the Moon, as in selenography, the description and mapping of the Moon's surface features. (From *Selēnē*, Greek for "Moon.")

semimajor axis (symbol *a*) Half the longest diameter of an **ellipse**. The semimajor axis of an elliptical **orbit** is the average distance of an object from its primary, and is one of the elements used when defining an orbit.

semiregular variable A pulsating giant or supergiant star, usually of late spectral type (typically M or C), with a period between 20 and 2000 days or more and an amplitude from a few hundredths of a magnitude to 1–2 magnitudes. Some semiregular variables show a definite periodicity, interrupted at times by irregularities. Others differ little from long-period variables except in having smaller amplitudes (less than 2.5 magnitudes). The various subtypes exhibit periodicities that cover a wide range from about 20 days to several thousand days.

separation The angular distance between two celestial objects, in particular the members of a visual binary or multiple star system.

Serpens A constellation in two parts, Caput (the head) and Cauda (the body), separated by the serpent-bearer Ophiuchus; Serpens represents a snake coiled around Ophiuchus. Alpha (Unukalhai), its brightest star, is magnitude 2.6. Theta is a wide, easy pair with components of magnitudes 4.6 and 5.0. Delta is also an easy telescopic double, magnitudes 4.2 and 5.2. The Mira-type variable R Serpentis can reach magnitude 5.2 (minimum 14.4, period 356 days). There are two bright Messier objects: M5, a globular cluster, and M16, an open cluster embedded in the **Eagle Nebula**.

Setebos A small outer **satellite** of Uranus, discovered in 1999.

SETI Abbreviation for "search for extraterrestrial intelligence," a term used for a variety of projects seeking signs of intelligent life elsewhere in the Universe. The first was Project Ozma, which in 1960 began to search for radio signals. None has met with success.

Seven Sisters Popular name for the **Pleiades**, an open star cluster in Taurus.

Sextans An obscure constellation introduced by **Hevelius** between Leo and Hydra representing a sextant. Its brightest star is only magnitude 4.5.

Seyfert galaxy A class of galaxies that have extremely bright, compact **active galactic nuclei** and whose spectra show strong emission lines. About 1% of all galaxies are Seyferts; M77 (NGC 1068) is one of the brightest. Seyferts emit strongly at ultraviolet and infrared wavelengths, and exhibit a degree of short-term variability. Some are strong X-ray sources but few are particularly strong radio emitters. Most are spirals. The gas clouds in Seyfert galaxies move at several thousand kilometers per second, probably because of the gravitational influence of a massive black hole at the nucleus. They are named after American astrophysicist Carl Seyfert (1911–60), who first studied them.

▲ **Seyfert galaxy** *M77 (NGC 1068) is the prototype type 2 Seyfert galaxy. Its spectrum indicates that there is high-velocity gas in its inner regions. Visually, Seyfert galaxies are characterized by their bright, small nuclei.*

shadow bands Gray ripples seen moving across light-colored surfaces in the minutes either side of a total eclipse of the Sun. Close to totality the bands are more ordered and show higher contrast. They are caused by turbulence in the atmosphere affecting light from the very thin crescent of the Sun, in the same way that it affects starlight to produce **scintillation**.

Shapley, Harlow (1885–1972) American astronomer. In 1914 he explained how the variability of **Cepheid variables** is caused by their pulsations. In the years that followed he calibrated the period–luminosity law discovered by Henrietta **Leavitt**, using Cepheid variables in globular clusters, which enabled him to make the first accurate estimate of the size of the Galaxy and the Sun's position in it. He later cataloged many thousands of galaxies, and discovered that galaxies are arranged in clusters.

shell star A hot B-type star with prominent absorption lines in its spectrum, from a shell of material surrounding the star. Some shell stars also exhibit an equatorial disk thrown out by the star's rapid rotation. Strong shell spectra, which include emission lines from the surrounding disk, are seen in B0e to B3e giants and dwarfs. Examples are Gamma Cassiopeiae, BU Tauri (Pleione) and 48 Librae (FX Librae).

shepherd moon A minor moon that through its gravitational influence keeps in check the particles in a planetary ring. Shepherd moons often act in pairs. Examples are Prometheus and Pandora, orbiting either side of Saturn's F Ring, and Cordelia and Ophelia, orbiting either side of Uranus's Epsilon Ring.

Shoemaker–Levy 9, Comet D/1993 F2 A **comet** discovered in 1993 jointly by Eugene and Carolyn Shoemaker, David Levy and Philippe Bendjoya. It was in a temporary orbit around Jupiter, and a close approach to the planet in 1992 had disrupted its nucleus into more than 20 separate fragments. Over a six-day period in July 1994 these fragments collided with Jupiter, creating fireballs and leaving temporary dark scars at the impact sites. This was a unique event in the history of observational astronomy, observed worldwide by amateur and professional astronomers.

shooting star A popular (chiefly British) name for a **meteor**.

short-period comet A **comet** whose apparitions are sufficiently frequent to permit correlation of orbital data; formally defined as one whose period is less than 200 years.

Siarnaq A small outer **satellite** of Saturn, discovered in 2000.

sidereal Of or pertaining to the stars; measured or determined with reference to the stars.

sidereal day The interval between two successive passages of a given star, or the vernal equinox, across the observer's meridian. It is equal to 23h 56m 4.091s of mean solar time.

sidereal month The time taken for one revolution of the Moon around the Earth relative to a fixed star. The sidereal month is equal to 27.32166 mean solar days.

sidereal period The orbital period of a planet or other celestial body with respect to a background star. It is the true orbital period. COMPARE **synodic period**.

sidereal time Local time reckoned according to the rotation of the Earth with respect to the stars. The time is 0h when the vernal equinox crosses the observer's meridian. The sidereal day is 23h 56m 4s of mean solar time – nearly 4 minutes shorter than the mean solar day. Sidereal time is equal to the **right ascension** of an object on the observer's meridian.

sidereal year The time required for the Earth to complete one revolution around the Sun relative to the stars. It is equal to 365.25636 mean solar days.

siderostat A mirror mounted equatorially and driven so as to counteract the apparent movement of a star or other celestial object and direct its light into a fixed telescope. A more sophisticated version is the **coelostat**.

Siding Spring Observatory An astronomical observatory in the Warrumbungle mountains of New South Wales, founded in 1962 by the Australian National University, which operates several tele-

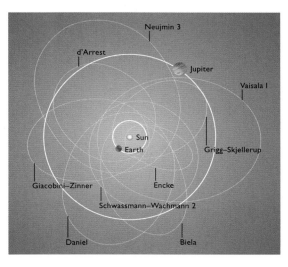

scopes there including the 2.3 m (90-inch) Advanced Technology Telescope opened in 1984. The site is shared by the **Australian Astronomical Observatory**.

singularity A point at which the known laws of physics break down. For example, the singularity at the center of a **black hole** is a point at which matter is compressed into an infinitely small volume and so has an infinitely large density. In the theory of the **big bang**, the Universe was born from a singularity of infinite density and temperature.

Sinope An outer **satellite** of Jupiter discovered in 1914 by Seth Nicholson. It is in a **retrograde** (1) orbit and may be a captured asteroid.

Sirius The star Alpha Canis Majoris. It is the brightest star in the sky, magnitude −1.44, distance 8.6 light years, luminosity 22 times that of the Sun. Sirius is a main-sequence star of spectral type A0, and is the sixth closest star system to us. It has a binary companion, Sirius B, whose existence was deduced by Friedrich **Bessel** in 1844 from its gravitational perturbations on Sirius, which affected the proper motion of Sirius. The companion, magnitude 8.4, was discovered optically by Alvan G. Clark in 1862. Sirius B is a **white dwarf** of 1.03 solar masses, but with a diameter only 0.92 times that of the Earth. Its orbital period is 50 years. It was the first star to be recognized as a white dwarf.

Sixty-one Cygni (61 Cygni) The first star to have its parallax measured (by Friedrich **Bessel** in 1838). It lies 11.4 light years away and is a binary consisting of two K-type main-sequence stars, magnitudes 5.2 and 6.0, with a period of 660 years.

SKA ABBREVIATION FOR **Square Kilometer Array**.

Slipher, Vesto Melvin (1875–1969) American astronomer. On the death of Percival **Lowell** he became director of the Lowell Observatory at Flagstaff, Arizona. He was a specialist in astronomical spectroscopy, especially of galaxies. In 1912 he became the first to obtain the spectrum of another galaxy, the Andromeda Galaxy, and in 1920 he discovered the **redshifts** in galactic spectra. By 1925, following Edwin **Hubble**'s finding that the redshifts of galaxies increase with distance, he had measured the radial velocities of more than 40 galaxies. Slipher also showed that light from certain nebulae was reflected from stars embedded in them, thus discovering reflection nebulae.

Sloan Digital Sky Survey (SDSS) An imaging and spectroscopic survey conducted over an eight-year period with a wide-field 2.5 m (100-inch) reflector at **Apache Point Observatory**. The telescope can image 1.5 square degrees of sky and measure the spectra of over 600 galaxies and quasars in a single observation. The survey, completed in 2008, produced spectra of 930,000 galaxies, 120,000 quasars and 460,000 stars covering one-quarter of the entire sky. Specialized follow-up surveys were then undertaken. It is named after Alfred Pritchard Sloan, Jr (1875–1966), whose foundation financed the project.

Small Magellanic Cloud SEE **Magellanic Clouds**.

◄ **short-period comet** Most short-period comets are thought to have originated in the Kuiper Belt and to have been perturbed into the inner Solar System by the gravitational influence of the gas giants. Encke has an orbital period of only 3.3 years, Grigg–Skjellerup 5.1 years and Giacobini–Zinner 6.5 years. Biela's Comet suffered the probable fate of all short-period comets: on one orbit, its nucleus was seen to have broken into at least two pieces and it never reappeared.

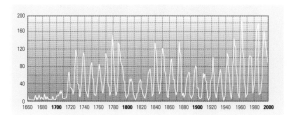

SNC meteorites A group of achondritic meteorites (SEE **achondrites**) believed to have originated on Mars. Their ages (mostly 1.3 billion years or less) indicate that they originated on a sizable planet that has been geologically active, and gases trapped in them are consistent with analyses of the Martian atmosphere made by spacecraft. It is unclear how they could have emerged as intact fragments from an impact of sufficient force to accelerate them to Mars' escape velocity. The name (pronounced "snick") comes from the three subtypes: *shergottites*, *nakhlites* and *chassignites*.

SNR ABBREVIATION FOR **supernova remnant**.

SOFIA ABBREVIATION FOR **Stratospheric Observatory for Infrared Astronomy**.

SOHO ABBREVIATION FOR **Solar and Heliospheric Observatory**.

Solar and Heliospheric Observatory (SOHO) A European Space Agency probe, operated jointly with NASA, to study the corona of the Sun, oscillations of the Sun's surface and the solar wind. It was launched on 2 December 1995, and placed at the L_1 **Lagrangian point**, 1.5 million km (900,000 miles) from the Earth in the direction of the Sun, in March 1996. A main aim of the mission is to probe the Sun's interior structure. SOHO discovered many **sungrazer** comets.

solar apex, antapex SEE **apex**.

solar constant A measure of the amount of solar energy received by a body a certain distance from the Sun. For the Earth, the solar constant is defined as the solar power received per unit area, at the top of the atmosphere, at the average Earth–Sun distance of 1 au; its value is about 1.35 kW/m². The "constant" varies from day to day with sunspot activity, and also in the longer term with the **solar cycle**.

solar cycle The periodic fluctuation in the number of **sunspots** and the levels of other kinds of solar activity; the cycle lasts about 11 years. Over the course of a cycle, sunspots vary both in number and in latitude in a way which is neatly illustrated in the **butterfly diagram**. The variation in latitude is described by **Spörer's law**. The diagram here shows the variations in sunspot activity going back to the 17th century, as reckoned by the **relative sunspot number**. At *solar maximum*, when sunspot numbers are greatest, astronomers refer to the *active Sun*; *solar minimum* is also called the *quiet Sun*. There is some variation in the level of activity at solar maximum, and in the period up to 1715 – during the **Maunder minimum** – activity was low for about 70 years.

solar day SEE **day**.

Solar Dynamics Observatory (SDO) A NASA satellite launched on 11 February 2010 which observes the Sun from geostationary orbit around the Earth. SDO's instruments take high-resolution images of the Sun in ten wavelengths every 10 seconds to study solar activity and variability.

solar eclipse An **eclipse** of the Sun by the Moon. Since the Moon's orbital plane is inclined to the plane of the ecliptic, a solar eclipse can occur only when the Moon is at conjunction (i.e. at new Moon) and at the same time is at or near one of its **nodes**; the Sun, Moon and Earth are then very nearly in a straight line. In addition, the Sun's angular distance from one of the Moon's nodes at conjunction will determine whether an eclipse can or cannot occur (SEE **ecliptic limits**).

The apparent diameters of the Sun and the Moon as seen in the sky are almost the same, but they do vary slightly, particularly the Moon's, whose maximum and minimum distances from the Earth differ by about 10%.

A *total solar eclipse* is seen at places where the umbra of the Moon's shadow-cone falls on and moves over the Earth's surface; at the same time the eclipse will appear *partial* to observers on either side of the central track of totality. Shortly before and after totality the phenomena of **shadow bands** and **Baily's beads** are observable. During the brief period of totality, the **corona** and any large **prominences** become visible. When the Moon is near apogee, the tip of its shadow-cone does not reach to the Earth and there is an *annular eclipse*, in which a rim of light is seen around the darkened disk of the Moon.

The overall duration of a solar eclipse from **first contact** to **fourth contact** can be as much as 4 hours; totality, from **second contact** to **third contact**, lasts at most 7½ minutes. There are from two to five solar eclipses each year; if there are five, they will all be partial. The total number of solar and lunar eclipses in a year varies from two to seven; if there are only two, they will both be solar.

solar mass The mass of the Sun, used as a unit of mass for other stars and celestial objects.

Solar Orbiter A European Space Agency probe due for launch in 2018. It will be placed into an elliptical orbit ranging between 0.9 and 0.3 au from the Sun. Gravity-assist passes of Venus will increase its inclination to the Sun's equator to 25° or more during the mission, affording it a better view of higher latitudes of the Sun than is possible from Earth.

solar parallax SEE **parallax**.

Solar System The Sun and all the celestial bodies which revolve around it: the eight **planets**, together with their **satellites** and ring systems, the **dwarf planets**, countless **asteroids** and **comets**, meteoroids and other interplanetary material. The boundaries of the Solar System extend beyond the orbit of Neptune, which orbits the Sun at an average distance of 30 au, to include the **Kuiper Belt**, which may extend to 1000 au, and the inferred **Oort Cloud** of comets, some of which may have aphelion distances of as much as 100,000 au (1½ light years). Interplanetary space contains cosmic dust and extremely tenuous ionized gas (particles of the **solar wind**); SEE **interplanetary matter**.

All the planets have direct orbits: they revolve around the Sun in the same direction as the Sun itself rotates; the orbital motion of most of the satellites is direct too. The Sun is by far the most massive component of the Solar System, its mass being 330,000 times that of the Earth. The eight planets have a total mass equal to 448 times that of the Earth, of which Jupiter accounts for 70%.

The Solar System came into being nearly 5000 million years ago as the end-product of a contracting cloud of interstellar gas and dust, the *solar nebula* (SEE **cosmogony**). Similar systems centered on other stars, referred to as "solar systems," without the capital Ss (SEE **extrasolar planet**), and the existence of disks of matter surrounding stars (SEE **protoplanetary disk**), suggest that planets are byproducts of star formation.

solar wind The steady flow of charged particles (mainly protons and electrons) from the solar corona into interplanetary space, controlled by the Sun's magnetic field. Solar wind particles are too energetic to be held back by the Sun's gravity. Some of them get trapped in planetary magnetic fields. At the Earth, some are trapped in the outer **Van Allen Belt**; others reach the Earth's upper atmosphere in oval regions surrounding the magnetic poles and cause **aurorae**. The solar wind carries away about 10^{-13} of the Sun's mass per year. Its intensity increases during periods of high solar activity. SEE ALSO **stellar wind**.

solstice The extreme northern or southern position of the Sun in its yearly path among the stars. At the solstices the Sun reaches its greatest declination, 23½°N or 23½°S. The *summer solstice* occurs around 21 June and the *winter solstice* is on or about 22 December.

solstitial colure The **great circle** that passes through the north and south celestial poles and the winter and summer **solstices**.

Sombrero Galaxy A spiral galaxy in the constellation Virgo, designated M104 or NGC 4594. The galaxy is seen nearly edge-on, and has the appearance of a sombrero hat. It is crossed by a thick dark lane of interstellar dust, and there is an unusually large central bulge containing numerous globular clusters. It lies about 30 million light years away.

South African Astronomical Observatory An observatory at Sutherland in the Karoo semidesert northeast of Cape Town, founded in 1972. Its main instrument is a 1.88 m (74-inch) reflector originally at the Radcliffe Observatory, Pretoria. On the same

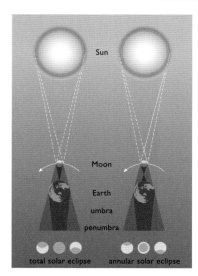

site is the 11 m (36 ft) Southern African Large Telescope (SALT), the largest single-mirror instrument in the southern hemisphere, opened in 2005. SALT, built in partnership with Poland, New Zealand, and universities in Germany and the USA, is a close copy of the **Hobby–Eberly Telescope**.

Southern Cross The popular name for the southern-hemisphere constellation **Crux**.

Southern Pleiades A bright open cluster in the southern constellation Carina, designated IC 2602. The third-magnitude star Theta Carinae is among its members.

space probe An uncrewed space vehicle sent to investigate the Moon, planets or interplanetary space.

spacetime A unified dimensional framework in which the three dimensions of space (length, breadth and height) and the dimension of time are considered together. An event in spacetime is specified by the three space coordinates and the time coordinate. The concept arose from **Einstein**'s special and general theories of relativity. In the general theory, gravity is a distortion of spacetime by matter.

space velocity The true velocity of a star in space with respect to the Sun. It is the hypotenuse of the right-angled triangle formed by its **tangential velocity** (obtained from observation of its proper motion) and its **radial velocity** (obtained by measurement of the Doppler shift in its spectrum).

Spacewatch A project to monitor the sky for **near-Earth asteroids** (NEAs), using the 0.9 m (36-inch) and 1.8 m (72-inch) telescopes at **Steward Observatory**. With most NEAs there is no risk of collision, but several hundred come close enough to be classified as *potentially hazardous asteroids* (PHAs). The Spacewatch project has discovered several of these objects, as well as many NEAs and main-belt asteroids.

speckle interferometer SEE **optical interferometer**.

spectral classification The categorization of stars based on the characteristics of their spectra. In the 1860s Angelo **Secchi** made the first spectral classification of stars by dividing them into four groups according to color and spectral lines. As spectroscopy improved, a more comprehensive system was developed at Harvard College Observatory. This was embodied in the **Henry Draper Catalog** of stellar spectra, published in 1918–24. The sequence of spectral types was arranged according to the prominence or absence of certain lines in the spectra. There were seven spectral types, labeled O, B, A, F, G, K and M (an order resulting from revisions of an earlier alphabetical sequence). This represents a temperature sequence, from the hottest stars, types O and B, which appear blue-white, to type M, the coolest stars, which appear orange-red. As a result of this ordering, O, B and A stars are called **early-type stars**; K and M stars are called **late-type stars**. Three new types were added when it was found that some cool stars had strong absorption bands not usually seen in other stars of the same color. These were classes R and N, with strong bands of molecular carbon, and class S, with bands of zirconium oxide.

Stellar spectra can be classified into even finer divisions within these seven types, so decimal subdivisions were introduced. G5, for example, indicates a star midway in type between G0 and K0. Further refinements include the use of additional letters as suffixes to the spectral type, giving more information about the star, for example the existence of emission lines (e), metallic lines (m), broad lines due to rotation (n and nn) or a peculiar spectrum (p).

However, the Harvard system could not deal with stars of different luminosities at a given temperature (that is, dwarfs, giants and supergiants). In 1943 William Morgan, Philip Keenan and Edith Kellman of Yerkes Observatory redefined the spectral types and added a classification scheme for the luminosity (absolute magnitude) of stars. This is now known as the **Morgan–Keenan system**, shortened to *MK system*, and is now used universally. For example, a star classified in the MK system as O9.5 IV–V has a spectral type (and therefore a temperature) midway between that of an O9 and a B0 star, and a luminosity between that of a dwarf and a subgiant. The spectral types and luminosity classes of the MK system are listed in the tables above right.

The MK system is applicable to stars of normal chemical composition, which is about 95% of all stars. The various types of peculiar star are given their own special classification schemes. The Harvard R and N types are now combined into one **carbon star** class, the designations for which include a temperature type and a carbon

SPECTRAL CLASSIFICATION: LUMINOSITY CLASSES	
Ia	Supergiants of high luminosity
Ib	Supergiants of lower luminosity
II	Bright giants
III	Normal giants
IV	Subgiants
V	Main-sequence stars (dwarfs)
VI	Subdwarfs

band strength, as for example in C2,4. A similar classification is used for the **S stars**. A class of **L stars** has been added for cool stars emitting in the infrared, and, most recently, a class of **T stars** for the coolest brown dwarfs. White dwarfs are usually classified on a Harvard-type scheme, with D preceding the type.

spectral type A series of divisions, indicating surface temperature and hence color, into which stars are classified according to the nature of their spectrum. SEE **spectral classification, A star, B star, carbon star, F star, G star, K star, L star, M star, O star, S star, T star**.

spectrogram A permanent record of the spectrum of an object, produced by a **spectrograph**.

spectrograph A **spectroscope**, fitted with a camera or an electronic detector such as a CCD, used to obtain a permanent record of the spectrum of a celestial object. Astronomical spectrographs are generally designed for use with a specific telescope. On telescopes with a **coudé** system or **Nasmyth focus** a large spectrograph can be mounted in a permanent position. Spectrographs generally operate in a band of wavelengths from the near-infrared to the near-ultraviolet. There are various designs for different purposes. *High-dispersion* instruments such as the **spectroheliograph** spread the spectral lines widely so that a narrow band of wavelengths can be studied in detail.

spectroheliograph A type of **spectrograph** used to image the Sun in the light of one particular wavelength only. A second slit placed in front of the photographic plate or detector images a narrow strip of the Sun's disk at a chosen wavelength. By moving the entrance slit and the second slit in tandem, the whole disk is scanned and an image of it, a *spectroheliogram*, is built up. The spectroheliograph was invented independently by George Ellery **Hale** and Henri Deslandres in the 1890s. The term is also used for a number of modern instruments of different design but working on the same principle.

spectrohelioscope A **spectroheliograph** adapted for visual use. The two slits are oscillated rapidly enough for the motion to be undetectable, so that by persistent vision the observer sees a steady, monochromatic image of the Sun's disk.

spectrometer A **spectroscope** equipped to measure accurately the positions and intensities of spectral lines, by means of a device such as a **Fabry–Pérot interferometer**. Spectrometers are also loosely referred to as **spectrographs**. Since spectra are now routinely recorded on CCDs, from which information in digital form can be

SPECTRAL CLASSIFICATION: MAIN SPECTRAL TYPES		
Type	Surface	Main features temperature (K)
O	50,000–30,000	Blue-white stars. Lines of ionized helium. Weak hydrogen lines.
B	30,000–10,000	Blue-white stars. Neutral helium lines, with hydrogen lines strengthening from B6 to B9.
A	9900–7500	White stars. Very strong hydrogen lines at A0, decreasing toward A9. Lines of ionized calcium increase in strength from A0 to A9.
F	7400–6100	Yellow-white stars. Ionized calcium continuing to increase in strength, and hydrogen weakening. Lines of other elements begin to strengthen.
G	6000–4600	Yellow stars. Strong lines of calcium, hydrogen weaker. Lines of iron become prominent.
K	4900–3250	Orange stars. Strong metallic lines. Molecular bands of CH and CN become prominent.
M	3500–2500	Orange-red stars. Strong absorption bands of titanium oxide and large numbers of metallic lines.

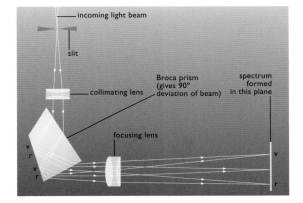

◀ *spectroscope* Shown here is one type of traditional spectroscope, using a prism and lenses. The study of spectra, known as spectroscopy, is the key technique by which the physical properties of astronomical bodies are revealed.

▶ *spectroscopic binary*
Redshift is used to ascertain
the relative motions of stars
in close binary pairs.

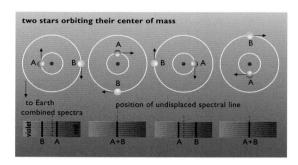

two stars orbiting their center of mass

to Earth
combined spectra

position of undisplaced spectral line

violet

B A   A+B   A B   A+B

analyzed by computer, the distinction between modern spectrometers and spectrographs lies in their application rather than their design.

spectrophotometer An instrument that scans the lines in a spectrum recorded by a spectrograph and measures their intensities. It outputs a graph of intensity against wavelength called a *line profile*.

spectroscope An instrument for producing spectra for the analysis of electromagnetic radiation (light or other wavelengths). *Spectroscopy* is the use of such an instrument to probe the chemical composition and physical conditions (such as temperature and motion) of an object. As such it is the most important investigative technique available to astronomers.

In the general form of the instrument, a focused beam of light or other radiation is passed through a narrow slit, to prevent images of different wavelengths from overlapping and to ensure a sharply defined spectrum, and collimated to produce a parallel beam. This beam is then split into its component wavelengths by a prism or – in most modern instruments – by a **diffraction grating**. After dispersion, the radiation is then focused on to the detector. In its original form, lenses were used for collimation and focusing, but modern equipment normally employs concave mirrors instead. In some instruments the dispersing element consists of a combined grating and prism (known as a *grism*).

All dispersing elements spread out the radiation into various *orders*; the overall dispersion varies between orders. Gratings can be designed to send most of the radiation into a specific chosen order. Particularly high spectral resolution over a wide range of wavelengths can be obtained by the use of a form known as an *echelle grating*, together with a *cross-disperser* (the two often combined in a grism), which separates the different orders and prevents them from overlapping.

Astronomical spectroscopes are known as **spectrographs** or **spectrometers**. Strictly, a spectrograph is a spectroscope equipped with a camera for recording a permanent record of a spectrum, whereas a spectrometer incorporates devices for accurately measuring the wavelengths and intensities of the spectral lines. In modern devices the detector is usually a **CCD**.

spectroscopic binary A **binary star** system whose two components are too close for them to be resolved visually as separate objects, but whose binary nature can be deduced from the periodic **Doppler shift** of the absorption lines in the combined spectrum. As the diagram above shows, when the two stars have reached positions in their orbits around each other where one is moving away from us and the other toward us, the spectral lines of the one are slightly redshifted, and those of the other slightly blueshifted, from their mean positions.

spectroscopic parallax A method of determining the distance of a star from its spectral type (SEE **spectral classification**) and apparent magnitude. Analysis of a star's spectrum reveals its spectral type. If the absolute magnitude of a star of that type is known, the star's distance can be calculated by comparing its absolute and apparent magnitudes. This is the commonest method of determining stellar distances. However, the scale must be calibrated against nearby stars whose distances can be measured directly by **trigonometric parallax**.

spectrum (plural **spectra**) The distribution of intensity of **electromagnetic radiation** with wavelength. When a beam of light from a source is deflected by a prism, or by another dispersing agent such as a diffraction grating, the radiation is fanned out into a spectrum because different wavelengths are deviated by different degrees.

There are various types of spectra. A *continuous spectrum* is an unbroken distribution of radiation over a broad range of wavelengths. White light, for example, is split into a continuous "rainbow" band of colors from red to violet. A *line spectrum* contains lines corresponding to only certain wavelengths or frequencies which are characteristic of the chemical elements present in the source. An *emission line*

spectrum contains lines emitted at particular wavelengths and is produced by substances in an incandescent state at low pressure. An *absorption line spectrum* contains dark lines at particular wavelengths superimposed on a continuous spectrum. It is produced when radiation from a hot source which is emitting a continuous spectrum passes through a layer of cooler gaseous material. The spectra of most stars are continuous spectra crossed by absorption lines, and sometimes by emission lines. Emission line spectra are typical of luminous nebulae. *Band spectra* contain absorption bands, wider than the spectral lines of atoms, and these are the spectral signatures of molecules.

The intensity of an emission line in a spectrum is a measure of the energy of the radiation at that wavelength. The width of the line indicates the wavelength band over which it is distributed. The intensity of the emission is greatest at the center of the line and decreases toward the edges; the reverse is true for an absorption line. The pressure of the gas producing the spectral line can affect its width; this effect is known as *pressure broadening*. SEE ALSO **flash spectrum, hydrogen spectrum, Zeeman effect.**

speculum A reflective alloy used for making early telescope mirrors. The abbreviation "spec." is occasionally used to refer to a telescope mirror, as in 6-inch spec.

spherical aberration An **aberration** (2) that results from the unequal deflection of light rays by different zones of a spherical lens or mirror. The rays are not brought to the same focus, causing a lack of definition about the edges of the image. Spherical aberration in lenses can be minimized by using a suitable combination of elements, and in reflecting telescopes it is overcome by the use of a **paraboloidal** mirror.

spheroidal Having the form of the surface formed by rotating an ellipse about one of its axes. If the minor axis is chosen, the result is an *oblate spheroid*. This is the shape of fluid bodies – stars and gas planets such as Jupiter – rotating at a sufficient speed, and also of solid but non-rigid bodies such as the Earth. Components of optical systems can also be spheroidal.

Spica The star Alpha Virginis, magnitude 0.98, distance 262 light years, luminosity 2200 times that of the Sun. It is a main-sequence star of type B1 and a spectroscopic binary.

spicule A narrow jet, like a tiny prominence, seen protruding from the Sun's chromosphere when it is viewed edge-on. Spicules are short-lived (about 5–10 minutes) and congregate at the edges of **supergranulation** cells. They are possibly caused by chromospheric gases shooting into the lower corona.

spiral galaxy A **galaxy** consisting of a nucleus of stars from which spiral arms emerge, winding around the nucleus and forming a flattened, disk-shaped region. The arms contain gas, dust and young stars, while the nucleus contains old stars.

Spirit SEE **Mars Exploration Rover.**

Spitzer Space Telescope A spaceborne telescope developed by NASA, formerly known as the Space Infrared Telescope Facility (SIRTF), launched on 25 August 2003 to study the sky at infrared

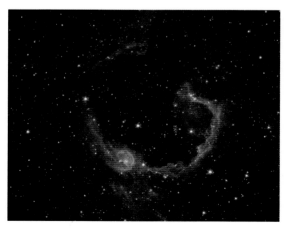

▲ *Spitzer Space Telescope* This image of a star-forming "bubble" of gas and dust (RCW 79) was captured by the Spitzer Telescope.

wavelengths between 3 and 180 μm with its 0.85 m (33-inch) telescope. It was placed into a solar orbit trailing behind the Earth.

sporadic meteor A meteor that is not part of a recognized meteor shower – one whose path cannot be traced back to a known **radiant**. The Earth is constantly sweeping up meteoroids, and 3–8 sporadic meteors should be visible each hour under ideal conditions. SEE ALSO **radar astronomy**.

Spörer's law The appearance of sunspots at lower latitudes over the course of the 11-year **solar cycle**. The average position drifts from midlatitudes (about 30–40°) toward the equator as the cycle progresses from one minimum to the next. The phenomenon was first studied in detail by the German astronomer Gustav Spörer (1822–95), after whom it was later named. Despite it being called a "law," it has no mathematical formulation. SEE ALSO **butterfly diagram**.

spring equinox SEE **vernal equinox**.

spring tide SEE **tides**.

Square Kilometer Array (SKA) A multi-national project to build the world's largest radio telescope with an effective collecting area of 1 sq km (0.4 sq mile) spread across two continents. It will consist of several thousand dishes of 15 m (49 ft) aperture and many more fixed panels known as aperture arrays. The SKA will be divided into two main sections, in South Africa and Australia, and will come into operation in phases. Final completion is expected by 2030.

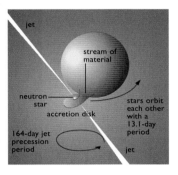

▲ **SS433** Shown here is a theoretical model of the unusual binary pair SS433, which is thought to consist of a hot, massive star orbited by a neutron star.

SS Cygni star A type of **U Geminorum star**.

SS433 An unusual binary star, number 433 in a catalog of stars with bright hydrogen emission lines compiled by American astronomers Bruce Stephenson and Nicholas Sanduleak. SS433 lies at the center of a supernova remnant in Aquila called W50. It is notable for the variable **Doppler shift** in certain emission lines of its spectrum and for having lines that are both red- and blueshifted. SS433 is thought to consist of a hot, massive star orbited every 13.1 days by a neutron star (presumably formed by the supernova that created W50). A stellar wind from the hot star, combined with orbital motion, produces the first set of emission lines, with variable Doppler shifts. Matter is transferred from the massive star to the neutron star via an **accretion disk**, but some of this material is ejected at high speed along two polar jets. The jets precess with a period of 164 days, producing the second set of variable lines with simultaneous red- and blueshifts.

S star A giant star whose surface temperature lies in the range for M stars, but whose spectrum contains absorption bands of zirconium oxide as well as the titanium oxide bands that are characteristic of M stars. The overabundance of zirconium, and also of carbon and many other heavy elements, is a result of convective mixing which brings to the surface the star products from nuclear reactions in the interior. As a result, many S stars show spectral lines of the radioactive element technetium. M stars with weak zirconium oxide bands are assigned the intermediate spectral type MS. S stars often show hydrogen emission lines in their spectra, and many are **long-period variables**.

standard epoch A fixed point in time, or **epoch**, to which star positions are referred in order to remove the effects of precession, proper motion and perturbation, together with a reference system in which the positions are measured. The standard epoch currently used for catalogs and star charts is designated 2000.0, meaning that all coordinates quoted are correct on noon 1 January 2000.

star A self-luminous ball of gas whose radiant energy is produced by nuclear reactions, mainly the conversion of hydrogen into helium. The temperatures and luminosities of stars depend on their masses (according to the **mass–luminosity relation**). The most massive known stars are about 130 solar masses (i.e. 130 times more massive than the Sun). Above about 150 solar masses a star is theoretically unstable and is likely to break up. Large stars can be very luminous and hot, and therefore appear blue. Medium-sized stars such as the Sun are yellow, while small stars are dull red. The smallest stars are about 0.08 solar masses; below this the temperature cannot become high enough for nuclear reactions to take place, and the object is instead a **brown dwarf**. The colors, and hence the temperatures, of stars give rise to the scheme of **spectral classification**. Most stars are born in pairs or groups (SEE **binary star**, **star cluster**). SEE ALSO **stellar evolution**, **stellar populations**.

STARS: THE NEAREST (CLOSER THAN 12 LIGHT YEARS)				
Star	Apparent magnitude	Spectral type	Absolute magnitude	Distance (light years)
Sun	226.78	G2	4.82	—
Proxima Centauri	11.01 v	M5	15.45	4.22
α Centauri A	20.01	G2	4.34	4.39
α Centauri B	1.35	K1	5.70	4.39
Barnard's Star	9.54	M4	13.24	5.94
Wolf 359	13.44 v	M6	16.50	7.8
Lalande 21185	7.49	M2	10.46	8.31
Sirius A	21.44	A0	1.45	8.60
Sirius B	8.44	DA2	11.34	8.60
UV Ceti A	12.54 v	M5.5	15.40	8.7
UV Ceti B	12.99 v	M6	15.85	8.7
Ross 154	10.37	M3.5	13.00	9.69
Ross 248	12.29	M5.5	14.79	10.3
ε Eridani	3.72	K2	6.18	10.50
HD 217987	7.35	M2	9.76	10.73
Ross 128	11.12 v	M4.5	13.50	10.89
L789–6	13.33	M5	15.64	11.2
61 Cygni A	5.20 v	K5	7.49	11.36
Procyon A	0.40	F5	2.68	11.41
Procyon B	10.7	DF	13.0	11.41
61 Cygni B	6.05 v	K7	8.33	11.43
HD 173740	9.70	M4	11.97	11.47
HD 173739	8.94	M3.5	11.18	11.64
GX Andromedae A	8.09 v	M1	10.33	11.64
GX Andromedae B	11.10	M4	13.35	11.64
G51–15	14.78	M6.5	16.98	11.8
ε Indi	4.69	K5	6.89	11.83
τ Ceti	3.49	G8	5.68	11.90
L372–58	13.03	M4.5	15.21	11.9

v variable

STARS: THE BRIGHTEST (FIRST MAGNITUDE OR BRIGHTER)				
Star	Name	Apparent magnitude	Spectral type	Distance (light years)
	Sun	226.78	G2	—
α Canis Majoris	Sirius	21.44	A0	8.60
α Carinae	Canopus	20.62	A9	313
α Centauri	Rigil Kentaurus	20.28 **	G2 I K1	4.39
α Boötis	Arcturus	20.05 v	K2	36.7
α Lyrae	Vega	0.03 v	A0	25.3
α Aurigae	Capella	0.08 v	G6 I G6	42.2
β Orionis	Rigel	0.18 v	B8	773
α Canis Minoris	Procyon	0.40	F5	11.4
α Eridani	Achernar	0.45 v	B3	144
α Orionis	Betelgeuse	0.45 v	M2	427
β Centauri	Hadar	0.61 v	B1	525
α Aquilae	Altair	0.76 v	A7	16.8
α Crucis	Acrux	0.77 **	B0.5 I B1	321
α Tauri	Aldebaran	0.87	K5	65.1
α Virginis	Spica	0.98 v	B1	262
α Scorpii	Antares	1.05 v **	M1 I B2.5	604
β Geminorum	Pollux	1.16	K0	33.7
α Piscis Austrini	Fomalhaut	1.16	A3	25.1
β Crucis	Mimosa	1.25 v	B0.5	353
α Cygni	Deneb	1.25 v	A2	3230
α Leonis	Regulus	1.36	B7	77.5
ε Canis Majoris	Adhara	1.50	B2	431

v variable   ** combined magnitude of double star

starburst galaxy A **galaxy** undergoing a strong burst of star formation. Some starburst galaxies are detected by their enhanced optical, ultraviolet or radio emission, but most are characterized by strong emission at far-infrared wavelengths. Hot young stars emit strongly in the ultraviolet. This radiation is absorbed by dust clouds and re-emitted as infrared. Tidal interactions between galaxies can trigger starburst, as can radiation from massive stars and supernovae.

star cluster A group of physically associated stars. Star clusters are classified as globular or open. A **globular cluster** contains an extremely large number (10^5 to 10^7) of ancient Population II stars (SEE **stellar populations**) in a roughly spherical space. An *open cluster* contains fewer stars than a globular, in an irregular volume of space. Their members are young (Population I) stars and they are found in or near the plane of the Galaxy. SEE ALSO **association, stellar; moving cluster.**

star diagonal SEE **diagonal** (1).

Stardust A space probe to Comet 81P/Wild 2, launched on 7 February 1999 as part of NASA's **Discovery** program. It encountered the comet in January 2004, making measurements and taking images. By using a collector covered with a special aerogel, it captured dust particles from the coma and brought them back to Earth, ejecting a sample-return capsule which was successfully recovered on 15 January 2006. Stardust was then sent on to Comet 9P/Tempel 1, the target of the earlier **Deep Impact** mission, which it flew past on 14 February 2011, photographing the nucleus.

Star of Bethlehem The "star seen in east" mentioned in the biblical Gospel of St Matthew. It has been ascribed to various astronomical phenomena, including comets, novae and the conjunction of two or three planets. The three separate conjunctions of Jupiter and Saturn in Pisces in 7 BC are one possibility. However, the most remarkable astronomical event about that time was the apparent merging in Leo of Venus and Jupiter, the two brightest planets, in June 2 BC, which would have created a brilliant temporary "star." It is also possible that the Star of Bethlehem was a literary invention inserted into St Matthew's Gospel at a later date.

stationary point A point in the apparent path of a planet when it changes between easterly and westerly motion – SEE **retrograde** (2).

steady-state theory A cosmological theory put forward by Hermann **Bondi** and Thomas **Gold** in 1948, and further developed by Fred **Hoyle** and others. According to this theory the Universe has always existed; it had no beginning and will continue for ever. Although the Universe is expanding, it maintains its average density – its steady state – through the continuous creation of new matter. The steady-state theory was rejected because, unlike the rival **big bang** theory, it cannot naturally explain the **cosmic microwave background**, the abundances of helium and other light elements, or the observation that the appearance of the Universe has changed with time.

stellar association SEE **OB association, T association.**

stellar evolution The development of a star over its lifetime, which can range from thousands of years to thousands of millions of years, depending on its mass. The theory of stellar evolution is based on mathematical models of stellar interiors. Its account of the evolution of stars can be checked by observations of stars at each of the predicted stages.

Stars form when a cloud of gas and dust collapses under its own gravity. As the cloud collapses, its temperature rises because the energy of infalling atoms is released as heat when they collide. At this stage it is termed a *protostar*. A protostar will continue to contract under gravity until its center is hot enough (about 15 million K) for nuclear reactions to commence, with the release of energy (SEE **carbon–nitrogen cycle, proton–proton reaction**). The outward flow of radiation from deep within halts the collapse (SEE **radiation pressure**). As this is happening, material is still being accreted. **T Tauri stars** are low-mass stars in this final stage of collapse.

Once nuclear fusion has commenced in the core of a star, the star adopts a stable structure with its gravity balanced by the radiation from its center. This is known as the main-sequence phase of a star's life (SEE **Hertzsprung–Russell diagram**). It is the longest phase in the life of a star. The Sun has a main-sequence lifetime of some 10,000 million years, of which half has expired. Less massive stars are redder and fainter than the Sun, and stay on the main sequence for longer. More massive stars are bluer and brighter and have a shorter lifetime. As a star ages on the main sequence, nuclear fusion changes the composition of its core from mainly hydrogen to mainly helium.

Eventually, the hydrogen supply in the core of the star runs out. With the central energy source removed, the core collapses under gravity and heats itself further until hydrogen fusion is initiated in a spherical shell surrounding the core. As this change occurs the outer layers of the star expand considerably and the star becomes a **red giant** or, in the case of the most massive stars, a **supergiant**. While it is a red giant, the star's core temperature will reach 100 million K, hot enough for the fusion of helium to carbon to begin (the **triple-alpha process**). In low-mass stars the onset of helium fusion is sudden (the *helium flash*), but in high-mass stars it gets under way more gradually. After this stage high-mass stars can fluctuate in size, becoming **long-period variable** stars such as Mira.

When all of the helium fuel in the core has been converted into carbon, the core of the star again collapses and heats up. In a low-mass star the central temperature cannot rise far enough to initiate carbon fusion. The red giant sheds its outer layers as a **planetary nebula**, after which only a **white dwarf** remains. In a high-mass star the contraction of the carbon core leads to further episodes of nuclear fusion involving heavier elements, during which the star remains a supergiant. Once fusion has converted all of the core into elements with atomic weights close to that of iron, no further fusion is possible. At this point the core implodes and the star throws off its outer layers in a **supernova** explosion. The explosion takes place not at the center but on the exterior of a dense core. The explosion expels several solar masses of gas at speeds of many thousands of kilometers per second, and also pushes inward on the core with immense force. The effect of this implosion is to crush the already extraordinarily dense core still further, forming a **neutron star** or a **black hole**, both effectively end-points of stellar evolution.

stellar interferometer SEE **optical interferometer.**

stellar nomenclature Some stars have proper names based on names given to them by Greek, Roman and Arab astronomers, such as Sirius, Capella and Aldebaran. In addition to any proper name, naked-eye stars in each constellation are identified by a letter or number followed by the name of the constellation in the genitive (possessive) form (SEE the table of **constellations** on page 26). The brightest stars are indicated by Greek letters, as in α (Alpha) Lyrae or ε (Epsilon) Eridani. These letters are known as *Bayer letters* because they were assigned by Johann **Bayer** in his star atlas *Uranometria* of 1603. Stars that do not have Bayer letters are prefixed by a number, as in 61 Cygni. These are known as *Flamsteed numbers* because they relate to the stars charted in John Flamsteed's *Historia coelestis britannica* of 1725. Fainter stars are referred to by their designation in any of a number of other catalogs. **Variable stars** have their own system of nomenclature.

stellar populations A classification of stars according to their age and location in the Galaxy. There are two principal populations. *Population I* consists of relatively young stars which are located in the plane of the Galaxy, in its spiral arms. The Sun is a Population I star. *Population II* consists of relatively old stars dispersed throughout the entire Galaxy, including **RR Lyrae stars** and subdwarfs, but prominently visible in its center and halo, particularly in globular clusters. There is also a gradation of stars intermediate between these two extremes.

The distribution of populations in our Galaxy is explained in terms of the contraction of the Galaxy during its formation. The oldest population of stars was made during the infall, which is why Population II is distributed throughout the galactic halo and central regions of the Galaxy. Stars in these regions, including those in globular clusters, continue to have elliptical orbits around the Galaxy. After the Galaxy had developed its flat rotating disk and spiral arms, stars of Population I formed, moving in circular orbits. Between these two extremes lie intermediate populations showing progressively flatter distributions, representing successive stages in the contraction of the Galaxy toward a disk.

Chemical elements produced in stars are recycled back into the interstellar material by stellar winds and supernova explosions, so Population I stars have greater concentrations of heavier elements (**metals**) than Population II. The Sun's content of heavy elements puts it among the younger Population I stars. The chemical history of the Galaxy suggests that a Population III once existed, created when the Galaxy first formed, but which has now disappeared. This early population would have manufactured the smattering of heavy elements present in Population II before dying to form compact objects such as neutron stars.

stellar wind A stream of charged particles, mostly protons and electrons, from the surface of a star. The strength of the wind depends on the type of star, and its velocity can range from a few hundred to

▲ **stellar wind** *The Bubble Nebula is caused by the strong stellar wind from the star at the bottom of the image. When young stars undergo a phase called the T Tauri phase their extremely strong stellar winds blow away much of the nebula in which they were formed.*

several thousand kilometers per second. Young stars evolving toward the main sequence have powerful stellar winds, up to a thousand times stronger than the Sun's **solar wind**. Old stars evolving into red giants also have strong stellar winds.

Stephano A small outer **satellite** of Uranus, discovered in 1999.

Stephan's Quintet A group of peculiar galaxies in the constellation Pegasus discovered in 1877 by the French astronomer Edouard Stephan (1837–1923). Four of the objects (NGC 7317, 7318A, 7318B and 7319) lie 270 million light years away, but the fifth (NGC 7320) is much closer, happening to lie in the same line of sight. The group, without NGC 7320, is sometimes called *Stephan's Quartet*.

STEREO The Solar Terrestrial Relations Observatory, a pair of NASA spacecraft to study coronal mass ejections (CMEs) and other solar phenomena from two directions simultaneously. Both spacecraft were launched on 26 October 2006. One, STEREO-A, orbits slightly closer to the Sun than the Earth, so it moves ahead of us, while the other, STEREO-B, has a slightly larger and slower orbit and hence lags behind the Earth. Combining their views with those from spacecraft such as the **Solar Dynamics Observatory** gives an all-round view of the Sun.

Steward Observatory The observatory of the University of Arizona, Tucson. Its main telescope is a 2.3 m (90-inch) reflector, sited on Kitt Peak.

stony-iron meteorite A meteorite with roughly equal proportions of silicates (stony material) and metals, mostly nickel–iron. There are two main subtypes: pallasites and mesosiderites. *Pallasites*, which consist of olivine (a magnesium–iron silicate) mixed with nickel–iron, may have originated near the core/mantle interface of a planetary body. *Mesosiderites* are a much coarser combination of chunks of various silicates and nickel–iron, and could have been produced by impacts on a solid planetary surface.

stony meteorite A **meteorite** that consists mostly of silicates (stony material), with only a small amount of chemically uncombined metals (typically 5% nickel–iron). There are two subtypes: **chondrites** and **achondrites**.

Stratospheric Observatory for Infrared Astronomy (SOFIA) An airborne observatory consisting of a 2.5 m (100-inch) telescope aboard a modified Boeing 747SP aircraft. SOFIA makes infrared observations from altitudes up to 14 km (45,000 ft), above most of the water vapor in the Earth's atmosphere. A joint project of NASA and the German Aerospace Center, it began operation in 2010.

strewnfield SEE **meteorite, tektites**.

Strömgren, Bengt George Daniel (1908–87) Swedish-born Danish astronomer. In 1940 he suggested the idea of what is now called a Strömgren sphere to explain how some nebulae shine. He also worked on photoelectric photometry, and the internal structure of stars.

Struve Russian–German family that produced four generations of astronomers. They included:

Struve, (Friedrich Georg) Wilhelm von (1793–1864) emigrated from Germany to Russia in 1833. In 1824 he had begun to observe double stars with a telescope made by Joseph von **Fraunhofer**, and in 1837 he published the first good catalog of double stars, containing nearly 2500 doubles he had discovered himself. In 1840 he made the third good measurement of the parallax of a star (Vega).

Struve, Otto (Wilhelm) (1819–1905), son of Wilhelm. He studied double stars, and determined an accurate value for the rate of **precession**, taking into account the motion of the Sun with respect to nearby stars.

Struve, Otto (1897–1963), a grandson of the above, emigrated and became a naturalized American. A leading observer and astrophysicist of his time, he applied spectroscopy to the study of binary stars and stellar rotation, and discovered interstellar hydrogen and calcium regions.

Subaru Telescope A Japanese 8.3 m (327-inch) reflecting telescope for use at visible and infrared wavelengths, sited at **Mauna Kea**. It commenced operation in 1999. Subaru is the Japanese name for the Pleiades.

subdwarf A star that is less luminous by 1 to 2 magnitudes than main-sequence stars of the same **spectral type**. Subdwarfs are mainly of types F, G and K, and lie below the main sequence on the **Hertzsprung–Russell diagram**. Most are Population II stars (SEE **stellar populations**). They are placed in luminosity class VI, although an alternative designation is to prefix their spectral type with the letters "sd".

subgiant A star of smaller radius and lower luminosity than a normal giant star of the same **spectral type**. They are mainly of types G and K, and lie between the main sequence and the giants on the **Hertzsprung–Russell diagram**. Subgiants are placed in luminosity class IV.

submillimeter-wave astronomy Observations of **electromagnetic radiation** at wavelengths of less than a millimeter; in practice, between 0.3 and 3 mm. The techniques for detecting submillimeter radiation are similar to those used for radio wavelengths, but as it is absorbed by water vapor, submillimeter telescopes have to be sited at high altitudes. Important targets at these wavelengths are molecular clouds and regions of star formation.

subsolar point The point on the surface of the Earth or other body in the Solar System at which, at any given moment, the Sun is directly overhead. Similar points are defined for other pairs of bodies.

substellar object SEE **brown dwarf, extrasolar planet**.

summer solstice SEE **solstice**.

Summer Triangle The prominent triangle that is formed by the first-magnitude stars Altair (in Aquila), Vega (in Lyra) and Deneb (in Cygnus). The triangle is overhead on summer nights in northern temperate latitudes.

Sun The star at the center of the Solar System, around which all other Solar System bodies revolve in their orbits. The apparent daily motion of the Sun across the sky and its annual motion along the **ecliptic** are caused by the Earth rotating on its axis and moving in its orbit. The Sun's light is occasionally blocked by the Moon (which, like the Sun, has an apparent diameter of about half a degree) in a **solar eclipse**. Data for the Sun are given in the table overleaf.

The Sun is a typical, average dwarf **G star**, on the **main sequence** of the Hertzsprung–Russell diagram. It consists of about 70% hydrogen (by weight) and 28% helium, with the remainder mostly oxygen and carbon. Its temperature, pressure and density increase toward the center, where the values are about 15×10^6 K, 10^{11} bar and 150 g/cm^3, respectively. As with all stars, the Sun's energy is generated by nuclear fusion reactions that take place under the extreme conditions in the core, chiefly the **proton–proton reaction**, which converts hydrogen into helium. The mass converted into energy is 4.3 million tonnes per second, but even at this enormous rate the loss amounts to only 0.07% of the total mass in 10^{10} years.

The amount of the Sun's energy normally reaching the Earth is quantified as the **solar constant**.

The Sun's core is about 400,000 km (250,000 miles) across. Energy released from the core passes up through the *radiative zone*, which is about 300,000 km (nearly 200,000 miles) thick, by a process of successive absorptions and re-emissions. It then passes through the 200,000 km (125,000-mile) thick *convective zone*, transported by rising and falling cells of gas, to the surface, the **photosphere**, from where it is radiated into space.

Most of the Sun's visible activity takes place in the 500 km (300-mile) thick photosphere. The pattern of convective cells that transport energy from below is visible as the **granulation** (SEE ALSO **supergranulation**). **Sunspots** are darker, cooler regions of the photosphere where the local magnetic field is enhanced. Their passage across the Sun's disk reveals **differential rotation** in the Sun's outer regions. Associated with sunspots are **flares** – sudden, violent releases of energy and high-speed atomic particles. Other phenomena occurring in the photosphere are **faculae** and **plages**. There is a periodicity in level of solar activity known as the **solar cycle**.

Above the photosphere lies the turbulent **chromosphere**, which consists of hot gases and extends for thousands of kilometers. This is the realm of **spicules** and **prominences**, huge magnetically constrained clouds of material above the Sun's limb. Extending outward from the chromosphere for millions of kilometers is the extremely tenuous solar **corona**. Material escapes continuously from the corona, producing the **solar wind**. Transient large-scale eruptions of material from the corona are known as **coronal mass ejections** which can occur with or without flares. The solar wind and the Sun's magnetic field dominate a region of space called the **heliosphere**, which extends to the boundaries of the Solar System.

The Sun must never be viewed directly through any optical instrument. There are two ways for amateurs to observe the Sun's disk. A *full-aperture filter* has a special metallic coating which cuts out harmful radiation before the Sun's light enters the telescope. Alternatively the Sun's image can be projected on to a white screen. Temporary or permanent blindness can result from direct viewing or from the use of unsuitable filters. Professional instruments for solar work include the **coronagraph** and the **spectroheliograph**.

Many probes and satellites, including the **Solar and Heliospheric Observatory**, **Solar Dynamics Orbiter** and **STEREO**, have been launched to study the Sun.

sundial A simple timekeeping device in which a shadow cast by the Sun falls on a dial graduated in hours. The shadow is cast by a short pillar called a *style* or *gnomon* standing out from the dial, which may be mounted vertically or horizontally. A sundial shows **apparent solar time**.

sundog SEE **parhelion**.

sungrazer A **comet** which at perihelion passes through the Sun's corona. They are the fragments of a large comet that broke up long ago. Often, such comets do not survive their close passage of the Sun. Over 2000 small sungrazers have been detected using the coronagraphs on the **SOHO** spacecraft.

sunrise The moment when the Sun's upper limb appears above the horizon in the morning. It is defined as the moment when the Sun's **zenith distance** is 90° 50′, and decreasing. This value is arrived at by making allowance for the Sun's semi-diameter (16′) and for atmospheric refraction (34′).

sunset The moment when the Sun's upper limb disappears below the horizon in the evening. Similarly to **sunrise**, it is defined as the moment when the Sun's zenith distance is 90° 50′, and increasing.

sunspot A region in the Sun's **photosphere** that is cooler than its surroundings and therefore appears darker. A sunspot consists of a dark central region, the *umbra*, and a lighter outer region,

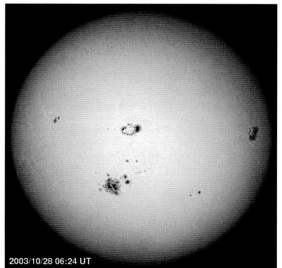

2003/10/28 06:24 UT

▼ **Sun** Although at a stable part of its life cycle, the Sun is not unchanging and undergoes a roughly 11-year cycle, during which its magnetic field "winds up" and then declines. Sunspots are the most obvious result of these changes.

SUN: DATA	
Distance from the Earth	
mean (the astronomical unit)	149.6 × 10⁶ km
maximum (at aphelion)	152.1 × 10⁶ km
minimum (at perihelion)	147.1 × 10⁶ km
Diameter	1.391 × 10⁶ km
Density (mean)	1.409 g/cm³
Mass	1.989 × 10³⁰ kg
Volume	1.412 × 10¹⁸ km³
Period of axial rotation	
at equator	25.4d
at poles	about 35d
Inclination of axis of rotation to pole of ecliptic	7° 15′
Surface gravity (Earth = 1)	28
Spectral type	G2V
Luminosity	3.86 × 10²⁶ W
Magnitude (mean visual)	−26.86 (apparent)
	+4.71 (absolute)
Rotational velocity (mean)	1.9 km/s
Escape velocity	617 km/s
Temperature of surface	5780 K
Temperature of core	15 million K

the *penumbra*. They vary in size from around 1000 to 50,000 km (600 to 30,000 miles), and occasionally up to about 200,000 km (125,000 miles). Their duration varies from a few hours to a few weeks, or months for the very biggest. Sunspots are seen to move across the face of the Sun as it rotates. The number of spots visible depends on the stage of the 11-year **solar cycle**.

Sunspots occur where there is a local strengthening of the Sun's magnetic field, which cools the area to around 4000 K. The stronger field is believed to suppress the convection of hotter gases from lower levels. Spots are usually found in pairs of opposite magnetic polarity, between latitudes 30° and 40° north or south of the equator (SEE ALSO **following**, **preceding**).

The spectrum of a sunspot differs from that of the photosphere as a result of the spot's lower temperature. The radial outflow of gases in the penumbra was detected by a Doppler shift of spectral lines (known as the *Evershed effect*). The **Zeeman effect** in the spectra of some sunspots demonstrates the presence of strong magnetic fields, the polarity of which reverses for each solar cycle.

supercluster SEE **cluster of galaxies**.

supergiant An extremely luminous star of large diameter and low density. Supergiants can be of any spectral type, from O to M. Rigel (type B) and Betelgeuse (type M) are examples. The luminosities of supergiants are several magnitudes greater than those of giant stars of the same spectral type, and so they lie at the top of the **Hertzsprung–Russell diagram**. Red (M-type) supergiants such as Betelgeuse have the largest diameters, around 1000 times that of the Sun. Supergiants are assigned to luminosity class I. The brightest are often given the separate class Ia, and the others placed in class Ib.

supergranulation A pattern of convective cells distributed fairly uniformly over the Sun's **photosphere**, and much larger – perhaps 30,000 km (19,000 miles) to a side – than ordinary photospheric granules (SEE **granulation**). Material has been detected flowing from the center to the edge of the cells, where most of the magnetic flux coming from the photosphere is concentrated, and it is believed that it is the magnetic field at the edges of the cells that leads to the formation of **spicules**.

superior conjunction SEE **conjunction**.

superior planet Any of the planets Mars, Jupiter, Saturn, Uranus and Neptune, all of whose orbits lie outside the Earth's.

supernova (plural **supernovae**) A stellar explosion in which virtually an entire star is disrupted. For a week or so, a supernova may outshine all the other stars in its galaxy. The luminosity (absolute magnitude up to about −19) is some 23 magnitudes (1000 million times) brighter than the Sun, and the energy released in the explosion is the same as is released over the star's entire previous life. A supernova is about 1000 times brighter than a **nova**. The diagram shows the light curve of Supernova 1989B, in the galaxy M66. Supernovae are designated by the year in which they are

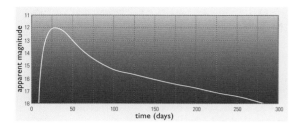

▲ **supernova** *The light curve of a supernova shows a sudden brightening as the outer atmosphere is thrown off after the stellar core has collapsed. This peak is followed by a steady decline for about two months, after which this fading slows further.*

observed, plus one or more letters to indicate the order of discovery in that year.

Analysis of ancient records has identified several supernovae in our own Galaxy, observed before the invention of the telescope. One was seen in 1054 by Chinese astronomers. The **Crab Nebula** and its associated pulsar are the remains of this event. In 1572 Tycho Brahe and others saw a new star as bright as Venus in the constellation Cassiopeia (SEE **Tycho's Star**). The last naked-eye supernova in our Galaxy erupted in 1604 in Ophiuchus, and was observed for over a year (SEE **Kepler's Star**). There is probably about one supernova every 30 years in a galaxy like our own, but most of them are concealed by dust.

Supernovae are of two main types, classified by their light curves and spectra. *Type II* supernovae have hydrogen lines in their spectra, whereas *Type I* supernovae do not. Type I supernovae have been found in all kinds of galaxies, but Type II have never been found in ellipticals. Type I supernovae are further classified into subtypes. Type Ia supernovae are believed to be formed by the explosion of the white dwarf component of a binary star. Hydrogen from the companion leaks on to the white dwarf, where it accumulates and drives the white dwarf over the critical mass of 1.4 solar masses (the **Chandrasekhar limit**), causing it to explode. Types Ib and Ic and Type II supernovae are thought to be caused by the core collapse and explosion of massive stars at the end of their life, when the nuclear fuel inside them is used up. In Types Ib and Ic, the progenitor star, probably a **Wolf–Rayet star**, is less than about 8 solar masses. In Type II supernovae, the collapse of the core of a star greater than about 8 solar masses forms a **neutron star** or a **black hole**; the collapse releases energy which is picked up by the outer layers, and these layers are ejected into space at about 5000 km/s (3000 miles/s).

After a couple of years the supernova has expanded so much that it becomes thin and transparent. For hundreds or thousands of years the ejected material remains visible as a **supernova remnant**.

Supernova 1987A A bright supernova that flared up in the Large Magellanic Cloud in February 1987, reaching magnitude 2.8 – the first naked-eye supernova since **Kepler's Star** of 1604. A blue supergiant designated Sanduleak −69° 202 was identified as the star that had exploded.

supernova remnant (SNR) A gaseous emission nebula, the expanding shell of matter thrown off into space by the outburst of a **supernova**. These remnants are often strong radio and X-ray sources. The ejected layers of the supernova collide with the surrounding interstellar gas and heat up to perhaps a million degrees K, emitting X-rays. Electrons spiral in the magnetic fields in the compressed gas and emit **synchrotron radiation** in the form of radio waves. The most intense radio source in the sky, **Cassiopeia A**, is a supernova remnant. Supernova remnants usually appear as hollow shells. However, some (including the **Crab Nebula**) are filled balls of radio and X-ray emission. Such a remnant is termed a *plerion*, meaning "filled." This implies that they contain electrons produced by an active pulsar, formed in the supernova explosion.

Surveyor A series of seven US spacecraft, five of which made successful soft landings on the Moon between May 1966 and January 1968 in preparation for the Apollo landings. They took photographs and made chemical analyses of the surface.

SU Ursae Majoris star SEE **U Geminorum star**.

Swan Nebula SEE **Omega Nebula**.

Swift Gamma Ray Burst Explorer A multi-wavelength observatory launched on 20 November 2004 and equipped to study gamma-ray bursts and their afterglows at gamma ray, X-ray, ultraviolet and optical wavelengths. Swift is a NASA mission with participation from the UK and Italy. It has helped confirm that long gamma-ray bursts come from supernova explosions, while short bursts are due to the merger of two compact objects, such as a pair of neutron stars or a neutron star and a black hole.

Swift–Tuttle, Comet 109P/ A **comet** discovered in 1862 by American astronomers Lewis Swift (1820–1913) and Horace Tuttle (1837–1923), subsequently identified with Comet Kegler of 1737. It is the second brightest short-period comet, **Halley's Comet** being the brightest. The large number of jets seen emerging from the nucleus at its return in 1992 suggest that the comet's changing period (at present 130 years) may be caused by **non-gravitational forces**. It is the parent comet of the **Perseid** meteor shower.

Sycorax An outer **satellite** of Uranus in an eccentric, **retrograde** (1) orbit, discovered in 1997.

symbiotic variable A **binary star** consisting of two stars of widely differing surface temperatures, such as a cool red giant and a hot dwarf. Gas from the cool star falls on to the smaller star and heats it. Symbiotic variables are also known as *Z Andromedae* stars, after their prototype, which has a range from magnitude 8.3 to 12.4.

synchronous orbit An orbit in which a satellite's period of revolution is the same as the primary's period of axial rotation. From the planet's surface the satellite appears to hover over one point, never rising nor setting. An example is **Charon**'s orbit around Pluto.

synchronous rotation The axial rotation of a celestial body in the same period as its period of revolution. Consequently, it always presents the same face toward the body about which it revolves, as in the case of the Moon orbiting the Earth. Tidal friction (SEE **tides**) has locked the Moon into this condition, which is also known as *captured rotation*. Other examples of bodies with synchronous rotation are the regular satellites of Jupiter and Saturn.

synchrotron radiation The electromagnetic radiation emitted by charged particles (usually electrons) that are accelerated by a strong magnetic field to speeds which are a significant fraction of the speed of light. The higher the energy of the particles, the shorter the wavelength of the radiation they emit. Synchrotron radiation is so called because it was first observed in particle accelerators called synchrotrons. Cosmic sources include **radio galaxies** and supernova remnants such as the **Crab Nebula**.

synodic month The period between two identical phases of the Moon; it is the same as the duration of one **lunation**. Its length is 29.53059 mean solar days.

synodic period The period of apparent revolution of one body about another as observed from the Earth, for example from one opposition or conjunction to the next. COMPARE **sidereal period**.

Syrtis Major A dark, triangular feature near the Martian equator, and the planet's most conspicuous feature in telescopic views from the Earth. It was first recorded by Christiaan **Huygens** in 1659. It is a sloping, cratered area, officially named Syrtis Major Planum.

syzygy The approximate alignment of three celestial bodies; in particular, the alignment of the Earth, the Sun, and the Moon or another planet. Thus syzygy occurs at full Moon and new Moon, and at planetary oppositions and conjunctions.

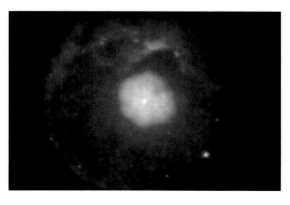

◀ **supernova remnant**
The central cloud of electrons and surrounding shell of hot gas are well seen in this image of G21.5-0.9 obtained by the Chandra X-ray Observatory.

TAI ABBREVIATION FOR **International Atomic Time** (from its name in French).

tail A long train of dust or gas particles expelled from the nucleus of a **comet**.

tangential velocity The component of a star's velocity at right angles to the line of sight. It is also known as *transverse velocity*. The tangential velocity can be found from measurements of the star's annual **proper motion** and its distance in parsecs. If the **radial velocity** is also known, then the **space velocity** of the star can be calculated.

Tarantula Nebula An emission nebula faintly visible to the naked eye on the southeastern edge of the Large Magellanic Cloud; also known as NGC 2070 or 30 Doradus. It is 1000 light years across and 500,000 solar masses, bigger and brighter than any nebula in our Galaxy or in any other nearby galaxy. It has a complex filamentary structure, and has a dense cluster of stars, designated R136, at its center.

T association A region of recent and active star formation consisting of low-mass stars known as **T Tauri stars**. The members of T associations offer a series of snapshots of stars like our Sun at various stages of its infancy. They are still surrounded by the dusty, obscuring material of the cloud from which they are forming, and consequently they are often brightest in the infrared. COMPARE **OB association**.

Taurids A meteor shower showing steady, low activity between mid October and late November, with a broad maximum during the first week of November. At maximum, the **ZHR** is little more than ten. In early November, the shower's two radiants lie close to the Pleiades and west of the Hyades star clusters in Taurus. The Taurids are produced by material from Comet **Encke**. Meteors from the shower are slow, and can sometimes be bright.

Taurus A constellation of the **zodiac**, representing the bull into which Zeus changed himself to carry off Princess Europa. It contains two of the most famous open clusters in the sky, the **Pleiades** and the **Hyades**, and also the **Crab Nebula**. Its brightest star is **Aldebaran**. Beta (Elnath), magnitude 1.7, and Zeta, magnitude 3.0, mark the tips of the bull's horns. Lambda is an eclipsing binary (range 3.4 to 3.9, period 3.95 days). There is also the prototype of the **T Tauri stars**.

tectonics Name given to the various processes by which the surface of a planetary body is deformed as a result of heating from within. Tectonic processes operate on bodies that have undergone **differentiation** and so possess a surface crust with a molten layer below. Large-scale movements of the molten layer affect the crust, producing faults and folds, and lifting mountains. Several satellites (e.g. **Ariel** and **Europa**) have surface features which appear to have been produced by tectonic activity. *Plate tectonics* operates, as far as is known uniquely, on the **Earth**.

tektites Small, glassy objects, typically centimeter-sized, found scattered across certain specific areas (*strewnfields*) of the Earth's surface. They consist mainly of silica, with small quantities of metallic oxides, and are from 600,000 to 65 million years old. All tektites have clearly solidified rapidly from a temperature of around 1500–2000 K. Their origin is uncertain, though their flight markings, which are characteristic of solidification while flying through the air at high speed, suggest that they originated as terrestrial rock which was vaporized by the impact of large meteorites. Tektites are named after the location of the strewnfield: for example, *Australites* are from Australia.

telescope An instrument for collecting and magnifying light or other electromagnetic radiation from a distant object. There are two basic types of optical telescope: the **refracting telescope**, in which the light-collecting element is a lens, and the **reflecting telescope**, in which it is a mirror; **catadioptric** systems are hybrids of the two. The image formed by the primary (main) lens or mirror is magnified by an **eyepiece**. The eyepiece may be replaced by a camera, photographic plate or **CCD** for imaging (SEE ALSO **Schmidt camera**); a **photometer** for measuring brightness; or a **spectrometer** for examining spectra. A telescope is supported (see **mounting**) in such a way that it is easily aimed at celestial objects.

The **focal ratio** of a telescope is equal to the focal length of the primary lens or mirror divided by its diameter. Refractors have longer focal ratios than reflectors, which, aperture for aperture, are thus shorter. The most compact designs are short-focus instruments like the **Schmidt-Cassegrain telescope**. The image produced by an astronomical telescope is inverted, as opposed to that formed in a terrestrial telescope, which is erect (upright).

The refractor installed in the Yerkes Observatory in 1897, which has an objective 1 m (40 inches) in diameter, is still the largest of its kind. Large telescopes built since then have all been reflectors, because large mirrors are easier to make and support than large lenses. The world's largest reflector is currently the **Gran Telescopio Canarias** in the Canary Islands, which has a segmented mirror 10.4 m (410 inches) across. Telescopes with mirrors up to 39 m across are in development (the **European Extremely Large Telescope**). With the development of spacecraft, it has become possible to send instruments such as the **Hubble Space Telescope** into space above the Earth's atmosphere, and so make observations unaffected by atmospheric absorption. SEE ALSO **drive**, **radio telescope**.

Telescopium A faint southern constellation representing a telescope, introduced by **Lacaille**. Alpha, its brightest star, is magnitude 3.5, but it contains little of interest.

Telesto A small **satellite** of Saturn, discovered by Bradford Smith and others in 1980 between the two Voyager encounters. Telesto is irregular in shape, and is **co-orbital** with Tethys and **Calypso**.

telluric line An **absorption line** in the spectrum of a celestial object produced by molecules such as oxygen and water in the Earth's atmosphere.

Tempel 1, Comet 9P/ A faint periodic comet discovered in 1867 by Wilhelm Tempel (1821–89). The comet has an orbital period of 5.5 years and was the target of the **Deep Impact** spacecraft in 2005.

Tempel–Tuttle, Comet 55P/ The parent comet of the **Leonid** meteor shower. It was discovered independently in December 1865 by Wilhelm Tempel (1821–89) and January 1866 by Horace Tuttle (1837–1923). It has an orbital period of 33 years and Leonid meteor activity reaches a peak every 33 years or so when the Earth encounters dust released from the comet near perihelion.

terminator The boundary between the illuminated and unilluminated hemispheres of a planet or satellite. Viewed from Earth, Mercury, Venus, Mars

► *Tarantula Nebula*
Located in the Large Magellanic Cloud, the Tarantula Nebula is bigger and brighter than any nebula in our own Galaxy.

▲ **Tethys** *Ithaca Chasma can be seen to the right in this Voyager 2 image of Saturn's satellite Tethys. Younger craters are superimposed on the giant trough, indicating that it is an ancient feature.*

and the Moon show phases and therefore their terminators are visible. The mountainous lunar surface gives the Moon a visibly irregular terminator.

terrestrial planets The planets Mercury, Venus, Earth and Mars, so called because they have rather similar characteristics in respect of size and density, and few or no satellites.

Tethys A medium-sized **satellite** of Saturn, diameter 1058 km (657 miles), discovered by Giovanni Cassini in 1684. The whole of Tethys' surface is densely cratered, but evidence of past geological activity is provided by some less heavily cratered areas which must have undergone resurfacing. The two dominant features are Odysseus, a crater 440 km (270 miles) in diameter, and Ithaca Chasma, a 100 km (60-mile) wide trough running three-quarters of the way round the satellite; their formation may be linked. Tethys shares its orbit with two **co-orbital** satellites, **Calypso** and **Telesto**.

Thalassa One of the small inner **satellites** of Neptune discovered in 1989 during the Voyager 2 mission.

Thales of Miletus (*c.*625–*c.*550 BC) Greek philosopher. He provided the first theory of the Universe's origin, in which everything developed from a primordial mass of water. He knew of the **saros** and is credited with having used it to predict the eclipse of the Sun that occurred probably on 28 May 585 BC, during a battle between the Lydians and the Medes.

Thebe One of the small inner **satellites** of Jupiter discovered in 1979 during the Voyager missions. It is irregular, measuring about 110 × 90 km (70 ×55 miles).

theory of everything (TOE) SEE **grand unified theory**.

third contact In a total **solar eclipse**, the moment when the trailing edge of the Moon's disk begins to uncover the Sun's disk, and totality ends. In a **lunar eclipse** it is the moment when the Moon begins to leave the umbra of the Earth's shadow. The term is also used for the corresponding stage in eclipses involving other bodies.

Thirty Meter Telescope (TMT) A joint project between the US, China, Japan and India to build a 30 m (100 ft) optical/infrared reflecting telescope on Mauna Kea, Hawaii. The primary mirror will consist of 492 hexagonal segments. The telescope is due for completion in 2024.

three-body problem A fundamental problem in celestial mechanics: to determine the motions of three bodies under the influence only of their mutual gravitational attractions. There is no exact general solution, only solutions for special cases, but highly accurate approximations can be achieved by using modern computers. Work on the three-body problem was stimulated originally by the need to understand the orbit of the Moon, under the gravitational influence of the Earth and the Sun, and more recently by the need to calculate the orbits of artificial satellites. SEE ALSO **many-body problem**.

tides Distortions induced in a celestial body by the gravitational attraction of one or more others. The gravitational attraction a body experiences is greatest on the side nearest the attracting body, and

least on the side farthest away, causing the body to elongate slightly in the direction of the attracting force and thus acquiring a *tidal bulge* on each side. If the deformed body is orbiting the attracting body, different parts of its surface periodically experience tidal distortion as it rotates. These tides have a significant effect on a gaseous atmosphere or a fluid ocean, but the solid crust of a planet such as the Earth, which is supported by a fluid mantle, is able to flex and so also experiences tidal distortion, but to a much lesser degree. The enormous gravitational force exerted by Jupiter on its satellite **Io** produces a large amount of flexure and interior heating, and this *tidal heating* is what makes Io so volcanically active. Smaller and completely rigid bodies can under certain circumstances be broken up by tidal forces.

The familiar ocean tides on the Earth are raised by the gravitational attraction of the Moon and the Sun, the Moon's influence being about three times the Sun's. There are two high tides and two low tides each day. When the Sun and the Moon are exerting a pull in the same direction (at new Moon and full Moon) their effects are additive and high tides are higher (*spring tides*). When the pull of the Sun is at right angles to that of the Moon (at first quarter and last quarter), high tides are lower (*neap tides*). *Tidal friction* caused by the rotation of the Earth beneath the oceanic bulges acts as a brake on the Earth's rotation, causing the length of the day to increase by about one millisecond per century. The angular momentum which is thus being lost is transferred to the Moon (SEE **secular acceleration**).

time The continual passage of existence, perceived as an ordering of events. Time may also be defined as the dimension of **spacetime** in which different events that have identical spatial dimensions are distinguished.

timescale A scale for measuring the passage of time by reference to some regularly recurring phenomenon. From remote antiquity, timescales have been based on the passage of the Earth once around its orbit, giving the unit of the **year**, and the rotation of the Earth on its axis, giving the unit of the **day**. However, the rotation of the Earth is not constant, and so it cannot be used as an accurate and invariable standard of time. Modern timescales are now based on a continuous count of time units defined by atomic resonators (**International Atomic Time**). The basis of civil timekeeping is **Greenwich Mean Time** (known technically as **Universal Time**). In addition, astronomers reckon time by the stars: **sidereal time**. SEE ALSO **apparent solar time**, **equation of time**, **local time**, **mean solar time**, **time zone**.

time zone One of 24 divisions of the Earth's surface, each 15° of longitude wide, within which the time of day is reckoned to be the same. The meridian of Greenwich serves as the zero of longitude. Standard time in each successive zone west of Greenwich is one hour behind that in the preceding zone (with some local variations). Large territories such as the USA and Russia span several time zones. There is a discrepancy at longitude 180° (the *International Date Line*), which is resolved by omitting one day from the calendar when crossing from west to east, or repeating one day if the crossing is from east to west.

Titan The largest **satellite** of Saturn, and the second largest in the Solar System, discovered by Christiaan **Huygens** in 1655. Titan is unique among planetary satellites in having a substantial atmosphere. At 1.88 g/cm³ Titan is the densest of Saturn's large satellites, and is composed of rock and water ice in roughly equal proportions. Titan's atmosphere consists mostly of nitrogen, with some methane and traces of other hydrocarbon compounds, and exerts a surface pressure 1.5 times that on the Earth. Photochemical reactions driven by sunlight produce a largely opaque reddish cloud layer 200 km (125 miles) above the surface. The surface temperature is 95 K, at which methane can exist as solid, liquid or gas; methane can play the role that water does on Earth, forming, clouds, rain, and pools or more substantial bodies of liquid.

During its descent to Titan's surface on 14 January 2005, the Huygens probe (SEE **Cassini–Huygens**) imaged features which are clearly river channels, caused by flows of liquid methane. The surface is believed to have been shaped by the same erosional processes that have operated on Earth. Huygens landed on a surface with the consistency of wet sand, covered by a thin harder crust. Substantial quantities of methane are apparently bound up in the topsoil as "groundwater"; heat generated by the landing released some of this methane. On the surface, Huygens' cameras imaged a landscape strewn with "pebbles" of dirty water ice over which a light fog of methane or ethane was present.

Titania The largest of the five main **satellites** of Uranus, discovered by William **Herschel** in 1787. It is 1580 km (980 miles) in diameter,

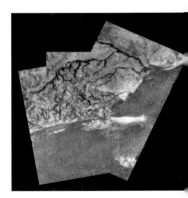

▲ **Titan** *River channels on Titan's surface are clearly visible in this composite of images obtained by Huygens as it descended to the surface on 14 January 2005.*

▲ **Titania** *The surface of Titania is heavily cratered and traversed by faults. This Voyager 2 picture shows details down to 7 km (4 miles) across.*

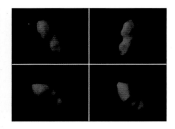

▲ **Toutatis** These four images of the asteroid Toutatis were constructed using radar information from NASA's Goldstone and Arecibo radar telescopes. Toutatis appears to consist of two irregularly shaped, cratered objects.

▼ **trigonometric parallax** If a nearby object (X) is observed from opposite sides of the Earth's orbit (A and B), it will appear in different positions against the stellar background (X₁ and X₂). As the distance between A and B is known, the distance, d, to X can be calculated.

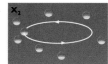

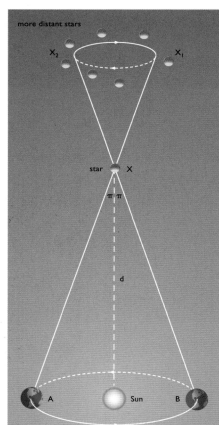

and its density of 1.68 g/cm³ indicates that it consists of a mixture of rock and water ice. Titania is Oberon's twin in size, but its surface resembles that of **Ariel**, with similar features and an apparently similar geological history.

Tombaugh, Clyde William (1906–97) American astronomer. In 1930, nearly a year into a search based on predictions by Percival **Lowell**, he discovered the planet **Pluto**. He continued to search for other planets for more than ten years, discovering in the process star clusters, clusters of galaxies, a comet and hundreds of asteroids. After World War II he developed telescopic cameras for tracking rockets after launch.

topocentric Describing observations made from a point on the surface of the Earth, e.g. topocentric coordinates as opposed to geocentric coordinates (which are measured from the center of the Earth).

total eclipse SEE **lunar eclipse**, **solar eclipse**.

Toutatis Asteroid no. 4197, discovered by Christian Pollas in 1989. Radar observations suggest that it consists of two bodies, 4 km (2.5 miles) and 2.5 km (1.5 miles) in diameter, in close contact. Toutatis is an **Apollo asteroid**, and one of the largest **near-Earth asteroids**.

transient lunar phenomenon (TLP) SEE **lunar transient phenomenon**.

transit (1) The passage of a celestial body across the observer's **meridian**. SEE ALSO **transit instrument**.

transit (2) The passage of a body directly between the Earth and the Sun. The planets Mercury and Venus do so on occasion, when they appear as a black spot crossing the solar disk. Accurate measurements of such transits helped to establish the scale of the Solar System.

transit (3) The passage of a planetary satellite across the planet's disk. Jupiter's four main satellites frequently transit its disk, and Saturn's inner satellites are occasionally seen in transit. Observations of transits of Jupiter's satellites by Ole **Römer** established the enormous but finite speed of light.

transit (4) The passage of a surface or atmospheric feature of a body across its **central meridian** as it rotates, the timing of which gives a means of measuring the body's rotation period.

transit instrument A telescope that is mounted on a horizontal east–west axis and movable only in the vertical plane of the meridian. Transit instruments are used to time the **transits** (1) of bodies across the local meridian and measure their altitudes. Such measurements are used for precision timekeeping and accurately measuring star positions (**astrometry**).

Transition Region and Coronal Explorer (TRACE) A NASA satellite launched on 2 April 1998 to study the Sun's **corona** and the transition region – the border between the chromosphere and the corona. Its telescope of 0.3 m (12-inch) aperture imaged the Sun at four ultraviolet wavelengths. TRACE ceased operations on 21 June 2010.

trans-Neptunian object SEE **Kuiper Belt**.

transverse velocity SEE **tangential velocity**.

Trapezium Popular name for the multiple star Theta Orionis in the Orion Nebula. Its four main components, magnitudes 5.1, 6.7, 6.7 and 8.0, are arranged in the shape of a trapezium. The stars are very hot, and light up the nebula.

Triangulum A small but easily found constellation between Andromeda and Aries. Its brightest star, Beta (magnitude 3.0), forms a well-marked triangle with Alpha (magnitude 3.4) and Gamma (magnitude 4.0). The main feature of interest is the spiral galaxy known as the Pinwheel or Triangulum Galaxy (M33), 2.8 million light years away in our **Local Group**, visible with binoculars under clear, dark skies.

Triangulum Australe A southern constellation representing a triangle, marked by the stars Alpha (magnitude 1.9), Beta and Gamma (both magnitude 2.9). The open cluster NGC 6025 is visible with binoculars.

Trifid Nebula An emission nebula in the constellation Sagittarius, apparently divided into three main sectors by dark lanes of dust. It is also known as M20 or NGC 6514. At its center is an eighth-magnitude double star. The Trifid lies 6700 light years away.

trigonometric parallax A means of determining the distance of a star by triangulation. As the Earth revolves around the Sun, a nearby star shifts its position slightly with respect to more distant stars. This change in position can be measured from images obtained six months apart, i.e. when the Earth is on opposite sides of the Sun. As the diagram shows, the angular displacement is a direct measure of the star's parallax (π). Since the baseline (the distance from the Sun to the Earth) is known, the distance d of the star can be determined by trigonometry.

The first star to have its distance measured by this method was 61 Cygni, by Friedrich **Bessel** in 1838. Parallax is measured in arc seconds. The inverse of the parallax is the distance in **parsecs**. Alpha Centauri is the star with the largest known parallax, 0.752 arc seconds, which corresponds to a distance of 1.3 parsecs. The error in measuring parallax from individual images is typically about 10 milliarcseconds, but by combining images this can be reduced to about 4 milliarcseconds. The parallax of a star at 25 parsecs is 40 milliarcseconds, so the uncertainty is around 10%. The most accurate parallax determinations over the entire sky have been made by the **Hipparcos** satellite.

triple-alpha process A nuclear reaction in which three alpha particles (helium nuclei) are transformed into carbon with the release of energy. It takes place after all of the hydrogen in a star's core has been exhausted. The core contracts and its temperature rises until it exceeds 100 million K, when the triple-alpha reaction begins. In each reaction two helium nuclei combine to form a beryllium nucleus which, in turn, captures a further helium nucleus to form a carbon nucleus. Additional helium captures can produce oxygen, neon and a number of heavier elements. The reaction is believed to be the dominant energy-producing process in red giants. It is also known as the *Salpeter process* after the US physicist Edwin Salpeter (1924–2008), who described it.

triplet A compound lens consisting of an assembly of three component lenses, which may be air-spaced or cemented.

Triton The largest **satellite** of Neptune, discovered in 1846 by William Lassell. It is in an inclined, **retrograde** (1) orbit and has almost certainly been captured by Neptune, probably from

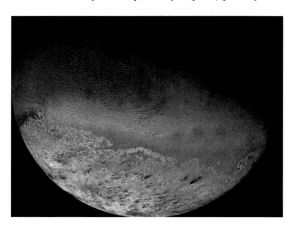

▲ **Triton** Voyager 2 imaged Triton in 1989, and this mosaic shows some of the varied terrains. The broad area at the bottom of the picture is the southern polar ice cap, which probably consists mainly of nitrogen ice.

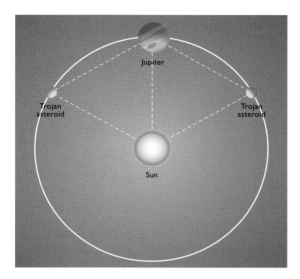

▲ **Trojan asteroid** *The asteroids known as the Trojans orbit at the Lagrangian points in front of or behind Jupiter. They oscillate around these points over a period of 150–200 years.*

the **Kuiper Belt**. Its diameter is 2705 km (1680 miles), and its density of 2.07 g/cm^3 is consistent with a composition largely of three parts rock to one of water ice. Triton is the coldest and most distant world so far visited: Voyager 2's instruments measured a surface temperature of 35 K (−238°C). Its surface is complex and varied. There is much evidence of very recent volcanic activity in a variety of features formed by the upwelling of fluid material. In the south polar region, which is covered with a thin polar cap of nitrogen ice, dark eruptive plumes send fine dark material rising to altitudes of 8 km (5 miles) before being carried downwind for 100 km (60 miles) or more in the tenuous nitrogen atmosphere.

Trojan asteroid A member of one of two groups of **asteroids** sharing Jupiter's orbit. One group lies ahead of Jupiter, and the other behind, oscillating about the L_4 and L_5 **Lagrangian points** of Jupiter's orbit around the Sun. The Trojans ahead of Jupiter, about two-thirds of the total, are known as the *Achilles group*, after the first one to be found, **Achilles** in 1906; those behind form the *Patroclus group*, named after the second Trojan to be discovered, also in 1906. Over 6000 were known by early 2015. The term is also used for asteroids sharing the orbits of other planets; 12 Neptunian Trojans, four Martian, and one each in the orbits of Earth and Uranus have been found.

tropical month The time taken for one revolution of the Moon around the Earth relative to the vernal **equinox**. It equals 27.32158 days.

tropical year The time taken for the Earth to orbit the Sun relative to the vernal equinox. Its length is 365.24219 mean solar days. Because of the precession of the vernal equinox, which moves along the ecliptic in a direction opposite to the Sun's motion, the tropical year is about 20 minutes shorter than the sidereal year. It is sometimes also known as the *solar year*.

Tropic of Cancer The parallel of latitude on Earth, $23\frac{1}{2}°$ north, that marks the most northerly declination reached by the Sun at the summer **solstice**, on or about 21 June.

Tropic of Capricorn The parallel of latitude on Earth, $23\frac{1}{2}°$ south, that marks the most southerly declination reached by the Sun at the winter **solstice**, on or about 22 December.

true anomaly SEE anomaly.

Trumpler classification The classification of open star clusters devised by the Swiss–American astronomer Robert Trumpler (1886–1956) on the basis of three characteristics: the number of stars in the cluster, the concentration of stars toward the center of the cluster, and the range of brightness within each cluster.

T star An extremely low-temperature **brown dwarf** star characterized by methane lines in its spectrum. Methane molecules can form only below about 1500 K. Like the **L stars**, no thermonuclear processes are occurring in T stars and they are detectable only because they derive some energy from their continued gravitational contraction.

T Tauri star A very young star still settling on to the main sequence and belonging to a class of irregular variables, named after the prototype, T Tauri. T Tauri stars are found in nebulae or young clusters and are characterized by high-velocity infall or outflow of gas which occurs as they adjust to the onset of nuclear reactions. Emission lines in their spectra indicate an extended atmosphere. Most known T Tauri stars are less massive than the Sun. Heavier stars either pass through the T Tauri stage while they are still obscured in the nebulae from which they were born, or have a different appearance at the same stage of life. SEE ALSO **Hertzsprung–Russell diagram**, **T association**.

Tucana A far southern constellation representing a toucan. Its overall faintness is redeemed by the presence of the Small Magellanic Cloud and the superb globular cluster 47 Tucanae (NGC 104). At magnitude 4 and half a degree across, 47 Tucanae is inferior only to Omega Centauri. It lies 15,000 light years away. NGC 362 is another globular, visible in binoculars.

Tunguska Event A huge aerial explosion on 30 June 1908, just north of the Stony Tunguska River in Siberia, which devastated the surrounding forest. The explosion, which was heard more than 800 km (500 miles) away and was recorded seismographically all around the world, was preceded by a fireball as bright as the Sun. Trees up to 40 km (25 miles) away were felled, and in the central 1000 sq km (400 sq miles) trees were incinerated.

The absence of an impact crater or meteoritic fragments puzzled earlier investigators. The best current theory is that a **near-Earth asteroid** roughly 50 m (150 ft) in diameter entered the atmosphere obliquely, and shattered and vaporized at an altitude of about 8 km (5 miles), blanketing the area with dust. Dust particles preserved in resin taken from trees in the area have the composition of known stony meteorites, and the damage pattern is consistent with the impactor having had an asteroidal density.

tuning fork diagram A diagram of **galaxy** types, originated by Edwin Hubble. The "handle" of the fork consists of elliptical galaxies (E), numbered according to their degree of elongation, while the two prongs of the fork are made up of ordinary (S) and barred (SB) spirals, designated a, b or c according to how tightly wound the spiral arms are. The classification was originally interpreted as a sequence of evolution, with ellipticals developing into spirals, but this is now known to be incorrect. SEE the diagram at **Hubble classification**.

twenty-one centimeter line The emission line of neutral hydrogen in interstellar clouds. It lies in the radio spectrum at a wavelength of about 21 cm; the frequency is 1420 MHz. Its existence was predicted by Hendrik van de **Hulst** in 1944 and discovered by him and others in 1951. SEE ALSO **H I region**.

twilight The period during which the illumination of the sky gradually increases before sunrise, and decreases after sunset. The phenomenon is caused by the scattering of sunlight by molecules and particles of dust in the Earth's atmosphere. The duration of twilight depends on the steepness of the Sun's apparent path with respect to the horizon, so that twilight lasts longer at higher latitudes. Three forms of twilight are distinguished. *Civil twilight* ends or begins when the center of the Sun's disk is 6° below the horizon, and is regarded as a period during which normal daytime activities cease to be possible. *Nautical twilight* ends or begins when the center of the Sun's disk is 12° below the horizon, and the marine horizon is no longer visible. *Astronomical twilight* ends or begins when the center of the Sun's disk is 18° below the horizon, and is the time when the faintest stars can be seen with the naked eye.

Tycho A large lunar crater, 85 km (53 miles) in diameter, in the Moon's southern uplands. It is the center of the most conspicuous system of lunar **rays**, which stretch across much of the Moon's near side, evidence that Tycho is a relatively young crater.

Tycho Brahe SEE **Brahe, Tycho**.

Tycho's Star A supernova in Cassiopeia in 1572 that was observed and described by Tycho **Brahe**. It was brighter than Venus at maximum and visible during daytime. Its remnant has been detected as a radio and X-ray source. Its light curve identifies it as a Type Ia supernova.

UBV system A system of three-color photometry devised by Harold Johnson and William Morgan of Yerkes Observatory. Stellar magnitudes are determined at three different wavelength bands through three color filters: ultraviolet (U) peaking at 360 nm; blue (B) peaking at 440 nm; and yellow (V for visual) peaking at 550 nm. SEE ALSO **color index**.

U Geminorum star A member of a class of dwarf novae which show sudden outbursts of between 2 and 6 magnitudes, followed by a slower return to minimum, where they remain until the next outburst. Periods range from 10 days to several years. They are sometimes known also as *SS Cygni stars*. The best-known examples are SS Cygni, whose light curve is shown in the diagram, and U Geminorum. All are close binaries, comprising a subgiant or dwarf star of type K or M, comparable in mass to the Sun, which has filled its **Roche lobe**, and a white dwarf surrounded by an **accretion disk** of infalling matter. Almost all show rapid irregular flickering of about half a magnitude at minimum. Because they are binaries, some show eclipses at minimum, such as U Geminorum itself. As well as normal maxima, the *SU Ursae Majoris* subclass have "supermaxima" which last about five times as long and are twice as bright. It has been suggested that they evolve from **W Ursae Majoris stars**. They are closely related to **Z Camelopardalis stars**.

Uhuru The first X-ray astronomy satellite, launched by the USA on 12 December 1970. Many X-ray sources bear numbers prefixed with the letter U, being their designation in the Uhuru catalog.

ultraviolet astronomy The study of **electromagnetic radiation** from space with wavelengths between those of the visible spectrum and X-rays, i.e. from about 350 nm down to about 90 nm. Apart from the longest wavelengths (called the *near ultraviolet*), ultraviolet (UV) radiation does not penetrate the Earth's atmosphere, so observations must be made from rockets and satellites. The **International Ultraviolet Explorer** (IUE) was launched in 1978. Coverage was extended to the shortest ultraviolet wavelengths by **Rosat** in 1990 and the **Extreme Ultraviolet Explorer** (EUVE) in 1992. Recent satellites include the **Far Ultraviolet Spectroscopic Explorer**, launched in 1999, and the **Galaxy Evolution Explorer**, in 2003.

Unlike explorations at other wavelengths, such as radio, infrared and X-rays, UV astronomy has discovered few new sources. Instead, the main application is in spectroscopy. Many of the abundant atoms and ions in celestial sources have their strongest spectral lines in the UV. Observations at UV wavelengths have been significant for studies of the composition and motions of interstellar gas. Hot stars of spectral types O and B emit strongly in the UV, and this emission excites surrounding nebulae. In the Solar System, UV spectroscopy is used to study planetary atmospheres and the gas in comets. Farther off in the Universe, UV astronomy is important in the study of luminous galaxies and quasars, which contain hot stars and large quantities of gas.

Ulugh Beg (1394–1449) Mongol ruler and astronomer. He established an observatory at Samarkand, and compiled the first star catalog to surpass those of Hipparchus and **Ptolemy** in precision.

Ulysses A European Space Agency probe launched on 6 October 1990 to study the polar regions of the Sun. After a flyby of Jupiter in February 1992, it entered a solar orbit which took it over the Sun's south pole 1994, and over the north pole a year later. In the late 1990s, Ulysses studied the evolution of **coronal mass ejections** from a position back near the orbit of Jupiter. It passed over the south and north poles again in 2000 and 2001. It ceased operation on 30 June 2009.

▲ **Umbriel** *This pole-on view of Umbriel was obtained by Voyager 2 on 24 January 1986. The smallest details that can be resolved are 10 km (6 miles) across, and the heavily cratered surface is clearly visible.*

▼ **U Geminorum star**
The light curve of a typical U Geminorum star shows repeated outbursts at intervals ranging from approximately ten days to several years. During outburst, the star's brightness climbs rapidly, with a slower decline. Outbursts occur as a result of mass transfer to an accretion disk around the smaller, more massive star in these close binary systems.

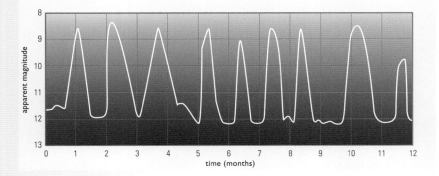

umbra (1) The inner part of the shadow cast by a celestial body illuminated by an extended source such as the Sun. An observer in the umbral region of a shadow sees all of the source obscured, as, for example, in a total **solar eclipse**. SEE ALSO **lunar eclipse**. COMPARE **penumbra** (1).

umbra (2) The dark central area of a sunspot. COMPARE **penumbra** (2).

Umbriel One of the five main **satellites** of Uranus, discovered by William **Lassell** in 1851. Its diameter is 1170 km (727 miles), and its density of 1.51 g/cm³ is consistent with a composition of mostly water ice and rock. Umbriel is a geologically inactive, heavily cratered satellite with a predominantly dark surface. The only bright spot is the floor of the 140 km (85-mile) diameter crater named Wunda.

United States Naval Observatory The US government observatory with headquarters at Washington, DC. Its main telescope there is a 0.66 m (26-inch) refractor. USNO has an observing station at Flagstaff, Arizona, with a 1.55 m (61-inch) reflector.

Universal Time (UT) Standard timescale used for scientific purposes throughout the world, popularly known as **Greenwich Mean Time** (GMT). There are several versions of Universal Time. *UT0* is based on the observed rotation of the Earth relative to the stars. *UT1* is UT0 corrected for a slight wandering of the Earth's geographical poles, which affects the longitude of the place of observation. *Coordinated Universal Time* (UTC; the abbreviation is of the term in French) is based on the caesium atomic clock, and is used for radio time signals. As the Earth's rotation is gradually slowing, the time shown by an atomic clock gradually diverges from UT1. To keep atomic time in step with UT1 to the nearest second, time signals are retarded when necessary by one **leap second** at midnight on 30 June or 31 December. UTC is what is generally known as GMT.

Universe All of space and everything contained in it. SEE **cosmology**.

upper culmination SEE **culmination**.

Uranus The seventh major planet from the Sun, and the second smallest of the four giant planets. With a mean magnitude of 5.5, Uranus is visible to the naked eye under good conditions. Its maximum apparent diameter is 4.1 arc seconds, and through a telescope it appears as a small, featureless, greenish-blue disk. In size, mass, atmosphere and color, Uranus resembles Neptune. The main data for Uranus are given in the table.

Uranus was the first planet to be discovered since ancient times. Although it had been observed on several occasions (and once had even been cataloged as a star), its non-stellar nature was first recognized by William **Herschel**, who observed its disk on 13 March 1781, during a telescopic survey of faint naked-eye stars. Uranus's axis of rotation is steeply inclined, and lies close to the ecliptic plane. Its poles thus spend around half of its 84-year orbit in sunlight, and half in darkness, and highly exaggerated seasonal variations are

URANUS: DATA	
Globe	
Diameter (equatorial)	51,118 km
Diameter (polar)	49,947 km
Density	1.29 g/cm³
Mass (Earth = 1)	14.53
Volume (Earth = 1)	62.18
Sidereal period of axial rotation (equatorial)	17h 14m (retrograde)
Escape velocity	21.3 km/s
Albedo	0.51
Inclination of equator to orbit	97° 52'
Temperature at cloud-tops	55 K
Surface gravity (Earth = 1)	0.79
Orbit	
Semimajor axis	19.22 au = 2871 × 10⁶ km
Eccentricity	0.046
Inclination to ecliptic	0° 46'
Sidereal period of revolution	84.01y
Mean orbital velocity	6.81 km/s
Satellites	27

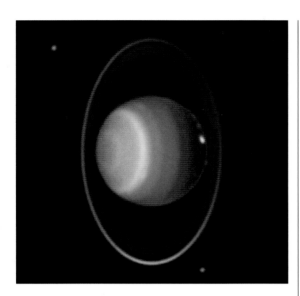

▲ **Uranus** *The rings of Uranus captured in infrared light by the Hubble Space Telescope. High-altitude clouds in the planet's atmosphere show up as orange. White dots around the planet are some of its moons.*

experienced by both the planet and its satellites. The flyby of the Voyager 2 probe in 1986 provided most of our current knowledge of the planet.

The upper atmosphere is about 83% molecular hydrogen and 15% helium, the other 2% being mostly methane. Strong enhancement of Voyager images revealed faint banding and a haze over the Sun-facing pole. Near-infrared images obtained from the ground and from the Hubble Space Telescope have since revealed dark spots and bright clouds similar to those visible when Voyager flew past **Neptune**. It may be that Uranus's "weather" was unusually calm at the time of the Voyager visit. The planet has insufficient internal heat to drive a dynamic meteorology like Neptune's. **Differential rotation** operates on Uranus, atmospheric features taking 16 to 17 hours to rotate, compared with the internal rotation period of over 17 hours. Wind speeds vary from about 150 to 600 km/h (90 to 370 miles/h).

Uranus's interior structure remains conjectural. There may be an iron–silicate core, surrounded by a deep, possibly semiliquid "mantle" consisting of water, ammonia and methane ices in which the planet's magnetic field (comparable in strength to Saturn's) originates. The magnetic axis is tilted by 59° to the axis of rotation, and is displaced from the planet's center. The magnetic field gives rise to an appreciable **magnetosphere**.

The five largest **satellites** were known before the Voyager encounter, which led to the discovery of 11 more. A total of 27 Uranian satellites were known as of mid-2015. The main components of Uranus's ring system were discovered in 1977 when the planet occulted a star (SEE **occultation**). Others were imaged by the Voyager spacecraft. The brightest and outermost is the Epsilon Ring, while the innermost is a very diffuse sheet of material. Between the two are nine narrow, darkish rings. SEE ALSO **ring, planetary**.

Ursa Major A famous northern constellation, the Great Bear, whose main pattern is known as the Big Dipper or Plough. This feature consists of seven stars: Alpha (Dubhe), magnitude 1.8; Beta (Merak), magnitude 2.3; Gamma (Phekda), magnitude 2.4; Delta (Megrez), magnitude 3.3; Epsilon (Alioth), magnitude 1.8; Zeta (**Mizar**, a famous double), magnitude 2.2; and Eta (Alkaid), magnitude 1.9. Dubhe and Merak are the "pointers" to the pole star. Five of the Big Dipper stars make up a **moving cluster**; the exceptions are Dubhe and Alkaid. There are many galaxies in Ursa Major, including several Messier objects (M81, M82, M101, M108 and M109), as well as the **Owl Nebula**.

Ursa Minor The constellation that contains the north celestial pole. Its brightest star is Alpha (**Polaris**), the north pole star, magnitude 2.0. The constellation's seven main stars make a pattern rather like a faint and distorted Big Dipper (Plough), and is known as the Little Dipper.

UT ABBREVIATION FOR **Universal Time**.

UV Ceti star Another name for a **flare star**.

Valhalla A 4000 km (2500-mile) diameter impact basin – the largest such structure in the Solar System – on Jupiter's satellite **Callisto**, surrounded by a series of concentric rings representing "ripples" from the impact.

Valles Marineris SEE **Mariner Valley**.

Van Allen Belts Two zones of plasma (high-energy charged particles) in the Earth's **magnetosphere**. The concentric, torus-shaped belts trap charged particles which then spiral around magnetic field lines, back and forth between the two magnetic poles. The diagram, which is roughly to scale, shows the form of the belts. The outer belt contains mainly electrons from the **solar wind**, and the inner belt mainly protons from the same source. Within the inner belt is a radiation belt consisting of particles produced by interactions between the solar wind and heavier **cosmic ray** particles. The intensity of the Van Allen Belts varies with the level of solar activity (SEE **solar cycle**). Because the Earth's magnetic axis is offset from its rotational axis, the belts are not a uniform distance from the Earth's surface. The inner belt comes very close to the surface over the South Atlantic Ocean; this is called the *South Atlantic Anomaly*. The two main belts were discovered by American physicist James Van Allen (1914–2006) and his collaborators during investigations using Geiger counters on board the first US Explorer satellites in 1958. The cosmic ray belt was found by Russian Cosmos satellites in 1991.

variable star A star whose brightness varies with time. *Intrinsic variables* are stars that vary because of some inherent process, such as pulsations in size or events in their atmosphere. In *extrinsic variables*, external factors such as eclipses or obscuring dust affect the amount of light reaching us from the star. There are five main classes of variable star. **Eclipsing binaries** are extrinsic variables in which members of a binary system periodically pass in front of each other as seen from Earth. **Pulsating stars** are intrinsic variables that expand and contract in size, either regularly or irregularly. **Cepheid variables** and **RR Lyrae stars** are examples of pulsating stars with regular cycles of variation. Less regular variation is found among the **long-period variables** (or *Mira stars*), **semiregular variables** and **irregular variables**. **Cataclysmic variables** include **supernovae**, **novae**, **dwarf novae** and **symbiotic variables**. **Eruptive variables** are a very diverse class, including **flare stars**, **R Coronae Borealis stars** and **shell stars**. Finally, **rotating variables** include various types where a star's shape or non-uniform surface brightness produces the variations in magnitude. Each main group also has its own subclassification.

Variable stars have their own peculiar naming system. The letter R is assigned to the first variable to be discovered in a particular constellation (unless it already has an existing designation with a Greek or Roman letter). Subsequent discoveries are given the letters S to Z. Then come RR to RZ, followed by SS to SZ, and similarly to ZZ. After that come AA to AZ, BB to BZ, and so on to QZ (but the letter J is never used). This scheme caters for 334 variable stars in a constellation; subsequently discovered variables are numbered V335, V336, and so on.

variation A perturbation in the Moon's motion caused by the Sun's changing gravitational pull on the Moon as the Moon orbits the Earth.

Varuna A large Kuiper Belt object discovered by R. S. McMillan on 28 November 2000. It has an estimated diameter of 900 km (560 miles) and takes 285 years to orbit the Sun at a mean distance of 43 au.

Vega The star Alpha Lyrae, fifth brightest in the sky, magnitude 0.03, distance 25 light years, luminosity 50 times that of the Sun. Vega is surrounded by a disk of gas and dust extending out to 85 au from it, possibly a **protoplanetary disk**.

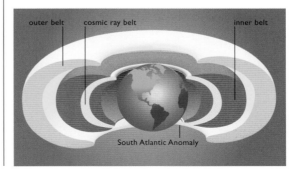

◄ **Van Allen Belts** *Earth is surrounded by radiation belts, which contain energetic particles trapped by the magnetic field. The inner belt dips low over the South Atlantic, where its particle population can present a hazard to satellites.*

Veil Nebula The brightest part of the **Cygnus Loop**, a large supernova remnant in Cygnus.

Vela A southern constellation, once part of the larger grouping of Argo Navis, the ship Argo, and representing the ship's sails. Its leading star is Gamma, magnitude 1.8 but slightly variable; this is the brightest **Wolf–Rayet star**, with a luminosity 40,000 times that of the Sun. Vela is a rich region; the stars Delta (magnitude 1.9) and Kappa (magnitude 2.5) make up the **False Cross** with Iota and Epsilon Carinae. Other features of the constellation are the **Gum Nebula** and the **Vela Pulsar**.

Vela Pulsar A pulsar with a period of 89 milliseconds, discovered in 1968 in the constellation Vela; known as PSR 0833-45. The optical counterpart, discovered in 1977, flashes with the same period.

Venera A series of space probes launched by the Soviet Union to the planet Venus between 1961 and 1983. Venera 7 in 1970 was the first craft to land successfully on Venus, although it transmitted for only 23 minutes before succumbing to the extreme temperatures and pressures at the surface. Venera 8 in 1972 proved more resilient and transmitted for 50 minutes. Veneras 9 and 10 in 1975 obtained the first images of the surface. Veneras 11 and 12 landed in 1978 and sent back information about the atmosphere, but no pictures. Veneras 13 and 14 landed in 1982, obtaining color images of the surface and sampling the rocks. Veneras 15 and 16, both radar mappers, went into orbit around the planet in October 1983 and operated until the following July.

Venus The second major planet from the Sun, and the second largest of the terrestrial planets. Visible around dawn or dusk as the so-called **Morning Star** or **Evening Star**, it is the most conspicuous celestial object after the Sun and the Moon. At its brightest (magnitude −4.7) it is even visible to the naked eye by day in a clear sky. At **greatest elongation** Venus is 45° to 47° from the Sun and has an apparent diameter of 25 arc seconds. Like the Moon and Mercury, it exhibits **phases**. **Transits** of Venus are rare, and occur in pairs eight years apart, with over a century between each pair. The most recent was on 5–6 June 2012 and the next is not until 11 December 2117. The main data for Venus are given in the table.

A telescope shows the planet's dazzling yellowish-white cloud cover. In ultraviolet images, visually elusive markings show clearly as a Y-shaped feature, enabling the clouds' rotation period of just under 4 days to be measured. The **Ashen Light**, **cusp caps** and **Schröter effect** may be watched for by amateur observers; their causes have been long debated, but they are most probably real atmospheric phenomena, and not optical effects.

Spacecraft, including the two US **Pioneer Venus** probes and **Magellan**, have revealed information about Venus's surface. Later Soviet **Venera** landers were built to withstand the extreme surface conditions – a temperature of 750 K and a pressure of over 90 bars – for long enough to transmit measurements and images. Magellan produced highly detailed radar maps of the surface. **Venus Express** reached the planet in 2006 to continue studies of its atmosphere from orbit.

A gently undulating plain covers two-thirds of Venus. Highlands account for a further quarter, and depressions and chasms the remainder. The two principal highland regions are Ishtar Terra, in

VENUS: DATA	
Globe	
Diameter	12,104 km
Density	5.20 g/cm^3
Mass (Earth = 1)	0.8149
Volume (Earth = 1)	0.8568
Sidereal period of axial rotation	243d 0h 30m (retrograde)
Escape velocity	10.4 km/s
Albedo	0.76
Inclination of equator to orbit	177° 20'
Surface temperature	750 K
Surface gravity (Earth = 1)	0.90
Orbit	
Semimajor axis	0.723 au = 108.2 × 10^6 km
Eccentricity	0.007
Inclination to ecliptic	3° 24'
Sidereal period of revolution	224.701d
Mean orbital velocity	35.02 km/s
Satellites	0

the northern hemisphere, and Aphrodite Terra, largely in the southern. The plains are cratered, but there are no craters less than a few kilometers across because smaller **meteoroids** would have been destroyed during passage through the dense atmosphere. Most of the surface features are volcanic in origin. There are large shield volcanoes such as Rhea Mons, and calderas. The volcanic peak Maat Mons is surrounded by lava flows that may be no more than two decades old, suggesting that Venus is still volcanically active. There are many other types of near-circular volcanic feature, which have been given names such as *arachnoids*, *coronae* and *ovoids*.

The atmosphere is 96% carbon dioxide and 3½% nitrogen, with traces of helium, argon, neon and krypton. Its great density has produced the very high surface temperature by an extreme **greenhouse effect**. At various altitudes there are haze layers of sulfuric acid, sulfur dioxide, water vapor and sulfur. A chemically active atmosphere is probably the only agent for "weathering" of surface features. Wind speeds in the atmosphere, which are 350 km/h (220 miles/h) at the cloud-tops, fall to 10 km/h (6 miles/h) at the surface.

The internal structure is largely unknown. Venera measurements indicate a low-density crust. By analogy with the similar-sized Earth, there may be a silicate mantle and iron–nickel core. Certainly there is no appreciable magnetic field, although the solar wind does interact strongly with Venus's ionosphere to produce a well-defined **bow shock** 2000 km (1300 miles) ahead of the planet.

Venus Express A European Space Agency Venus orbiter launched on 9 November 2005. It reached the planet in April 2006 to conduct the first global investigation of the Venusian atmosphere and its plasma environment.

vernal equinox (spring equinox) The point at which the Sun crosses the celestial equator from south to north, or the time at which this occurs. This point, also known as the *First Point of Aries*, is the zero point of the celestial coordinate called right ascension. The vernal equinox moves westward by about one-seventh of an arc second daily because of the effect of **precession**.

Very Large Array (VLA) The most complex radio telescope on a single site in the world. It consists of 27 dishes, each 25 m (82 ft) in diameter, which can be moved along a giant Y-shaped track on the plains near Socorro, New Mexico. Each arm of the VLA can be made up to 21 km (13 miles) long, using the technique of **aperture synthesis** to produce a virtual dish some 34 km (21 miles) wide. In 2012 it was renamed the Karl G. Jansky Very Large Array.

Very Large Telescope (VLT) A group of four 8.2 m (232-inch) telescopes at Paranal Observatory, Chile, built by the **European Southern Observatory**, the last of which was completed in 2001. With all four telescopes linked together, the VLT has a light-gathering power equivalent to that of a single 16 m (630-inch) mirror. The four main instruments can be linked with several smaller ones to form an **optical interferometer**.

Very Long Baseline Array (VLBA) An array of ten 25 m (82 ft) radio telescopes spread across the continental USA and Hawaii, giving a baseline for **very long baseline interferometry** 8000 km

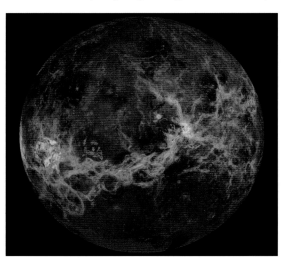

▶ **Venus** *Centered on longitude 180°, this image is a composite view of Venus prepared from Magellan Imaging Radar results in 1991.*

(5000 miles) long. Resolutions of 1 milliarcsecond are possible. The instrument began operation in 1993. SEE ALSO **radio interferometer**.

very long baseline interferometry (VLBI) A technique in radio astronomy that extends the concept of **aperture synthesis** by combining signals from radio telescopes which may be thousands of kilometers apart to achieve a high accuracy of positional measurements, of 1 milliarcsecond for the **Very Long Baseline Array**, for example. By combining signals from radio astronomy satellites and ground-based radio telescopes, baselines of around 40,000 km (25,000 miles) have been achieved.

Vesta Asteroid no. 4, diameter 501 km (311 miles), discovered in 1807 by Heinrich Olbers. At magnitude 6.4 it is the brightest asteroid and the only one ever visible (under ideal conditions) to the naked eye. Vesta's spectrum suggests that its surface, unique among the larger asteroids, was once molten (SEE **achondrite**). There is a huge, 500 km (310-mile) impact basin at the south pole called Rheasilvia.

Viking Two NASA space probes to Mars, launched on 20 August and 9 September 1975. Each was a combined orbiter and lander. Viking 1 went into orbit around Mars in June 1976 and the lander landed on 20 July 1976; Viking 2 went into orbit in August 1976 and its lander reached the surface on 3 September. Both landers photographed their surroundings, studied the weather and soil, and conducted experiments which found no indications of life at the landing sites. The orbiters photographed the surface of the planet and its two satellites.

Virgo The largest constellation of the **zodiac**, lying on the celestial equator, and representing the goddess of justice. Its brightest star is **Spica**. Gamma is a splendid binary with equal components of magnitude 3.6, period 169 years, at their closest in the year 2005. Virgo is one of the richest areas for faint galaxies since it includes the **Virgo Cluster**. Closer to us than this cluster is the **Sombrero Galaxy**.

Virgo Cluster (Coma–Virgo Cluster) A rich concentration of galaxies in the constellation Virgo, although some lie across the border in Coma Berenices (hence the alternative name). The center of the cluster is about 55 million light years away, and is marked by the giant elliptical galaxy M87, also known as the radio source Virgo A, which is ejecting a jet of gas. In all there are 16 Messier objects in the cluster, which contains over 2000 galaxies in all.

visible spectrum The range of wavelengths in the electromagnetic spectrum perceptible by the human eye. It varies slightly from one person to another, but extends typically from 385 nm (violet) to 700 nm (red). SEE ALSO **electromagnetic radiation**.

visual binary A **binary** system that can be observed as a double star with a telescope, as distinct from a spectroscopic binary.

visual magnitude (symbol m_v) The **apparent magnitude** of a celestial object in the color region to which the human eye is most sensitive, i.e. about 560 nm.

VLA ABBREVIATION FOR **Very Large Array**.

VLBA ABBREVIATION FOR **Very Long Baseline Array**.

VLBI ABBREVIATION FOR **very long baseline interferometry**.

VLT ABBREVIATION FOR **Very Large Telescope**.

Volans A small, faint southern constellation adjacent to Carina, representing a flying fish. Its brightest stars are of fourth magnitude.

volatile A substance that solidifies at a low temperature. Substances such as water and carbon dioxide, familiar as gases on Earth, are

◀ **Very Large Array** One of the world's premier radio observatories, the Very Large Array is located at an elevation of 2120 km (6970 ft) on the plains of San Agustin, near Socorro, New Mexico.

examples. Bodies in the Solar System that formed nearer the Sun have a lower volatile content than those that formed farther away. Volatiles in comets, which formed in the outer reaches of the Solar System, evaporate to form the gas tail as the comet approaches the Sun.

volcanism The eruption of molten material at the surface of a planetary body. Volcanism in which molten silicate rock erupts, cools, and becomes new surface material has taken place on Venus and Mars as well as the Earth, and similar material produced the lunar maria (SEE **mare**) by volcanic flooding. In the outer Solar System a mixture of water ice and ammonia ice can behave as lava in a process called *cryovolcanism*, and has produced, for example, the volcanic floodplains on **Ariel**. Sulfur volcanism operates on **Io**, the most volcanically active world known.

Voyager Two NASA space probes to the outer planets. Launched in 1977, on 20 August (Voyager 2) and 5 September (Voyager 1), they passed Jupiter in March and July 1979, respectively, photographed the planet and satellites, and discovered Jupiter's ring system. Voyager 1 reached Saturn in November 1980, followed by Voyager 2 in August 1981; they photographed the rings, as well as the satellites. Voyager 1's path was bent out of the ecliptic by its close approach to Titan, but Voyager 2 proceeded to Uranus, reaching it in January 1986, and Neptune, in August 1989. Both Voyager probes are now on their way out of the Solar System, their last task to locate the heliopause, the boundary of the **heliosphere**. Voyager 1 crossed the heliopause in 2013, some 125 au from the Sun.

Vulpecula A dim constellation next to Cygnus, representing a fox. Its brightest stars are of fourth magnitude. It contains the **Dumbbell Nebula**, reputedly the easiest planetary nebula to see in the sky.

▼ **Viking** Compiled from images obtained by the Viking orbiters, this composite shows the Amphitrites Patera region of Mars. Radial ridges extend northward for about 400 km (250 miles) from this old volcano.

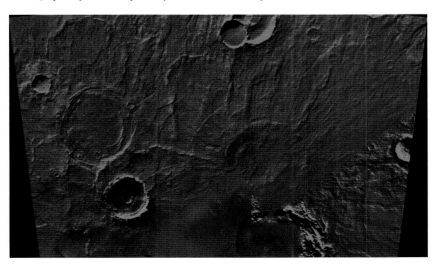

W

weakly interacting massive particle SEE **WIMP**.

Weizsäcker, Carl Friedrich von (1912–2007) German theoretical physicist and astrophysicist. In 1938 he and Hans **Bethe** independently proposed a detailed theory for the production of energy in the Sun and other stars (SEE **carbon–nitrogen cycle**) in which hydrogen is converted into helium by nuclear fusion. In 1944 he revived the *nebular hypothesis* of the origin of the Solar System originally proposed by Immanuel Kant and Pierre Simon de **Laplace**.

West, Comet C/1975 V1 A bright comet discovered by Danish astronomer Richard West (1941–) in November 1975. The following March it became a prominent naked-eye object with a fan-shaped tail. The comet was extremely active, and the nucleus broke into at least four fragments as it passed within 30 million km (20 million miles) of the Sun.

Westerbork Radio Observatory A radio astronomy observatory located near Groningen in the Netherlands, operated by the Netherlands Institute for Radio Astronomy. Its major instrument is an **aperture synthesis** telescope consisting of 14 dishes 25 m (82 ft) in diameter, along an east–west baseline 2.7 km (1.7 miles) long.

Whipple, Fred Lawrence (1906– 2004) American astronomer. He was best known for his "dirty snowball" theory of comets, proposed in 1949 and shown to be correct in 1986 when space probes were sent to **Halley's Comet**. Whipple discovered several comets and worked on cometary orbits; he also studied planetary nebulae, stellar evolution and the Earth's upper atmosphere.

Whirlpool Galaxy A well-defined spiral galaxy of type Sc which appears face-on and has very prominent arms; also known as M51 and NGC 5194. It lies in Canes Venatici, not far from the end of the tail of Ursa Major, and was the first galaxy in which spiral structure was noted, by Lord **Rosse** in 1845. A small companion galaxy, NGC 5195, is connected to it by an extension of one of the spiral arms. The Whirlpool Galaxy is 25 million light years away.

white dwarf A type of star about the size of the Earth, but with a mass about that of the Sun. As a result, its density is enormously greater than that of any terrestrial material (0.1 to 100 tonnes/cm³). This is because the normal atomic structure is broken down completely, with electrons and nuclei packed tightly together (the state known as **degenerate matter**). A white dwarf cannot have a mass of more than about 1.4 solar masses (the **Chandrasekhar limit**). For larger masses, gravity overwhelms the pressure of the electrons and the star collapses under its own weight, forming a **neutron star**

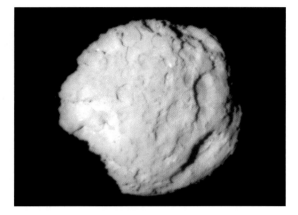

▲ **Wild 2, Comet 81P/** *The heavily cratered nucleus of Comet 81P/Wild 2, as imaged by the Stardust spacecraft on 2 January 2004.*

or **black hole**. White dwarfs are of low luminosity and gradually cool down to become cold, dark objects. They represent the final stage in the evolution of low-mass stars after they have lost their outer layers. The first white dwarf to be discovered was the companion of **Sirius**.

Wide Field Infrared Survey Explorer (WISE) A NASA satellite launched on 14 December 2009 that made an all-sky survey at four infrared wavelengths with its 0.4 m (16-inch) telescope. One of its aims was to provide a source catalog for the **James Webb Space Telescope**. Its main mission ended in February 2011 but the spacecraft was reactivated in September 2013 to begin a search for near-Earth objects, and was renamed NEOWISE.

Widmanstätten pattern The distinctive pattern on a sectioned and polished surface of an **iron meteorite** that has been etched with acid. It is named after Count Aloys Joseph von Widmanstätten (1754–1849), who noted it in 1808. The pattern is only seen in meteoritic material, and consists of intersecting plates of the iron–nickel minerals kamacite and taenite.

Wild 2, Comet 81P/ A periodic comet in a 6.4-year orbit, discovered by the Swiss astronomer Paul Wild in 1978. The **Stardust** spacecraft made a close (236 km; 148 miles) flyby in January 2004, imaging the comet's nucleus and collecting material for return to Earth. The nucleus was found to be a 5 km (3-mile) diameter, more or less spherical body with heavy cratering; its surface topography was suggestive of greater mechanical strength than had previously been assumed for comets. Numerous gas jets were seen to be emerging from the nucleus.

Wild Duck Cluster An open cluster in the constellation Scutum, also known as M11 and NGC 6705. Its wedge shape resembles wild ducks in flight. The cluster lies 6200 light years away.

Wilkinson Microwave Anisotropy Probe (WMAP) A NASA spacecraft launched on 30 June 2001 to study the small-scale variations (anisotropy) in the **cosmic microwave background** (CMB). It was positioned at the Earth's L_2 **Lagrangian point**, from where it mapped the CMB at high resolution. It is named after David Wilkinson, a member of the mission's science team who died in 2002. WMAP made its last observations on 19 August 2010.

William Herschel Telescope (WHT) A reflecting telescope with a 4.2 m (165-inch) mirror at the **Roque de los Muchachos Observatory** in the Canary Islands. It has facilities for high-dispersion and multi-object spectroscopy. The WHT began operation in 1987.

Wilson, Robert Woodrow (1936–) American physicist. In 1964, with Arno **Penzias**, he detected the **cosmic microwave background** radiation that provides the strongest evidence for the **big bang**. They shared the 1978 Nobel Prize for Physics for their discovery.

Wilson effect The foreshortening of a **sunspot** when it is near the Sun's limb, accompanied by a widening of the penumbra on the side nearest to the limb, and a narrowing on the side farthest from the limb. The phenomenon was discovered by the Scottish astronomer Alexander Wilson (1714–86), who took it as an effect of perspective, indicating that sunspots are saucer-shaped depressions. However, not all spots show the Wilson effect. The effect is now thought to

▲ **white dwarf** *This close-up view from the Hubble Space Telescope shows part of the globular cluster M4 in Scorpius. While most of the objects seen here are old, red stars, several white dwarfs (circled in blue) have also been found.*

▶ **Whirlpool Galaxy**
The magnificent Whirlpool Galaxy takes its name from its pronounced spiral structure. The small companion galaxy at top, NGC 5195, appears to be interacting with the Whirlpool.

arise from sunspots being more transparent than the surrounding photosphere, and the umbra more transparent than the penumbra.

WIMP Abbreviation for "weakly interacting massive particle," a hypothetical atomic particle whose existence has been invoked to account for the **missing mass** of the Universe. WIMPs would interact with each other and with ordinary matter only via the weak nuclear force and gravity, and so would take no part in nuclear fusion reactions.

winter solstice SEE **solstice**.

WISE ABBREVIATION FOR **Wide Field Infrared Survey Explorer**.

WMAP ABBREVIATION FOR **Wilkinson Microwave Anisotropy Probe**.

Wolf, Maximilian Franz Joseph Cornelius ("Max") (1863–1932) German astronomer. He was a pioneer of photographic methods in astronomy. In 1891 he made the first photographic discovery of an asteroid (Brucia, no. 323); in all he made 232 asteroid discoveries. He detected several new nebulae, including the **North America Nebula**, the nebulosity around the Pleiades, numerous dark nebulae and the first cluster of galaxies (in Coma Berenices) to be identified as such, as well as the **Virgo Cluster**.

Wolf number SEE **relative sunspot number**.

Wolf–Rayet star (WR star) A highly luminous star whose spectrum contains extremely strong emission lines. WR stars are divided into two kinds. In the WN type, emission lines from nitrogen dominate the spectrum. In the WC type, emission lines of carbon and oxygen predominate. Both types have strong lines of helium and a few have moderate or weak lines of hydrogen as well. All the emission lines are broad. WR stars are very hot, with surface temperatures between 25,000 and 50,000 K, luminosities between 100,000 and 1 million times the Sun's, and masses from 10 to 50 solar masses. They have strong **stellar winds** that typically carry away 3 solar masses per million years. Many central stars of planetary nebulae are WR stars. The type of WR star that emerges after the original hydrogen-rich envelopes have been lost depends on how far the star's evolution had progressed. Wolf–Rayet stars are named after their French discoverers, Charles Wolf (1827–1918) and Georges Rayet (1839–1906).

wormhole A hypothetical tunnel in **spacetime**. In 1916 Albert Einstein discovered that one of the solutions to equations describing a black hole showed that there could be a connection between different regions of spacetime. In principle, this would make time travel possible. Recent cosmological theories suggest that on very small scales spacetime has a foam-like structure, pervaded by wormholes. These wormholes could lead to the formation of "baby universes," which would have profound implications for our understanding of the **big bang**.

wrinkle ridge A winding ridge on the surface of a lunar mare, having sloping sides rising to a typical height of 200 m (650 ft), and several hundred kilometers long. They are often associated with **rilles**, and may have originated when the lava that formed the maria cooled and contracted. Wrinkle ridges are also found on Venus.

W Ursae Majoris star A member of a class of eclipsing binary stars, almost in contact as they orbit each other. They differ from **Beta Lyrae stars** in that the two components are smaller, less luminous stars of nearly identical brightness.

W Virginis star A member of a group of pulsating variable stars superficially similar to **Cepheid variables**. They are giant stars, typically in the spectral range G0 to M0 and with absolute magnitudes from −1 to −4, on the instability strip in the **Hertzsprung–Russell diagram**. Their masses can be as low as 0.5 solar masses, suggesting that they have evolved from low-mass main-sequence stars. Light curves of W Virginis stars can be distinguished from those of classical Cepheids by their less regular shape and a double-peaked maximum. They are Population II stars (SEE **stellar populations**) with periods that range from about 1 to about 100 days, and their **period–luminosity law** is distinctly different from that of normal Cepheids; however, they can still be used as distance estimators for galactic and extragalactic objects. Occasionally W Virginis stars show small period changes but in one case, RU Camelopardalis, the variations suddenly stopped in 1964 for about three years. It is not understood how pulsations can stop, or start, in such a short time.

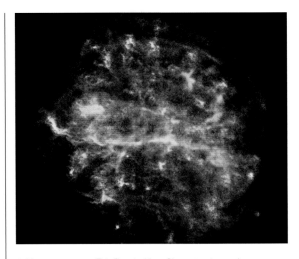

▲ **X-ray astronomy** This Chandra X-ray Observatory image shows a young supernova remnant in Centaurus. The rapidly expanding shell of gas contains large amounts of oxygen, neon, magnesium, sulfur and silicon. It has a pulsar at its center.

XMM-Newton The ESA X-Ray Multi-Mirror Mission satellite, launched on 10 December 1999. XMM-Newton carries X-ray telescopes with a large effective collecting area, enabling long uninterrupted exposures, and an optical monitor, the first to be flown on an X-ray observatory.

X-ray astronomy The study of celestial objects emitting **electromagnetic radiation** of short wavelength (about 0.01 to 10 nm), shorter than ultraviolet radiation but longer than gamma rays. The Earth's atmosphere is opaque to X-rays, so this radiation must be studied by means of instruments aboard rockets or satellites. For astronomical purposes X-rays are usually described according to the energy associated with the X-ray photons; this energy is expressed in **electronvolts**.

X-ray astronomy began in 1949 with the discovery that the Sun emits X-rays. Other celestial X-ray sources were detected by rocket flights in the 1960s, followed by the first all-sky X-ray surveys by the US satellite Uhuru, launched in 1970, and later the UK satellite Ariel V. These surveys revealed the existence of **X-ray binaries**, in which one component is a neutron star or a black hole, as well as X-ray emission from supernova remnants, active galaxies, quasars and hot gas pervading clusters of galaxies. An **X-ray telescope** aboard the Einstein satellite, launched in 1978, produced the first true images of X-ray sources, rather than simply counting X-rays as previous instruments had done. Objects in nearby galaxies such as M31 and the Magellanic Clouds were resolved and studied for the first time. It was also discovered that almost all types of ordinary star emit X-rays. Subsequent X-ray satellites have included **Rosat**, **Yohkoh**, the **Chandra X-ray Observatory** and **XMM-Newton**.

X-ray binary A binary system consisting of a normal star and a collapsed star – a **neutron star**, **black hole** or, in less intense sources, a **white dwarf**. The two components are very close together. Where the normal star is a giant or supergiant, a stellar wind blows directly on to the compact companion, causing the X-ray emission. In other cases, expansion of the normal star in the course of its evolution results in a flow of gas toward the collapsed star. This matter forms an **accretion disk** of hot material which emits X-rays. SEE ALSO **X-ray burster**.

X-ray burster A source of intense flashes of X-rays. These X-ray bursts have rise times of approximately 1 second, fall times of approximately 60 sec-

▼ **X-ray binary** The strong X-ray source Cygnus X-3 is a close binary system in which matter from a normal star is being drawn into a neutron star or black hole. The X-ray emission from Cygnus X-3 varies regularly with a 4.8-hour period, as the compact star circles its companion star.

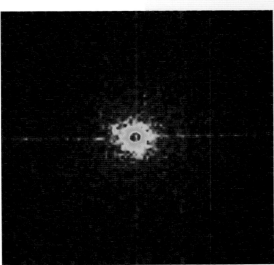

TYPES OF YEAR	
anomalistic year	365.25964 mean solar days
eclipse year	346.62003 mean solar days
sidereal year	365.25636 mean solar days
tropical year	365.24219 mean solar days

onds and total luminosities equivalent to one week's energy output of the Sun. The intervals between bursts are irregular, ranging from hours to days, while many sources undergo burst-inactive phases that can last for weeks or even months. Burst sources are associated with old Population II objects (SEE **stellar populations**) and emit no pulsations. They are thought to be binaries containing an old neutron star with no magnetic field, which explains the absence of pulsations. The neutron star is paired with an old low-mass star, and individual bursts come from thermonuclear explosions after material accreted from the companion star has exceeded a critical mass on the neutron star's surface. *X-ray transients* are nova-like outbursts of X-rays believed to come from similar binaries in which the mass transfer is very uneven. SEE ALSO **pulsar**.

X-ray telescope An instrument for imaging X-ray sources. All X-ray telescopes are satellite-borne because X-rays do not penetrate the Earth's atmosphere. Unlike less energetic electromagnetic radiation, X-rays cannot be focused by reflection from a conventional concave mirror (as in a **reflecting telescope**) because, for all but the shallowest angles of incidence, they penetrate the surface. In *grazing incidence* X-ray telescopes the focusing element is a pair of coaxial surfaces, one **paraboloidal** and the other **hyperboloidal**, from which incoming X-rays are reflected at a very low "grazing" angle toward a focus. The detecting element is usually a **CCD** adapted for X-ray wavelengths. An alternative instrument is the *microchannel plate detector*, in which the incident radiation falls on a plate which is made up of many fine tubes, rather like a short, wide fiber-optic bundle. The plate is charged, so that the radiation generates electrons which are accelerated down the tubes and together form an image that can be read off.

X-ray transient SEE **X-ray burster**.

Yagi antenna A basic form of antenna used in simple **radio telescopes**. It consists of several parallel elements mounted on a straight member, and often forms the basis of cheap arrays used in **aperture synthesis**; it is also a familiar form of TV antenna. It was developed by the Japanese engineer Hidetsugu Yagi (1886–1976).

year The time taken by the Earth to complete one revolution around the Sun. Various years, defined according to the choice of reference point, are given in the table. The *civil year* (**calendar** year) averages 365.2425 **mean solar days**. SEE ALSO the entries for each type of year listed in the table.

Yerkes Observatory The observatory of the University of Chicago, at Williams Bay, Wisconsin. It was founded by George Ellery **Hale**. Its main instrument is a 1 m (40-inch) refractor, opened in 1897 and still the largest in the world.

Ymir A small outer **satellite** of Saturn, discovered in 2000.

Yohkoh A satellite launched on 30 August 1991 to study the Sun, in particular solar flares, at X-ray and gamma-ray wavelengths. The mission was a Japanese–British–American collaboration; Yohkoh is Japanese for "sunbeam." The satellite returned many X-ray images of the Sun, the first obtained from orbit since Skylab missions in 1973. Its instruments revealed much about the rapid changes that occur in the corona, and how flares originate and evolve. Yohkoh ceased operations on 14 December 2001.

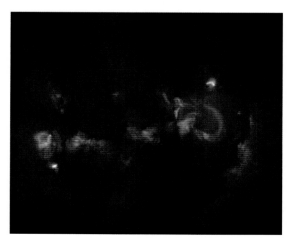

▶ **Yohkoh** *This Yohkoh image of the active Sun at X-ray wavelengths shows hot coronal plasma in magnetic loops above active regions. Cooler, dark regions above the poles are described as coronal holes; these extend to lower solar latitudes at sunspot minimum.*

Z Andromedae star SEE **symbiotic variable**.

Z Camelopardalis star A member of a small subgroup of **dwarf novae** similar to **U Geminorum stars**, except that they experience occasional "standstills" (called "stillstands" in North America), remaining more or less constant at some intermediate brightness. The onset of the standstills, and their duration, ranging from a few days to many months, are unpredictable.

Zeeman effect The splitting of a spectral line into several components by a strong magnetic field. Where these components cannot be resolved, the effect is apparent as a widening of the original line. The Zeeman effect occurs in the spectra of sunspots and stars. It demonstrates the existence of magnetic fields in celestial bodies and, since the field strength depends on the degree of splitting, it allows the fields to be measured. The phenomenon was predicted by Hendrik Lorentz, and discovered in 1896 by Pieter Zeeman (1865–1943).

Zelenchukskaya Observatory An observatory, also known as the Special Astrophysical Observatory, on Mount Pastukhov in the Caucasus Mountains of southern Russia, the site of the 6 m (236-inch) Large Azimuthal Telescope, which was opened in 1975. Also at Zelenchukskaya is the RATAN-600 radio telescope, which consists of a ring of reflecting panels 600 m (2000 ft) in diameter.

zenith The point on the **celestial sphere** that is vertically above the observer, 90° from the horizon. The point diametrically opposite, beneath the observer's feet, is called the **nadir**.

zenithal hourly rate (ZHR) The number of meteors per hour that would be seen at the maximum of a particular **meteor shower** under ideal conditions if the radiant were immediately overhead. The ZHR is obtained by applying a number of correction factors to the observed hourly rate of meteors in the shower. The observed hourly rate is always less than the ZHR. There are short-term and long-term variations in showers' ZHRs as the associated **meteor streams** evolve.

zenith distance The angular distance of a celestial body from the zenith. It is equal to 90° minus the altitude of the body above the horizon.

zero-age main sequence (ZAMS) The **main sequence** on the Hertzsprung–Russell diagram as defined by stars of zero age, i.e. before they have undergone any substantial evolution.

ZHR ABBREVIATION FOR **zenithal hourly rate**.

zodiac A belt on the celestial sphere, about 8° either side of the ecliptic, which forms the background for the motions of the Sun, Moon and planets (except Pluto). The zodiac is divided into 12 *signs*, each 30° long, which are named after the constellations they contained at the time of the ancient Greeks: Aries, Taurus, Gemini, Cancer, Leo, Virgo, Libra, Scorpio, Sagittarius, Capricorn, Aquarius and Pisces. Because of **precession** these signs no longer coincide with the constellations of the same names. The ecliptic also passes through the constellation Ophiuchus, which is not a zodiacal sign.

zodiacal light A cone of faint light, usually fainter than the Milky Way, visible at all seasons from dark locations in the tropics in the absence of moonlight. It stretches along the ecliptic from the western horizon after evening twilight, or from the eastern horizon before morning twilight. At the point in the sky opposite the Sun, the zodiacal light brightens a little; this is the **gegenschein**. The spectrum of the zodiacal light resembles the Sun's, indicating that the phenomenon results from the scattering of sunlight by particles in the plane of the ecliptic. These particles constitute the *zodiacal dust cloud*, and probably originate partly from matter ejected by the Sun, and partly from the decay of comets and asteroids.

zone of avoidance A region of the sky along the plane of the Milky Way where almost no galaxies can be seen because of absorption by interstellar gas and dust in our Galaxy. It is between 10° and 40° wide.

Zwicky, Fritz (1898–1974) Swiss–American astronomer, born in Bulgaria. In 1934 he and Walter **Baade** suggested that what is left after a supernova explosion is a neutron star; this was confirmed in 1968 with the discovery of the pulsar in the **Crab Nebula**. In the mid-1960s, using the newly developed **Schmidt camera**, Zwicky discovered and examined many supernovae in other galaxies, and began a long study of **clusters of galaxies**.

SYMBOLS FOR UNITS, CONSTANTS AND QUANTITIES

a	semimajor axis
Å	angstrom unit
au	astronomical unit
c	speed of light
d	distance
e	eccentricity
E	energy
eV	electron-volt
f	following
F	focal length, force
g	acceleration due to gravity
G	gauss
G	gravitational constant
h	hour
h	Planck constant
H_0	Hubble constant
Hz	hertz
i	inclination
IC	*Index Catalogue*
Jy	jansky
k	Boltzmann constant
K	degrees kelvin

L	luminosity
L_n	Lagrangian points ($n = 1$ to 5)
l.y.	light year
m	meter, minute
m	apparent magnitude, mass
m_{bol}	bolometric magnitude
m_{pg}	photographic magnitude
m_{pv}	photovisual magnitude
m_V	visual magnitude
M	absolute magnitude, mass (stellar)
N	newton
p	preceeding
P	orbital period
pc	parsec
q	perihelion distance
q_0	deceleration parameter
Q	aphelion distance
r	radius, distance
R	Roche limit
s	second
t	time

T	temperature (absolute), epoch (time of perihelion passage)
T_{eff}	effective temperature
v	velocity
W	watt
y	year
z	redshift
α	constant of aberration, right ascension
δ	declination
λ	wavelength
μ	proper motion
ν	frequency
π	parallax
ω	longitude of perihelion
Ω	observed/critical density ratio, longitude of ascending node
°	degree
'	arc minute
"	arc second

MULTIPLES AND SUBMULTIPLES USED WITH SI UNITS

Multiple	Prefix	Symbol	Submultiple	Prefix	Symbol
10^3	kilo-	k	10^{-3}	milli-	m
10^6	mega-	M	10^{-6}	micro-	μ
10^9	giga-	G	10^{-9}	nano-	n
10^{12}	tera-	T	10^{-12}	pico-	p
10^{15}	peta-	P	10^{-15}	femto-	f
10^{18}	exa-	E	10^{-18}	atto-	a

CONVERSION FACTORS

Distances

1 nm = 10 Å
1 inch = 25.4 mm
1 mm = 0.03937 inch
1 ft = 0.3048 m
1 m = 39.37 inches = 3.2808 ft
1 mile = 1.6093 km
1 km = 0.6214 mile
1 km/s = 2237 miles/h
1 pc = 3.0857×10^{13} km = 3.2616 l.y. = 206,265 au
1 l.y. = 9.4607×10^{12} km = 0.3066 pc = 63,240 au

Temperatures (to the nearest degree)

°C to °F :	×1.8, +32
°C to K :	+273
°F to °C :	−32, ÷1.8
°F to K :	÷1.8, +255
K to °C :	−273
K to °F :	×1.8, −460

Note: To convert temperature *differences*, rather than points on the temperature scale, ignore the additive or subtractive figure and just multiply or divide.

THE GREEK ALPHABET

α	A	alpha		ν	N	nu
β	B	beta		ξ	Ξ	xi
γ	Γ	gamma		o	O	omicron
δ	Δ	delta		π	Π	pi
ε	E	epsilon		ρ	P	rho
ζ	Z	zeta		σ	Σ	sigma
η	H	eta		τ	T	tau
θ	Θ	theta		υ	Y	upsilon
ι	I	iota		ϕ	Φ	phi
κ	K	kappa		χ	X	chi
λ	Λ	lambda		ψ	Ψ	psi
μ	M	mu		ω	Ω	omega

ACKNOWLEDGMENTS

ABBREVIATIONS

AURA – Association of Universities for Research in Astronomy, Inc.
CXC – Chandra X-ray Observatory Center, Harvard-Smithsonian Center for Astrophysics
DLR – Deutschen Zentrum für Luft- und Raumfahrt
ESA – European Space Agency
ESO – European Southern Observatory
GSFC – Goddard Space Flight Center
JHU – Johns Hopkins University
JPL – Jet Propulsion Laboratory
JSC – Johnson Space Center
MPE – Max-Planck-Institut für extraterrestrische Physik, Garching
NASA – National Aeronautics and Space Administration
NOAO – National Optical Astronomy Observatory
NRAO – National Radio Astronomy Observatory
NSF – National Science Foundation
SOHO – Solar and Heliospheric Observatory. SOHO is a project of international cooperation between ESA and NASA
STScI – Space Telescope Science Institute
ST-ECF – Space Telescope European Coordinating Facility
USGS – US Geological Survey

ACKNOWLEDGMENTS

Contributors to the "A to Z of Astronomy" section (1st edition): Neil Bone, Storm Dunlop, Professor Chris Kitchin, Tim Furniss, Professor Fred Watson, John Woodruff. Editorial consultant Ian Ridpath (1st and 2nd editions).

All photographs supplied by Galaxy Picture Library.

Front/rear endpapers: NASA/JPL-Caltech/University of Colorado (*Eta Carinae star-forming region*)

Robin Scagell 8, 9bl, 11, 13, 17b, 18b, 22, 24, 26, 28, 29, 30, 31t, 32, 33, 36b, 38t, 40tl, 41, 42, 43r, 44, 45, 48t, 54c, 54bl, 63, 67b, 70t, 72tr, 84t, 92t, 92c, 94b, 96tl, 97b, 99, 110r, 112, 122, 124, 131br, 132, 134, 139tl, 139bl, 140t, 141, 147cr, 154b, 155b, 160cr, 162bl, 162bc, 163, 166c, 166b, 168tl, 168cr, 168br, 169, 171cr, 175, 176br, 178bl, 182c, 185, 186l, 190r, 191cr, 193l, 233.

4–5 (*Milky Way*) NASA/JPL-Caltech/University of Wisconsin; 6–7 (*AE Aurigae*) T. A. Rector and B. A. Wolpa/ NOAO/AURA/NSF; 9tr ESO; 10t Rainer Girnstein/ING Telescopes; 10b Nando Patat; 12 Nick King; 14t Maurice Gavin; 14b University of Michigan; 15 Galaxy Picture Library; 16t NOAO/AURA/NSF; 16b STScI; 17t STScI; 18t IDA; 19t NASA; 19b NASA/WMAP Science Team; 21tl Michael Stecker; 21tr Darren Bushnall; 31b Jon Harper; 34 Celestron International; 35t Orion Optics; 35b Optical Vision Ltd (www.opticalvision.co.uk); 36t Celestron International; 37tl Optical Vision Ltd; 37tr Ian Palmer; 37b Meade; 38c Ian Palmer; 38b, 39 Dave Tyler; 40br Optical Vision; 43l Celestron International; 46 Thierry Legault; 47 NASA; 48bl, 48br, 51 Thierry Legault; 53bl ESO; 53t, 53br, 54t Thierry Legault; 56, 57t, 57l, 57r NASA; 57c Thierry Legault; 57b Damian Peach; 58, 59tl, 59tr NASA; 59b Thierry Legault; 60tr Jamie Cooper; 60tl, 60b, 61 NASA; 62 SOHO/NASA/ESA; 64t Ian Phelps; 64bl Damian Peach/Dave Tyler; 65t ESA/NASA; 65c Coronado; 65bl G. Giuliani/P. Confino/P. Chavalley; 65br SOHO, NASA/ESA; 66t SDO/NASA/Galaxy; 66b Swedish Solar Vacuum Telescope; 67t David Graham; 67c Erwin van der Velden; 68 NASA/JHU-APL/Carnegie Institute of Washington; 69t JPL; 69c David Graham; 69b Damian Peach; 70b NASA; 71t, 71c Carle Pieters, Brown University; 71cl, 71b JPL; 72tl David Graham; 72b, 73t, 73br Damian Peach; 73bl NASA/JPL/Cornell; 74t USGS; 74b NASA; 75t Calvin J. Hamilton; 75c NASA/JPL/Malin Space Science Systems; 75b Malin Space Science Systems; 76t, 76cl, 76b NASA; 76r NASA/JPL/Malin Space Science Systems; 77l, 77t NASA/JPL/Cornell; 77cr ESA/DLR/FU Berlin (G. Neukum); 77b NASA/JPL-Caltech/MSSS/Galaxy; 78tl ESA/DLR/FU Berlin (G. Neukum); 78tr NASA/JPL-Caltech/MSSS/Galaxy; 78c NASA/JPL-Caltech/Cornell; 78b JPL; 79t STScI; 79b, 80t Damian Peach; 80b Matthew Boulton; 81t NASA/JPL/University of Arizona; 81b STScI; 82, 83 JPL; 84b NASA/JPL/STScI; 85tl Matthew Boulton; 85tr STScI; 85bl NASA/JPL-Caltech/University of Arizona/University of Idaho/Galaxy; 85br ESA/NASA/University of Arizona; 86t Damian Peach; 86b JPL; 87 NASA/JPL/Space Science Institute; 88t Ed Grafton; 88c Maurice Gavin; 88b NASA/E. Karkoschka/University of Arizona; 89t JPL; 89b NASA/E. Karkoschka/University of Arizona; 90 JPL; 91t STScI; 91c NASA/Johns Hopkins University Applied Physics Laboratory/Southwest Research Institute; 91b Maurice Gavin; 92b NASA/JPL-Caltech/JAXA/ESA/Galaxy; 93t NASA/ESA/Galaxy; 93b Martin Lewis; 94t JHU/NASA; 95t Paul Andrew; 95c Nigel Evans; 95b John Gillett; 96tr David Arditti/Galaxy; 96b Paul Sutherland; 97t Nigel Evans; 100t STScI; 100b Hubble Heritage STScI/AURA; 101l C. R. O'Dell/Rice University; 101r STScI; 102t NASA/Greg Bacon (STScI/AVL); 102b STScI; 103 Andrea Dupree, Ronald Gilliland (STScI) NASA/ESA; 104t NASA/SAO/CXC; 104b Yoji Hirose; 105t, 105b Bruce Balick/STScI; 105cl STScI; 105cr Howard Bond (STScI), Robin Ciardullo (PSU); 106t Tom Boles; 106b STScI; 107t ESA/NASA/Felix Mirabel; 107b Michael Stecker; 108 Celestron International; 109 Damian Peach; 110l Sandy Tweddle; 111t Hillary Mathis, REU program/NOAO/AURA/NSF; 111b Philip Perkins; 113l Michael Stecker; 113r Darren Bushnall; 114t STScI; 114b Bill Schoening, Vanessa Harvey/REU program/AURA/NOAO/NSF; 115 Hubble Heritage/AURA/STScI; 116t Local Group Galaxies Survey Team/NOAO/AURA/NSF; 116b NASA/ESA; 117t, 117c, 117br STScI; 117bl P. Wilkinson et al; 118l Faith Jordan; 118r NOAO/AURA/NSF; 119 Peter Shah; 120 Martin Lewis; 121 Philip Perkins; 125l Michael Stecker; 125tr Duncan Radbourne; 126l NOAO/AURA/NSF; 126tr, 126br ESO; 127l Axel Mellinger; 127tr Michael Stecker; 127b NASA, C. R. O'Dell and S. K. Wong (Rice University); 128l STScI; 128r, 129t, 129b NOAO/AURA/NSF; 130t N. A. Sharp/NOAO/AURA/NSF; 130c Martin Lewis; 130b NASA/AURA/STScI; 131bl Digital Sky Survey; 133t STScI; 133c Martin Lewis; 133b JPL-Caltech; 135t, 135c Eddie Guscott; 135b Martin Lewis; 137l REU program/NOAO/AURA/NSF; 137r Damian Peach; 138l NOAO/AURA/NSF; 138r Lee Macdonald; 139tr NASA/ESA/STScI/AURA; 139br Giovanni dal Lago; 140tl,140cl Martin Lewis; 140bl NOAO/ AURA/NSF; 140cr Adrian Catterall; 142tl Martin Lewis; 142b Bill Schoening/NOAO/AURA/NSF; 143t Gordon Garradd; 143b Hubble Heritage Team (STScI/AURA/NASA); 144tr NOAO/AURA/NSF; 144bl Alex Mellinger; 144br, 145c Hubble Heritage; 145t Chandra X-Ray; 145br Spitzer Space Telescope; 146l NOAO/AURA/NSF; 146r, 147b Gordon Garradd; 147t Digital Sky Survey; 147cl John Gleason; 149 NOAO/AURA/NSF; 150bl, 150tr NASA/STScI/AURA; 150cr STScI; 150br National Radio Astronomy Observatory/National Science Foundation; 151t NOAO/AURA/NSF; 151b STScI; 152tl, 152cl Martin Lewis; 152bl Todd Boroson/NOAO/AURA/NSF; 152r N. A. Sharp, Vanessa Harvey/REU program/NOAO/AURA/NSF; 153tr, 153cr, 153bl Martin Lewis; 153br Hubble Heritage; 154t N. A. Sharp/NOAO/AURA/NSF; 155tl STScI; 155tr Martin Lewis; 156bl Gordon Garradd; 156br ESO; 157tl Martin Ratcliffe; 157c Howard Bond; 157b ESO; 158tl, 158br Yoji Hirose; 158bl AURA/NOAO/NSF; 160cl Damian Peach; 160b REU program/NOAO/AURA/NSF; 161r STScI; 161b Damian Peach; 162r Eddie Guscott; 164cl Michael Stecker; 164b Bill Schoening/NOAO/ AURA/NSF; 165 NASA and ESA; 166t Michael Stecker; 167 Philip Perkins; 168cl Darren Bushnall; 168bl Michael Stecker; 169t Canadian Galactic Plane Survey/Univ. of Calgary; 169c ESO; 169b NASA/UMass/D. Wang; 170bl N. A. Sharp, REU program/NOAO/AURA/NSF; 170br AURA/NOAO/NSF; 171t, 171br N. A. Sharp, Mark Hanna, REU program/NOAO/AURA/NSF; 171bl, 172bl, 172br, 173c Gordon Garradd; 173b Michael Stecker; 176t Michael Stecker; 176cl Martin Lewis; 176bl Darren Bushnall; 177l Michael Stecker; 177r NASA/JPL-Caltech/W. Reach (SSC-Caltech); 177cb CFHT and DSS; 178br Pedro Ré; 179bl NASA/NOAO/M. Meixner (STScI) and T. A. Rector (NRAO); 179cl Faith Jordan; 179cr Martin Lewis; 179br Bruce Balick et al/STScI; 180tl Doug Williams, REU program/NOAO/AURA/NSF; 180tr REU program/NOAO/AURA/NSF; 180bl Pedro Ré; 181bl Chris Livingstone; 181cr STScI; 181br Faith Jordan; 182t Chris Livingstone; 182b Yoji Hirose; 182–3tc NASA, ESA, Y. Nazé (University of Liège, Belgium) and Y.-H. Chu (University of Illinois, Urbana); 183tr NASA/ESA/HEIC/STScI/AURA; 183bl NASA, ESA and A. Nota (STScI/ESA); 183br NASA/ESA/STScI/AURA; 186r Michael Stecker; 187l Jay Gallagher (University of Wisconsin)/WIYN/NOAO/NSF; 187r Martin Lewis; 188 Margarita Karovska (Harvard-Smithsonian CfA)/NASA; 189cr ESO; 189b Howard Bond (STScI), Robin Ciardullo (Pennsylvania State University)/NASA; 190l Nick Hewitt; 191tl Local Group Survey Team and T. A. Rector (University of Alaska Anchorage); 191c, 191br Michael Lewis; 192t NASA/CXC/IoA/A. Fabian et al; 192bl Digitized Sky Survey; 192br NASA/STScI/AURA; 193r N. A. Sharp/AURA/NOAO/NSF; 194tr John Gillett; 194tl Martin Lewis; 194b N. A. Sharp/WIYN/NOAO/NSF; 195cr Michael Stecker; 195bl N. A. Sharp, REU program/AURA/NOAO/NSF; 195br Hillary Mathis, N. A. Sharp/ NOAO/AURA/NSF; 212 N. A. Sharp/NOAO/AURA/NSF; 213t ESO; 213b NOAO/AURA/NSF; 214t NASA/JPL; 216t NASA/David R. Scott; 216b National Astronomy and Ionosphere Center/Cornell University/NSF; 217 NASA/JPL; 219 ALMA (ESO/NAOJ/NRAO)/W. Garnier (ALMA); 220 NASA/ESA/Hubble Heritage Team (STScI/AURA); 221 Big Bear Solar Observatory; 222 ESO; 225t NASA/JPL-Caltech/Space Science Institute; 225b NASA/CXC/GSFC/U. Hwang et al; 226 ESO/WFI (Optical); MPIfR/ESO/APEX/A. Weiss et al (Submillimeter); NASA/CXC/CfA/R. Kraft et al (X-ray); 228 ESO/E. Slawik; 230 Courtesy of the TRACE consortium, Stanford-Lockheed Institute for Space Research/NASA; 231 ESA and the Planck Collaboration; 232 ESO; 234 NASA/JPL-Caltech/UCAL/MPS/DLR/IDA; 235 NASA/JPL; 236 ESO; 237l ESA/Hubble and NASA; 237r NASA/ESA/Hubble Heritage Team (STScI/AURA); 240t, 240b NASA/JPL/Space Science Institute; 241 NASA/JHU Applied Physics Laboratory; 242 NASA/JPL/DLR; 243cl Solar Dynamics Observatory/NASA; 243cr Courtesy SOHO/MDI Consortium; 243b Courtesy SOHO EIT Consortium; 244 N. A. Sharp/NOAO/AURA/NSF; 246t NASA/JPL/DLR; 246c NASA/JPL/University of Arizona; 246b ESO; 247t NASA/JPL/USGS; 247b Gemini Observatory; 248t N. A. Sharp/NOAO/AURA/NSF; 248bl ESA/MPE; 248br Michael Rich, Kenneth Mighell and James D. Neill (Columbia University)/Wendy Freedman (Carnegie Observatories)/NASA; 249c A. Fruchter and the ERO Team (STScI, ST-ECF)/NASA; 249b ESO; 250 NASA/JPL/Cornell University; 251, 252t, 252b ESO; 253 ESA/Herschel/PACS/SPIRE/J. Fritz, U. Gent; 254 T. A. Rector (NOAO/AURA/NSF) and Hubble Heritage Team (STScI/AURA/NASA); 255 NASA/JSC; 256t NASA/JPL/Space Science Institute; 256bl NASA/JPL; 256br NASA/GSFC/Langley Research Center/JPL/Multi-angle Imaging Spectro Radiometer Team; 257t ESA/ISOCAM/ISOGAL Team; 257b IRAS/IPAC/JPL-Caltech/NASA; 258t, 258b NASA; 259tr NASA/JPL/Lunar and Planetary Laboratory; 259b NASA/JPL/Cornell; 260t NASA/JPL/University of Arizona; 260b Purepix/Alamy; 262t NOAO/AURA/NSF; 262b David Jewitt; 263 NASA/JPL; 264 NOAO/AURA/NSF; 267 S. Deiries/ESO; 268 NASA/JPL/USGS; 269tl NASA, J. Bell (Cornell U.) and M. Wolff (SSI); 269tr NASA/JHU Applied Physics Team; 269c NASA/JPL/Malin Space Science Systems; 269b NASA/ASU/Cornell; 270 NASA/Johns Hopkins University Applied Physics Laboratory/Arizona State University/Carnegie Institution of Washington. Image reproduced courtesy of Science/AAAS; 272t NASA/JPL/Space Science Institute; 272b, 273 NASA/JPL/USGS; 276t NASA/JPL; 276b NASA/CXC/SAO/T. Temim et al (X-ray); NASA/JPL-Caltech (IR); 278t Todd Boroson/NOAO/AURA/NSF; 278b NASA/JPL; 280t Mark McCaughrean/ESO; 280b Karen Kwitter (Williams College), Ron Downes (STScI), You-Hua Chu (University of Illinois) and NOAO/AURA/NSF; 281 Mir 27 Crew © Centre National d'Études Spatiales, France; 282 ESA/DLR/FU Berlin (G. Neukum); 283t NASA/JPL/Space Science Institute; 283c ESO/A.-M. Lagrange et al; 283b NASA, ESA, K. Kuntz (JHU), F. Bresolin (University of Hawaii), J. Trauger (Jet Propulsion Lab), J. Mould (NOAO), Y.-H. Chu (University of Illinois, Urbana) and STScI; 284t NASA/ESA/Hubble Heritage Team (STScI/AURA); 284b NASA/ESA/AURA/Caltech; 285 NASA/JHUAPL/SwRI/Galaxy; 286c, 286b NASA/JPL; 287c NASA/CXC/SAO/F. Seward et al; 287b John Bahcall, Institute for Advanced Study, Princeton/Mike Disney (University of Wales)/NASA; 288bl NASA/JPL; 288br Image courtesy of NRAO/AUI; 289 NASA/JPL/USGS; 290 NASA/Hubble Heritage Team (STScI/AURA); 291 NASA/JPL/Space Science Institute; 292 NASA/ESA/Hubble Heritage Team (AURA/STScI); 296 NOAO/AURA/NSF; 300 NASA/JPL-Caltech/E. Churchwell (University of Wisconsin-Madison); 303 NASA/Donald Walter (South Carolina State University)/Paul Scowen and Brian Moore (Arizona State University); 304 SOHO (ESA and NASA); 305 NASA/CXC/U. Manitoba/H. Matheson and S. Safi-Harb; 306 ESO; 307t NASA/JPL; 307c ESA/NASA/JPL/University of Arizona; 307b NASA/JPL; 308t NASA/NSSDC; 308b NASA/JPL/USGS; 310 NASA/JPL; 311 Erich Karkoschka (University of Arizona) and NASA; 312 NASA/JPL; 313t NRAO/AU/NSF; 313b NASA/JPL/USGS; 314t NASA/JPL-Caltech; 314c Harvey Richer (University of British Columbia)/STScI/NASA; 314b T. A. Rector and Monica Ramirez/NOAO/AURA/NSF; 315t NASA/CXC/Rutgers/J. Hughes et al; 315b NASA/SRON/MPE; 316 Institute of Space and Astronautical Science, Japan.

All artwork © Philip's, except as noted below.

Star maps by Wil Tirion (© Philip's): 20–21, 122, 123, 124, 128, 129, 131, 132, 134, 136, 137, 138, 141, 142, 143, 146, 148, 149, 152, 153, 154, 156, 158, 159, 160, 161, 162, 163, 164, 165, 167, 170, 172, 173, 174, 175, 177, 178, 179, 180, 181, 182, 184, 185, 186, 188, 189, 190, 193, 195, 196–211.

Moon maps by John Murray (© Philip's): 49, 50, 52, 55.

Illustrations by Jonathan Bell (© Philip's): 220, 223b.

Illustrations © Robin Scagell: 22t, 110t, 150, 163c, 164cr, 177tr, 193bl, 197, 199, 201, 203, 205, 207, 209, 211.

GENERAL INDEX

Entries in *italics* refer to illustrations

Airy disk 108, *108*
albedo *73*, 74
altazimuth mount *35*, 36–37, *36*, *37*
altitude 25, 36
American Association of Variable
    Star Observers 110
angular measurements 24
aperture 32, 34, 40
Apollo *47*, 56, 57–61, *57*, *59*, *60*
apparition 29
arc minute 25
arc second 25
Association of Lunar and Planetary
    Observers 84
asteroid belt 92
asteroids 47, 92
    observing 92
aurora 98–99, *99*
    corona 99
    rays 98
    streamers 98
averted vision 32, 118, 120, 176, 187,
    193
azimuth 25, 36, 42

Bayer, Johann 28
Bayer letters 28
binoculars 10, 32–33, *32*, *33*
    adjusting 33, *33*
    choosing 32–33
    field of view 32–33
    using 20, 32–33
black holes 107, *107*, 116, 145, 151,
    169, 188, 192
bolide 98
brightness of stars *see magnitude*
British Astronomical Association
    (BAA) 84, 110

Caldwell Catalogue 121
Callisto *82*, 83, *83*
captured rotation 44
catadioptric telescopes 33, 35–36, 43,
    122
    Schmidt-Cassegrain (SCT) 35, 36,
        36, 39, 43
    Maksutov-Cassegrain 35, 43
    Schmidt-Newtonian 36
CCDs *see imaging*
celestial equator 23, *23*, 24
celestial pole 23, *23*
celestial sphere 23, *23*, 24
Cepheids *see variable stars*
chromatic aberration (false color) 35
chromosphere 63
circumpolar stars 25
Clerke, Agnes M. 167
collimation 35
comets 92–93, *93*, 97–98
    coma of 92
    Hale–Bopp 92, *92*
    Halley's 92, *157*
    ISON *65*
    NEAT *93*
    nucleus of 92, *93*
    observing 32
    orbits of 92
    pseudo-nucleus of 92, *93*
    with binoculars 92
conjunction 29
constellations 20, 26–27
    learning the 28
    list of 26
continuous emission 112
corona, solar *65*, 95

Daylight Saving Time 22
declination (dec.) 23, 24, 122
deep-sky objects 118–121
    finding 120
    observing 32, 38, 40, 42, 119, 120
Doppler effect 18
Doppler shift 14, 17, 108, 109, 121
double stars 14, 17, 108, 109, 121
    eclipsing binary 109, 134, 145, 161
    optical 109
    separation of 109
    true 109
drawing
    deep-sky objects 120
    Jupiter 80, *80*
    Mars *72*
    Moon 44
    Saturn 84, *85*
    star clusters 120
    the Sun *64*
    Venus 69
Dreyer, J. L. E. 121

Earthshine 45, *45*
eccentricity 93
eclipse 94–95
    lunar 94
        observing 95
        partial 95
        total 95, *95*
    solar 63, 94
        annular *96*
        diamond-ring effect *96*
        observing 94–95
        partial 94
        total 95, *95*
ecliptic 20, 23, *23*, 24, 123, 132, 148,
    149, 167, 170, 174, 178, 186, 188,
    197
ephemeris 93
Equation of Time 23
equatorial mount *35*, 37–38, *38*, 40, 41,
    41
equinox 23, 24
Europa 82
evening star *see Venus*
exit pupil 32, 39
eye 20, 32
eyepieces 8, 34, 39, 42, 119, 120
    Plössl 34, *34*
    Huygenian 34, 62
    for solar observing 62

faculae, solar 63, 64, *64*
field of view 24, 33
filters 35, 63
    hydrogen-alpha 63, *65*
    hydrogen-beta 127, 135, 194
    light-pollution rejection (LPR) 113,
        176
    Moon 46
    narrowband 63, 113
    O III 113, *113*, *176*, *179*, *179*, *181*,
        182, 194
    solar 63–64, 95
finder
    telescope 36, 43, *43*, 120
    alignment of 43
    red-dot 36
fireball 98
First Point of Aries (vernal equinox)
    23, 24
Flamsteed, John 28
flares, solar 63, 64

galaxies 113–118, 124–126, 137, 138,
    138, 139, *139*, 140, *140*, 142, 149
    barred spiral 115, *118*
    distance to 113
    elliptical *114*, 115, 116
    irregular 115, *116*, 181
    lenticular 115
    Local Group of 189
    magnitude of 118
    observing 118
        with binoculars 118
        with telescopes 118
    Seyfert *117*
    spiral 115, *115*, 116, 118, *118*
    starburst 139, *139*
Galaxy, the 168–169, *168*
Galileo Galilei 62, 81
Ganymede 80, 83, *83*
Geminids 98
globular clusters 110, *110*
    with binoculars 111
    with telescopes 111
    see also star clusters
Go To telescope 21, 28, 36, *36*, 37,
    42–43, *43*, 70, 88, 89, 91, 92, 118,
    119, 120, 121, 139, 140, 149, 152,
    153
greatest elongation 29
Greek alphabet 27
Greenwich Observatory 24

H II regions *see nebulae*
Henry Draper Catalog 28
Herschel, William 88, 177
Hubble, Edwin 16

image processing 22, *39*
image rotation 36
imaging 8, 11, 21, 38–39
    afocal method 38
    digital 38
    of aurorae 99
    of constellations 38
    of deep-sky objects 11, 38, 121
    of double stars 121
    of galaxies 121
    of globular clusters 121
    of meteors 98
    of planets 38, 67, 69, 72, 79, 84, 88,
        91, 108

of planetary nebulae 121
    with piggy-back cameras 39
    with webcams 38–39, *38*, 39, 67, 69,
        71, 85, *86*, 121
    with CCDs 39, 121
    with SLR cameras 39
Index Catalogue 121, 166
International Astronomical Union 93
International Occultation Timing
    Association 97
Io 81

Jupiter 9, 11, 28, *28*, 30, *31*, 79–83, 137
    belts of 79, 80, *80*, 81
    clouds of 81, 83
    Great Red Spot 80, *80*, 81
    moons of 79, 80, 81–83, *82*, *83*
    orbit of 79
    rotation of 79
    with binoculars 79
    with telescopes 79–80
    zones of 79, *79*

leap year 23
lens, Barlow 34
lens, objective
    apochromatic 35
    in binoculars 32, 33
    in telescopes 42
light
    distribution 14
    speed of 9
    wavelength 13, 14
light pollution 10, 20, 29, 32, 39, 62, 99,
    108, 112, 113, 118–120, 176, 179,
    181, 187, 194
light year 9, 30
limb darkening 63
line emission 112
lunar eclipse *see under eclipse*
lunation 44, *45*

magnification (power) 44, 46, 119, 121,
    122, 125
    binoculars 32–33
    telescopes 34, 36, 39, 72, 79
magnitude 25
maps
    constellation 20, *26*, 122–195
    ecliptic 20
    equatorial regions 200–207
    photo-realistic 20, 25, 196–211
    polar regions 196–199, 208, 211
    seasonal 25, 28, 123, 136, 148, 159,
        174, 184
Mars 28, 29, 72–78, 170
    canals 74
    dust storms 73
    map of *73*
    markings on 73
    movement of 29
    oppositions of 29, 72
    orbit of 73
    phases of 72
    polar caps 73, *75*
    rotation of 73
    with a telescope 72–74
    with binoculars 72
Mercury 28, 67–68
    elongation of 67
    phases of 67
    telescopes for 67
    transits of 67
meridian 22
Messier, Charles 111, 152
Messier Catalogue 121, 167, 190, 194
meteorites 98
meteoroids 97
meteors 97–98, *97*
    orbit of 98
    sporadic 98
meteor shower 98
    list of 98
    radiant of 98
    zenithal hourly rate 98
Milky Way 9, 29, *29*, 32, 123, 128, 129,
    132, 134, 136, 137, 143, 146, 148,
    156, 158, 159, 160, 162, 163, 164,
    166, 167, 168, 170, 172–177, 180,
    181, 183, 193–195
mirror 34, 35
Moon 44–61
    craters 44, 46, 47, *47*, 48
    ejecta 51
    far side 46, *56*
    first quarter 45, *46*
    formation of 44
    full 44, 45, *45*
    gibbous 45, *45*
    last quarter 45, *45*
    librations 44, 46
    magnitude of 25

mare 47
    movement of 44–45
    names of lunar features 47
    near side 46
    new 45
    northeast quadrant 48–49
    northwest quadrant 50–51
    observing 39
    Old 45, *45*, 46
    orbit of 44
    phases of 45, *45*
    position of 24
    ray craters 48, 54
    rille 48, 54, *60*
    scarp fault 53
    southeast quadrant 54–55
    southwest quadrant 52–53
    volcanic features 48
    walled plain 48, *51*
    waning 45
    waxing 45, *45*
    wrinkle ridge 48, *48*
Moore, Sir Patrick 121
motor drives 38
multiple stars 14, 100

naked-eye limit 25, *31*
naked-eye observing 32
    see also eye
nebulae 11, 30, 111–113
    dark 111, 112, 127, 158, 163, *163*,
        166, *168*, 178
    diffuse 113
    H II regions *111*, 112, 113, 115, 130,
        181, 182
    observing 32, 113
    reflection *111*, 112, *121*
    see also planetary nebulae
Neptune 88–90, *89*
    satellites of 90
    with binoculars 88
    with telescope 88
neutron stars 107
New General Catalogue 121, 166
Newton's Law 15
noctilucent clouds 99, *99*
Northern Lights *see aurora*
novae 109

occultations 96
off-axis guider 38
open star cluster 30, 110
    with binoculars 110
    with telescopes 110
    see also star clusters
opposition 29, *29*
orbit, Earth's 22, *23*
orbital elements 93

parallax 13
perihelion 93
Perseids 98
photography *see imaging*
photosphere 63
planets 28, 67–91
    movement of 29, *29*
    observing 34, 36
    position of 24, 25, 28
    superior 29
    visibility of 20, 29
    see also individual planets
planetary nebulae 112, *112*, 121
Pluto 91, *91*
    position of 91
    telescope for 91
position angle 109
precession 24, 156
prominence, solar 63, 64, *65*, 94, 96, *96*
pulsars 107

quasar 151, 188

radio waves 18
rays
    gamma 17
    X-ray 17
recording *see drawing and imaging*
Red Planet *see Mars*
reflector 33, 35, *35*, 39
    Newtonian *35*, 40, *40*
refractor 33, 34–35, *35*, 36, 42, 122, 185
    apochromatic 35
retrograde loop 29, *29*
right ascension (RA) *23*, 24
Rosse, Lord 152
Saturn 28, 84–87
    belts and zones 84
    imaging 85
    inclination of 84
    moons of 84
    orbit of 84
    rings of 84, 85, *85*

seasons 23
seeing 11, 38, 39, 64, 72, 81, 88, 108,
    108, 109, 121
    seeing disk 108, *108*
semimajor axis 93
setting circles 38
shooting stars *see meteors*
sky, movement of 23
Smyth, Admiral 135
software 73, 93
solar filters *see filters*
solar cycle 64
solar eclipse *see eclipse*
solar maximum 63, *64*, 95, 96
solar minimum 64, 95, 96, *96*
Southern Lights *see aurora*
spectroscopy 14, 112
stars 14, 100–110
    carbon 185
    classification of 15, 100–103
    distances of 108
    formation of 100, *101*, 106, 116
    masses of 15, 103
    observing 108–109
    red dwarfs 15, 16, 109
    red giants 15, 103–105, *103*, *104*,
        109, 129, 147, 158, 177
    supergiants 104
    white dwarfs *104*, 105, 106
    Wolf-Rayet 143
    *see also double stars, multiple stars,
        variable stars*
star clusters 30, 110–111, *110*, 120
    observing 32
    see also globular clusters, open clusters
star-hopping 118, 120
star map, computer-plotted 91, 92
star names 26–28
star novae 109
star trails *22*
Summer Time 22
Sun
    active region 64
    magnitude of 25
    movement of 23
    observing the 62–65
        filters for *see filters, solar*
        telescopes for 62–65, *65*
    projecting image of 63, *63*
    rotation of 64
sunspots 64
    penumbra 64
    umbra 64
supernovae 17, *106*, 107, 112, *113*, 115,
    124, 169, 183, 187
supernova remnant 106, 112, *113*

telescope
    choice 33, 36
    catadioptric 33, 35–36, 43, 122
        Schmidt-Cassegrain (SCT)
            8, 35, 36, *36*, 39, 43
        Maksutov-Cassegrain 35, 43
        Schmidt-Newtonian 36
    collimation 35
    dewing 119
    eyepieces 34, 39, 42, 120
        Plössl 34, *34*
        Huygenian 34, 62
        for solar observing 62
    field of view 24, 34
    finder 36, 43, *43*, 120
        alignment of 43
        red-dot 36
    focal length 34, 39
    focal ratio (f-number) 34
    focusing 34
    Go To 21, 28, 36, *36*, 37, 42, *43*,
        70, 88, 89, 91, 92, 118, 119, 120,
        121, 139, 140, 149, 152, 153
        catalogs 27, 121
    large 149, 153, 157, 162, 180, 181,
        182, 187, 191, 195
    magnitude limits 25
    medium 122
    observing tips 39
    reflector 34, 35, *35*, 39
        Newtonian *35*, 40, *40*
    refractor 34, 35, *35*, 36, 42, 122
        apochromatic 35
    resolving power 40, 108, 109
    small 122
    solar 65, *65*
    star diagonal 35, *35*, 42, *42*
    testing 108
    using 22–23
telescope mounts 36–38
    altazimuth *35*, 36–37, *36*, *37*
    Dobsonian 37, 42
    equatorial *35*, 37–38, *38*, 40, 41,
        41
    fork 36, *36*, 37, *37*
    German *35*, 37, 41

Go To see Go To telescope
yoke 36, *36*
terminator 46, 47, 51, 53, 54, 70
time zone 23
Titan 84, *85*
transparency 39, 109, 155
Triton 88, 89, 90

Uranus 88–90, *88*
  satellites of 89
  with binoculars 88
  with telescope 88

variable stars 109, 110, *110*, 134, 161, 163, 164, 175,
  177, 185, 188, 193
  Algol-type 161, 177
  Cepheids 15–17, *16*, 109, 154, 155, 163, 177
  comparison charts 109, 110, *110*, *164*, *193*
  comparison stars 110, *110*, *163*
  eruptive 109
  estimates of 110, *110*
  Mira-type 142, 175
  observing 32
  RV Tauri 164
Venus 28, 69–71, *69*
  apparitions of 69
  ashen light 70
  atmosphere of 69
  cusp caps 70
  dichotomy 70
  finding 69–71
  greatest elongation of 69
  magnitude of 24, 25, 69, *69*
  phases of 69, *69*
  transits of 69
vernal equinox see First Point of Aries

webcams see imaging
wedge 37, *37*

zenith 24
zodiac 20, 137, 141, 159, 165, 174

INDEX OF CONSTELLATIONS

Andromeda 123, 149, 174, 180, 181, 184, 185, 189,
  190, 191, 193, 194, 196, 200, 206
Antlia 204
Apus 210
Aquarius 174, 179, 180, 184, 200
Aquila 160, 163, *163*, 164, 174, 178, 200
Ara 173, 200, 202
Aries 126, 174, 185, 186, 206
Auriga 123, 127, 134, *135*, *193*, 196, 198, 204, 206
Boötes 136, 148, 152, 202
Caelum 206
Camelopardalis 196, 198
Cancer 123, 141, 142, 204
Canes Venatici 152, 202
Canis Major 122, *123*, 124, 128, 129, 130, 136, 204
Canis Minor 123, *123*, 124, 129, 204
Capricornus 170, 174, 178, 179, 200
Carina 143, 146, *146*, 148, 210
Cassiopeia 123, 156, 165, 177, 184, 190, 193, *193*,
  195, 196
Centaurus 148, 156, 157, *157*, 158, 202, 204, 210
Cepheus 177, 196
Cetus 180, 184, 188, 190, 193, 196
Chamaeleon 210
Circinus 202, 210
Columba 204, 206
Coma Berenices 150, 153, 202
Corona Australis 200
Corona Borealis 202
Corvus 202
Crater 204
Crux 148, 157, 158, 172, 195, 210
Cygnus 148, 158, 159, 160, 161, 174, 175, 176, *176*,
  177, 194, 200
Delphinus 200
Dorado 181, 206, 208
Draco 154, 162, 200
Equuleus 200
Eridanus 206
Fornax 206
Gemini 123, 132, 136, 186, 204
Grus 200
Hercules 154, 159, 162, 168, 200, 202
Horologium 206, 208
Hydra 136, 142, 149, 202, 204
Hydrus 208
Indus 208
Lacerta 196, 200
Leo 27, 28, 123, 136, 137, 148, 149, 153, 204
Leo Minor 137, 204
Lepus 124, 128, 129, 204, 206
Libra 20, 202
Lupus 172, 202, 210
Lynx 198, 204
Lyra 143, 155, 161, 175, 179, 200
Mensa 181, 208

Microscopium 200, 208
Monoceros 129, 204
Musca 158, 210
Norma 172, 210
Northern Cross see Cygnus
Octans 208
Ophiuchus 165, 176, 200, 202
Orion 167, *169*, 170, 204, 206
Pavo 200, 208, 210
Pegasus 123, 174, 180, 184, 185, 190, *193*, 200
Perseus 177, 184, 190, 191, *192*, 193, 194, 196, 206
Phoenix 200, 206
Pictor 204, 206, 208, 210
Pisces 180, 185, 188, 200, 206
Piscis Austrinus 200
Puppis 131, 143, 144, 204, 210
Pyxis 204
Reticulum 206, 208
Sagitta 160, 174, 200
Sagittarius 159, 164, 167, 168, *168*, 169, *169*, 174,
  176, 200, 202
Scorpius 148, 159, 165, 166, *168*, 170, *171*, 173,
  174, 202
Sculptor 200, 206
Scutum 163, 164, 165, 174
Serpens 165, 168, 200, 202
Sextans 137, 138, 204
Southern Cross see Crux
Taurus 123, 124, 134, 184, 186, *193*, 206
Telescopium 200, 208
Triangulum 189, 206
Triangulum Australe 148, 210
Tucana 181, 200, 208
Ursa Major 138, 155, 198, 202, 204
Ursa Minor 138, 155, 196, 198
Vela 143, 144, *144*, *146*, 148, 204, 210
Virgo 17, 136, 148, 149, 150, *150*, *151*, 153, 154,
  202
Volans 146, 210
Vulpecula 160, *160*, 174, 200

INDEX OF OBJECTS

30 Doradus 181
47 Tucanae 182, *182*
61 Cygni 175

Achernar 181, *182*, 184
AE Aurigae 126, 135
Albireo 160, 161, *175*, 178
Alcor 140, *140*
Aldebaran 123, 184, 186, *186*
Algol 145, 161, 177, 192, 193, *193*
Alpha Capricorni 178
Alpha Centauri 156, *156*
Alpha Crucis 158
Alpha Persei Cluster (Melotte 20) 193, *193*
Alphard 136, 142
Alrescha 185
Altair 129, 159, 160, 161, *163*, 174, 175, 178, 184
Andromeda Galaxy (M31) 149, 181, 190, *190*
Antares 159, 166, *166*, 170, 172
Arcturus 136, 148, 159

Barnard's Loop 124, 126
Beta Capricorni 178, *178*
Beta Lyrae 161
Betelgeuse 28, 124, 129
Big Dipper 136, 138, *140*, 148, 152, 174
Black Eye Galaxy (M64) 153
Blinking Planetary (NGC 6826) 175, *176*
Blue Planetary (NGC 3918) 157, *157*
Blue Snowball (NGC 7662) 191, *191*
Bubble Nebula (NGC 7635) 195, *195*

California Nebula (NGC 1499) 127, 194, *194*
Canopus 181
Capella 123, 132, 134
Castor 9, 132, *132*
Centaurus A 148, 156, 157, *157*, 167
Circlet 185, *185*
Coalsack Nebula 158
Coathanger 160, *160*
Coma Star Cluster 153
Cone Nebula
Cor Caroli 152
Crab Nebula (M1) 21, 187, *187*
Cygnus Rift 29, 158, 176

Delta Cephei 177, *177*
Deneb 129, 137, 148, 158, 159, 161, 175, *175*, 176,
  184, 188
Denebola 28, 137, 149, 153
Double Cluster (NGC 869 and 884) 193, *193*
Dumbbell Nebula (M27) 160

Eagle Nebula (M16) 164, 165, *165*, 182
Eight-Burst Nebula (NGC 3132) 143, *143*
"Engagement Ring" 155, *155*
Epsilon Aurigae 134, *134*
Epsilon Lyrae (Double Double) 161, *161*

Eskimo Nebula (Clown Face, NGC 2392) *133*
Eta Aquilae 163, *163*
Eta Carinae Nebula *144*, 145, *145*, 146, *146*, 147

False Cross 146, *146*
Fish's Mouth 125
Flame Nebula (NGC 2024) 125, 127, *127*
Flaming Star Nebula 127, 135, *135*
Fomalhaut 84, 174, 184

Gamma Andromedae 190, *190*, 194
Gamma Leonis 137, *137*
Gamma Velorum 143, 144
Garnet Star (Mu Cephei) 177
Ghost of Jupiter (NGC 3242) 142, *142*
Great Rift see Cygnus Rift

Haedi 134, *134*
Helix Nebula (NGC 7293) 179, *179*
Horsehead Nebula 127, 173
Hyades 14, 184, 186, *186*

IC 410 *135*
IC 434 127, *127*
IC 1805 195
IC 1848 195
IC 2602 (Southern Pleiades) 146, *146*
IC 4665 166, *166*
IC 4756 165
Izar 109

Jewel Box (Kappa Crucis Cluster, NGC 4755) 112,
  158, *158*, 195

Keyhole Nebula 144, *144*
Keystone 159, 162
Kochab 155
Lagoon Nebula (M8) 21, 115, 167
Large Magellanic Cloud see Magellanic Clouds
Little Dipper 155
Little Dumbbell (M76) 194

M2 179, 180, *180*
M3 110, 152, *152*
M4 166, 170, *170*
M6 (Butterfly Cluster) 170, 171, *171*
M7 (Ptolemy's Cluster) 170, 171, *171*
M10 166, *166*
M12 166, *166*
M13 162, 168
M15 180, *180*
M16 164, 165, *165*, 167, 168
M18 168
M22 112, 168, *168*
M23 168, *168*
M24 *164*, 167
M25 112, 168, *168*
M32 *114*, 120, 190, 191, *191*
M35 112, 132, *132*
M36 134, 135, *135*
M37 134, 135, *135*, 194
M38 135, *135*
M39 175, *175*
M41 *128*, 129
M43 125, *125*
M44 (Beehive, Praesepe) 141, *141*
M46 131, *131*
M47 131, *131*
M49 150, *150*
M50 129, 129, 130
M51 (Whirlpool Galaxy) *12*, 140, 152
M52 195, *195*
M58 150, *150*
M59 150, *150*
M60 150, *150*
M65 *137*, *137*
M66 *137*, 138
M67 141, *141*
M71 160, *160*
M72 180, *180*
M73 179
M74 185
M77 188
M81 119, 139, 140
M82 139, *139*
M83 (Southern Pinwheel) 142
M84 149, 150, *150*
M85 153
M86 149, *149*, 150, *150*
M87 149, 150, *150*, 151, *151*
M89 149, 150, *150*
M90 150, *150*
M92 *109*, 162, *162*
M94 152, *152*
M95 138, *138*
M96 138
M99 150, *150*, 154
M101 140, *140*
M103 195, *195*
M105 138
M109 *118*

Magellanic Clouds 119, 120, 149, 181, 183
  Large Magellanic Cloud 181, *181*, *182*, *183*
  Small Magellanic Cloud 181, 182, *182*
Merope Nebula 186, *186*, 187
Mira 188, *188*
Mizar 139, 140, *140*

NGC 205 (M110) 190, 191
NGC 253 *118*
NGC 362 182
NGC 663 195, *195*
NGC 752 191, *191*
NGC 891 191, *191*
NGC 1977 125, *127*
NGC 1981 127
NGC 2264 *111*
NGC 2451 131, *131*
NGC 2516 147, *147*
NGC 2547 143, *143*
NGC 3114 146, 147, *147*
NGC 3115 (Spindle Galaxy) 137, *138*
NGC 3532 147, *147*
NGC 3628 *137*, 138
NGC 4365 137
NGC 4526 *17*
NGC 4565 153, *153*
NGC 4833 158
NGC 5128 157, *157*
NGC 5195 152
NGC 5822 172, *172*
NGC 6067 172, *172*
NGC 6087 172, *172*
NGC 6188 173, *173*
NGC 6193 173, *173*
NGC 6200 173
NGC 6204 173
NGC 6231 171, *171*
NGC 6397 173, *173*
NGC 6543 (Cat's Eye Nebula) 154, 155, *155*
NGC 6603 168, *168*
Norma Star Cloud 172
North America Nebula (NGC 7000) 116, 176, *176*
Nu Draconis 154, *154*

Omega Centauri 157, *157*
Orion Nebula (M42) 100, *101*, *104*, 115, 124, 125,
  167, 181
Owl Cluster (ET Cluster, NGC 457) 195, *195*
Owl Nebula 140

Pelican Nebula 176
Pillars of Creation 165, *165*
Pinwheel Galaxy (M33) 189, *189*
Pleiades (M45) *38*, *110*, 112, 123, 141, 147, 171,
  186, *186*, 187
Plough 136, 138, *140*, 148, 152, 174
Pointers 24, 138, 156, *158*
Polaris 25, *139*, 155, *155*, 158
Pole Star see Polaris
Pollux *9*, 132
Procyon 123, 129, 132
Proxima Centauri 16, 156

Regulus 136, 137, 138, 148
Rho Ophiuchi 166, *166*
R Hydrae 142
Rigel *104*, 105, 124
Ring Nebula (M57) 144, 155, 161, *161*, 175, 179
Rosette Nebula (NGC 2244) 130, *130*
R Scuti 164, *164*

Saturn Nebula (NGC 7009) 179, *179*, 180
Seven Sisters see Pleiades
Sigma Orionis 125, 125, *127*
Sirius 123, 128, *128*, 129
Sirius B 128, *128*
Small Magellanic Cloud see Magellanic Clouds
SN1987A 182, *182*
Sombrero Hat (M104) *149*, 150
Spica *72*, 148, 149
Square of Pegasus 174, 180, 184, 185, 188, 190
Summer Triangle 129, 161
Swan Nebula (Omega Nebula, M17) 167, 168,
  168

Tarantula Nebula (NGC 2070) 181, *181*
Trapezium 125, 127, *127*
Trifid Nebula (M20) 167, *167*
TX Piscium (19 Piscium) 185

Vega 129, 148, 159, 161, 162
Veil Nebula (NGC 6960 and 6992) 107, *113*, 114,
  144, 176, *176*
Virgo–Coma Cluster see Virgo Cluster
Virgo Cluster 16, 17, 117, 119, 120, 149, 149,
  151, 153, 154, 192

Water Jar 174, 179
Wild Duck Cluster (M11) 112, 164, *164*
Winter Triangle 129

Zosma 28